WELLS'S

NATURAL PHILOSOPHY;

FOR THE

USE OF SCHOOLS, ACADEMIES, AND PRIVATE STUDENTS.

INTRODUCING

THE LATEST RESULTS OF SCIENTIFIC DISCOVERY AND RESEARCH; ARRANGED WITH SPECIAL REFERENCE TO THE PRACTICAL APPLICATION OF PHYSICAL SCIENCE TO THE ARTS AND THE EXPERIENCES OF EVERY-DAY LIFE.

WITH THREE HUNDRED AND SEVENTY-FIVE ENGRAVINGS

BY

DAVID A. WELLS, A.M.,

AUTHOR OF "THE SCIENCE OF COMMON THINGS," EDITOR OF THE "ANNUAL OF SCIENTIFIC DISCOVERY," "KNOWLEDGE IS POWER," ETC.

FIFTEENTH EDITION REVISED.

NEW YORK:
IVISON, PHINNEY, BLAKEMAN & CO.,
CHICAGO: S. C. GRIGGS & CO.

1864.

ELECTROTYPED BY
THOMAS B SMITH,
82 & 84 Beekman-street, N. Y.

PREFACE.

THE constant progress made in every department of physical science, is a sufficient apology for the preparation and publication of a new elementary text-book on Natural Philosophy.

The principles of physical science are so intimately connected with the arts and occupations of every-day life, with our very existence and continuance as sentient beings, that public opinion, at the present time, imperatively demands that the course of instruction on this subject should be as full, thorough, and complete as opportunity and time will permit. With this view, the author has endeavored to render the work, in all its arrangements and details, eminently practical, and, at the same time, interesting to the student. The illustrations and examples have been multiplied to a greater extent than is usual in works of like character, and have been derived, in most cases, from familiar and common objects.

Great care has been also taken to render the work complete and accurate, and in full accordance with the latest results of scientific discovery and research.

In the arrangement of the subjects treated of, and in the incorporation of questions with the text, the most approved methods, it is believed, have been followed. The teacher will also observe that the principles and important propositions are presented in large and prominent type, and the observations and illustrations in smaller letters. The advantage of this to the learner is most evident.

HEAT, which is often considered as belonging more especially to chemistry, has been discussed at length, and the familiar application of its principles in the industrial arts, in warming and ventilation, in the production of dew, etc., carefully explained. A full and complete outline of the subject of Meteorology has also been given. On the other hand, ASTRONOMY, which is often included in text-books on Natural Philosophy, has been omitted, as rightfully and properly forming the subject of a separate treatise.

An elementary work on physical science can have little claim to originality, except in the arrangement and classification of subjects, and the selection of illustrations. In this respect the author makes no pretensions, and acknowledges his indebtedness to the very superior French treatises of Ganot, Delaunay, Archambault, and to the writings of Müller, Arnott, Lardner, Brewster, and others.

The engravings in the present volume are of a superior character, and have been prepared, in part, from new and original designs.

NEW YORK, August, 1857.

CONTENTS.

CHAPTER VI.

CHAPTER VII.

CHAPTER VIII.

CHAPTER IX.

CHAPTER X.

CHAPTER XI.

CHAPTER XII.

CHAPTER XIII.

CHAPTER XIV.

CHAPTER XV.

CHAPTER XVI.

CHAPTER XVII.

CHAPTER XVIII.

CHAPTER XIX.

NATURAL PHILOSOPHY.

INTRODUCTION.

What is Natural Philosophy?

1. Natural Philosophy, or Physics, is that department of science which treats of all those phenomena observed in masses of matter, in which there is a sensible change of place.

What is Chemistry?

2. Chemistry, on the contrary, treats of all those phenomena observed to take place in minute particles, or portions of matter, in which there is a change in the character and composition of the matter itself, and not merely a change of place.

What are examples of the phenomena of Natural Philosophy?

3. A falling body, the motion of our limbs or of machinery, the flow of liquids, the occurrence of sound, the changes occasioned by the action of heat, light, and electricity, are all examples of phenomena which come under the consideration of Natural Philosophy.

Strictly speaking, we have no right, in Natural Philosophy, to conceive or imagine any thing, for the truths of all its laws and principles may be proved by direct observation,—that is, by the use of our senses. When we conceive, reason, or imagine concerning the properties of matter, we have in reality passed beyond the limits of Natural Philosophy, and entered upon the application of the laws of mind or of mathematics to the principles of Natural Philosophy. Practically, however, no such division of the subject is ever made.

The truths and operations of Chemistry, in contradistinction to the truths and operations of Natural Philosophy, can not all be proved and made evident by direct observation. Thus, when we unite two pieces of machinery, as two wheels, or when we lift a weight with our hands, or move a heavy body by a lever, we are enabled to see exactly how the different substances come in contact, how they press upon one another, and how the power is transmitted from one point to another: these are experiments in Natural Philosophy, in which every part of the operation is clear to our senses. But when we mix alcohol and water together, or burn a piece of coal in a fire, we see merely the result of these processes, and our senses give us no direct information of the manner in which one particle of alcohol acts upon another particle of water, or how the oxygen of the air acts upon the coal. These are experiments in Chemistry, in which we can not perceive every part of the operation by means of our senses, but only the results. Had there been but one kind of substance or matter in the universe, the laws of Natural Philosophy would have explained all the phenomena or changes which could possibly take place; and as the character, or composition of this one substance, could not be changed by the action of any different substance upon it, there could be no such department of knowledge as Chemistry.

What is meant by the term Physics?

4. The term PHYSICS is often used instead of the term Natural Philosophy, both having the same general meaning and signification. It is also customary to speak of "PHYSICAL LAWS," "Physical Phenomena," and "Physical Theories," instead of saying the laws, phenomena, and theories of Natural Philosophy.

What are Physical Laws and Theories?

5. A PHYSICAL LAW is the constant relation which exists between any phenomenon and its cause. A PHYSICAL THEORY is an exposition of all the laws which relate to a particular class of phenomena.

Thus, when we speak of the "theory" of heat, or of electricity, we have reference to a general consideration of the whole subject of heat, or light, or electricity; but when we use the expression a "law" of heat, of light, or of electricity, we have reference to a particular department of the whole subject.

CHAPTER I.

MATTER, AND ITS GENERAL PROPERTIES.

What is Matter?

1. MATTER is the general name which has been given to that substance which, under an infinite variety of forms, affects our senses. We apply the term matter to every thing that occupies space, or that has length, breadth, and thickness.

How do we know that any thing exists?

2. It is only through the agency of our FIVE senses (hearing, seeing, smelling, tasting, and feeling), that we are enabled to know that any matter exists. A person deprived of all sensation, could not be conscious that he had any material existence.

What is a body?

3. A BODY is any distinct portion of matter existing in space.

What are the properties of matter?

4. The properties, or the qualities of matter, are the powers belonging to it, which are capable of exciting in our mind certain sensations.

It is only through the different sensations which different substances excite in our minds, or, in other words, it is by means of their different properties, that we are enabled to distinguish one form or variety of matter from another.

The forms and combinations of matter seen in the animal, vegetable, and mineral kingdoms of nature, are numberless, yet they are all composed of a very few simple substances or elements.

What is a simple substance?

5. By a simple substance we mean one which has never been derived from, or separated into any other kind of matter.

Gold, silver, iron, oxygen, and hydrogen, are examples of simple substances or elements, because we are unable to decompose them, convert them into, or create them from, other bodies.

What is the number of the elements?

6. The number of the elements or simple substances with which we are at present acquainted, is sixty-two.

How distributed?

7. These substances are not all equally distributed over the surface of the earth: most of them are exceedingly rare, and only known to chemists. Some ten or twelve only make up the great bulk or mass of all the objects we see around us.

All the different forms and varieties of matter are in some respects alike—that is, they all possess certain general properties. Some of these properties are essential to the very existence of a body; others are non-essential, or a body may exist without them. Thus it is essential to the existence of a body that it should occupy a certain amount of space, and that no other body should occupy the same space at the same time; but it is not necessary for its existence that it should possess color, hardness, elasticity, malleability, and the like non-essential properties.

What are the most important properties of matter?

8. The following are the most important of the general properties of matter—MAGNITUDE or EXTENSION, IMPENETRABILITY, DIVISIBILITY, POROSITY, INERTIA, ATTRACTION, AND INDESTRUCTIBILITY.

What is Magnitude?

9. By MAGNITUDE we mean the property of occupying space. We can not conceive that a portion of matter should exist so minute as to have no magnitude, or, in other words, to occupy no space.

The SURFACES of a body are the external limits of its magnitude; the SIZE of a body is the quantity of space it occupies; the AREA of a body is its quantity, or extent of surface.

The FIGURE of a body is its form or shape, as expressed by its boundaries or terminating extremities. The VOLUME of a body is the quantity of space included within its external surfaces. The figure and volume of a body are entirely independent of each other. Bodies having very different figures may have the same volume, or bodies of the same figure may have very different volumes. Thus a globe may have ten times the volume of another globe and yet have the same figure, or a globe and a cylinder may have the same volume, that is, may contain the same amount of matter within their surfaces, but possess very different figures.

What is Impenetrability?

10. By IMPENETRABILITY we mean that property or quality of matter, which renders it impossible for two separate bodies to occupy the same space at the same time.

There are many instances of apparent penetration of matter, but in all of them the particles of the body which seem to be penetrated are merely displaced. When a nail is driven into a piece of wood, the particles of wood are not penetrated, but merely displaced. If a needle be plunged into a vessel of water, all the water which previously filled the space into which it entered, will be displaced, and the level of the water in the vessel will rise to the same height as it would have done, had we added a quantity of water equal in volume to the bulk of the needle. When we walk through the atmosphere, we do not penetrate into any of the particles of which the air is composed, but we merely push them aside, or displace them. If we plunge an inverted tumbler into a vessel of water, the air contained in it will prevent the water from rising in the glass—and notwithstanding the amount of pressure we may exert upon the tumbler, it cannot be filled with water until the air is removed from it.

What is Divisibility?

11. By DIVISIBILITY we mean that property which matter possesses of being divided, or separated into parts.

It has until quite recently been taught that matter was infinitely divisible; that is, a body could be separated into smaller and still smaller particles without limit. So far as our senses inform us, this is true. So long as we can perceive the existence of a portion of matter by our sense of sight, of feeling, of taste, or of smell, so long we can continue to divide it. Beyond this our senses give us no information. But the recent discoveries and investigations in chemistry, have proved beyond a doubt, that all bodies are ultimately composed of exceedingly minute particles, which can not be subdivided.

What is an Atom?

12. To such an ultimate portion of matter as is no longer separable into parts, we apply the term ATOM.

Extent to which matter can be divided.

The extent to which matter can be divided and yet perceived by the senses is most wonderful.

A grain of musk has been kept freely exposed to the air of a room, of which the door and windows were constantly kept open, for a period of two years, during all which time the air, though constantly changed, was densely impregnated with the odor of musk, and yet at the end of that time the particle was found not to have greatly diminished in weight. During all this period, every particle of the atmosphere which produced the sense of odor must have contained a certain quantity of musk.

In the manufacture of silver-gilt wire, used for embroidery, the amount of gold employed to cover a foot of wire does not exceed the 720,000th part of an ounce. The manufacturers know this to be a fact, and regulate the price of their wire accordingly. But if the gold which covers one foot is the 720,000th part of an ounce, the gold on an inch of the same wire will be only the 8,640,000th part of an ounce. We may divide this inch into one hundred pieces, and yet see each piece distinctly without the aid of a microscope: in other words, we see the 864,000,000th part of an ounce. If we now use a microscope, magnifying five hundred times, we may clearly distinguish the 432,000,000,000th part of an ounce of gold, each of which parts will be found to have all the characters and qualities which are found in the largest masses of gold.

Some years since, a distinguished English chemist made a series of experiments to determine how small a quantity of matter could be rendered visible to the eye, and by selecting a peculiar chemical compound, small portions of which were easily discernible, he came to the conclusion that he could distinctly see the billionth part of a grain.

In order to form some conception of the extent of this subdivision of matter, let us consider what a billion is. We may say a billion is a million of millions, and represent it thus, 1,000,000,000,000; but the mind is incapable of conceiving any such number. If a person were to count at the rate of 200 in a minute, and work without intermission twelve hours in a day, he would take, to count a billion, 6,944,944 days, or more than 19,000 years. But this may be nothing to the division of matter. There are living creatures so minute, that a hundred millions of them may be comprehended in the space of a cubic inch. But these creatures, until they are lost to the sense of sight, aided by the most powerful instruments, are seen to possess arrangements fitted for collecting their food, and even capturing their prey. They are therefore supplied with organs, and these organs must consist of parts corresponding to those in larger animals, which in turn must consist of atoms, or little particles, if we please so to term them. In reckoning the size of such atoms, we must not speak of billions, but of billions of billions. Such a number can be represented thus, 1,000,000,000,000,000,000,000,000, but the mind can form no rational conception of it.*

What are Molecules, or Particles of Matter?

13. We use the term MOLECULES, or PARTICLES of matter to designate very small quantities of a substance, not meaning, however, the ultimate atoms. A molecule, or particle of matter may be supposed to be formed of several atoms united together.

What are Pores?

14. No two atoms of matter are supposed to touch, or be in actual contact with each other, and the openings or spaces which exist between them are called PORES. This property of bodies, according to which their atoms are thus separated by vacant places, we call POROSITY.

What is Porosity?

* The billion is here used according to the English notation.—*Vide Webster.*

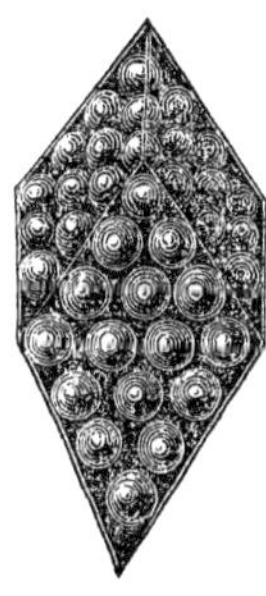
FIG. 1.

If we suppose the atoms of matter to consist of minute spheres or globes, it is obvious that it will be impossible for them to come into perfect contact at all points: so that there must be small spaces between them, where they do not touch each other. Fig. 1 represents the manner in which we may imagine a collection of such atoms to be arranged to form a crystal.

What is the evidence of the existence of Pores in all matter?

15. The reasons for believing that the atoms or particles of matter do not actually touch each other, are, that every form of matter, so far as we are acquainted with it, can by pressure be made to occupy a smaller space than it originally filled. Therefore, as no two particles of matter can occupy the same space at the same time, the space, by which the size or volume of a body may be diminished by pressure, must, before such diminution took place, have been filled with openings, or pores. Again, all bodies expand or contract under the influence of heat and cold. Now, if the atoms were in absolute contact with each other, no such movements could take place.

What is generally meant by the term Pores?

The porosity of bodies is sometimes illustrated and explained by reference to a sponge, which allows the cavities which pervade it to be filled with water, or some other fluid. Such an illustration is not strictly correct. The cavities of a sponge are not really its pores, any more than the cells of a honey-comb are the pores of wax. In common speech, however, the term pore is often used to designate those openings which exist naturally in the substance of a body, which are sufficiently large to admit of the passage of fluids like water, and gases like air.

Several very important properties of matter are dependent on porosity; or, in other words, they owe their existence to the fact, that the particles of matter do not actually touch each other. The principal of these are DENSITY, COMPRESSIBILITY, and EXPANSIBILITY. These properties of matter belong to all bodies, but not to all alike.

What is Density?

16. By DENSITY we mean the proportion which exists between the quantity of matter contained in a body and its magnitude, or size. Thus, if of two substances, one contains twice as much

matter in a given space as the other, it is said to be twice as dense.

There is a direct connection between the density of a body and its porosity. A body will be more or less dense, according as its particles are arranged closely together, or are separated from each other; and hence it is clear, that the greater the density the less the porosity, and the greater the porosity the less the density.

17. If the particles of a body do not touch each other, then, if it is subjected to pressure, they may be forced nearer, and made to occupy less space.

This we find to be the fact. All matter may be compressed. The most solid stone, when loaded with a considerable weight, is found to be compressed. The foundations of buildings, and the columns which sustain great weights in architecture, are proofs of this. Metals, by pressure and hammering, are made more compact and dense. Air, and all gases, are susceptible of great compression. Water, and all liquids, are much less easily compressed than either solid or gaseous bodies.

What is Compressibility?

18. By COMPRESSIBILITY, therefore, we mean that property of matter in virtue of which a body allows its volume or size to be diminished, without diminishing the number of the atoms or particles of which it is composed.

What is Expansibility?

19. Again, if the particles of matter of which a body is composed do not touch each other, it is clear that they may be forced further apart. This we find to be the case with all matter. Expansibility is, therefore, that property of matter in virtue of which a body allows its volume or size to be increased, without increasing the number of the atoms or particles of which it is composed.

Illustrations of Expansibility?

All bodies, when submitted to the action of heat, expand, and occupy a larger space than before. To this increase in dimensions there is no limit. Water, when sufficiently heated, passes into steam, and the hotter the steam the greater the space it will occupy. All bodies, if subjected to a sufficient degree of heat, will pass from the state of solids or liquids, into the state of vapor, or gases.

What is Inertia?

20. INERTIA signifies the total absence in a body of all power to change its state. If a body is at rest, it can not of itself commence moving; and if a body be in motion, it can not of itself stop, or come to rest. The motion, or cessation of

motion in a body, requires a power to exist independent of itself.

It is obvious, from the definition given, that when a body is once put in motion, its inertia will cause it to continue to move until its movement is destroyed, or stopped, by some other force.

A ball fired from a cannon would move on forever, were it not for the resistance or friction of the air, and the attraction of the earth.

What is Friction?

21. By FRICTION, we mean the resistance which a moving body meets with from the surface on which it moves.

A marble rolled upon a carpet will move but a short distance, on account of the roughness and unevenness of the surface. Its motion would be continued much longer on a flat pavement and longer still on fine, smooth ice. If friction, the attraction of the earth, and the resistance of the air, were entirely removed, the marble would move on forever.

What are Examples of Inertia?

Owing to the property of inertia, or the indifference of matter to change its state, we find it difficult, in running, to stop all at once. The body tends to go on, even after we have exerted the force of our muscles to stop. We take advantage of this property, by running a short distance when we wish to leap over a ditch or chasm, in order that the tendency to move on, which we acquire by running, may help us in the jump. For the same reason, a running-leap is always longer than a standing one.

Many of the most frightful railroad accidents which have happened, are due to the laws of inertia. The locomotive, moving rapidly, is suddenly checked by an obstruction, collision, or breakage of machinery; but the train of cars, in virtue of the velocity previously acquired, continue to move, and in consequence are driven into, or piled upon each other.

For the same reasons the wheel of an engine continues to pursue its course for a time after the driving force has stopped. This property is taken advantage of to regulate the motions of machinery. A large, heavy wheel is used in connection with the machinery, called a FLY-WHEEL. This heavy wheel, when once set in motion, revolves with great force, and its inertia causes it to move after the force which has been imparted to it has ceased to act. A water-wheel or a steam-engine rarely moves perfectly uniformly, but as it is not easy, on the instant, either to check or increase the movement of the heavy wheel, its motion is steady, and causes the machinery to which it is attached to work smoothly and without jerking, even if the action of the driving force be less at one moment than at another.

What is Attraction?

22. ATTRACTION is that tendency which all the particles of matter in the universe have to approach to each other.*

* As Attraction, in its various forms and relations to matter, is so comprehensive and important, it is treated separately in advance.

What are Examples of Attraction?

The force which holds the particles of a stone, a piece of wood, or metal together, the falling of a body to the earth, the tendency which a piece of iron or steel has to adhere to a magnet, are all familiar examples of the different forms of attraction.

Is Matter indestructible?

23. All the researches and investigations of modern science teach us, that it is impossible for any finite agent to either create or destroy a single particle of matter. The power to create and destroy matter belongs to the DEITY alone. The quantity of matter which exists, in and upon the earth has never been diminished by the annihilation of a single atom.

When a body is consumed by fire, there is no destruction of matter: it has only changed its form and position. When an animal or vegetable dies and decays, the original form vanishes, but the particles of matter, of which it was once composed, have merely passed off to form new bodies and enter into new combinations.

PRACTICAL QUESTIONS ON THE PROPERTIES OF MATTER.

1. Why will water, or any other liquid, when poured into a tunnel closely inserted into the mouth of a bottle, run over the sides of the bottle?

Because the bottle is filled with air, which, having no means of escape, prevents the water from entering, since no two bodies can occupy the same space at the same time. If, however, the tunnel be lifted from the bottle a little, so as to afford the air an opportunity to escape, the water will then flow into the bottle in an uninterrupted stream.

2. Are the pores of a body entirely empty, vacant spaces?

The pores of a body are often filled with another substance of a different nature. Thus, if the pores of a body be greater than the atoms of air, such a body being surrounded by the atmosphere, the air will enter and fill its pores.

3. When a sponge is placed in water, that liquid appears to penetrate it. Does the water really enter the SOLID particles of the sponge?

It does not; it only enters the *pores*, or vacant spaces between the particles.

4. When we plunge the hand into a mass of sand, do we PENETRATE the sand?

We do not; we only *displace* the particles.

5. Why do bubbles RISE to the surface when a piece of sugar, wood, or chalk is plunged under water?

Because the *air* previously existing in the pores becomes displaced by the water, and rises to the surface as bubbles.

6. What occasions the SNAPPING of wood or coal when laid upon the fire?

Because the air or liquid contained in the pores becomes expanded by heat, and bursts the covering in which it is confined.

7. Why does LIGHT, POROUS WOOD, like chestnut or pine, make more snapping in burning than any OTHER kind?

Because the pores are *very large*, and contain *more air* than wood of a *closer grain*, like oak, etc.

8. How is water, or any other liquid, made PURE by filtering through paper, cloth, a layer of sand, rock, etc.?

The process of filtration depends on the presence of pores in the substance used as a filter, of such magnitude as to allow the particles of liquid to pass freely, but not the particles of the matter contained in it, which we wish to separate.

9. Why is not the substance suitable for the filtration of ONE liquid equally adapted for the filtration of ALL liquids?

Because the magnitude of the pores in different substances and of the impurities in liquids is different; and no substance can be separated from a liquid by filtration, except one whose particles are larger than those of the liquid.

10. Gold and lead are metals of great density; their pores are not visible. Is there any PROOF of their existence beside the fact that they can be compressed?

Water can be forced mechanically through a plate of lead or gold without rupturing any portion of the metal. Mercury, or quicksilver, confined in a dish of lead or gold, will soak through the pores, and escape at the bottom.

An interesting experiment was tried at Florence, Italy, nearly two centuries ago, which furnished a striking illustration of the porosity of so dense a substance as gold. A hollow ball of this metal was filled with water, and the aperture exactly and firmly closed. The globe was then submitted to a very severe pressure, by which its figure was slightly changed. Now, it is proved in geometry, that a globe has this peculiar property—that any change whatever in its figure necessarily diminishes its volume, or capacity. The result was, that the water oozed through the pores, and covered the surface of the globe, presenting the appearance of dew, or steam cooled by the metal. This experiment also proved that the pores of the gold are larger than the elementary particles of water, since the latter are capable of passing through them.

11. When a CARRIAGE is in motion, drawn by HORSES, why is the same exertion of power in the horses required to STOP IT, as would be necessary to BACK IT, if it were at rest?

Because, according to the laws of inertia, the *force* required to destroy motion in one direction is *equal* to that required to produce as *much motion in the opposite direction.*

12. If a carriage, railroad-car, or boat, moving with speed, be suddenly STOPPED or RETARDED, from any cause, why are the passengers, or the baggage carried, precipitated from their places in the DIRECTION OF THE MOTION?

Because, by reason of their *inertia*, they *persevere* in the motion which they shared in common with the body that transported them, and are not deprived of that motion by the same cause.

13. Why will a PERSON, leaping from a carriage in rapid motion, fall in the direction in which the carriage is moving at the MOMENT his feet meet the ground?

Because his entire *body*, on quitting the vehicle and descending to the ground, *retains*, by its *inertia*, the progressive motion which it has in common with it. When his feet reach the ground, they, and they alone, will be suddenly deprived of this progressive motion, by the resistance of the earth, but the remainder of his body will retain it, and he will fall as if he were tripped.

14. Why is a man standing carelessly in the STERN of a boat liable to fall into the water behind, when the boat begins to move?

Because his *feet* are pulled forward while the *inertia of his body* keeps it in the same position, and, therefore, behind its support. For a similar reason, when the boat stops, the man is liable to fall forward.

15. When the sails of a ship are first spread to receive the FORCE OF IMPULSE of the wind, why does not the vessel acquire her full speed at once?

Because it requires a little time for the *impelling force* to overcome the *inertia* of the *mass* of the ship, or its disposition to remain at rest.

16. Why, when the sails are taken in, does the vessel continue to move for a considerable time?

Because the *inertia of the mass* is opposed to a change of state, and the vessel will continue to move until the resistance of the water overcomes the opposition.

17. Why do we KICK against the door-post to SHAKE the snow or dust from our SHOES?

The forward motion of the foot is arrested by the impact against the post; but this is not the case with respect to the particles of dust or snow which are not attached to the foot, and are free to move. According to the laws of inertia, they tend to persevere in the direction of the original motion, and when the foot stops, they move on, or fly off.

18. Why do we BEAT a coat or carpet to EXPEL the dust?

The cause which arrests the motion imparted to the coat or carpet by the blow does not arrest the particles of dust, and their motion being continued, they fly off.

CHAPTER II.

FORCE.

Is matter constantly changing?

23. MATTER is constantly changing its form and place. The most solid substance will in time wear away. The air about us is never perfectly still. We see water sometimes as ice, sometimes as a liquid, sometimes as a vapor, in steam or clouds. The earth moves sixty-eight thousand miles every hour. An animal or vegetable dies, decays, and its form vanishes from our sight.

To what cause do we attribute the changes observed in matter?

24. As the cause of all the changes observed to take place in the material world, we admit the existence of certain forces, or agents, which govern and control all matter.

What is Force?

25. FORCE is whatever produces, or opposes motion in matter.

What is Mobility?

26. MOBILITY, or the susceptibility of motion, is that property whereby a body admits of change of place.

What are the great forces in nature?

27. All the great forces, or agents in nature, those which produce, or are the cause of all the changes which take place in matter, may be enumerated as follows: INTERNAL, or MOLECULAR FORCES, the ATTRACTION of GRAVITATION, HEAT, LIGHT, the ATTRACTIVE and REPULSIVE FORCES of MAGNETISM and ELECTRICITY, and, finally, a force or power which only exists in living animals and plants, which is called, VITAL FORCE.

What do we know of the nature of these forces?

Concerning the real nature of these forces, we are entirely ignorant. We suppose, or say, they exist, because we see their effects upon matter. In the present state of science, it is impossible to know whether they are merely properties of matter, or whether they are forms of matter itself, existing in an exceedingly minute, subtile condition, without weight, and diffused throughout the whole universe. The general opinion, however, among scientific men, at the present day, is, that these forces, or agents, are not matter, but properties, or qualities, of matter.

We see a stone fall to the ground, and say that the cause of it is the attraction of gravitation;—we observe an object at a distance, and say that we see it through the action of light on the eye;—we notice a tree shattered by lightning, and say it is the effect of electricity;—we observe an animal or plant to grow and flourish, and ascribe this to the action of the vital force. But if it is asked, What is the original cause of gravitation, light, electricity, and vital force?—the wisest man can give no satisfactory answer. If the Creator governs matter through the agency of instruments, these forces may be called his agents, or his instruments.

CHAPTER III.

INTERNAL, OR MOLECULAR FORCES.

What is an Internal, or Molecular Force?

28. An Internal, or Molecular Force, is one that acts upon the particles of matter *only* at insensible distances. This variety of force differs from all others in this respect.

What is Attraction and Repulsion?

29. The various changes which matter undergoes, render it certain that the atoms, or particles of all bodies are acted upon by two distinct and opposite forces, one of which tends to draw the atoms, or particles, close together, while the other tends to separate them from one another. The first of these forces we call Attraction, the second Repulsion, both acting at insensible distances.

Give an example of Attraction as acting at an insensible distance.

A blade of steel, or a thin piece of wood, when bent within a certain limit, will, when the restraint is removed, restore itself to its original form. This takes place through the agency of an internal force, attracting the particles together, and tending to keep them in their original place.

What is Elasticity?

30. Elasticity is that property of matter which disposes it to resume its original form and shape, after having been bent or compressed by some external force.

Elasticity, therefore, is not so much a distinct property of matter, as is usually stated, as it is a phenomenon of attractive and repulsive forces.

Do all bodies possess elasticity?

All bodies possess the property of elasticity, but in very different degrees. There are some in which the atoms, after bending, or displacement, almost perfectly resume their former position. Such bodies are especially termed elastic, as tempered steel, India-rubber, ivory, etc. Other bodies, like iron, lead, etc., are elastic in a limited degree, not being able to bear any great displacement of their atoms without breaking, or permanent disarrangement. Putty, moist clay, and similar bodies, possess a very slight degree of elasticity.

Give an example of repulsion acting at an insensible distance.

31. If we compress a certain quantity of gas, as common air, and then allow it to dilate, by removing all restraint, it will expand without limit, and fill every really empty space which is open to it. This takes place through the agency of an internal force which tends to drive the particles from one another. There are many reasons which lead us to suppose that the repulsive force which tends to keep the particles of matter asunder is the agent known as heat. Gases may be considered as perfectly elastic.

In what three forms or conditions does all matter exist?

32. According as the attractive or repulsive forces prevail, all bodies will assume one of three forms or conditions—the SOLID, the LIQUID, or the AERIFORM,* or GASEOUS condition.

What is a Solid?

33. A SOLID body is one in which the particles of matter are attracted so strongly together, that the body maintains its form, or figure, under all ordinary circumstances.

What is a Liquid?

34. A LIQUID body is one in which the particles of matter are so feebly attracted together, that they move upon each another with the greatest facility.

Hence a liquid can never be made to assume any particular form, except that of the vessel in which it is inclosed.

What is a Gaseous Body?

35. An AERIFORM, or GASEOUS body is one in which the particles of matter are not held together by any force of attraction, but have a tendency to separate and move off from one another.

What are the properties of a Gaseous Body?

A gaseous body is generally invisible, and, like the air surrounding us, affords to the sense of touch no evidence of its existence when in a state of complete repose. Gaseous bodies may be confined in vessels, from whence they exclude liquids,

* Aeriform, having the form, or resemblance, of air.

or other bodies, thus demonstrating their existence, though invisible, and also their impenetrability.

Under what circumstances will a body assume the form of a Solid, a Liquid, or a Gas?

36. Most substances can be made to assume successively the form of a solid, a liquid, or a gas. In solids, the attractive force is the strongest; the particles keep their places, and the solid retains its form. But if we heat the solid to a sufficient degree, as, for example, a piece of iron, we gradually destroy the attractive force, and the repulsive force increases; the particles become movable, and we say the body melts, or becomes a liquid. In liquids, the attractive and repulsive forces are nearly balanced, but if we supply an additional quantity of heat, we destroy the attractive force altogether, and the liquid changes to a gas, in which the repulsive force prevails, and the particles tend to fly off from each other. By the withdrawal of heat (*i. e.*, by the application of cold), we can diminish, or destroy the repulsive force, and allow the attractive force to again predominate.

Fig. 2.

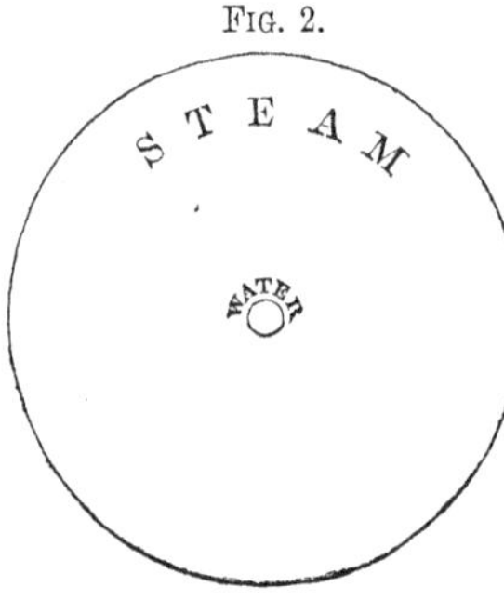

Thus steam, when cooled, becomes a liquid, water; and this in turn, by the withdrawal of an additional amount of heat, becomes a solid, ice.

The power of the repulsive force is strikingly illustrated by the conversion of water into steam. In a cubic inch of water converted into steam, the particles will repel each other to such an extent, that the space occupied by the steam will be 1700 times greater than that occupied by the water. Fig. 2 illustrates the comparative difference between the bulk of steam and the bulk of water.

What are Fluids?

37. The term Fluid is applied to those bodies whose particles move easily among themselves. It is used to designate either liquids or gases.

What are the four kinds of Molecular Attraction?

38. We distinguish FOUR kinds of molecular attraction, or attraction acting upon the particles of bodies at insensible distances. These

are, COHESION, ADHESION, CAPILLARY ATTRACTION, and AFFINITY.

What is Cohesive Attraction?

39. COHESION, or COHESIVE ATTRACTION, is that force which binds together atoms of the same kind to form one uniform mass.

The force which holds together the atoms of a mass of iron, wood, or stone, is cohesion, and the atoms are said to cohere to each other.

What is Adhesion?

40. ADHESION is that form of attraction which exists between unlike atoms, or particles of matter, when in contact with each other.

Dust floating in the air sticks to the wall or ceiling, through the force of adhesion. When we write on a wall with a piece of chalk, or charcoal, the particles, worn off from the material, stick to the wall and leave a mark, through the force of adhesion. Two pieces of wood may be fastened together by means of glue, in consequence of the adhesive attraction between the particles of the wood and the particles of glue.

What is Capillary Attraction?

41. CAPILLARY ATTRACTION is that form of attraction which exists between a liquid and the interior of a solid, which is tubular, or porous.

When one end of a sponge, or a lump of sugar is brought into contact with water, the liquid, by capillary attraction, will rise, or soak up above its level, into the interior of the sponge, or sugar, until all its pores are filled.*

What is Affinity?

42. AFFINITY is that form of attraction which unites atoms of unlike substances into compounds possessing new and distinct properties.

Oxygen, for example, unites with iron, and forms iron-rust, a substance different from either oxygen or iron. The consideration of the attraction of Affinity belongs wholly to Chemistry.

How does the force of Cohesive Attraction vary?

43. The force, or strength of Cohesive Attraction varies greatly in different substances, according as the nature, form, and arrangement of the atoms of which they are composed vary.

What properties of bodies depend on the variation of Attraction?

44. These modifications of the force of Attraction, acting at insensible distances between the atoms of different substances, give rise to certain important properties in bodies, which are designated under the names of MALLEABILITY, DUC-

* Capillary Attraction is treated of more fully under the department of Hydrostatics and Hydraulics.

TILITY, PLIABILITY, FLEXIBILITY, TENACITY, HARDNESS, and BRITTLENESS.

These are not, as is often taught, distinct, independent properties of matter, like magnitude, porosity, inertia, etc., but modifications of the force of attraction.

What is Malleability?

45. MALLEABILITY is that property in virtue of which a substance can be reduced to the form of thin leaves, or plates, by hammering, or by means of the intense pressure of rollers.

In malleable bodies, the atoms seem to cohere equally in whatever relative situations they happen to be, and therefore readily yield to force, and change their positions without fracture, almost like the atoms of a fluid.

What are examples of Malleability?

The property of malleability is possessed in the most eminent degree by the metals; gold, silver, iron, and copper being the most malleable. Gold may be hammered to such a degree of thinness, as to require 360,000 leaves to equal an inch in thickness.

What is Ductility?

46. DUCTILITY is that property in virtue of which a substance admits of being drawn into wire.

We might suppose that ductility and malleability would belong to the same substances, and to the same degree, but they do not. Tin and lead are highly malleable, and are capable of being reduced to extremely thin leaves, but they are not ductile, since they can not be drawn into fine wire. Some substances are both ductile and malleable in the highest degree. Gold has been drawn into wire so fine, that an ounce of it would extend fifty miles.

What are Flexibility and Pliability?

47. FLEXIBILITY and PLIABILITY are those properties which permit considerable motion of the particles of a body on each other, without breaking.

What is Tenacity?

48. TENACITY is that property in virtue of which a body resists separation of its parts, by extension in the direction of its length.

What is Hardness?

49. HARDNESS is a property in virtue of which the particles of a body resist impression, separation, or the action of any force which tends to change their form, or arrangement.

When is a body Soft?

50. A body, whose particles can be removed, and changed in position, by a slight degree of force, is said to be soft. SOFTNESS is, therefore, the opposite of hardness.

The property of Hardness is quite distinct from Density. Gold and lead possess great density, yet they are among the softest of metals.

What is Brittleness?

51. BRITTLENESS is a property in virtue of which bodies are easily broken into fragments. It is a characteristic of most hard substances.

In a brittle body, the attractive force between the atoms exists within such narrow limits, that a very slight change of position, or increase of distance among them, is sufficient to overcome it, and the body breaks.

52. The modifications of the force of cohesive attraction between the particles of matter, which give rise to the properties of malleability, ductility, flexibility, pliability, hardness, and brittleness, seem to be intimately connected with, or depend upon the particular form of the atoms of the substance, and the particular manner in which they are arranged.

Every one knows that it is easier to split wood lengthwise than across the fibers; hence, the force which binds the particles of the wood together is exerted in a less degree in one direction than in the other.

Explain how the force of attraction depends on the arrangement of the atoms.

By changing the form or arrangement of the atoms of a substance, we can in many instances apparently renew or destroy the various modifications of the attractive force. The following is a familiar illustration of this principle:

Steel, when heated and suddenly cooled, is rendered not only very hard, but very brittle; but if heated and cooled gradually, it becomes soft and flexible. We may suppose that when the atoms of steel are expanded—forced apart from each other by the action of heat, and then suddenly caused to contract—forced in upon each other—by cooling, that no opportunity is afforded them for arrangement in a natural manner. But when the steel is cooled slowly, each atom has an opportunity to take the place best adapted for it, without interfering with its neighbor. According to one arrangement of the atoms, the steel is brittle, or the atoms will not admit of any motion among themselves without breaking; but according to a different arrangement, the attractive force is modified, and the steel is soft and flexible. In a similar manner, bricks stacked up irregularly, may be made to fall easily, but if piled in a regular manner, they retain their stability.

It is a very singular circumstance, that the same operation of heating and cooling suddenly, which hardens steel, should soften copper. A piece of steel which has been hardened in this way is not condensed—made smaller—as we might have supposed it would be, but is actually expanded, or made larger. This proves that the arrangement of the atoms, or particles, has been changed. Any one may satisfy himself of this by taking a piece of steel, fitting it exactly into a guage, or between two fixed points, and then hardening it. It will then be found that the steel will not go into the guage, or between the fixed points.

What is Annealing?

53. The process of rendering metals, glass, etc., soft and flexible by heating and gradually cooling, is called ANNEALING, and is of great importance in the arts.

For example, the workman, in fashioning and shaping a steel instrument, requires it to be soft and flexible; but in using it after it has been constructed, as for the cutting of stone, wood, etc., it is necessary that it should be hard. This is accomplished by making the steel soft by annealing, and then rendering it hard by heating and cooling quickly.*

When will a body, bent or compressed, break?

54. When we bend or compress a body so that its particles are separated beyond a certain limited distance, the force of cohesive attraction existing between them ceases to act, or is destroyed, and the body falls apart, or breaks.

Can we restore the attraction of cohesion when destroyed?

55. When the Attraction of Cohesion between the particles of a substance is once destroyed, it is generally impossible to restore it. Having once reduced a mass of wood or stone to powder, we can not make the minute particles cohere again by pushing them into their former position.

In some instances, however, this can be accomplished by resorting to various expedients. The particles of the metals may be made to again cohere by melting. Two pieces of perfectly smooth plate-glass, or marble, laid upon each other, unite together with such force, that it is impossible to separate them without breakage. In the manufacture of looking-glass plates, this attraction between two smooth surfaces is particularly guarded against.

* There are many practical illustrations in the arts, of the principle, that the modifications of the attractive force which unites the atoms of solid bodies together, are dependent in a great degree upon the forms, or arrangement of the atoms themselves. If we submit a piece of metal to repeated hammering, or jarring, the atoms, or particles of which it is composed, seem to take on a new arrangement, and the metal gradually loses all its tenacity, flexibility, malleability, and ductility, and becomes brittle. The coppersmith who forms vessels of brass and copper by the hammer alone, can work on them only for a short time before they require annealing; otherwise they would crack and fly into pieces.

For this reason, also, a cannon can only be fired a certain number of times before it will burst, and a cannon which has been long in use, although apparently sound, is always condemned and broken up.

A more important illustration, and one that more closely affects our interests, is the liability of railroad car-axles and wheels to break from the same cause. A car-axle, after a long lapse of time and use, is almost certain to break.

That these phenomena are due to changes in the manner of the arrangement and the form of the particles, or atoms, of matter, was conclusively proved by an experiment made a few years since in France:—An accident having occurred upon a railroad, by the breaking of an axle, by which many lives were lost, the attention of scientific men was called to the fact, that the iron composing the axle, when first used, was strong, and capable of standing a test, but after use in locomotion for a certain period, could be broken by a force far inferior to that by which it had formerly been tested. Many suppositions were made to account for this phenomenon, when finally a person took a series of rods about the size of pipe-stems, all strong and tough, and, with great patience, allowed them to fall for hours and hours upon an anvil, thus producing rapid strokes and vibrations. After subjecting them for a long time to this treatment, he found that the rods could be snapped and broken into fragments almost as easily as rotten wood.

What is Welding?

56. Iron may be made to cohere to iron by heating the metal to a high degree, and hammering the two pieces together. The particles are thus driven into such intimate contact, that they cohere and form one uniform mass. This property is called WELDING, and only belongs to two metals, iron and platinum.

PRACTICAL QUESTIONS ON THE INTERNAL, OR MOLECULAR FORCES.

1. In what respect does a gas DIFFER from a liquid?

A liquid, like water, milk, syrup, etc., can be made to flow regularly *down a slope, or an inclined plane,* but a gas can not.

2. Why is a bar of IRON stronger than a bar of WOOD of the same size?

Because the cohesion existing between the particles of iron is *greater* than that existing between the particles of wood.

3. Why are the particles of a LIQUID more easily separated than those of a SOLID?

Because the cohesive attraction which binds together the particles of a liquid is much less strong than that which binds together the particles of a solid.

4. Why will a small needle, carefully laid upon the surface of water, FLOAT?

Because its weight is not sufficient to overcome the cohesion of the particles of water constituting the surface; consequently, it can not pass through them and sink.

5. If you drop water and laudanum from the same vessel, why will SIXTY drops of the water fill the same measure as ONE HUNDRED drops of laudanum?

The cohesion between the particles of the two liquids is different, being greatest in the water. Consequently, the number of particles which will adhere together to constitute a drop of water, is greater than in the drop of laudanum.

6. Why is the prescription of medicine by DROPS an unsafe method?

Because, not only do drops of fluid from the same vessel, and often of the same fluid from different vessels, differ in size, but also drops of the same fluid, to the extent of a third, from different parts of the lip of the same vessel.

7. Why are cements and mortars used to fasten bricks and stone together?

Because the adhesive attraction between the particles of brick and stone and the particles of mortar, is so strong, that they unite to form one solid mass.

8. How may the efficacy of a locomotive engine be said to depend upon the force of adhesion?

If there were no adhesion, or even insufficient adhesion, between the tire of the driving-wheel of the locomotive, and the rails upon which it presses, the wheel would turn without advancing.

This actually happens when the rails are greasy, or covered with frost and ice. The contact is thus interrupted, and the adhesion between the rail and wheel is impaired.

9. When a liquid adheres to a solid, what term do we apply to designate the act of adhesion?

Wetting. It is necessary that a liquid should adhere to the surface of a solid before it can be wet. Water falling upon an oiled surface does not wet it, because there is no adhesion between the particles of the oil and the particles of the water.

10. Why are drops of rain, of tears, and of dew upon the leaves of plants, generally spherical, or globular?

The force of cohesion always tends to cause the particles of a liquid, when unsupported, or supported on a surface having little attraction for it, to assume the form of a sphere—a globe, or sphere, being the figure which will contain the greatest amount of matter within a given surface.

This property of fluids is taken advantage of in the arts, in the manufacture of shot. The melted lead is made to fall in a shower, from a great elevation. In its descent the drops become globular, and before they reach the end of their fall become hardened by cooling, and retain their form.

CHAPTER IV.

ATTRACTION OF GRAVITATION.

What is Attraction of Gravitation?

57. The Attraction of Gravitation is that form of attraction, by which *all* bodies at sensible distances, tend to approach each other.

How does Gravitation differ from other forms of attraction?

Electricity and Magnetism attract bodies at sensible distances also, but their influence upon different classes of bodies varies, and is limited by distance. Molecular, or Internal Attraction, acts only at insensible distances. The Attraction of Gravitation acts at all distances, and upon all bodies.

What is the great law of the attraction of Gravitation?

58. Every portion of matter in the universe attracts every other portion, with a force proportioned directly to its mass, or quantity, and inversely as the square of the distance. This is the great general law of the Attraction of Gravitation.

By the Attraction of Gravitation being directly proportional to the mass of a body, we mean, that if of two bodies, the mass of one be twice as large as that of the other, its force of attraction will be twice as great: if it is only half as large, its attraction will be only half as great.

By the Attraction of Gravitation being inversely proportioned to the square of the distance, we mean, that if one body, or substance, attracts another body with a certain force at the distance of a mile, it will attract with four times that force at half a mile, nine times the force at one third of a mile, and so on, in like proportion. On the contrary, it will attract with but one fourth of the force at two miles, one ninth of the force at three miles, one sixteenth of the force at four miles, and so on, as the distance increases.

FIG. 3.

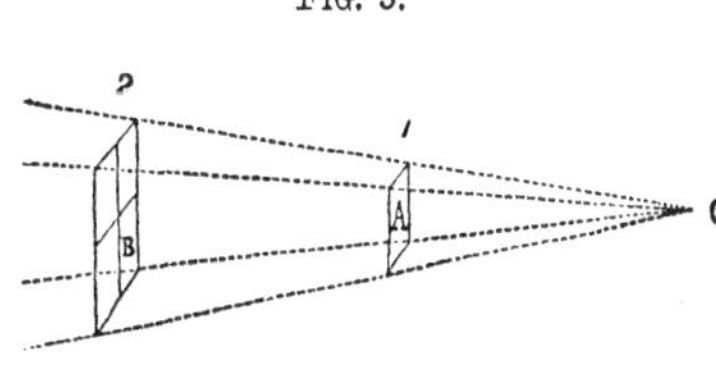

This law may be further illustrated by reference to Fig. 3. Let C be the center of attraction, and let the four dotted lines diverging from C represent lines of attraction. At a certain distance from C they will comprehend the small square A; at twice that distance they will include the large square B, four times the size of A; and since there is only a certain definite amount of attraction included within these lines, it is clear that as B is four times as great as A, the attraction exerted upon a portion of B equal to A, will be only one fourth that which it would experience when in the position marked 1, just half as far from C.

Why do not all bodies upon the earth's surface come in contact?

As gravitative attraction is the common property of all bodies, it may be asked, why all bodies not fastened to the earth's surface do not come in contact? They would do so, were it not for the overpowering influence of the earth's attraction, which in a great measure neutralizes, or overcomes, the mutual attraction of smaller bodies on its surface.

Does a feather attract the earth?

We throw up a feather into the air, and it falls through the influence of the earth's attraction; but as all bodies attract each other, the feather must also attract, or draw up, the earth, in some degree, toward itself. This it really does, with a force proportioned to its mass; but as the mass of the earth is infinitely greater than the mass of the feather, the influence of the feather is infinitely small, and we are unable to perceive it.

What are illustrations of Mutual Attraction?

In some instances, where bodies are free to move, the mutual attraction of all matter exhibits itself. If we place upon water, in a smooth pond, two floating bodies at certain distances from each other, they will eventually approach, the conditions affecting the experiment being alike for each. Two leaden balls suspended by a string near each other, are found, by delicate tests, to attract each other, and therefore not to hang quite perpendicular. A leaden weight suspended near the side of a mountain, inclines toward it to an extent proportionate to the magnitude of the mountain.

What is the cause of Tides?

The earth attracts the moon, and this in turn attracts the earth. The solid particles of matter upon the earth's surface, not being free to move, do not sensibly show the influence of the moon's attraction; but the particles of water composing the ocean, being free to move, furnish us evidence of this attraction, in the phenomena of the tides. When, by the revolution of the earth, a certain portion of its surface is brought within the direct influence of the moon's attraction, the surface of the ocean is attracted, or drawn up, to form a wave. This wave, or elevation of the surface of the water, occurring uniformly, is called a tide; when the moon is the nearest to the earth, its attraction is the greatest, and at these periods we have high tides, or "high water."

What is Terrestrial Gravity?

59. All bodies upon the earth are attracted toward its center. This we call Terrestrial Gravitation.

What is the law of the earth's attraction?

The attraction of the earth is not the same at all distances from the center, being greatest at the surface, and decreasing upward as the square of the distance from the center increases, and downward simply as the distance from the center decreases.

SECTION I.

WEIGHT.

How is a body at rest upon the surface of the earth attracted?

60. When a body falls to the earth, it descends because it is attracted toward the center of the earth. When it reaches the surface of the earth, and rests upon it, its tendency to continue to descend toward the center is not destroyed, and it presses downwards with a force proportioned to the degree by which it is attracted in this direction. This pressure we call Weight.

What is Weight?

61. Weight is, therefore, the measure of force with which a body is attracted by the earth. In ordinary language, it is the quantity of matter contained in a body, as ascertained by the balance.

How does Weight vary?

Weight being, then, the measure of the earth's attraction, it follows that as the attraction of the earth varies, weight must also vary, or a body will not have the same weight at all places.

Where will a body weigh the most, and where the least?

The weight of a body will be greatest at the surface of the earth, and greatest at those points upon the surface which are nearest the center.

As the earth is not a perfect sphere, but flattened at the poles, the poles are nearer the center than the equator. A body, therefore, will be attracted most strongly, that is, will weigh the most, at the poles, or at that portion of the earth's surface which is nearest the center, and weigh the least at the equator, or at that portion of the earth's surface which is most remote from the center.

A ball of iron weighing one thousand pounds in the latitude of the city of New York, at the level of the sea, will gain three pounds in weight, if removed to the north pole, and lose about four pounds if conveyed to the equator.

How does weight vary as we ascend from the earth's surface?

62. If a body be lifted above the surface of the earth, its weight will decrease in accordance with the law, that the attraction of gravitation decreases upward from the surface, as the square of the distance from the center of the earth increases.

The weight of a body, therefore, will be four times greater at the earth's surface, than at double the distance of the surface from the center; or a body weighing one pound at the earth's surface, will have only one fourth of that weight, if removed as far from the surface of the earth, as the surface is from the center.

How does weight vary as we descend from the surface?

63. As the attraction of gravitation decreases downward from the surface to the center of the earth, simply as the distance decreases, weight will decrease in like manner.

A body weighing a pound at the surface of the earth, will weigh only half a pound at one half the distance from the surface to the center.

Where will a body have no weight?

64. At the center of the earth a body will necessarily lose all weight, since, being surrounded on all sides by an equal quantity of matter, it will be attracted equally in all directions, and, therefore, can not exert a pressure greater in one direction than in another.

What are heavy and light bodies?

As the attractive force which the earth exerts upon a body is proportioned to its mass, or to the quantity of matter contained in it, and as weight is merely the measure of such attraction, it follows that a body of a large mass will be attracted strongly, and possess great weight, while, on the contrary, a body made up of a small quantity of matter, will be attracted in a less degree, and possess less weight. We recognize this difference of attraction by calling the one body heavy and the other light.

If, as is represented in Fig. 4, we place a mass of lead, *a*, at one extremity of a well-balanced beam, and a feather, *b*, at the other, we shall find that the lead is drawn to the earth with a force exactly equal to the superiority of its mass over that of the feather. If, however, we tie on a sufficient number of feathers to make up a quantity of matter equal to that of the lead, the equilibrium is restored—the two quantities are attracted with equal force, and the beam is supported in a horizontal position.

Fig. 4.

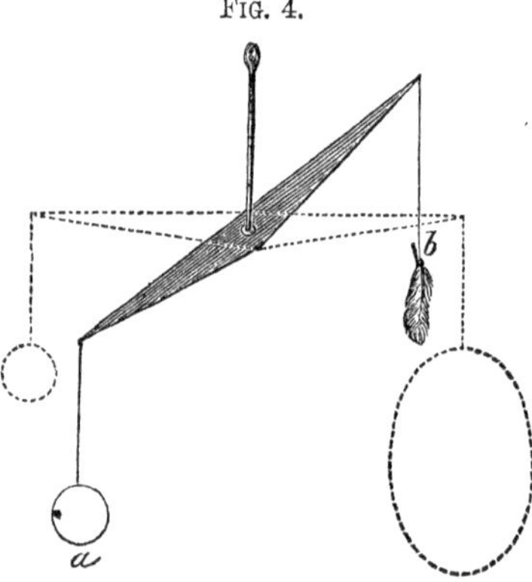

What is a System of Weights and Measures?

65. In all the operations of trade and commerce, we sell, or exchange a given quantity of one article or substance for a certain quantity of some other article or substance—so much flour for so much sugar, or so much sugar and flour for so much gold. Hence the necessity, which has existed from the earliest ages, of having some fixed rules or standards, according to which different quantities of different substances may be compared. A set, or series, of such rules or standards of comparison, is called a System of Weights and Measures.

What are the two great Systems of Weights and Measures?

Various nations adopt different standards, but in the civilized and commercial world, but two great Systems of Weights and Measures are generally recognized. These are known as the English, and the French Systems.

What are the peculiarities of the English System?

In the English System, which is the one used in the United States, there are two systems of weights—Troy and Avoirdupois Weight. Troy Weight is principally used for weighing gold and silver; Avoirdupois for weighing merchandize, other than the precious metals. It derives its name from the French *avoirs* (*averia*), goods or chattels, and *poids*, weight. The smallest weight made use of in the English System is a grain. By a law of England enacted in 1286, it was ordered that 32 grains of wheat, well dried, should weigh a pennyweight. Hence the name *grain* applied to this measure of weight. It was afterward ordered that a pennyweight should be divided into only 24 grains. Grain weights for practical purposes, are made by weighing a thin plate of metal of uniform thickness, and cutting out, by measurement, such a proportion of the whole as should give one grain. In this way, weights may be obtained for chemical purposes, which weigh only the 1,000th part of a grain.

How do we obtain a Standard of Weights and Measures?

66. In constructing a System of Weights and Measures, it is necessary, in the first place, to fix upon some dimension which shall forever serve as a standard from which all other weights and measures may be derived, and by which they may be compared and verified. If an artificial standard were taken, it is evident that it might be falsified, or even entirely lost or destroyed, thus creating great confusion. It is, therefore, necessary to fix upon some unchanging and invariable space or size in nature, which will always serve as a standard, and which the art of man can not affect. In the English System of Weights and Measures, such an unvarying dimension, or standard, is found in the length of a pendulum.

Describe the Pendulum.

67. A pendulum is a heavy body, suspended from a fixed point by a wire or cord, in such a manner that it may swing freely backward and forward. The alternate movements of a pendulum in opposite directions are called its VIBRATIONS, or OSCILLATIONS, and the part of a circle over which it moves is called its ARC.

In Fig. 5, A B represents a pendulum; D C, the arc in which it vibrates.

FIG. 5.

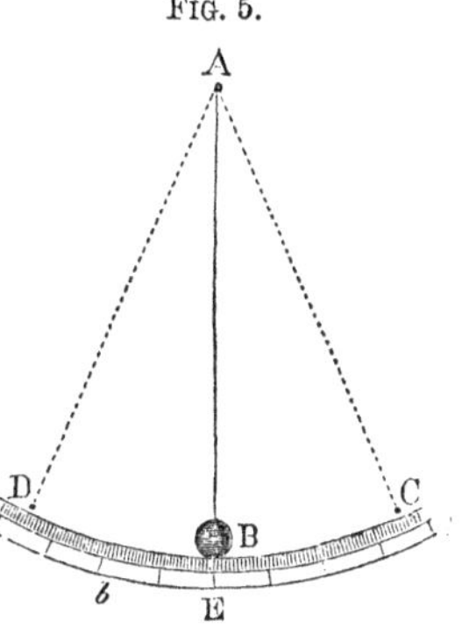

How does the Pendulum furnish a Standard of Measures of Length?

Now, it has been found that a pendulum, of any weight, which in the latitude of London will vibrate, or swing over the same arc, or from the highest point on one side, to the highest point on the other side, in one second of time, will always, under the same circumstances, have the same length. The length of this pendulum (the part A B, Fig. 5) is divided into 391,393 equal parts. Of these parts, 10,000 are called an inch, twelve of which make one foot, thirty-six of them one yard. Thus we obtain standards of linear measure.

How do we obtain a Standard of Weight?

To obtain a Standard of Weight, a cubic inch (*accurately obtained from the pendulum*) of distilled water, of the temperature of 62° Fahrenheit's thermometer, is taken and weighed. This weight is divided into 252,458 equal parts; and of these, 1,000 will be a *grain*. The grain multiplied, gives ounces, pounds, etc.

How do we obtain Standards of Liquid Measures?

To obtain standards of Liquid Measure, ten pounds, or 7,000 grains of distilled water, at the same temperature, are made to constitute a gallon. The gallon, by division, gives quarts, pints, and gills.

68. The French System of Weights and Measures is constructed on a different plan, and originated in the following manner :

Explain the construction of the French System of Weights and Measures.

In 1788, the *French Government*, feeling the necessity of having some *standard* by which all weights and measures might be compared and made uniform, ordered a scientific inquiry to be made; the result of which was the establishment of the present system of *French Weights and Measures*, which, from its perfect accuracy and simplicity is superior to all other systems. It is sometimes called the Decimal System, all its divisions being made by ten.

The French standard is based on an *invariable dimension of the globe*, viz., a *fourth part of the earth's meridian*, or the fourth part of the largest circle passing through the poles of the earth.

Fig. 6.

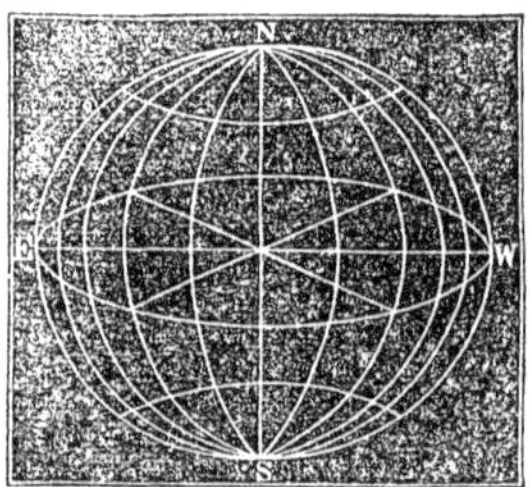

In Fig. 6, the circle N E S W represents a meridian of the earth; and a fourth part of this circle, or the distance N E, constitutes the dimension on which the French System is founded. This distance, which was accurately measured, is divided into ten million equal parts; and a single ten millionth part adopted as a measure of length, and called a *metre*. The length of the metre is about 39 English inches. By multiplying or dividing this quantity by ten, the other varieties of weights and measures are obtained.

69. In the United States, Standards of Weights and Measures, prepared according to the English System by order of the Government, are to be found at Washington, and at the capital of every State.

PRACTICAL PROBLEMS ON THE ATTRACTION OF GRAVITATION.

1. Suppose two bodies, one weighing 30 and the other 90 pounds, situated ten miles apart, were free to move toward each other, under the influence of mutual attraction: what space would each pass over before they came in contact?

The mutual attraction of any two bodies for each other is proportional to the quantity of matter they contain.

2. A body upon the surface of the earth weighs one pound, or sixteen ounces: if by

any means we could carry it 4,000 miles above the earth's surface, what would be its weight?

Solution: The force of gravity decreases upward, as the square of the distance from the center increases: weight, therefore, will decrease in like proportion. The distance of the body upon the surface of the earth, from the center, is 4,000 miles. Its distance from the center, at a point 4,000 miles above the surface, is 8,000. The square of 4,000 is 16,000,000; the square of 8,000 is 64,000,000. The weight, therefore, will be diminished in the proportion that sixty-four bears to sixteen; that is, it will be diminished $\frac{3}{4}$ths, or weigh $\frac{1}{4}$th of a pound, or 4 ounces.

3. What will be the weight of the same body removed 8,000 miles from the earth's surface?

4. A body on the surface of the earth weighs ten tons: what would be its weight if elevated 2,000 miles above the surface?

5. How far above the surface of the earth must a pound weight be carried, to make it weigh one ounce avoirdupois?

6. What would a body weighing 800 pounds upon the earth's surface, weigh 1,000 miles below the surface?

The force of gravity decreases as we descend from the surface into the earth, simply as the distance downward increases,—weight being the measure of gravity, it therefore decreases in the same proportion. The distance from the surface of the earth to the center may be assumed to be 4,000 miles: 1,000 miles is one fourth of 4,000. The distance being decreased one fourth, the weight is diminished in like proportion, and the body will lose 200 pounds, or its total weight would be 600 pounds.

7. Suppose a body weighing 800 pounds upon the surface of the earth were sunk 3,000 miles below the surface: what would be its loss in weight?

8. If a mass of iron ore weighs ten tons upon the earth's surface, what would it weigh at the bottom of a mine a mile below the surface?

9. What will be the weight of the same mass at the bottom of a mine one half a mile below the earth's surface?

SECTION II.

SPECIFIC GRAVITY, OR WEIGHT.

In what two senses may the term Weight be used?

70. A piece of iron sinks in water, and floats upon quicksilver. In the first instance, we say the iron sinks because it is heavier than water; and in the second, it floats, because it is lighter than quicksilver. Iron, therefore, is a heavy body compared with water, and a light body compared with mercury. But in ordinary language, we always consider iron as a heavy body. The term weight may, therefore, be used in two very different senses, and a body may be at once very light or very heavy according to the sense in which the terms are used. A mass of cork which weighs a ton is very heavy, because its absolute weight as indicated by the balance, viz., 2,000 pounds, is considerable. It is, however, in another sense, a light body, because if compared, bulk for bulk, with most other solid substances, its weight is very small. Hence we make a distinction between the absolute, or real weight of a body, and its specific, or comparative weight.

What is Absolute Weight? 71. The ABSOLUTE WEIGHT of a body is that of its entire mass, without any reference to its bulk, or volume.

What is Specific Weight? 72. The SPECIFIC WEIGHT, or the SPECIFIC GRAVITY of a body, is the weight of a given bulk, or volume of the substance, compared with the weight of the same bulk, or volume, of some other substance.

Why do we use the term "Specific," as applied to Weight? The term "Specific" Weight, or Gravity, is used, because bodies of different species of matter have different weights under equal bulks, or volumes. Thus, a cubic inch of cork, has a different weight from a cubic inch of oak, or of gold, and a cubic inch of water contains a less weight than a cubic inch of mercury. Hence we say that the specific gravity, or specific weight, of cork is less than that of oak or gold, and the specific gravity of mercury is greater than that of water.

What is the Standard for estimating the Specific Gravity of bodies? 73. Specific Gravity, or Weight, being merely the comparative gravity, or weight, it is convenient that some standard should be selected, to which all other substances may be referred for comparison. Distilled water has accordingly been taken, by common consent, as the standard for comparing the weights of all bodies in the solid, or liquid form. The reason for using distilled water is, that we may be certain of its purity.

Water, therefore, being fixed upon as the standard, we determine the specific gravity of a body, or we ascertain how much heavier or lighter a substance is than water, by the following rule:—

How do we find the Specific Gravity of bodies? 74. Divide the weight of a given bulk of the substance, by the weight of an equal bulk of water.

Explain the application of this rule. Suppose we take five vessels, each of which would contain exactly one hundred grains of water, and fill them respectively with spirits, ice, water, iron, and quicksilver. The following differences in weight will be found:—The vessel filled with spirits would weigh 80 grains; with ice, 90 grains; with water, 100 grains; with iron, 750 grains; with quicksilver, 1,350 grains.

Water having been selected as the standard for comparing these different weights, the question to be settled is simply this: How much lighter than water are spirits and ice, and how much heavier than water are iron and quicksilver; or, in other words, how many times is 100 contained in 80, 90, 750, and 1,350? The weights of the different substances filling the vessel are, therefore, to be divided by 100, the weight of the water; and there is found for spirits the weight 0·80, one fifth lighter than water; for the ice, 0·90, one tenth lighter than water; for the iron, 7·50, or seven and a half times heavier than water; for the quicksilver, 13·50, or thirteen and a half times heavier than water. *These numbers*, therefore, are the *specific gravities* of the *spirits*, *ice*, *iron*, and *quicksilver*.

How do we obtain the Specific Gravity of Liquid bodies?

For obtaining the specific gravity of Liquids the method above described is substantially the one usually adopted in the arts. A bottle capable of holding exactly 1,000 grains of distilled water, at a temperature of 60° Fahrenheit, is obtained, filled with water, and balanced upon the scales. The water is then removed, and its place supplied with the fluid whose specific gravity we wish to determine, and the bottle and contents again weighed. The weight of the fluid, divided by the weight of the water, gives the specific gravity required. Thus a bottle holding 1,000 grains of distilled water, will hold 1,845 grains of sulphuric acid; 1,845÷1,000=1.845, or, the sulphuric acid is 1.845 times heavier than an equal bulk of water.

When we immerse a body in water, what occurs?

For obtaining the specific gravity of solid bodies, a different method is adopted. When we immerse a body in water, it displaces a quantity of water equal to its own bulk. (In Fig. 7, the space occupied by the cube A B is obviously equal to a cube of water of the same size.) The water that before occupied the space which the body now fills was supported by the pressure of the other particles of water around it. The same pressure is exerted on the substance which we have immersed in the water, and, consequently, it will be supported in a like degree.

FIG. 7.

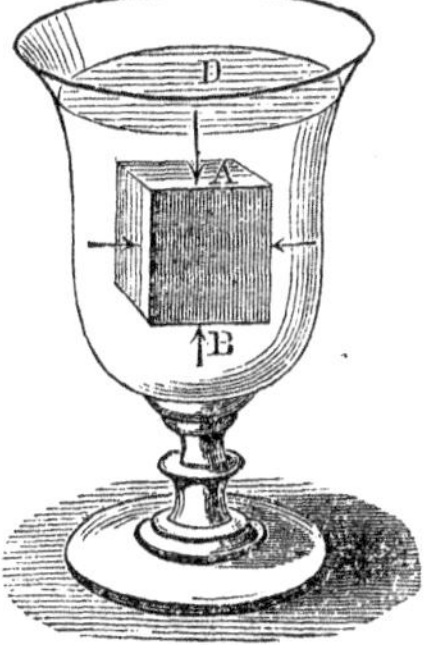

When will a body sink, and when float, in water?

If the body weighs less than an equal bulk of water, the pressure of the water will sustain it entirely, and the body will float; if, on the contrary, it is heavier than an equal bulk of water, the pressure of the particles of water will be unable wholly to sustain it, and, yielding to the attraction of gravitation, it descends, or sinks.

But to whatever extent a body may be supported in water, to the same extent it will cease to press downward, or its weight will diminish. We accordingly find, that a solid body, when immersed in water and weighed, will weigh less than when weighed in air, and the difference between these two weights will be equal to the weight of a quantity of water of the same size or bulk as the solid body; all bodies of the same size, therefore, lose the same quantity of their weight in water. To find the Specific Gravity of Solids heavier than water, or their weight compared with the weight of an equal bulk of water, we have the following rule:

How do we determine the Specific Gravity of Solids heavier than water?

75. Ascertain the weight of the body in water, and also in air. Divide the weight in air by the loss of weight in water, and the quotient will be the specific gravity required.

Fig. 8.

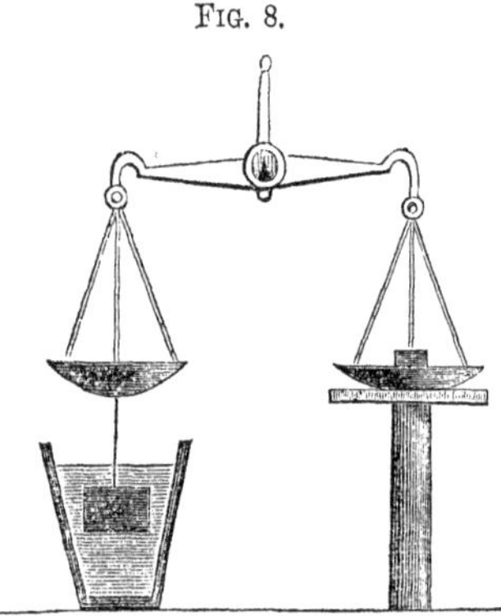

Suppose a piece of gold weighs in the air 19 grains, and in water 18 grains; the loss of weight in water will be 1; $19 \div 1 = 19$, the specific gravity of gold.

Fig. 8 represents the arrangement of the balance for taking specific gravities, and the manner of suspending the body in water from the scale pan, or beam, by means of a fine thread, or hair.

How do we find the Specific Gravity of a body lighter than water?

76. To find the specific gravity of a body lighter than water, tie it to some substance sufficiently heavy to sink it, whose weight in air and water is known. Weigh the two together, both in air and water, and ascertain the loss in weight. This loss will be the weight of as much water as is equal in bulk to the two solids taken together. Subtract the loss of the heavy body weighed by itself in water, previously known, from the loss sustained by the combined solids. The remainder will be the weight of as much water as is equal in bulk to the lighter body. Divide the weight of the lighter body in air by this remainder, and the quotient will be the specific gravity required.

Thus, for example, let the weight of the lighter solid be 3 ounces, and that of the heavier solid 15 ounces. Let the weight which the two together lose when submerged in water, be 5 ounces, and let the weight which the heavier alone loses when immersed be 1 ounce. Subtracting the loss of weight of the heavier body, in water, 1 ounce, from the combined loss of the two in water, 5 ounces, we have 4 ounces as the weight of a mass of water equal in bulk to the lighter body. But the weight of the lighter body in air is 3 ounces; $3 \div 4 = 0.75 = \frac{3}{4}$. It will, therefore, weigh three quarters of its own volume of water, or have a specific gravity 0.75.

How may we find the Specific Gravity directly by the balance?

77. The specific gravity of Liquids may also be found by the balance in the following manner: Weigh a solid body in water, as well as in the liquid whose specific gravity is to be determined; then the loss in each case will be the respective weights of equal bulks of water and liquid. We have therefore, the following rule:

78. Divide the loss of weight in the liquid by the loss

of weight in water; the quotient will give the specific gravity of the liquid.

Thus a solid body (a piece of glass is generally used) loses 20 grains when weighed in water, and 30 grains when weighed in acid; $30 \div 20 = 1.5$, the specific gravity of the acid.

79. There are various other methods of obtaining the specific gravity of solids and liquids.* Those we have described are the ones most generally adopted.

How do we obtain the Specific Gravity of a Gas?

80. For obtaining the specific gravity of gases, air instead of water is adopted as the standard of comparison. The weight of a given volume or measure of a gas is compared with the weight of an equal volume of pure atmospheric air, and the weight of the gas divided by the weight of the air, will express the specific gravity of the gas.

81. The following table exhibits the specific gravity of various solid, liquid, and gaseous bodies; pure water, having a temperature of 60 degrees Fahrenheit's thermometer, being assumed as the standard of comparison for solids and liquids, and pure, dry air, having the same temperature, being assumed as the standard of comparison for gases. The metal platinum has the greatest specific gravity of any solid body, being 21.50 times heavier than an equal bulk of water; and hydrogen gas the least specific gravity of any of the gases, being 14.4 lighter than an equal bulk of air, and 12.000 lighter than an equal bulk of water. These two substances are respectively the heaviest and lightest forms of matter with which we are acquainted.

SOLIDS AND LIQUIDS.

Substance	Specific gravity
Distilled water	1.000
Platinum	21.500
Gold	19.360
Mercury	13.600
Lead	11.450
Silver	10.500
Copper	8.870
Iron	7.800
Flint Glass	3.320
Marble	2.830
Anthracite coal	1.800
Box-wood	1.320
Sea-water	1.020
Whale oil	0.920
Pitch-pine wood	0.660

* See Hydrometer.

White pine	0.420
Alcohol	0.800
Ether	0.720
Cork	0.240
GASES.	
Pure, dry atmospheric air	1.000
Carbonic acid gas	1.520
Oxygen	1.100
Nitrogen	0.970
Ammoniacal gas	0.580
Hydrogen	0.070

How can we determine the absolute weight of a body from its Specific Gravity?

82. A cubic foot of water weighs almost exactly 1,000 ounces avoirdupois, or 62½ pounds. If, therefore, the specific gravity of water be represented by the number 1,000, the numbers which express the specific gravity of all other solids and liquids, will also express the number of ounces contained in a cubic foot of their dimensions. Thus, the specific gravity of gold being 19.360, it follows that a cubic foot of gold will weigh 19,360 ounces; and the specific gravity of cork being 0.240, the weight of a cubic foot of cork will be 240 ounces. By means of a table of specific gravities, therefore, the weight of any mass of matter can be ascertained, provided we know its cubical contents, by the following rule:

83. Multiply the weight of a cubic foot of water by the specific gravity of a substance; the product will be the weight of a cubic foot of that substance.

Thus, anthracite coal has a specific gravity of 1.800. This, multiplied by the weight of a cubic foot of water, 1,000 ounces, gives 1,800 ounces, which is the weight of a cubic foot of coal.

How can we ascertain the bulk of a substance from its Specific Gravity?

84. The volume, or bulk, of any given weight of a substance can also be readily calculated, by dividing the number expressing the weight in ounces by the number expressing the specific gravity of the substance, omitting the decimal points; the quotient will express the number of cubic feet in the volume, or bulk.

Thus, for example, if it be desired to ascertain the bulk of a ton of iron, it is only necessary to reduce the ton weight to ounces, and divide the number of ounces by 7.800, the specific gravity of iron; the quotient will be the number of cubic feet in the ton weight.

If the particles of matter were free to move, how would they arrange themselves?

85. If the particles of all matter were perfectly free to move among themselves, their arrangement in space would always be in ex-

act accordance with their different specific gravities: in other words, light bodies, or those having a small specific gravity, would rest upon, or rise above all heavier bodies, or those possessing a greater specific gravity.

What are illustrations of this principle?

In the case of different liquids, the particles of which are free to move among themselves, this arrangement always exists, so long as the different substances do not combine together, by the force of chemical attraction, to form a compound substance. Thus, water floats upon sulphuric acid, oil upon water, and alcohol upon oil, and by carefully pouring each of these liquids successively upon the surface of the other, they may be arranged in a glass in layers.

Carbonic acid gas is heavier than atmospheric air. We accordingly find that it accumulates at the bottom of deep pits, wells, caverns and mines.

Why does a balloon ascend, or a cork rise to the surface of water?

This principle also explains certain phenomena which at first seem opposed to the law of terrestrial gravity, that all matter is attracted toward the center of the earth. We observe a balloon, a soap-bubble, or a cloud of smoke or steam to ascend; and a cork, or other light body, placed at the bottom of a vessel of water, rises through it, and swims on the surface. These phenomena are a direct consequence of gravitation; the attraction of which, increasing with the quantity of matter, draws down the denser air and water to occupy the place filled by the lighter bodies, which are thus pushed up, and compelled to ascend.

Fig. 9.

Suppose *a*, Fig. 9, a ball of wood so loaded with lead that it will float exactly in the middle of a vessel of water. The weight of the wood and the upward pressure of the water have such a relation to each other, that the ball is balanced in this position. If now we add a few drops of strong salt and water, we shall see, as it sinks and mixes with the water, that the ball, *a*, is forced to the top of the fluid, because the attraction of gravitation on the denser fluid draws it down, and compels it to occupy the place of *a*.

The principle that the particles of liquids arrange themselves according to their specific gravities, has been taken advantage of in the West Indies by the slaves, in order to enable them to steal rum from casks. The long neck of a bottle filled with water, is inserted through the bung of the cask into the rum. The water falls out of the bottle into the cask, while the lighter rum rises to take its place.

Mention some of the practical applications of specific gravity.

The principle of specific gravity admits of many valuable applications in the arts. It offers a very sure and quick method of determining whether a substance is pure or adulterated. Thus, silver may be mixed with gold to a considerable extent, without changing, to any great degree, the appearance of the gold. The specific gravity of pure gold being 19, and of pure silver 10, it is obvious that a mixture of the two will have a specific gravity less than pure gold, and greater than pure silver, the difference being proportioned to the amount of adulteration. In the same way we can determine whether cheap oils have been mixed with expensive oils, cheap and poor illuminating gas, with expensive and brilliant gas. In any case it enables us to ascertain the exact size or solid bulk of a mass, however irregular—even of a bundle of twigs.*

PRACTICAL PROBLEMS RELATING TO SPECIFIC GRAVITY.

1. The weight of a solid body is 200 grains, but its weight in water is only 150 grains; what is the specific gravity of the body?

Solution: 50 grains = loss of weight in water; 200 grains (weight in air) ÷ 50 = 4, specific gravity required.

2. A body weighed in the air 28 pounds, and in water 24 pounds; what is its specific gravity?

3. An irregular fragment of stone weighed in air 78 grains, but lost 30 upon being weighed in water; what was the specific gravity of the stone?

4. A piece of cork weighed in the air 48 grains, and a piece of brass 560 grains; the brass weighed in water 488 grains, and the brass and cork when tied together weighed in water 336 grains. What was the specific gravity of the cork?

5. How much more matter is there in a cubic foot of sea-water, than in a cubic foot of fresh water?

6. Would a piece of steel sink or swim in melted copper?

7. When alcohol and whale-oil are put in the same vessel, which of these two substances will occupy the top, and which the bottom part of the vessel?

8. If a cubic foot of water weigh 1,000 ounces, what will be the weight of a cubic foot of lead?

9. What will be the weight of a cubic foot of cork, in ounces and in pounds?

---

* The attempt to ascertain whether a particular body had been adulterated led Archimedes, it is said, to the discovery of the principle of specific gravity. Hiero, King of Syracuse, having bought a crown of gold, desired to know if it were formed of pure metal; and as the workmanship was costly, he wished to accomplish this without defacing it. The problem was referred to Archimedes. The philosopher for some time was unable to solve it, but being in the bath one day, he observed that the water rose in the bath in exact proportion to the bulk of his body beneath the surface of the water. He instantly perceived that any other substance of equal size, would raise the water just as much, though one of equal weight and less size, or bulk, could not produce the same effect. Convinced that he could, by the application of this principle, determine whether Hiero's crown had been adulterated, and moved with admiration and delight, he is said to have leaped from the water and rushed naked into the street, crying "Ευρηκα! Ευρηκα!" "I have found it! I have found it!" In order to apply his theory to practice, he procured a mass of pure gold and another of pure silver, each having the same weight as the crown; then plunging the three metallic bodies successively into a vessel quite filled with water, and having carefully collected and weighed the quantity of liquid which was displaced in each instance, he ascertained that the mass of pure gold, of the same weight as the crown, displaced less water than the crown; the crown was, therefore, not pure gold. The mass of pure silver of the same weight as the crown, displaced more water than the crown; the crown, therefore, was not pure silver, but a mixture of gold and silver.

10. How many cubic feet in a ton of gold?

11. How many cubic feet in two tons of anthracite coal?

12. How many cubic feet in a ton of cork?

13. A fragment of metal lost 5 ounces when weighed in water; what were its dimensions, supposing a cubic foot of water to weigh 1,000 ounces?

Solution: The loss of weight in water, 5 ounces, is the weight of a bulk of water equal to that of the body. As we know the weight of a cubic foot of water, we can determine the number of cubic inches or feet in any given weight, thus: as 1,000 (the weight of a cubic foot of water in ounces) is to 5 ounces, so is 1,728 (the number of cubic inches in a cubic foot) to 8.64 cubic inches, the dimensions of the fragment.

14. Wishing to ascertain the number of cubic inches in an irregular fragment of stone, it was weighed in water, and its loss of weight observed to be 4.25 ounces. What were its dimensions?

SECTION III.

CENTER OF GRAVITY.

What is the Center of Gravity in a body?

86. The CENTER of GRAVITY in a body, is that point about which, if supported, the whole body will balance itself.

FIG. 10.

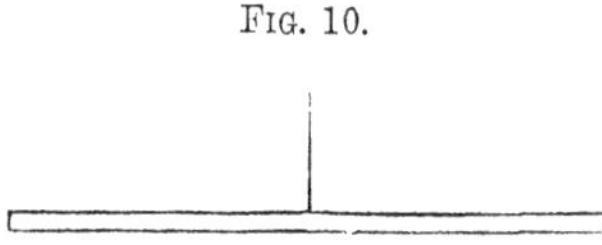

If we take a rod, or beam, of equal size throughout, and suspend it from the middle, Fig. 10, the two sides will exactly balance each other, and it will remain at rest in a horizontal position. There being as much matter similarly situated on one side of the support as on the other, the force of attraction exerted on both sides will be alike, and therefore one side can not overpower, or outweigh the other.

How may we consider the whole attraction exerted on a body concentrated at its Center of Gravity?

In every body, of whatever size or form, a point may be found, about which, if supported, all the parts of the body will balance, or remain at rest. Every body may be considered as made up of separate particles, each acted upon separately by gravity, but as by supporting this one point we support the whole, as by lifting it we lift the whole, and as by stopping it we cause the whole body to rest, the whole attraction exerted on the entire mass may be considered as concentrated at this one point, and this point we call the CENTER of GRAVITY.

What is the Center of Magnitude?

87. The CENTER OF MAGNITUDE of a body, is the central point of the bulk, or mass of the body.

Where is the Center of Gravity of a body?

88. When a body is of uniform density, the CENTER OF GRAVITY will coincide with its center of magnitude; but when one part of a body is composed of heavier materials than another part,

the center of gravity no longer corresponds with the center of magnitude, or the central point of the bulk of the body.

Fig. 11.

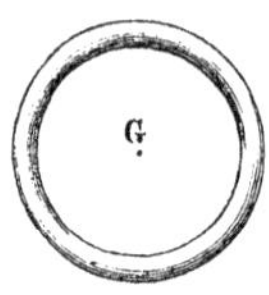

Thus, in a sphere, a cube, or a cylinder, the center of gravity is the same as the center of the body. In a ring of uniform size and density, the center of gravity is the center of the space inclosed in the ring (see Fig. 11). This example shows that the center of gravity is not necessarily included in that portion of space occupied by the matter of the body.

In a wheel of wood of uniform density and thickness the center of gravity will be the center of the wheel, but if a part of the rim be made of iron, the center of gravity will be removed to some point aside from the center.

When two bodies are connected together, they may be regarded as one body, having but one center of gravity. If the two bodies be of equal weight, the center of gravity will be in the middle of the line which unites them; but if one be heavier than the other, the center of gravity will be as much nearer the heavier body, as the heavier exceeds the lighter one in weight.

Fig. 12.

Thus, if two balls, each weighing four pounds, be connected together by a bar, the center of gravity will be a point on the bar equally distant from each. But if one of the balls be heavier than the other, then the center of gravity will, in proportion, approach the larger ball. This is illustrated by reference to Fig. 12, in which the center of gravity about which the two balls support themselves, is seen to be nearest to the heavier and larger ball.

When will the Center of Gravity be in permanent rest, or equilibrium?

89. The center of gravity of a body being regarded as the point in which the sum of all the forces of gravity acting upon the separate particles of the body are concentrated, it may be considered as influenced by the attraction of the earth in a greater degree than any other portion of the body. It follows, therefore, that if a body has freedom of motion, it can not be brought into a position of permanent equilibrium, until its center of gravity occupies the lowest situation which the support of the body will allow; that is, the center of gravity will descend as far toward the center of the earth as possible.

What do we mean by Equilibrium?

90. By Equilibrium we mean a state of rest produced by the counterpoise, or balancing, of opposite forces.

Thus when one force tending to produce motion in one direction, is opposed by an equal force tending to produce motion in an exactly opposite direction, the two balance each other, and no motion results. To produce any action, there must be an inequality in the condition of one of the forces.

By what experiment can you illustrate this principle?

The truth of this principle may be illustrated by certain experiments which at first seem to be contradictory to it. Thus a cylinder may be made to roll up an inclined plane. Fix a piece of lead, *l*, Fig. 13, on one side of the cylinder *a*, so that the center of gravity of the cylinder will be at the point *l*, while its center of magnitude is at *c*. The cylinder will then roll up the inclined plane to the position *a l*, because the center of gravity of the mass, *l*, will endeavor to descend to its lowest point.

Fig. 13.

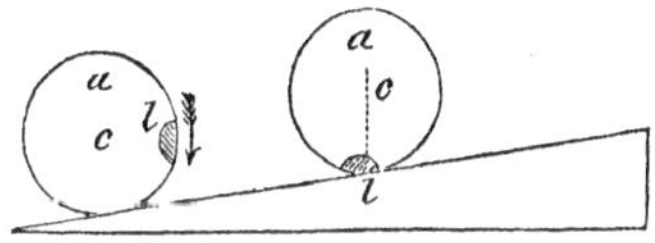

91. A prop that supports the center of gravity supports the whole body. This support may be applied in three different ways :

In what three ways may the Center of Gravity be supported?

1. The point of support may be applied directly to the center of gravity of the body.

2. The point of support may have the center of gravity immediately below it.

3. The point of support may have the center of gravity immediately above it.

Illustrate the first case.

In the first case, where the point of support is applied directly to the center of gravity, the body will remain at rest in any position; this is illustrated in the case of a common wheel, where the center of gravity is also the center of the figure, and this being supported on the axle, the wheel rests indifferently in any position. In Fig. 14, let *a*, the center of the wheel, which is also its center of gravity, be supported by an axle;—the wheel rests, no matter to what extent we turn it.

Fig. 14.

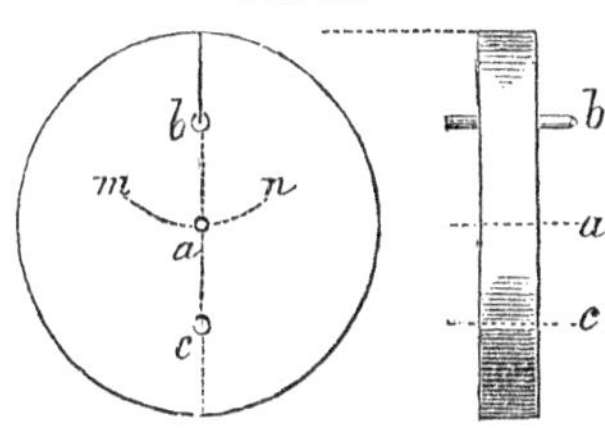

In the second case, where the point of support is above the center of gravity, the body, if it is allowed freedom of motion, will not rest in perfect equilibrio until its center of gravity has descended to the lowest position, which in all cases will be immediately beneath the point of suspension.

Illustrate the second case.

Thus, in Fig. 14, let the wheel, the center of gravity of which is at *a*, be suspended from the point *b*, by a thread, or hung upon an axle, having freedom of motion on that point. However much we may move it, either right or left, toward *m* or *n*, as shown by the dotted lines, *am* and *an*, it swings back again, and is only at rest when *b* and *a* are in the same perpendicular line.

Illustrate the third case.

In the third case, where the point of support has the center of gravity above it, a body will remain at rest only so long as the center of gravity is in a vertical line, above the point of support. In Fig. 14, suppose the wheel to be supported at the point *c*, situated in a vertical line *a c*, immediately below the center of gravity, *a*; so long as this position is maintained, the wheel will remain at rest, but the moment the center of gravity, a, is moved a little to the right or left, so as to throw it out of the vertical line joining a and c, the wheel will turn over, and assume such a position as to bring the center of gravity immediately beneath the point of support, as in the second case.

Upon what does the stability of a body depend?

92. The stability of a body, therefore, depends upon the manner in which it is supported, or in other words, upon the position of its center of gravity.

What are the three conditions of Equilibrium?

93. As a body may be supported in three positions, we have, as a consequence, three conditions of equilibrium, viz., Indifferent, Stable, and Unstable Equilibrium.

What is Indifferent Equilibrium?

INDIFFERENT EQUILIBRIUM occurs when a body is supported upon its center of gravity; for then it remains at rest indifferently in every position.

What is Stable Equilibrium?

STABLE EQUILIBRIUM occurs when the point of support is above the center of gravity. If a body be moved from this position, it swings backward and forward for a time, and finally returns to its original situation.

What is Unstable Equilibrium?

UNSTABLE EQUILIBRIUM occurs when the point of support is beneath the center of gravity. The tendency of the center of gravity in such cases is to change, and take the lowest situation the support of the body will allow.

How may we determine the center of gravity in irregular bodies?

94. The principle that when a body is suspended freely, it will have its center of gravity in a vertical line, immediately below the point of support, has been taken advantage of to determine experimentally the position of the center of gravity, in irregular shaped bodies. Suppose we suspend, as in Fig. 15, an irregular piece of board by means of cord. A plumb-line let fall from the point of support, or the prolongation of the cord, will pass through the center of gravity, G. If we now attach the cord to another point, and suspend the body anew, the prolongation of the cord in this instance, also, will pass through the center of gravity, G. The intersection of these two lines will be the center of gravity, and the board, if suspended by a cord attached to this point, will hang evenly balanced.

FIG. 15.

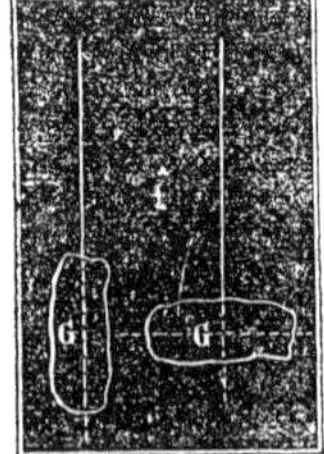

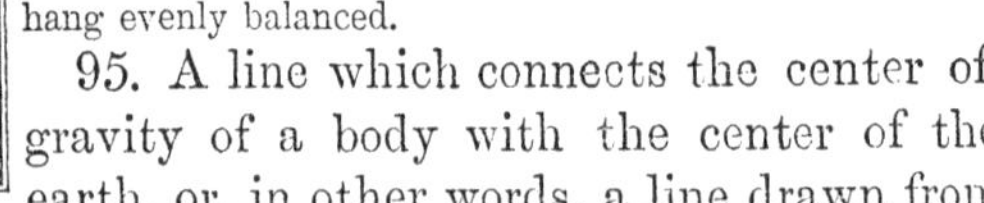

95. A line which connects the center of gravity of a body with the center of the earth, or, in other words, a line drawn from the center of gravity perpendicularly downward, is called the LINE OF DIRECTION. It is called the Line of Direction,

What is the Line of Direction?

because when a solid body falls, its center of gravity moves along this line until it reaches the ground. When bodies are supported upon a basis, their stability depends on the position of their Line of Direction.

When will a body stand, and when will it fall?

96. If the line of direction falls within the base upon which the body stands, the body remains supported; but if it falls without the base, the body overturns.

Fig. 16.

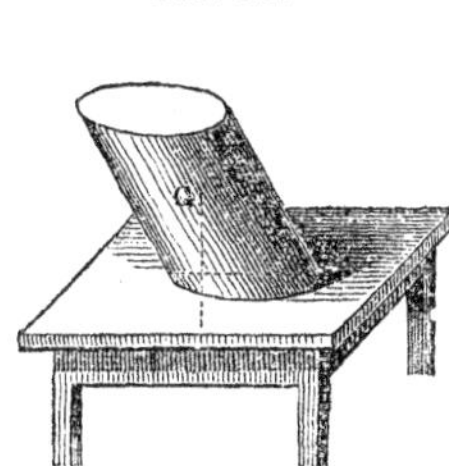

Fig. 17.

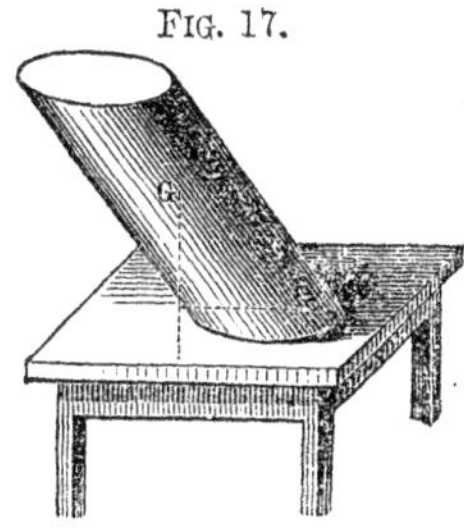

Thus, in Fig. 16, the line directed vertically from the center of gravity, G, falls within the base of the body, and it remains standing; but in Fig. 17 a similar line falls without the base, and the body, consequently, can not be maintained in an upright position, and must fall.

A wall, or tower stands securely, so long as the perpendicular line drawn through its center of gravity falls within its base. The celebrated leaning-tower of Pisa, 315 feet high, which inclines 12 feet from a perfectly upright position, is an example of this principle. For instance, the line in Fig. 18, falling from the top of the tower to the ground, and passing through the center of gravity, falls within the base, and the tower stands securely. If, however, an attempt had been made to build the tower a little higher, so that the perpendicular line passing through the center of gravity, would have fallen beyond the base, the structure could no longer have supported itself.

Fig. 18.

97. The broader, or larger

3

the base of a body, and the nearer its principal mass is to the base, or, in other words, the lower its center of gravity is, the firmer it will stand.

When will a body stand most firmly?

A pyramid, for this reason, is the firmest of all structures.

What is the advantage of turning out the toes in walking?

The base upon which the human body rests, or is supported, is the two feet and the space included between them. The advantage of turning out the toes when we walk is, that it increases the breadth of the base supporting the body, and enables us to stand more securely.

In every movement of the body, a man adjusts his position unconsciously, in such a way as to support the center of gravity, and cause the line of direction to fall within the base.

Why does a person carrying a load upon his back bend over?

A person carrying a load upon his back, bends forward in order to bring the center of gravity and his load over his feet.

Fig. 19.

Fig. 20.

If he carried the load in the position of A, Fig. 19, he would be liable to fall backward, as the direction of the center of gravity would fall beyond his heels; to bring the center of gravity over his feet, he assumes the position indicated by B, Fig. 20.

Why does a person lean forward in ascending a hill, and backward in descending?

For the same reason, when a man ascends a hill he leans forward, and when he descends he leans backward. See Fig. 21.

Fig. 21.

Why is a high carriage more liable to overturn than a low one?

A high carriage is much more liable to be overset by an irregularity in the road than a low one; because the center of gravity being high, the line of direction is easily thrown without the base. This will appear evident from the following illustration, Fig. 22.

FIG. 22.

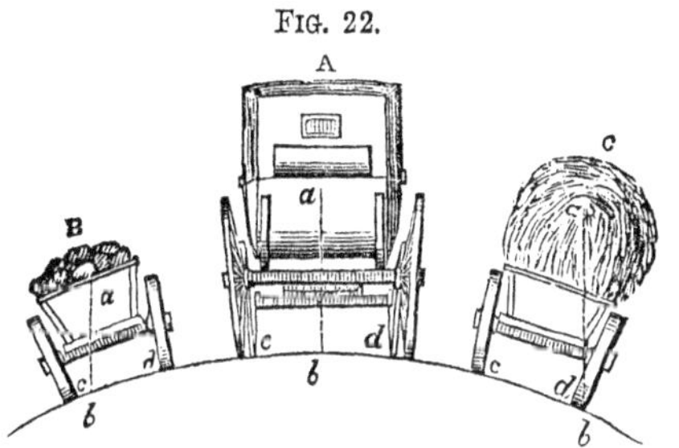

Let A represent a coach standing on a level; B, a cart loaded with stones on a slope; C, a wagon loaded with hay on a slope; *a a a* the centers of gravity; *a b*, line of direction; *c d*, base.

Here it is obvious that the hay-wagon must upset, because the line of direction falls without the base; that the coach is very secure, because the line of direction falls far within the base; and the stone-cart, though the center of gravity is low down, is not very secure, because the line of direction falls very near the outside of the base.

FIG. 23.

The effect on the stability of a body occasioned by placing its center of gravity in a very low position, is shown in an amusing toy for children, represented by Fig. 23. The horse, with his rider, is firmly supported on his hind feet, because, by means of a leaden ball attached to the bent wire, the center of gravity is brought below the point of support.

When will a body slide and when roll down a slope?

If a body be placed on an inclined surface, it will slide down when its line of direction falls within the base; but it will roll down when it falls without the base. Thus the body, *e*, Fig. 24, having its line of direction *e a*, within the base, will slide down the inclined surface, *c d;* but the body *b a*, will roll down, since its line of direction, *b a*, falls without the base.

FIG. 24.

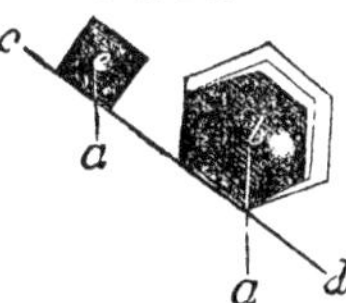

PRACTICAL QUESTIONS ON THE CENTER OF GRAVITY.

1. Why does a person in rising from a chair bend forward?

When a person is sitting, the center of gravity is supported by the *seat;* in an erect position, the center of gravity is supported by the *feet;* therefore, before rising it is necessary to change the center of gravity, and, by bending forward, we transfer it from the chair to a point over the feet.

2. **Why is a turtle placed on its back unable to move?**

Because the center of gravity of the turtle is, *in this position, at the lowest point*, and the animal is unable to change it; therefore it is obliged to remain at rest.

3. **Why do very fat people throw back their head and shoulders when they walk?**

In order that they may effectually keep the center of gravity of the body over the base formed by the soles of the feet.

4. **Why can not a man, standing with his heels close to a perpendicular wall, bend over sufficiently to pick up any object that lies before him on the ground, without falling?**

Because the wall prevents him from throwing part of his body backward, *to counterbalance the head and arms* that must project forward.

5. **What is the reason that persons walking arm-in-arm shake and jostle each other, unless they make the movements of their feet to correspond, as soldiers do in marching?**

When we walk at a moderate rate, the center of gravity comes alternately over the right and over the left foot. The body advances, therefore, in a *waving line;* and unless two persons walking together keep step, the waving motion of the two fails to coincide.

6. **In what does the art of balancing or walking upon a rope consist?**

In keeping the center of gravity in a line over the base upon which the body rests.

7. **Why is it a very difficult thing for children to learn to walk?**

In consequence of the natural upright position of the human body, it is constantly necessary to employ some exertion to keep our balance, or to prevent ourselves from falling, when we place one foot before the other. Children, after they acquire strength to stand, are obliged to acquire this knowledge of preserving the balance by experience. When the art is once acquired, the necessary actions are performed involuntarily.

8. **Why do young quadrupeds learn to walk much sooner than children?**

Because a body is tottering in proportion to its great *altitude* and *narrow base.* A child has a body thus constituted, and learns to walk but slowly because of this difficulty (perhaps in ten or twelve months), while the young of quadrupeds, having a *broad supporting base*, are able to stand and move about almost immediately.

9. **Are all the limbs of a tall tree arranged in such a manner, that the line directed from the center of gravity is caused to fall within the base of the tree?**

Nature causes the various limbs to shoot out and grow from the sides with as much exactness, in respect of keeping the center of gravity within the base, as though they had been all arranged artificially. Each limb grows, in respect to all the others, in such a manner as to preserve a due balance between the whole.

SECTION IV.

EFFECTS OF GRAVITY AS DISPLAYED BY FALLING BODIES.

What is a Vertical Line? 98. When an unsupported body falls, its motion will be in a straight line toward the center of the earth. This line is called a VERTICAL LINE.

What is a Plumb Line? 99. If a body be suspended by a thread, the thread will always assume a vertical direction, or it will represent that path in which the body would have fallen. A weight thus suspended by a thread, is called a PLUMB-LINE,* Fig. 25, and is used by carpenters, masons, etc., to ascertain by comparison, whether their work stands in a vertical or perpendicular position.

FIG. 25.

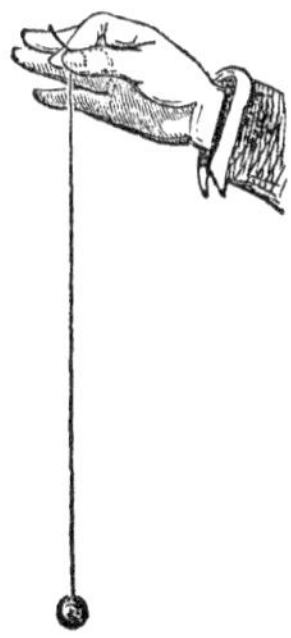

What is a Level Surface? 100. A plumb-line is always perpendicular to the surface of water at rest. The position of such a surface we call LEVEL.

No two plumb-lines upon the earth's surface will be parallel, but will incline toward each other, since no two bodies from different points can approach the center of a sphere in a parallel direction. If their distance apart be one mile, this inclination will amount to one minute, and if it be sixty miles, to one degree. In Fig. 26, let E E be a portion of the earth's surface, and D its center; the bodies A, B, and C, when allowed to drop, will fall in the direction A D, B D, and C D.

FIG. 26.

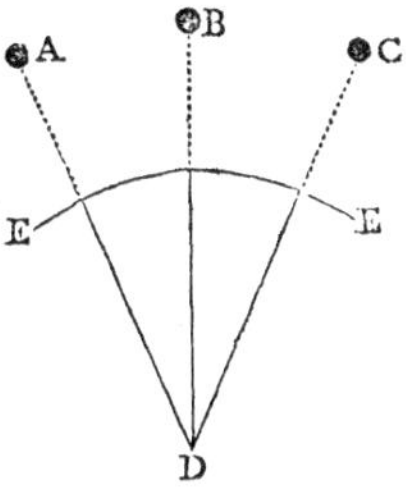

Will all bodies, under the influence of gravity alone, fall with equal velocities? 101. As the attraction of the earth acts equally and independently on all the particles composing a body, it is clear that they must all fall with equal velocities. It makes no difference whether the several particles fall singly, or whether they fall compacted together, in the form of a large or a small body.

* Plumb Line, so called from the Latin word *plumbum*, lead, the weight usually attached to the string.

If ten or a hundred leaden balls be disengaged together, they will fall in the same time, and if they be molded into one ball of great magnitude, it will still fall in the same manner.

102. Hence all bodies under the influence of gravity alone, must fall with equal velocities.*

By what experiment can you prove this law?

There are some familiar facts which seem to be opposed to this law. When we let go a feather and a mass of lead, the one floats in the air, and the other falls to the ground very rapidly. But in this case, the operation of gravity is modified by the resistance of the air; the feather floats because the air opposes its descent, and it can not overcome the resistance offered. But if we place a mass of lead and a feather in a vessel exhausted of air, and liberate them at the same time, they will fall in equal periods. The experiment is easily shown by taking a glass tube, Fig. 27, closed at one end, and supplied with an air-tight cap and screw-cock at the other. A feather and a piece of metal are previously inclosed in the tube. The tube being filled with air, and inverted, the metal will fall with greater speed than the feather, as might be expected. If the tube be now exhausted of air by means of an air-pump and the screw-cock, and in this condition inverted, the feather and the metal will fall from end to end of the tube with equal velocity.

Fig. 27.

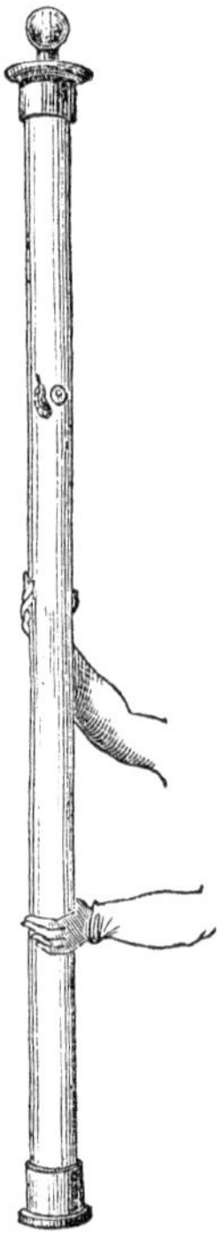

Upon what do the force and velocities of falling bodies depend?

103. If a man leap from a chair or table, he will strike the ground without injury. If the same man leap from the top of a high house, he will probably be killed. These, and many like instances, prove that the force with which a falling body strikes the ground depends upon the height from which it falls. But the force depends on the velocity of the body the moment it touches the ground; therefore, the velocity with which a body falls depends also upon the height from which it descends.

* Previous to the time of Galileo, the philosophers maintained that the velocity of a falling body was in proportion to its weight, and that if two bodies of unequal weights, were let fall from an elevation, at the same moment, the heavier would reach the ground as much sooner than the lighter, as its weight exceeded it. In other words, a body weighing two pounds would fall in half the time that would be required by a body weighing one pound. Galileo, on the contrary, asserted that the velocity of a falling body is independant of its weight, and not affected by it. The dispute running high, and the opinion of the public being generally averse to the views of Galileo, he challenged his opponents to test the matter by a public experiment. The challenge was accepted, and the celebrated leaning-tower of Pisa agreed upon as the place of trial. In the presence of a large concourse, two balls were selected, one having exactly twice the weight of the other. The two were then dropped from the summit of the tower at the same moment, and, in exact accordance with the assertions of Galileo, they both struck the ground at the same instant.

How does gravity act on a falling body?

104. When a body falls, it is attracted by gravity during the whole time of its falling. Gravity does not merely set the body in motion and then cease, but it continues to act. During the first second of time, the force of gravity will cause the body to descend through a certain space. At the end of this time, the body would continue to move, with the motion it has acquired, without the action of any further force, merely on account of its inertia. But gravity continues to act, and will add as much more motion to the falling body during the second second of time, as it did during the first second, and as much again during the third second, and so on.

What is the law of falling bodies?

105. Falling bodies, therefore, descend to the earth with a uniform accelerated motion. A body falling from a height will fall 16 feet in the first second of time,* three times that distance in the second, five times in the third, seven in the fourth, the spaces passed over in each second increasing as the odd numbers 1, 3, 5, 7, 9, 11, etc.

How does the space passed over and the time of a falling body compare?

106. The entire space passed over by a body in falling is as the square of the time; that is, in twice the time it will fall through four times the space; in thrice the time, nine times the space.†

The time occupied in falling, therefore, being known, the height from which a body falls may be calculated by the following rule:

Time being given, how can the height from which a body falls be found?

107. Multiply the square of the number of seconds of time consumed in falling, by the distance which a body will fall in one second of time.

Thus, a stone is five seconds in falling from the top of a precipice; the square of five seconds is 25; this multiplied by 16, the number of feet a body will fall in one second, gives 400—the height of the precipice.

How do the velocities and times of falling compare?

108. As the effect of gravity is to produce a uniform accelerated motion, the velocity of a falling body will increase as the time increases.

* The spaces described by falling bodies are here given in round numbers, the fractions being omitted. The space described by a falling body during the first second is 16 1-10th feet.

† The resistance of the air essentially modifies the laws of the motions of falling bodies, as here stated, and with a certain velocity, will become equal to the weight of the falling body. After this takes place, the body will descend with a uniform velocity. There is, therefore, a limit to the velocity which a body can acquire by falling through the atmosphere.

Thus, at the end of two seconds, the velocity acquired by a falling body will be twice as great as at the end of one second, thrice as great at the end of the third second, and so on.

How are bodies projected upward influenced by gravity?

109. Bodies projected directly upward, will be influenced by gravitation in their ascent, as well as in their descent, but in a reversed order; producing continually retarded motion while they are rising, and continually increasing motion during their fall.

Thus, a body projected up perpendicularly into the air, if not influenced by the resistance of the air, would rise to a height exactly equal to that from which it must have fallen to acquire a final velocity the same as it had at the first instant of its ascent.

How can we determine the height which a body projected upward with a given velocity will ascend?

110. To determine the height to which a body projected upward will rise, with a given velocity, ascertain the height from which a body would fall to acquire the same velocity. The answer in one case will be the answer in the other.

How do the times of ascent and descent compare?

111. The time, also, which the ascending body would require to attain its greatest height, would be just equal to the time it would require to fall to the ground from that height.

The following table exhibits an analysis of the motions of a falling body; the spaces passed over in each interval of time of falling, increasing as the odd numbers 1, 3, 5, 7, 9, etc.; the velocities acquired at the end of each interval increasing directly as the times; and the whole space passed over being as the squares of the times.

Number of Seconds in the Fall, counted from a State of Rest.	Spaces fallen through in each successive Second.	Velocities acquired at the End of Number of Seconds expressed in First Column.	Total Height fallen through from Rest in the Number of Seconds expressed in First Column.
1	1	2	1
2	3	4	4
3	5	6	9
4	7	8	16
5	9	10	25
6	11	12	36
7	13	14	49
8	15	16	64
9	17	18	81
10	19	20	100

Where extreme accuracy is not required, most of the problems connected with the descent of falling bodies, may be worked with great readiness—16 feet, the space passed through by a falling body in one second, being taken as the common multiple of distances and velocities.

Thus, to ascertain the height from which a body would fall in 5 seconds, take in the fourth column of the table the number opposite 5 seconds, which is 25, and multiply it by 16; the product, 400, will be the height required. Problems of this character may also be worked by the rule given (§ 107).

In the same manner, if it be required to determine the space a falling body would descend through in any particular second of its motion, as, for example, the 5th second, we take in the second column of the table the number opposite five seconds, which is 9, and multiply it by 16; the product, 144, is the space required.

In like manner, if it be required to determine with what velocity a body would strike the ground after falling during an interval of 5 seconds, we take the number in the third column of the table opposite 5 seconds, which we find to be 10, and multiply this by 16. The product, 160 feet, will be the velocity required; and a body thus falling for 5 seconds would have, when it strikes the ground, a velocity of 160 feet.

What will be the velocity of a body falling down an inclined plane?

112. If a body, instead of falling perpendicularly, be made to roll down an inclined plane, free from friction, the velocity acquired at the termination of its descent, will be equal to that it would acquire in falling through the perpendicular height of the inclined plane.

FIG. 28.

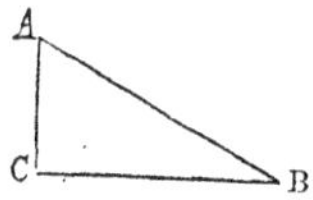

Thus, the velocity acquired in rolling down the whole length of A B, Fig. 28, is equal to that it would acquire by falling down the perpendicular height A C.

How, and by whom was the pendulum discovered?

113. The great Italian philosopher Galileo, during the early part of the 17th century, had his attention directed, while in a church at Florence, to the swinging of the chandeliers suspended from the lofty ceiling. He noticed that when they were moved from their natural position by any disturbing cause, they swung backward and forward in a curve, for a long time, and with great uniformity, rising and falling alternately in opposite directions. His inquiry into the cause of these motions led to the invention of the pendulum, the theory of which may be explained as follows:

Explain the theory of the pendulum.

114. All bodies will have their motion as much accelerated whilst descending a curve, as retarded whilst ascending. Let C A B be a curve, Fig. 29. If a ball be placed at C, the attraction of gravitation will cause it to descend to A, and in so doing it will acquire velocity sufficient to carry it to B, all opposing obstacles being removed, such as friction and resistance of the air. Gravitation

FIG. 29.

will once more bring it down to A; it will then rise again to C, and so continue to oscillate backward and forward.

If we now suspend the ball by a string, or wire, in such a manner that it will swing freely, its motions will be the same as that of the ball rolling upon the curve. A body thus suspended is called a PENDULUM. In Fig. 30, D C, the part of the circle through which the pendulum moves, is called its *arc*, and the whole movement of the ball from D to C is called an *oscillation*.

FIG. 30.

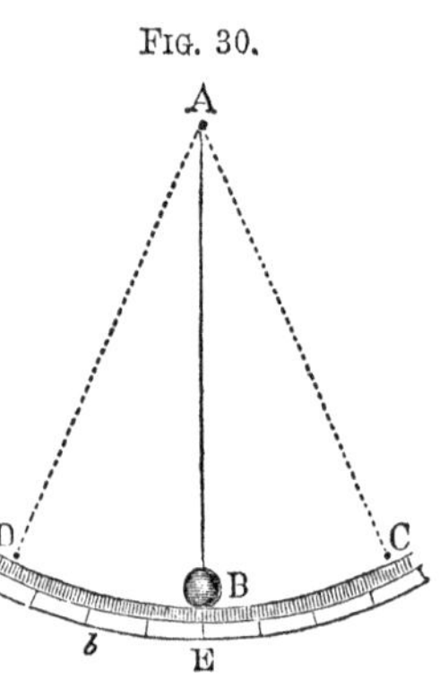

How do the times of the vibrations of a pendulum compare with each other?

115. The times of the vibrations of a pendulum, are very nearly equal, whether it moves much or little; or, in other words, through a greater, or less part of its arc.

Explain the reason of this law.

The reason that a large vibration is performed in the same time as a small one, or, in other words, the reason the pendulum always moves faster in proportion as its journey is longer, is, that in proportion as the arc described is more extended, the steeper are the declivities through which it falls, and the more its motion is accelerated. Thus, if a pendulum, Fig. 30, begins its motion at D, the accelerating force is twice as great as when it is set free at *b*; and if we take two pendulums of equal lengths, and liberate one at D and another at *b* at the same time, they will arrive at the same moment at E.

How does this property of the pendulum enable us to register time?

116. This remarkable property of the pendulum enables us to employ it as a register, or keeper of time. A pendulum of invariable length, and in the same location, will always make the same number of oscillations in the same time. Thus, if we arrange it so that it will oscillate once in a second, sixty of these oscillations will mark the lapse of a minute, and 3,600 an hour.

What is a common clock?

A common clock is, therefore, merely an arrangement for registering the number of oscillations which a pendulum makes, and at the same time of communicating to the pendulum, by means of a weight, an amount of motion sufficient to make up for what it is continually losing by friction on its points of support, and by the resistance of the air.

The wheels of the clock turn round by the action of the weight, but they are so connected with the pendulum, that with every double oscillation a tooth of the last wheel is allowed to pass. If, now, this wheel has thirty teeth, as is common in clocks, it will turn round once for every sixty vibrations. And, if the axis of this wheel project through the dial-plate or face of a clock, with a hand fastened on it, this hand will be the second hand of the clock. The other wheels are so connected with the first, and the number of teeth so proportioned, that the second one turns sixty times slower than the first, and this will be the minute hand; a third wheel moving twelve times slower than the last will constitute the hour hand.

How does a watch differ from a clock?

A watch differs from a clock in having a *vibrating wheel* instead of a *vibrating pendulum.* This wheel, called the *balance-wheel,* is moved by a *spring,* which is always forcing it to a middle position of rest, but does not fix it there, because the velocity acquired during its approach from either side to the middle position, carries it just as far past on the other side, and the spring has to begin its work again. The *balance-wheel* at each vibration allows *one tooth* of the adjoining wheel to pass, as the pendulum does in a clock, and the record of the beats is preserved by the wheels which follow, as already explained for the clock.

FIG. 31.

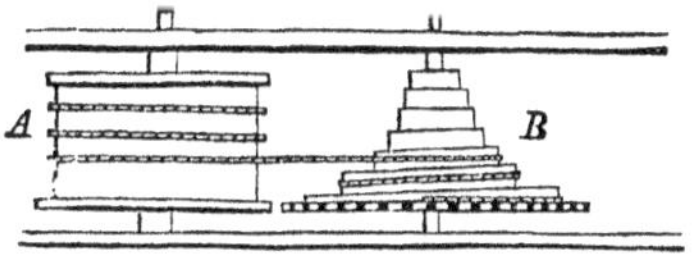

Fig. 31 represents the arrangement used to keep up the motion in a watch. The barrel, or wheel A, incloses a spring, which, when compressed by winding up, tends to liberate itself, or unwind, in virtue of its elasticity. This effort to unwind, turns the barrel upon its axis, and thus, by means of a chain coiled round it, motion is communicated to the other wheels of the watch.

What influence has the length of a pendulum on its time of vibration?

117. The length of a pendulum influences the time of its vibration ; the longer the pendulum the slower are its vibrations.

The reason why long pendulums vibrate more slowly than short ones is, that in corresponding arcs, or paths, the ball of the long pendulum has a greater journey to perform, without having a steeper line of descent.

What is the center of oscillation in a pendulum?

118. If we take a pendulum rod, Fig. 32, A D, having balls upon it at C and D, and cause it to vibrate, the ball, B, being nearer to the point of suspension, will tend to perform its oscillations more quickly than the ball C. In like manner, every other point on the pendulum rod tends to complete its oscillations in a different time; but as they are all connected together inflexibly, all are compelled to perform their oscillations in the same time. But the action of the portions of the rod near to the ball, B, is to accelerate the motion of the pendulum, and the action of the portions of the rod near to the ball C, is to retard it; therefore a point may be found where all these counteractions will balance one another, or be neutralized, and this point is termed the CENTER OF OSCILLATION, and the sum of the momenta of all the portions of the rod on each side of this point will balance. The center of oscillation does not correspond with the center of gravity, but is always a little below it; the practical method of bringing them near together, is to make the rod light, and the termination of the pendulum heavy.

FIG. 32.

Why do clocks go faster in winter than in summer?

119. As heat expands, and cold contracts all metals, a pendulum rod is longer in warm than in cold weather; hence, clocks gain time in winter, and lose in the summer.

How are the changes in the length of pendulums counteracted?

Fig. 33.

As the smallest change in the length of a pendulum alters the rate of a clock, it is highly important, for the maintaining of uniform time, that the expansion and contraction of pendulums, caused by changes in temperature, should be counteracted. For this purpose various contrivances have been employed. The one most commonly employed at the present time is the mercurial pendulum, which is constructed as follows: The pendulum rod, A B, Fig. 33, supports a glass jar, G H, containing mercury, inclosed in a steel frame-work, F C D E. When the weather is warm, the steel rod and frame-work expand, and thus increase the length of the pendulum, and depress the center of oscillation. But, at the same time, the mercury contained in the jar also expands, and rises upward; and thus, by a proper adjustment, the center of oscillation is carried as far upward in one direction, as downward in the opposite direction, or the expansion in both directions is equal, and the vibrations of the pendulum remain unaltered. Another form of pendulum, called the "gridiron pendulum," Fig. 34, is composed of rods of different metals, which expand unequally under the same changes of temperature, and, by counteraction, keep the length of the pendulum constant.

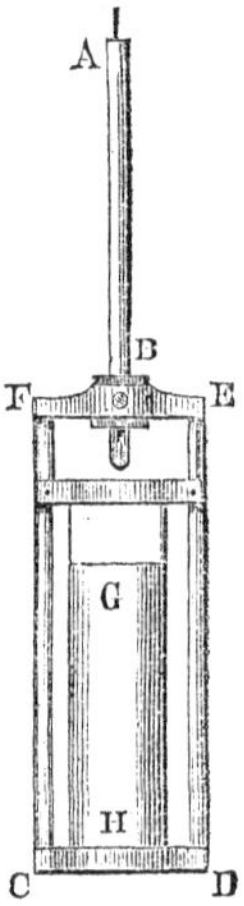

Fig. 34.

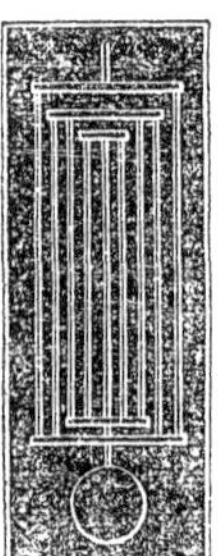

How do the variations in the force of gravity affect the vibrations of a pendulum?

120. As the force of gravity determines how long the pendulum shall be in falling down its arc, and the time also of its rising in the opposite direction (since the ball of the pendulum, as already stated, may be considered as a body descending by its weight on a slope), it follows, that the time of vibration of a pendulum will vary as the attraction of gravity varies.

Where will a pendulum of a given length vibrate slowest, and where the fastest?

The same pendulum will vibrate more slowly at the equator than at the poles, because the attraction of gravitation is less powerful at the equator. Therefore a pendulum to vibrate once in a second, must be shorter at the equator than at the poles. Corresponding results take place when a pendulum is carried to a mountain-top, away from the center of the earth, which is the center of attraction, or when carried to the bottom of a mine, where it is attracted both by matter above it and below it.

What is the length of a seconds pendulum?

121. The length of a pendulum that will describe sixty oscillations in a minute, each oscillation having the duration of a second, is, in the latitude of Greenwich, England, 39.1393 inches in length; one to vibrate in half seconds must measure 9.7848, or rather more than 9¾ inches.

At the pole it would require to be somewhat longer; at the equator somewhat shorter. A pendulum that vibrated seconds at Paris, was found to require lengthening .09 of an inch in order to perform its vibrations in the same time at Spitzbergen.

How may the length of a seconds pendulum be used as a standard of measure?

122. The length of a pendulum vibrating seconds being always invariable at the same place, since the attraction under the same circumstances is always the same, it may be used as a standard of measure.

This application has already been described under the section Weight (§ 67).

The duration of the oscillation of a pendulum is not affected by altering the weight of the ball, since all bodies moving over the same space, under the influence of gravitation, acquire equal velocities.

How do the lengths of pendulums vibrating in different times compare?

123. The lengths of different pendulums, vibrating in unequal times, are to each other as the squares of the times of their vibration.

Thus a pendulum, to vibrate once in two seconds, must have four times the length of one that vibrates once in one second; to vibrate once in three seconds, it must have nine times the length, etc.—the duration of the oscillation being as the whole numbers,

1, 2, 3, 4, 5, 6, 7, 8, 9.

The length of the pendulum will be as their squares.

1, 4, 9, 16, 25, 36, 49, 64, 81.

A pendulum, therefore, that will vibrate once in nine seconds, must have a length of 81 times greater than one vibrating once in one second.

PRACTICAL PROBLEMS ON THE THEORY OF FALLING BODIES.

1. A stone let fall from the top of a tower struck the earth in two seconds; how high was the tower?

2. How far will a body acted upon by gravity alone, fall in ten seconds?

3. How deep is a well, into which a stone being dropped, reaches the surface of the water in two seconds, the depth of the water in the well being ten feet?

4. If a body be projected downward with a velocity of twenty-two feet in the first second of time, how far will it fall in eight seconds?

The multiple in this case will be the distance fallen through in the first second.

5. What space will a body pass through in the fourth second of its time of falling?

6. A body falls to the ground in eight seconds; how large a space did it pass over during the last second of its descent?

7. A body falls from a height in eight seconds; with what velocity did it strike the ground?

8. A cannon-ball fired upward, continued to rise for nine seconds; what was its velocity during the first second, or with what force was it projected?

9. Suppose a bullet fired upward from a gun returned to the earth in sixteen seconds; how high did it ascend?

The time occupied in ascending and descending being equal, the body rose to such a height that it required eight seconds to descend from it. The square of 8=64. This multiplied by the space it would fall in the first second, 16 feet = 924 feet.

10. A bird was shot while flying in the air, and fell to the ground in three seconds. How high up was the bird when it was shot?

11. What must be the length of a pendulum to vibrate once in seven seconds?

12. If the length of a pendulum to vibrate seconds at Washington is 39.101 inches, how long must it be to vibrate half seconds? How long to vibrate quarter seconds?

CHAPTER V.

MOTION.

What is Motion?

124. MOTION is the act of changing place.

If no motion existed, the universe would be dead. There would be no alternation of the seasons, and of day and night; no flow of water, or change of air; no sound, light, heat, or animal existence.

What is Absolute and Relative Motion?

125. MOTION is ABSOLUTE or RELATIVE. ABSOLUTE MOTION is a change of position in space, considered without reference to any other body. RELATIVE MOTION is motion considered in relation to some other body, which is either in motion or at rest.

Thus the motions of the planets in space are examples of Absolute Motion, but the motion of a man sitting upon the deck of a vessel, while sailing, is an example of Relative Motion, since he is in motion as respects the land, but at rest as regards the parts of the vessel. Rest, which is the opposite of motion, so far as we know, exists only relatively. We say a body on the surface of the earth is at rest, when it maintains a constant position as regards some other body; but at the same time that it is thus at rest, it partakes of the motion of the earth, which is always revolving. We do not, therefore, really know any body to be in a state of absolute rest.

Define Uniform and Variable Motion.

126. A moving body may have a UNIFORM or a VARIABLE MOTION. UNIFORM MOTION is the motion of a body moving over equal spaces in equal times. VARIABLE MOTION is the motion of a body moving over unequal spaces in equal times.

What is Accelerated and Retarded Motion?

127. When the spaces passed over in equal times increase, the body is said to possess ACCELERATED MOTION; when they diminish, the body is said to possess RETARDED MOTION.

A stone falling through the air is an example of Accelerated Motion, since, acted upon by the force of gravity, its rate of motion constantly increases; while the ascent of a stone projected from the hand, is an example of Retarded Motion, its upward motion continually decreasing.

What is Power and Resistance?

128. When a body commences to move from a state of rest, we assign some force as the cause of its motion; and a force acting in such a manner as to produce motion, is generally termed "POWER." On the contrary, a force acting in such a way as to retard a moving body, destroy its motion, or drive it in a contrary direction, is termed RESISTANCE. The chief forces which tend to retard or destroy the motion of a body are GRAVITATION, FRICTION, and RESISTANCE OF THE AIR.

What is Velocity?

129. The speed, or rate, at which a body moves, is termed its VELOCITY.

Moving bodies pass over their paths with different degrees of speed; one may pass through ten feet in a second of time, and another through a hundred feet in the same period. We say, therefore, that they have different velocities.

The velocity of a moving body is estimated by the time it occupies in moving over a given space, or by the space passed over in a given time. The less the time and the greater the space moved over in that time, the greater the velocity.

How do we ascertain the Velocity of a moving body?

130. To ascertain the VELOCITY of a moving body, divide the space passed over by the time consumed in moving over it.

Thus, if a body moves 10 miles in 2 hours, its velocity is found by dividing the space, 10, by the time, 2; the answer, 5, gives the velocity per hour.

How can we ascertain the space passed over by a body in motion?

131. To ascertain the SPACE passed over by a moving body, multiply the velocity by the time.

Thus, if the velocity be 10 miles per hour, and the time 15 hours, the space will be 10 multiplied by 15, or 150 miles.

How is the time occupied by a body in motion ascertained?

132. To ascertain the TIME employed by a body in motion, divide the space passed over by the velocity.

Thus, if the space passed over be 150 miles, and the velocity 10 miles per hour, the whole time employed will be 150 divided by 10 = 15 hours.

What is Momentum?

133. The MOMENTUM of a body is its quantity of motion. Momentum expresses the force with which one body in motion would strike against another.

Illustrations of Momentum.

That a mass of matter moving in any manner exerts a certain force against any object with which it may come in contact, is a principle of Natural Philosophy which experience teaches us most frequently and most readily. The child has hardly emerged from the nurse's arms, before it becomes conscious of the force with which it would strike the ground if it fell. We take advantage of momentum, or the force of a moving body, in almost all mechanical operations. The moving mass of a hammer-head drives or forces in the nail, shapes the iron, breaks the stone; the force of a moving mass of water gives strength to a torrent, and turns the wheel; the force of a moving mass of air gives strength to the wind, carries the ship over the ocean, forces round the arms of a wind-mill.

Is motion imparted to all the particles of a body at the same instant?

134. When a body is caused to move, the motion is not imparted simultaneously to every particle of the body, but at first only to the particles which are directly exposed to the influence of the force—for instance, of a blow. From these particles, it spreads to the rest.

How can you illustrate this fact?

A slight blow is sufficient to smash a whole pane of glass, while a bullet from a gun will only make a small round hole in it, because, in the latter case, the particles of glass that receive the blow are torn away from the remainder with such rapidity, that the motion imparted to them has no time to spread further. A door standing open, which would readily yield on its hinges to a gentle push, is not moved by a cannon-ball passing through it. The ball, in passing through, overcomes the whole force of cohesion among the atoms of wood, but its force acts for so short a time, owing to its rapid passage, that it is not sufficient to affect the inertia of the door to an extent to produce motion. The cohesion of the part of the wood cut out by the ball would have borne a very great weight laid quietly upon it; but supposing the ball to fly at the rate of 1200 feet in a second, and the door to be one inch thick, the cohesion being allowed to act for only the minute fraction of a second, its influence is not perceived.

It is an effect of this same principle, that the iron head of a hammer may be driven down on its wooden handle, by striking the opposite end of the handle against any hard substance with force and speed. In this very simple operation, the motion is propagated so suddenly through the wood of the handle, that it is over before it can reach the iron head, which therefore, by its own inertia, sinks lower on the handle at every blow, which drives the handle up.

How is the Momentum of a body calculated?

135. The MOMENTUM, or force, which a moving body exerts, is estimated by multiplying its mass or quantity of matter by its velocity.

Thus, a body weighing 10 pounds, and moving with a velocity of 500 feet in a second, will have a momentum of (10×500) 5,000.

What connection is there between the Momentum of a body and its weight and velocity?

136. The velocity being the same, the momentum, or moving force of a body, will be directly proportionate to the mass, or weight; and the mass or weight remaining the same, the momentum will be directly proportionate to the velocity.

Thus, if 2 leaden balls, each of 5 pounds' weight, move with a velocity of 5 miles per minute, the momentum, or striking force of each, will be 25; if now the two balls, molded into one of 10 pounds' weight, move with the same velocity of 5 miles per minute, the momentum, or striking force, will be 50, since with the same velocity the mass, or weight, will be doubled. If, on the contrary, we double the velocity, allowing the weight to remain the same, the same effect will be produced; a ball of 5 pounds, with a velocity of 5, will have a momentum, or striking force, of 25; but a ball of 5, with a velocity of 10, will have a momentum of 50.

How can a small body in motion be made to exert the same force as a large one?

137. A small, or light body, may be made to strike with a greater force than a heavier body, by giving to the small body a sufficient velocity.

Illustrations of these principles are most familiar. Hail-stones, of small mass and great velocity, strike with sufficient force to break glass, and destroy standing grain; a ship of huge mass, moving with a scarcely perceptible velocity, crushes in the side of the pier with which it comes in contact.

SECTION I.

ACTION AND REACTION.

What is meant by Action and Reaction?

138. When a body communicates motion to another body, it loses as much of its own momentum, or force, as it gives to the other body. We apply the term ACTION to designate the power which a body in motion has to impart motion, or force, to another body; and the term REACTION to express the power which the body acted upon has of depriving the acting body of its force, or motion.

What is the great law of Action and Reaction?

139. There is no motion, or action, in the universe without a corresponding and opposite action of equal amount; or, in other words, ACTION and REACTION are always equal and opposed to each other.

What are Illustrations of Action and Reaction?

If a person presses the table with his finger, he feels a resistance arising from the reaction of the table, and this counter-pressure is equal and contrary to the downward pressure. When a cannon or gun is fired, the explosion of the powder which gives a forward motion to the ball, gives at the same time a backward motion, or "recoil," to the gun. A man in rowing a boat, drives the water astern with the same force that he impels the boat forward.

To what is the quantity of motion in a body proportionate?

140. The quantity of motion in a body is measured by the velocity and the quantity of matter it contains.

A cannon-ball of a thousand ounces, moving one foot per second, has the same quantity of motion in it as a musket-ball of one ounce, leaving the gun with a velocity of a thousand feet per second. The momentum, or quantity of motion, in the musket-ball being, however, concentrated in a very small mass, the effect it will produce will be apparently much greater than that of the cannon-ball, whose motion is diffused through a very large mass. This explanation will enable us to understand some phenomena which at first appear to contradict the law, that action and reaction are always equal, and opposed to each other.

Thus, when we fire a bullet from a gun, the gun recoils back with as much force as the bullet possesses, proceeding in an opposite direction. The reason the effects of the gun are not equally apparent with those of the ball, is that the motion of the gun is diffused through a great mass of matter, with a small velocity, and is, therefore, easily checked; but in the ball the motion is concentrated in a very small compass, with a great velocity. A gun recoils more with a charge of fine shot, or sand, than with a bullet. The explanation of this is, that with a ball the velocity is communicated to the whole mass *at once*, but with small shot, or sand, the velocity communicated by the explosion *to those particles of the substance immediately in contact with the powder*, is greater than that received *at the same instant* by the *outer particles;* consequently, a larger proportion of explosive force acts momentarily in an opposite direction.

Fig. 35.

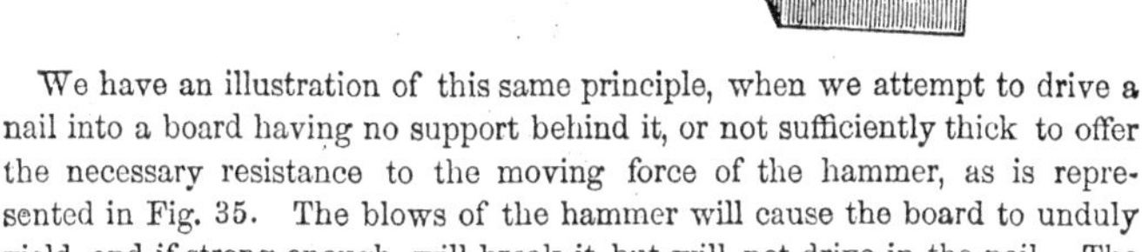

We have an illustration of this same principle, when we attempt to drive a nail into a board having no support behind it, or not sufficiently thick to offer the necessary resistance to the moving force of the hammer, as is represented in Fig. 35. The blows of the hammer will cause the board to unduly yield, and if strong enough, will break it, but will not drive in the nail. The object is attained by applying behind the board, as in Fig. 36, a block of wood,

Fig. 36.

or metal, against which the blows of the hammer will be directed. By adopting this plan, however, no increased resistance is opposed to the blows of the hammer, the momentum, or moving force of which is equally imparted in both cases; but in the first case, the momentum is received by the board alone, which, having little weight, is driven by it through so great a space as to produce considerable flexure, or even fracture; but in the second case, the same momentum being shared between the board and the block behind it, will produce a flexure of the board as much less as the weight of the board and block applied to it together, is greater than the weight of the board alone.

The same principle serves to explain a trick sometimes exhibited in feats of strength, where a man in a horizontal position, his legs and shoulders being supported, sustains a heavy anvil upon his chest, which is then struck by sledge-hammers. The reason the exhibitor sustains no injury from the blows, is that the momentum of the sledge is distributed equally through the great mass of the anvil, and gives to the anvil a downward motion, just as much less than the motion of the sledge, as the mass of the sledge is less than the mass of the anvil. Thus, if the weight of the anvil be 100 times greater than the weight of the sledge, its downward motion upon the body of the exhibitor will be 100 times less than the motion with which the sledge strikes it, and the body of the exhibitor easily yielding to so slight a movement, and also resisting it by means of the elasticity of the body, derived from its peculiar position, escapes without injury.

When is the collision of two bodies said to be direct?

141. When two bodies come in contact, the collision is said to be direct, when a right line passing through their centers of gravity passes also through the point of contact.

The center of gravity in such cases corresponds with the center of collision; and if such a center come against an obstacle, the whole momentum of the body acts there, and is destroyed; but if any other part hit, the body only loses a portion of its momentum, and revolves round the obstacle as a pivot, or center of motion.

When two inelastic bodies come into collision, what occurs?

142. When two non-elastic bodies, moving in opposite directions, come into direct collision, they will each lose an equal amount of momentum.

Hence, the momentum of both after contact, will be equal to the difference of the momenta of the two before contact, and the velocity after contact will be equal to the difference of the momenta divided by the whole quantity of matter. Let the quantity of matter in A be 2, and its velocity 12; its momentum is, therefore, 24. Let the quantity of matter in B be 4, its velocity 3; its momentum will be 12. The momentum of the mass after contact, on the supposition they move in opposite directions and come in direct collision, will be the difference of the two momenta, or 12, and the velocity of the mass will be its momentum divided by the quantity of matter, or 12 divided by 6, which is 2.*

If two non-elastic bodies, as A and B, Fig. 37, be suspended from a fixed point, and the one be raised toward Y, and the other toward X, an equal amount, they will acquire an equal force, or momentum, in falling down the arc, provided their masses are equal; and will by contact destroy each other's motion, and come to rest. If their momenta are unequal, they will, after contact move on together, in the direction of the body having the largest quantity of motion with a momentum equal to the difference of the momenta of the two before collision.

Explain the results of the collision of inelastic bodies.

Fig. 37.

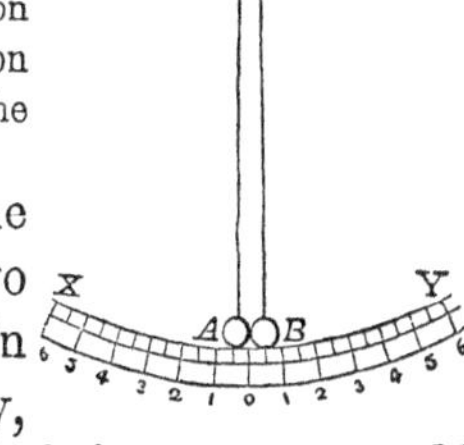

To what will the shock of collision of two bodies coming in contact be equivalent?

143. The force of the shock produced by two equal bodies coming in contact with equal velocity, will be equal to the force which either, being at rest, would sustain, if struck by the other moving with double the velocity ; for reaction and action being equal, each of the two will sustain as much shock from reaction as from action.

Illustrate this principle.

If a person running, come in contact with another who is standing, both receive a certain shock. If both be running at the same rate in opposite directions, the shock is doubled. In combats of pugilists, the most severe blows are those struck by fist against fist, for the force sustained by each in such cases, is equal to the sum of the forces exerted by the two arms. If two ships, moving in contrary directions at the rate of 20 miles per hour, come in collision, the shock will be the same as if one of them, being at rest, were struck by the other, moving at 40 miles per hour.

Fig. 38.

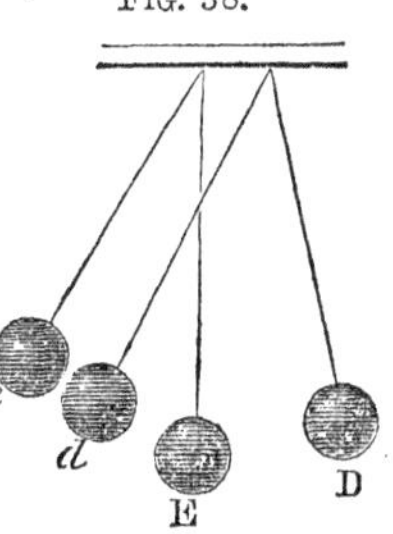

If one inelastic body comes in contact with another at rest, what occurs?

144. If we suspend two balls of some non-elastic substance, as clay or putty, by strings, so that they can move freely, and allow one of the balls to fall upon the other at rest, it will communicate to it a part of its motion, and both balls, after collision, will move on together.

* This whole subject, usually considered dry and uninteresting, will be found to possess a new interest, if the student will make himself a few simple experiments, by suspending leaden balls by the side of a graduated arc, as in Fig. 37, and allow them to fall under different conditions. The length of the arc through which they fall will be found to be an exact measure of the force with which they will strike.

The quantity of motion will remain unchanged, the one having gained as much as the other has lost; so that the two, if equal, will have half the velocity after collision that the moving one had when alone. Fig. 38 represents two balls of clay, E and D, non-elastic, of equal-weight, suspended by strings. If the ball D be raised and let fall against the ball E, a part of its motion will be communicated to E, and both together will move on to *e d*.

When two elastic bodies come into collision, what occurs?

145. If we suspend two balls, A and B, Fig. 39, of some elastic substance, as ivory, and allow them to fall with equal masses and velocities from the points X and Y on the arc, they will not come to rest after collision, but will recede from each other with the same velocity which each had before contact.

Fig. 39.

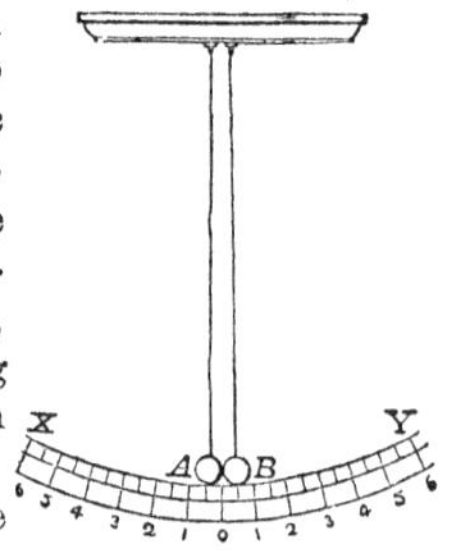

What occasions the difference in the results of the collision of elastic and non-elastic bodies?

The reason of this movement in highly elastic bodies, contrary to what takes place in non-elastic bodies, is this: the elastic substances are compressed by the force of the shock, but instantly recovering their former shape in virtue of their elasticity, they spring back, as it were, and react, each giving to the other an impulse equal to the force which caused its compression.

Suppose the ball A, however, to strike upon the ball B at rest; then, after impact, A will remain at rest, but B will move on with the same velocity as A had at the moment of contact. In this case the reaction of elasticity causes the ball A to stop, and the ball B to move forward with the motion which A had at the instant of contact.

Fig. 40.

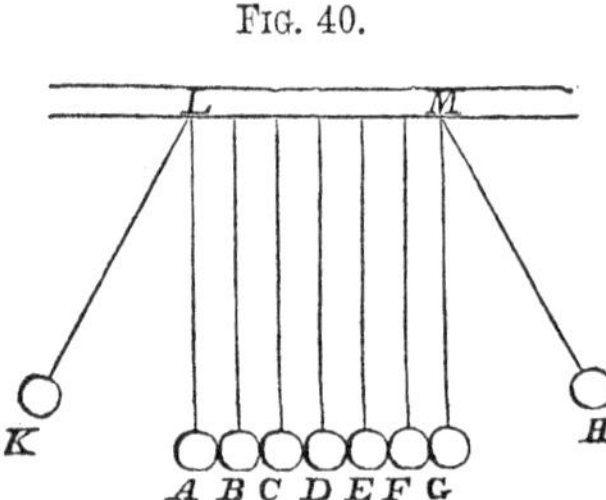

The same fact may be illustrated by suspending a number of elastic balls of equal weight, as represented in Fig. 40. If the ball H be drawn out a certain distance, and let fall upon G, the next in order, it will communicate its motion to G, and receive a reaction from it, which will destroy its own motion. But the ball B can not move without moving F; it will, therefore, communicate the motion it received from G to F, and receive from F a reaction which will stop its motion. In like manner, the motion and reaction are received by each of the balls E, D, C, B, A, until the last ball, K, is reached; but there being no ball beyond K to act upon it, K will fly off as far from A, as H was drawn apart from G.

SECTION II.

REFLECTED MOTION.

What is Reflected Motion?

146. When any elastic body, as an ivory ball, is thrown against a hard smooth surface, the reaction will cause it to rebound from such surface, and the motion it receives is called REFLECTED MOTION.

In what manner may a moving body be reflected?

147. If the ball be projected perpendicularly, it will rebound in the same direction; if it be projected obliquely, it will rebound obliquely in an opposite direction, making the angle of incidence equal to the angle of reflection.

What is the Angle of Incidence?

148. The ANGLE of INCIDENCE is the angle formed by the line of incidence with a perpendicular to any given surface.

What is the Angle of Reflection?

149. The ANGLE of REFLECTION is the angle formed by the line of reflection with a perpendicular to any given surface.

FIG. 41.

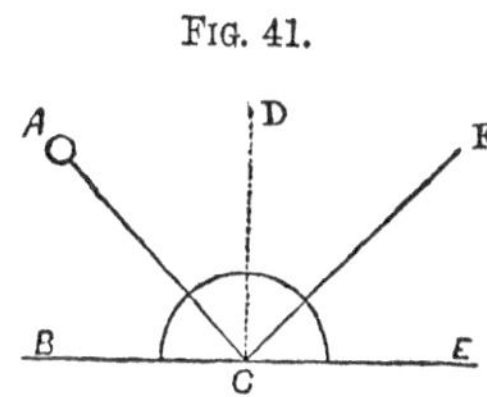

Thus, in Fig. 41, let B E be a smooth, flat surface. If the ball, A, be projected, or thrown upon this surface, in the direction A C, it will rebound, or be reflected, in the direction C F. In this case, the line A C is the line of incidence, and the angle A C D, which it makes with a perpendicular D C, is the angle of incidence. In like manner the line C F is the line of reflection, and the angle D C F the angle of reflection. If the ball be projected against the surface, B C, in the direction D C, perpendicular to the surface, it will be reflected, or rebound back in the same straight line.

What proportion exists between the angles of incidence and reflection?

150. The ANGLES of INCIDENCE and REFLECTION are always equal to one another.

Thus, in Fig. 41, the angles A C D and F C D are equal.

What is an Angle, and upon what does its size depend?

151. An ANGLE is simply the inclination of the lines which meet each other in a point. The size of the angle depends upon the opening, or inclination, of the lines, and not upon their length.

In what consists the skill of the Game of Billiards?

The skill of the player of billiards and bagatelle depends upon his dexterous application of the principles of incident and reflected motion, which he has learned by long-continued experience, viz., that the angle of incidence is always equal to the angle of reflection, and that action and reaction are equal and contrary. An illustration of the skillful reflection of billiard balls is given in Fig. 42, which represents the top of a billiard-table. The ball, P, when struck by the stick, Q,

FIG. 42.

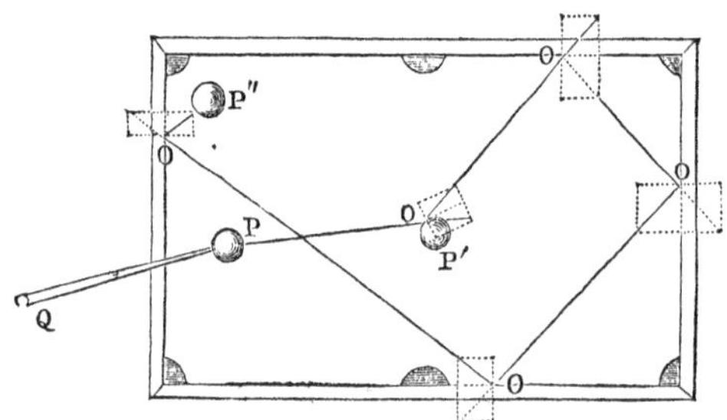

is first directed in the line P O, upon the ball P′, in such a manner, that being reflected from it, it strikes the four sides of the table successively, at the points marked O, and is finally reflected so as to strike the third ball, P″. At each of the reflections from the ball P′, and the four points on the side of the table, the angle of incidence is exactly equal to the angle of reflection.

Why are imperfectly elastic bodies peculiarly fitted to oppose and destroy momentum?

152. Imperfectly elastic bodies oppose the momentum of bodies in motion more perfectly than any others, in consequence of their yielding to the force of collision without reacting; opposing a gradual resistance instead of a sudden one.

Hence a feather-bed, or a sack of wool, will stop a bullet much more effectually than a plate of iron, from its *deadening*, as it is popularly called, the force of the blow.

SECTION III.

COMPOUND MOTION.

What is Simple Motion?

153. A body acted upon by a single force, moves in a straight line, and in the direction of that force. Such motion is designated as SIMPLE MOTION.

Illustrate Simple Motion.

A body floating upon the water is driven exactly south by a wind blowing south. A ball fired from a cannon takes the exact direction of the bore of the cannon, or of the force which impels it.

What is Compound Motion?

154. When a body is acted upon by two forces at the same time, and in different directions, as it can not move two ways at once, it takes a middle course between the two. Such motion is termed COMPOUND MOTION.

What is the course of a body acted upon by two forces called?

155. The course in which a body, acted upon by two or more forces, acting in different directions, will move, is called the RESULTANT, or the Resulting Direction.

FIG. 43.

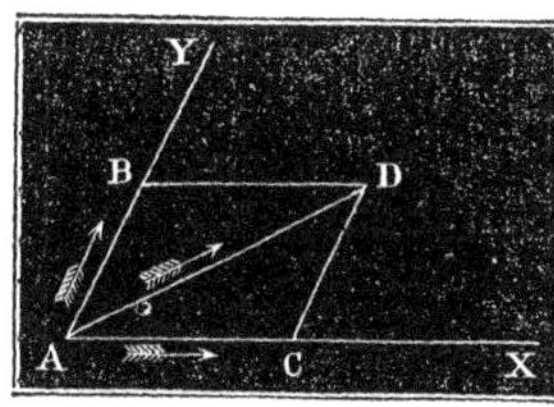

In Fig. 43, if a body, A, be acted upon at the same time by two forces, one of which would cause it to move in the direction A Y, over the space A B, in one second of time, and the other cause it to move in the direction A X, over the space A C, in one second; then the two forces, acting upon it at the same instant, will cause it to move in a Resultant Direction, A D, in one second. This direction is the diagonal of a parallelogram, which has for its sides the lines A B, A C, over which the body would move if acted upon by each of the forces separately.

What are familiar Examples of Resultant Motion?

156. The operations of every-day life afford numerous examples of Resultant Motion. If we attempt to row a boat across a rapid river, the boat will be subjected to action of two forces; viz., the action of the oars, which tend to drive it across the river in a certain time, as ten minutes, in a straight line, as from A to B, Fig. 43, and the action of the current, which tends to carry it down the stream a certain distance in the same time, as from A to C. It will, therefore, under the influence of both these forces, move diagonally across the river, or in the direction A D, and arrive at D at the expiration of the ten minutes. When we throw a body from the deck of a boat in motion, or from a railroad car, the body partakes of the motion of the boat or the car, and does not strike at the point intended, but is carried some distance beyond it. For the same reason, in firing a rifle from the deck of a vessel moving rapidly, at some object at rest upon the bank, allowance must be made for the motion of the vessel, and aim directed behind the object.

What is the Science of Projectiles?

157. The principles of the composition and resolution of different forces acting upon a body to produce motion, constitute the basis

of the SCIENCE of PROJECTILES, or that department of Natural Philosophy which considers the motion of bodies, thrown or driven by an impelling force above the surface of the earth.

What is a Projectile?

158. A PROJECTILE is a body thrown into the air in any direction; as a stone from the hand, or a ball from a gun, or cannon.

What is the direction of a body thrown obliquely?

If we project a body perpendicularly downward, or upward, it will move in a perpendicular line with a uniform accelerated or retarded motion, since the force of gravity and that of projection are in the same line of direction. But if a body is thrown in a direction oblique to the perpendicular, it is acted upon by two forces,* the projectile force which tends to impel it forward in a straight line, and the force of gravity, which tends to bring it to the earth. Instead, therefore, of following the direction of the projectile force, the path of the body will be a curve, the resultant of the two forces. Such a curve is called a PARABOLA.

FIG. 44.

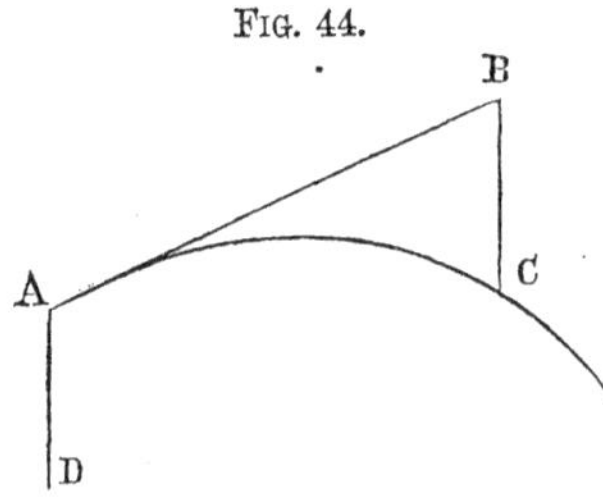

If a cannon-ball is fired from A toward B, Fig. 44, in an upward direction, instead of moving along the line A B, it is, by the influence of the earth's attraction, continually drawn downward, and its path is along a line which is indicated by the parabolic curve A C; and although it has been moving onward from the impulse it has received from the force of the gunpowder, it occupies exactly the same time in falling to the point C, as if the ball had been allowed to drop from the hand at A, and fall to D.

What effect has the projectile force on the action of gravity?

159. If a ball be projected from the mouth of a cannon in a horizontal direction, it will reach the earth in precisely the same time as a ball dropped from the mouth of the gun. The force of gravity is neither increased or diminished by the force of projection.

The same fact may be strikingly illustrated by placing a number of marbles at unequal distances from the edge of a table and sweeping them off with a ruler, or stick: those which are rolled along the farthest will be projected the farthest; yet all will strike the floor at the same time.

* The theoretical laws governing the motion of projectiles, as herewith given, are in practice essentially modified by the resistance of the air.

Fig. 45.

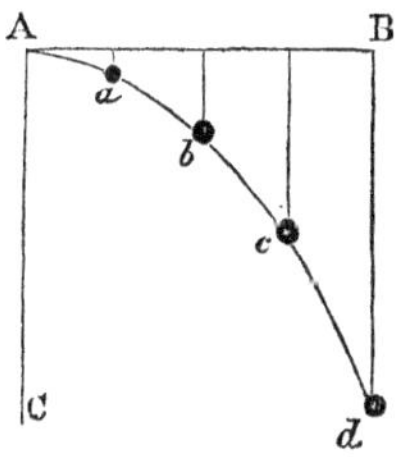

Suppose from the point A, Fig. 45, about 240 feet above the earth, a ball to be projected in a perfectly horizontal line, A B; instead of traversing this line, it would, at the end of the first second, be found that the ball had fallen 15 feet, at the same time it had moved onward in the direction of B. Its true position would be, therefore, at *a*; at the end of the second second, it would have passed onward, but have fallen to *b*, 60 feet below the horizontal line; and at the end of the third second, it would have fallen 135 feet below the line, and be at *c*; and thus it would move forward and reach the earth at *d* 240 feet, in precisely the same time it would have occupied in falling from A to C.

An oblique, or horizontal jet of water, is an instance of the curve described by a body acted upon by gravity and the force of projection. See Fig. 46.

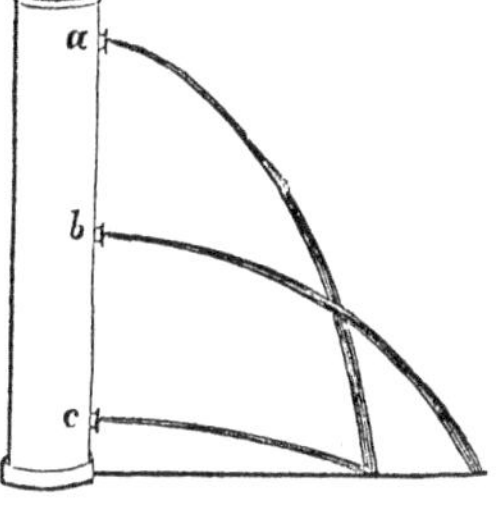

Fig. 46.

What is the Range of a Projectile?

160. The Range of a projectile, is the horizontal distance to which it can be thrown.

How can the greatest Range be obtained?

161. According to theory, the range is greatest when the angle of elevation is 45°; and is the same for elevations equally above and below 45°; as for example 70° and 20°. See Fig. 47.

These conclusions are, however, found to be essentially modified in practice by the resistance of the air, which not only changes the path but the velocity of the projectile. With great velocities, as in the case of a cannon-ball, the greatest range corresponds with an elevation of about 30°, but for slow motions it is near 45°.

Fig. 47.

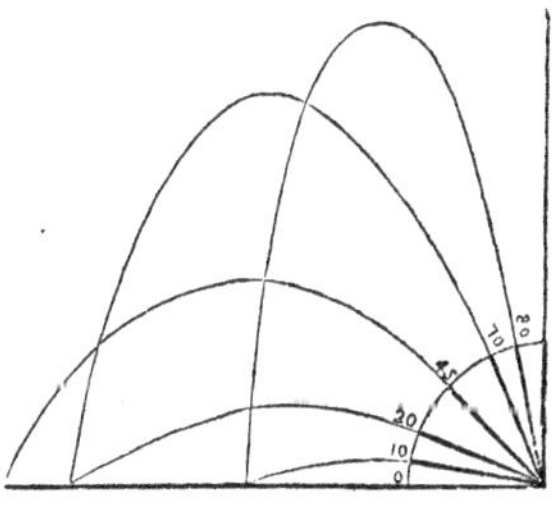

How are the Laws of Projectiles practically applied in military engineering?

162. The laws of projectiles are especially regarded in the art of gunnery. By knowing the force of the powder which drives the ball, the engineer is enabled to direct the cannon, or mortar, in such a manner as to cause the ball, or bomb, to fall

upon a particular spot in the distance; thus producing a desired effect without a wasteful expenditure of ammunition.

Fig. 48.

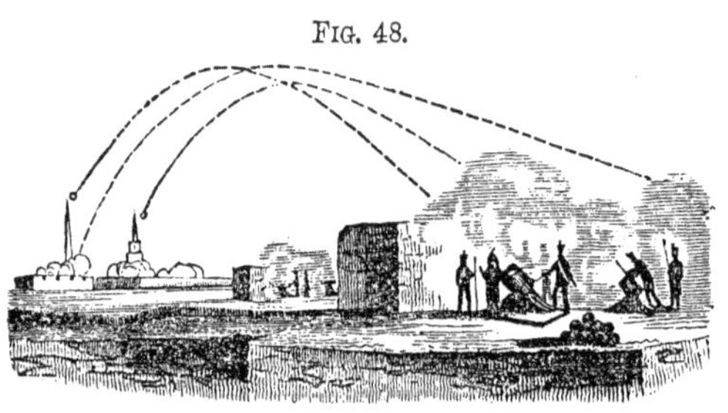

Fig. 48 represents a bombardment, and the three lines indicate the curves made by the balls. If the bombardment had been conducted from an elevation, instead of the level surface, the balls would have gone beyond the city, as shown by the familiar fact, that we can throw a heavy body to a greater distance from an elevation, as the steep bank of a river, than on a plain, or level ground. It was on this principle that Napoleon bombarded Cadiz, at the distance of five miles, and from a greater elevation, the balls could have been thrown to a still greater distance.*

* The following facts respecting the explosive force of gunpowder, and its application to projectiles, will be found interesting and instructive in this connection. The estimated force of gunpowder when exploded, is at least 14,750 pounds upon every square inch of the surface which confines it. Count Rumford showed, by his experiments made about 60 years ago, that if the powder were placed in a close cavity, and the cavity two thirds filled, its dimensions being at the same time restricted, the force of explosion would exceed 150,000 pounds upon the square inch.

The force of gunpowder depends upon the fact, that when brought in contact with any ignited substance, it explodes with great violence. A vast quantity of *gas*, or *elastic fluid*, is emitted, the *sudden production* of which, at a *high temperature*, is the cause of the violent effects which are produced.

The reason that gunpowder is manufactured in little grains, is that it may explode more quickly, by facilitating the passage of the flame among the particles. In the form of dust, the particles would be too compact.

The velocity of balls impelled by gunpowder from a musket with a common charge, has been estimated at about 1,650 feet in a second of time, when first discharged. The utmost velocity that can be given to a cannon-ball is 2,000 feet per second, and this only at the moment of its leaving the gun.

In order to increase the velocity from 1,650 to 2,000 feet, one half more powder is required; and even then, at a long shot, no advantage is gained, since, at the distance of 500 yards, the greatest velocity that can be obtained is only 1,200 or 1,300 feet per second. Great charges of powder are, therefore, not only useless, but dangerous; for, though they give little additional force to the ball, they hazard the lives of many by their liability to burst the gun. The velocity is greater with long than with short guns, because the influence of the powder upon the ball is longer continued.

The essential properties of a gun are to *confine the elastic fluid* generated by the explosion of the powder as completely as possible, and to *direct the course of the ball* in a

How should a gun be aimed to hit an object at a great distance?

According to the laws which govern the motion of projectiles, it is evident that a gun must be aimed, in order to hit an object, in a direction above that of the object, more or less, according to the distance of the object and the force of the charge. With an aim directed, as in Fig. 49, at the object, the ball, moving in a curved path, must necessarily fall below it.

straight, or rectilinear path. A rifle sends a ball more accurately than a musket, because the ball is in more accurate contact with the sides of the barrel than in the case of a common musket. The *space* produced by the *difference of diameter* between the *ball* and the *bore of the gun*, greatly diminishes the effect of the powder, by allowing a part of the elastic fluid to escape before the ball, and also permits the ball to deviate from a straight line. The peculiarity and superiority of the new rifle, called the "Minié rifle," is to be found in the construction of the ball, which, by the act of firing, is made to fit completely the barrel, or bore, of the gun. This is accomplished by making the ball of an oblong shape and a conical point, with an opening in the base extending up for two thirds the length of the ball. Into the opening of this internal cylinder there is placed a small *concave section of iron*, which the powder, at the moment of firing, *forces into the leaden ball* with great power, *spreading it open*, and causing it to fit tightly to the cavity of the barrel in its course out, thus giving it a perfect direction.

Cannon of different sizes are named according to the weight of the ball which they are capable of discharging. Thus, we have 68-pounders, 24-pounders, 18-pounders, and the lighter field-pieces, from 4 to 12-pounders. The quantity of powder generally used for discharging common iron or brass cannon, is one third the weight of the ball. In general warfare, the effective distance at which artillery can be used is from 500 to 600 *yards*, or from *a quarter to half a mile*. At the battle of Waterloo, the brigades of artillery were stationed about half a mile from each other. Cannon-balls and shells can be thrown with effect to the distance of a mile and a half to two miles.

The distance to which a ball may be thrown by a 24-pounder, with a quantity of powder equal to two thirds the weight of the ball, is about four miles. Its effective range is, however, much less. Were the resistance of the air entirely removed, the same ball would be thrown to about five times that distance, or twenty miles.

It has been found that, by the firing of an 18-pound shot into a butt, or target, made of beams of oak, when the charges were 6 *pounds of powder*, 3 *pounds*, 2¼ *pounds*, *and* 1 *pound*, the respective depths of the penetration *were* 42 *inches*, 30 *inches*, 28 *inches*, *and* 15 *inches;* and the velocities at which the balls flew, were 1,600 *feet in a second*, 1,140 *feet*, 1,024 *feet*, *and* 656 *feet.*

When the cannon is so pointed that the ball goes *perfectly straight* toward the object aimed at, the direction is said to be point-blank. Ricochet firing is when the ball is discharged in such a manner that it goes *bounding and skipping* along the surface of the ground. In this way a ball can be thrown more effectively, and for a greater distance, than in any other way.

There are several substances known to chemists which possess a greater explosive power than gunpowder. It has not, however, been considered possible to increase the range and effect of a projectile fired from a gun, or cannon, by using any of them. Supposing that the guns could be made indefinitely strong, and the gunpowder indefinitely powerful, the point would soon be reached where the *resistance which the air opposes* to a body moving very rapidly would balance the force derived from the explosive compound, which drives the projectile forward. Beyond this point no increase of impulsive force would urge the projectile farther; and this limit is considerably within the range of power that can be exercised by common gunpowder. Beside this, the strength of materials of which guns are made is limited. Practical experience has fully demonstrated that the largest piece of ordnance which can be cast perfect, sound, and free from flaws, *is a mortar* 13 *inches diameter;* and even this weighs five tons. The French, at the siege of Antwerp, constructed a mortar having a bore of no less than 20 inches diameter, but it burst on the ninth time of firing.

FIG. 49.

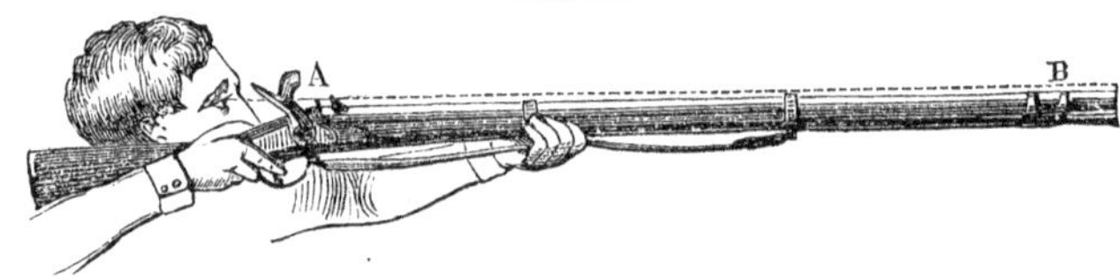

Until quite recently, the muskets placed in the hands of soldiers were usually aimed so that the line of sight was parallel to the barrel, and directed to the object, as in Fig 49. So long as the range of the musket was of limited extent, and great precision was not expected, the deviation of the ball from a straight line was not taken into account; but with the introduction of rifles throwing a ball to a great distance, the drop of the ball occasioned by the curvature of the path of the projectile, was found to deprive the weapon of the necessary precision. On all modern guns, therefore, a double sight is provided, by which the elevation necessary to secure accurate aim can always be given to the barrel. This is exhibited in Fig. 50, where one of the sights, B, is fixed, in the usual manner, at one extremity of the barrel, while the other is located nearer the breach. This last sight is often graduated and provided with an adjustment, by which it can be adapted to objects at different distances, so as to hit them exactly.

FIG. 50.

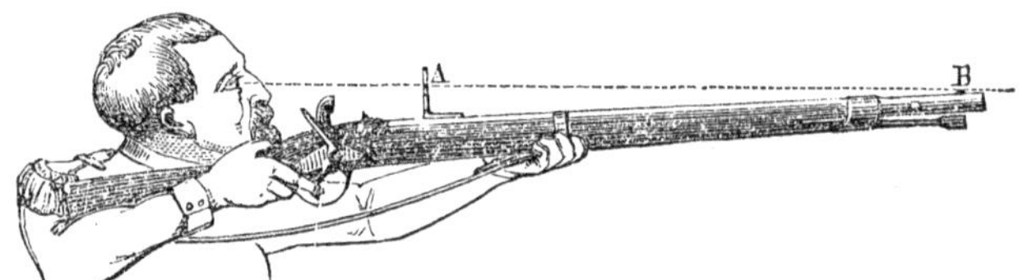

What is Circular Motion?

163. CIRCULAR MOTION is the motion produced by the revolution of a body about a central point.

How is Circular Motion produced?

164. Circular Motion is a species of compound motion, and is caused by the continued operation of two forces;—one the force of projection, which gives the body motion, tends to cause it to move in a straight line; while the other is continually deflecting it from a straight course toward a fixed point.

Illustrate the production of Circular Motion.

This fact is illustrated by the common sling, or by swinging a heavy body attached to a string round the head. The body, in this case, moves through the influence of two forces, the force of projection, and the string which confines it to the hand. These two forces act at right angles to one another, and according to the statements already made (§ 155), the path of the moving body will be a resultant of the two forces, or the diagonal of a parallelogram.

How may the curve of a circle be considered as equivalent to the diagonal of a parallelogram?

How then, it may be asked, does the body attached to the string and whirled round the head, move in a circle? This will be clear, if we consider that a circle is made of an infinite number of little straight lines (diagonals of parallelograms) and that the body moving in it, has its motion bent at every step of its progress by the action of the force which confines it to the hand. This force, however, only keeps it within a certain distance, without drawing it nearer to the hand. The two forces exactly balancing each other, the course of the whirling body will be circular.

What are the two forces which produce Circular Motion called?

165. The two forces by which circular motion is produced, are called the CENTRIFUGAL* and CENTRIPETAL Forces.†

What is Centrifugal Force?

166. The CENTRIFUGAL FORCE is that force which impels a body moving in a curve to move outward, or fly off from a center.

What is Centripetal Force?

167. The CENTRIPETAL FORCE is that force which draws a body moving in a curve toward the center, and assists it to move in a bent, or curvelinear course. In Circular Motion the Centrifugal and Centripetal Forces are equal, and constantly balance each other.

What follows if the Centrifugal or Centripetal Forces are destroyed?

If the Centrifugal Force of a body revolving in a circular path be destroyed, the body will immediately approach the center; but if the Centripetal Force be destroyed, the body will fly off in a straight line, called a tangent.

Thus, in whirling a ball attached by a string to the finger, the propelling force, or the force of projection, is given by the hand, and the Centripetal Force is exhibited in the stretching, or tension of the string. If the string breaks in whirling, the Centripetal Force no longer acts, and the ball, by the action of the Centrifugal Force, generated by the whirling motion, flies off in a tangent, or straight line, as is represented in Fig. 51. If, on the contrary, the whirling motion is too slow, the Centripetal Force preponderates, and the ball falls in toward the finger.

FIG. 51.

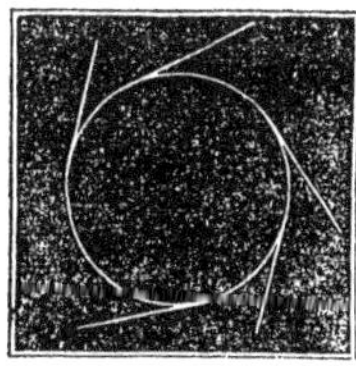

Familiar examples of the effects of Centrifugal Force are common in the experience of every-day life.

What are familiar illustrations of Centrifugal Force?

The motion of mud flying from the rim of a coach-wheel, moving rapidly, is an illustration of Centrifugal Force. Fig. 52 represents a coach-wheel throwing off mud; *a* the point at which the mud flies off; *a b*, the straight line in which it

* Centrifugal, compounded of center, and "*fugio*," to fly off.

† Centripetal, compounded of center and "*peto*," to seek.

would move but for the action of the two forces, which compel it to follow the parabolic curve, *a c*.

FIG. 52.

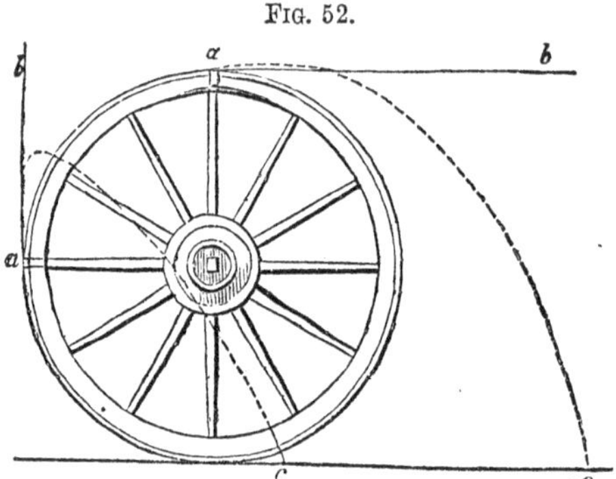

The mud sticks to the wheel, in the first instance, through the force of adhesion; but this force, being very weak, is overcome by the Centrifugal Force, and the particles of mud fly off. The particles which compose the wheel itself would also fly off in the same manner, were not the force of cohesion which holds them together stronger than the Centrifugal Force.

Under what circumstances will the Centrifugal Force overcome the Force of Cohesion?

The Centrifugal Force, however, increases with the velocity of revolution, so that if the velocity of the wheel were continually increased, a point would at last be reached, when the Centrifugal Force would be more powerful than the force of cohesion, and the wheel would then fly in pieces. In this way almost all bodies can be broken by a sufficient rotative velocity. Large wheels and grindstones, revolving rapidly, not infrequently break from this cause, and the pieces fly off with immense force and velocity.

FIG. 53.

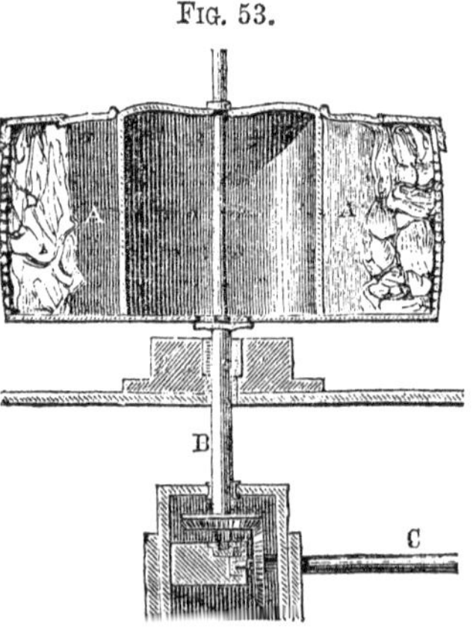

When we whirl a mop, the water flies off from it by the action of the Centrifugal Force. The fibers, or threads, which compose the mop, also tend to fly off, but being confined at one end, they are unable so to do. They, therefore, assume a spherical form, or shape.

The fact that water can be made to fly off from a mop, by the action of the Centrifugal Force produced by whirling it, has been most ingeniously applied in a machine for drying cloth, called the hydro-extractor (water-extractor), Fig. 53. The machine consists of a large hollow wheel, or cylinder, A A, turning upon an axis, B. The sides and bottom of the wheel are pierced with holes like a sieve. The wet cloths being in and around the sides, A, the wheel is caused to revolve with great rapidity, and the water contained in the material, by the action of the Centrifugal Force, flies out, and escapes through the apertures left in the sides of the wheel. A rotation of 1500 times per minute, is sufficient to almost entirely dry the cloth, no matter how wet it may have been originally.

FIG. 54.

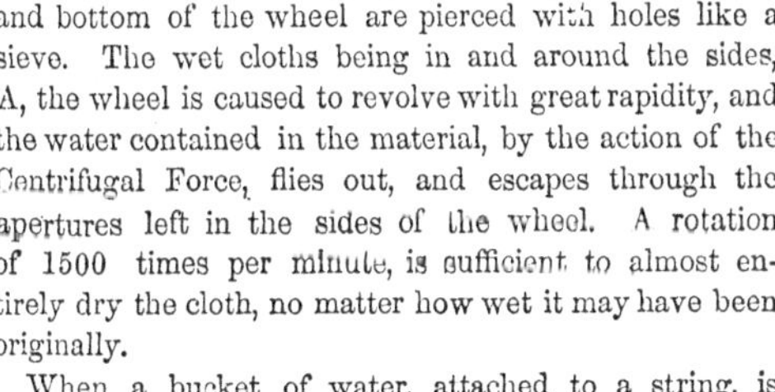

When a bucket of water, attached to a string, is whirled rapidly round, the water does not fall out when the mouth is presented downward, since the Centrifugal Force imparted to the water by rotation, tends to cause it to fly off from the center, and this overcomes, or balances, the attraction of gravitation, which tends to cause the water to fall out, or toward the center. Thus, in Fig 54, the water contained in the bucket which is upside down, has no support under it, and if the bucket were kept still in its inverted position for a single moment the water would fall out by its own weight, or, in other words, by the attraction of gravitation, which represents a Centripetal Force; but the Centrifugal Force, which is caused by the whirling of the bucket in the direction of the arrow, tends to drive the water out through the bottom and side of the vessel, and as this last force overcomes, or balances the other, the water retains its place, and not a drop is spilled.

When a carriage is moved rapidly round a corner, it is very liable to be overturned by the Centrifugal Force called into action. The inertia carries the body of the vehicle forward in the same line of direction, while the wheels are suddenly pulled around by the horses into a new one. Thus a loaded stage running south, and suddenly turned to the east, throws out the luggage and passengers on the south side of the road. When railways form a rapid curve, the outer rail is laid higher than the inner, in order to counteract the Centrifugal Force.

An animal, or man, turning a corner rapidly, leans in toward the corner or center of the curve in which he is moving, in order to resist the action of the Centrifugal Force, which tends to throw him away from the center.

In all equestrian feats exhibited in the circus, it will be observed that not only the horse, but the rider, inclines his body toward the center, Fig. 55, and according as the speed of the horse round the ring is increased, this inclination becomes more considerable. When the horse walks slowly round a large ring this inclination of his body is imperceptible; if he trot, there is a visible inclination inward, and if he gallop, he inclines still more, and when urged to full speed he leans very far over on his side, and his feet will be heard to strike against the partition which defines the ring. The explanation of all this is, that the Centrifugal Force caused by the rapid motion around the ring tends to throw the horse out of, and away from, the circular course, and this he counteracts by leaning inward.

Fig. 55.

How do the motions of the solar system illustrate the action of Centrifugal and Centripetal Forces?

The most magnificent exhibition of Centrifugal and Centripetal Forces balancing each other, is to be found in the arrangements of the solar system. The earth and other planets are moving around a center—the sun, with immense velocities, and are constantly tending to rush off into space, by the action of the Centrifugal Force. They are, however, restrained within exactly determined limits by the attraction of the sun, which acts as a centripetal power drawing them toward the center.

What is the Axis of a body?

168. The Axis of a body is the straight line, real or imaginary, passing through it, on which it revolves, or may revolve.

When a body revolves round its Axis, what peculiarities do its several parts exhibit?

169. When a body rotates upon an axis, all its parts revolve in equal times. The velocity of each particle of a revolving body increases with its perpendicular distance from the axis, and as its velocity increases, its Centrifugal Force increases.

A moment's reflection will show, that a point on the outer part, or rim, of a wheel, moves round the axis in the same time as a point nearer the center, as upon the hub. But the circle described by the revolution of the outer part of the wheel is much larger than that described by the inner part, and as both move round the center in the same time, the outer part must move with a greater velocity.

What effect does the action of Centrifugal Force have upon the figure of a body?

170. If the particles of a rotating body have freedom of motion among themselves, a change in the figure of the body may be occasioned by the difference of the Centrifugal Force in the different parts.

A ball of soft clay, with a wire for an axis, forced through its center, if made to turn quickly, soon ceases to be a perfect ball. It bulges out in the middle, where the Centrifugal Force is, and becomes flattened toward the ends, or where the wire issues.

Fig. 56.

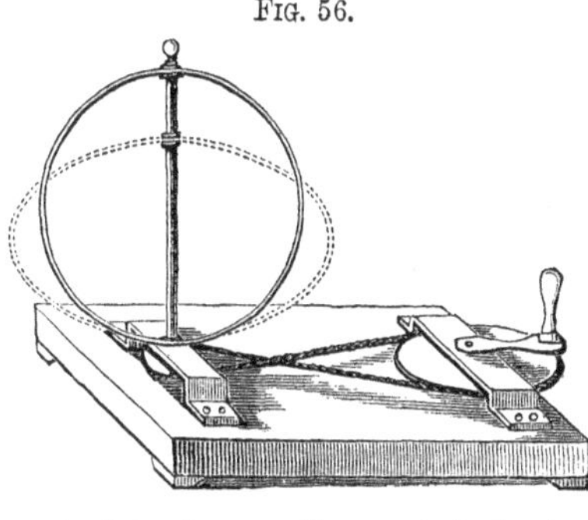

This change in the form of revolving bodies may be illustrated by an apparatus represented in Fig. 56. This consists of an elastic circle, or hoop, fastened at the lower side on a vertical shaft, while the upper side is free to move. On turning the wheels, so arranged as to impart a very rapid motion to the shaft and hoop, the hoop will be observed to bulge out in the middle (owing to the Centrifugal Force acting with greater intensity upon those parts furthest removed from the axis) and to become flattened at the ends.

What is the cause of the present form of the earth?

171. The earth itself is an example of the operation of this force. Its diameter at the equator is about twenty-six miles greater than its polar diameter. The earth is supposed to have assumed this form at the commencement of its revolution, through the action of the Centrifugal Force, while its particles were in a semi-fluid, or plastic state. In Fig. 57 we have a representation of the general figure of the earth, in which N S is the polar diameter, and also the axis of rotation, and E W the equatorial diameter.

Fig. 57.

What is the amount of Centripetal Force at the equator?

172. At the equator the Centrifugal Force of a particle of matter is 1-290ths of its gravity. This diminishes as we approach the poles, where it becomes 0.

If the earth revolved 17 times faster than it now does, or in 84 minutes instead of 24

What would be the effect if the velocity of rotation of the earth were increased?

hours, the Centrifugal Force would be equal to the attraction of gravitation, which may be considered as the Centripetal Force, and all bodies at the earth's equator would be deprived of weight, since they would have as great a tendency to leave the surface of the earth as to descend toward its center. If the earth revolved on its axis in less time than 84 minutes, terrestrial gravitation would be completely overpowered, and all fluids and loose substances would fly from its surface.

173. There appears to be a constant tendency to rotary motion in moving bodies free to turn upon their axes. The earth turns upon its axis, as it moves in its orbit; a ball projected from a cannon, a rounded stone thrown from the hand, all revolve around their axes as they move.

FIG. 58.

This phenomenon may be very prettily illustrated by placing a watch-glass upon a smooth plate of glass, Fig. 58, moistened sufficiently to insure slight adhesion, and fixed at any angle. As it begins to move toward the bottom of the inclined plane, it will exhibit a revolving motion, which uniformly increases with the acceleration of its downward movement.

PRACTICAL QUESTIONS AND PROBLEMS ON THE PRINCIPLES AND COMPOSITION OF MOTION.

1. The SURFACE of the EARTH at the EQUATOR moves at the rate of about a THOUSAND MILES in an HOUR: why are MEN not sensible of this rapid movement of the earth?

Because *all objects* about the observer *are moving in common* with him. It is the natural uniformity of the undisturbed motion which causes the earth and all the bodies moving together with it upon its surface to appear at rest.

2. How can you easily see that the EARTH is in motion?

By looking at some object that is entirely *unconnected* with it, as the *sun* or the *stars*. We are here, however, liable to the mistake that the sun or stars are in motion, and not we ourselves with the earth.

3. Does the SUN really RISE and SET each day?

The sun maintains very nearly a constant position; but the earth revolves, and is constantly changing its position. *Really, therefore, the sun neither rises nor sets.*

4. Why, to a PERSON SAILING in a BOAT on a smooth stream, or GOING SWIFTLY in a CARRIAGE on a smooth road, do the trees or buildings on the banks or roadside appear to move in an OPPOSITE DIRECTION?

The *relative situation* of the trees and buildings to the person, and to each other, is actually changed by the motion of the observer; but the mind, in judging of the real change in place by the difference in the position of the objects observed, *unconsciously* confounds the real and apparent motion.

5. Why will a tallow candle fired from a gun pierce a board, or target, in the same manner as a leaden bullet will, under the same circumstances?

When a candle starts from the breach of a gun, its motion is gradually increased, until it leaves the muzzle at a high velocity; and when it reaches the board, or target, every particle of matter composing it is in a state of great velocity. At the moment of contact, the particles of matter composing the target are at rest; and as the density of the candle, multiplied by the velocity of its motion, is greater than the density of the target at rest, the greater force overcomes the weaker, and the candle breaks through and pierces a hole in the board.

6. Why, with an enormous pressure and slow motion, can you not force a candle through a board?

Because the candle, on account of its slow motion, does not possess sufficient momentum to enable the density of its particles to overcome the greater density of the board; consequently the candle itself is mashed, instead of piercing the board.

7. Why will a large ship, moving toward a wharf with a motion hardly perceptible, crush with great force a boat intervening?

Because the great mass and weight of the vessel compensates for its want of velocity.

8. Why can a person safely skate with great rapidity over ice which would not bear his weight standing quietly?

Because time is required to produce a fracture of the ice; as soon as the weight of the skater begins to act upon any point, the ice, supported by the water, bends slowly under him; but if the skater's velocity be great, he passes off from the spot which was loaded before the bending has reached the point at which the ice would break.

9. A HEAVY COACH and a LIGHT WAGON came in collision on the road. A suit for damages was brought by the proprietor of the wagon. How was it shown that ONE of the VEHICLES was moving at an UNSAFE VELOCITY?

On trial, the persons in the wagon deposed *that the shock*, occasioned by coming in contact, was so great, that it *threw them over the head of their horse;* and thus lost their case by proving that the faulty velocity was their own.

10. Why did the FACT that they were THROWN over the HEAD OF THE HORSE by coming in contact with the coach, prove that their velocity was GREATER than it ought to have been?

The coach stopped the wagon by contact with it, but the bodies of the persons in the wagon, *having the same velocity as the wagon, and not fastened to it, continued to move on.* Had the wagon moved slowly, the distance to which they would have been thrown would have been slight. To cause them to be thrown *as far as over the head of the horse*, would require a great velocity of motion.

11. When TWO PERSONS STRIKE their HEADS together, one being in MOTION and the other at REST, why are both equally hurt?

Because, when bodies strike each other, *action and reaction are equal;* the head that is at rest returns the blow with equal force to the head that strikes.

12. When an elastic BALL is thrown against the side of a house with a CERTAIN FORCE, why does it rebound?

Because the *side of the house* resists the ball with the *same force*, and the ball being elastic, *rebounds.*

13. When the SAME BALL is thrown against a PANE of GLASS with the same force, it goes through, breaking the glass; why does it not rebound as before?

Because the glass has not sufficient power to resist the full force of the ball: it destroys a part of the force of the ball, but the remainder continuing to act, the ball goes through, shattering the glass.

14. Why did not the MAN succeed who undertook to make a FAIR WIND for his PLEASURE-BOAT, by erecting an IMMENSE BELLOWS in the STERN, and blowing against the SAILS?

Because the action of the stream of wind and the reaction of the sails were exactly equal, and, consequently, the boat remained at rest.

15. If he had blown in a CONTRARY DIRECTION from the sails, instead of against them, would the boat have moved?

It would, with the *same force* that the air issued from the bellows-pipe.

16. Why can not a MAN raise himself over a FENCE by pulling upon the STRAPS of his BOOTS?

Because the action of the force exerted by the muscles of his arms is counteracted by the reaction of the force, or, in other words, the resistance of his whole body, which tends to keep him down.

17. Why do WATER-DOGS give a SEMI-ROTARY MOVEMENT to free themselves from water?

Because in this way a *centrifugal force* is generated, which causes the drops of water adherent to them to fly off.

18. Why is the COURSE of rivers rarely STRAIGHT, but SERPENTINE and WINDING?

When, from any obstruction, the river is obliged to bend, *the centrifugal force tends to throw away the water from the center of the curvature,* so that when a bend has once commenced, it increases, and is soon followed by others. Thus, for instance, the water being thrown by any cause to the left side, it wears that part into a curve, or elbow, and, by its centrifugal force, acts constantly on the outside of the bend, until the rock, or higher land, resists its gradual progress; from this limit, being thrown back again, it wears a similar bend to the right hand, and after that another to the left, and so on.

19. A locomotive passes over a railroad, 200 miles in length, in 5 hours; what is its velocity per hour?

20. If a bird, in flying, passes over a distance of 45 miles in an hour, what is its velocity per minute?

21. The flash of a cannon three miles off was seen, and in 14 seconds afterward the sound was heard. How many feet did the sound travel in one second?

22. The sun is 95 millions of miles from the earth, and it requires 8¼ minutes for its light to reach the earth; with what velocity per second does light move?

23. If a vessel sail 90 miles a day for 8 days, how far will it sail in that time?

24 A gentle wind is observed to move 1,250 feet in 15 minutes: how far would it move in 2 hours, allowing 5,000 feet to the mile?

25. What distance would a bird flying uniformly at the velocity of 60 miles per hour, pass over in 12½ hours?

26. Suppose light to move at the rate of 192,000 miles in a second of time, how long a time will elapse in the passage of light from the sun to the earth, the distance being 95 millions of miles?

27. What is the momentum of a body weighing 25 pounds moving with the velocity of 30 feet per second?

28. A cannon-ball weighing 520 pounds, struck a wall with a velocity of 45 feet per second: what was its momentum, or with what force did it strike?

29. A locomotive and train of cars weighing 180 tons (403,200 pounds), and moving at the rate of 40 miles per hour, came in collision with another train weighing 160 tons, and moving at the rate of 25 miles per hour: what was the momentum, or force of collision?

30. A stone thrown directly at an object from a locomotive, moving at the rate of 3,520 feet per minute, was 2 seconds in the air; at what distance beyond the object did it strike?

CHAPTER VI.

APPLICATION OF FORCE.

What are the principal agents of power in the arts?

174. THE principal agents from whence we obtain power for practical purposes, are MEN and ANIMALS, WATER, WIND, STEAM, and GUNPOWDER.

From what great natural forces are these agents of power derived?

The power of all these may be ultimately resolved into some one or more of the great natural forces, or primary sources of power, viz., vital force, producing muscular energy, or strength in man and animals; gravitation, causing the flow of water; heat and molecular forces, the agents producing the power exhibited by wind, steam, and gunpowder.

Are there any other agents of power?

Magnetism and electricity when called into action, and capillary attraction, are also agents of power; but none of these are capable, as yet, of being used to any great extent for the production of motion.

How is muscular energy exerted?

175. Muscular energy in men and animals is exerted by means of the contraction of the fibers which constitute the muscles of the

body; the bones of the body facilitate and direct the application of this force.

Beasts of prey possess the greatest amount of muscular power; but some very small animals possess muscular power in proportion to their bulk, incomparably greater than the largest of the brute creation. A flea, considered relatively to its size, is stronger than an elephant, or a lion.

How can a man exert his greatest strength? A man can exert his greatest active strength in pulling upward from his feet, because the strong muscles of the back, and those of the upper and lower extremities, are then brought most advantageously into action.

The comparative effect produced in the different methods of applying the force of a man, may be indicated as follows: in the action of turning a crank, or handle, his force may be represented by the number 17; in working a pump, by 20; in pulling downward, as in ringing a bell, by 39; and in pulling upward from the feet, as in the action of rowing, by 41.

What is the estimated strength of a man? 176. The estimate of the uniform strength of an ordinary man, for the performance of ordinary daily mechanical labor is, that he can raise a weight of 10 pounds to the height of 10 feet once in a second, and continue to do so for 10 hours in the day.

What is the estimated strength of a horse, or a "horse-power?" 177. The estimated strength of a horse is, that he can raise a weight of 33,000 pounds to the height of one foot in a minute. Such a measure of force is called a "HORSE-POWER."

The strength of a horse is considered to be equal to that of five men. The average strength which a horse can exert in drawing is about 1600 pounds.

What is water-power? 178. WATER-POWER is the power obtained by the action of water falling perpendicularly, or running down a slope, by the influence of gravity.

What is the standard for comparing the amount of work performed by different forces? 179. When work is performed by any agent, there is always a certain weight moved over a certain space, or a resistance overcome; the amount of work performed, therefore, will depend on the weight, or resistance that is moved, and the space over which it is moved. For comparing different quantities of work, done by any force, it is necessary to have some standard; and this standard is the power, or labor, expended in raising a pound weight one foot high, in opposition to gravity.

How is the effect of a moving power expressed? 180. The effect produced by a moving power is always expressed by a certain weight raised a certain height.

To find, therefore, the effect of a moving power, or to find the power expended in performing a certain work, we have the following rule:—

How may the power expended in work be ascertained? 181. Multiply the weight of the body moved in pounds by the vertical space through which it is moved.

Thus, for example, if a horse draw a loaded wagon, with a force by which the traces are stretched to as great a degree as if 200 pounds were suspended vertically from them, and if the horse thus acting draws the wagon over a space of 100 feet, the mechanical effect produced is said to be 200 pounds raised 100 feet; or, what is the same thing, 20,000 pounds raised 1 foot. When a horse draws a carriage, the work he performs is expended in overcoming the resistance of friction of the road which opposes the motion of the carriage; but friction increases and diminishes as the weight of the load increases or diminishes. The work performed will, therefore, be estimated by multiplying the total resistance of friction, as expressed in pounds, by the space over which the carriage is moved.

Illustrate the manner of estimating power? The following examples will illustrate how we are enabled, by the above rules, to calculate the amount of power required to perform a certain amount of work:—Suppose we wish to know the amount of horse-power required to lift 224 pounds of coal from the bottom of a mine 600 feet deep. The weight, 224, multiplied into space moved over, 600 feet, equals 134,400, the amount of work to be performed each minute; a horse-power equals 33,000 pounds raised 1 foot per minute: therefore, 134,400÷33,000=4.07, horse-power required. If we wish to perform the same work by a steam-engine, we would order an engine of 4.07 horse-power, and the engine-builder, knowing the dimensions of the parts of an engine essential to give one horse-power, can build an engine capable of performing the requisite work.

Again. Suppose a locomotive to move a train of cars, on a level, at the rate of 30 miles per hour, the whole weighing 25 tons, with a constant resistance from friction of 200 pounds, what is the horse-power of the engine? 30 miles per hour equals 2,640 feet per minute; this space multiplied by 200 pounds, the resistance to be overcome, equals 528,000, the work to be done every minute; which, divided by 33,000 (one horse-power), equals 16, the horse-power of the locomotive.

What is a Dynamometer? 182. An instrument for measuring the relative strength of men and animals, and also of the force exerted by machinery, is called a DYNAMOMETER.

FIG. 59.

Fig. 59 represents one of the most common forms of the dynamometer, consisting of a band of steel, bent in the middle, so as to have a certain degree of flexibility. To the expanded extremity of each limb is fixed an *arc* of iron, which passes freely through an opening in the other limb, and terminates outside in a hook or ring. One of these arcs is graduated, and represents in pounds the force required to bring the two limbs nearer together. Thus, if a horse were pulling a rope attached to a body which he had to move, we may imagine the rope to be cut at a certain point, and the two ends attached to the ends of the *arcs*, as represented in Fig. 59; the force of traction exerted by the animal would be seen by the greater or less bringing together of the ends of the instrument.

FIG. 60.

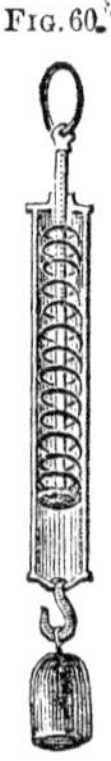

In another form of dynamometer, Fig. 60, which is also used as a spring balance in weighing, the force is measured by the collapsing of a steel spring, contained within a cylindrical case. The construction and operation of this instrument will be easily understood from an examination of the figure.

What is a Machine?

183. A MACHINE is an instrument, or apparatus, adapted to receive, distribute, and apply motion derived from some external force, in such a way as to produce a desired result.

A steam-engine and a water-wheel are examples of machines. They receive the power of steam in the one case, and the power of falling water in the other, and apply it for locomotion, sawing, hammering, etc.

Do we produce force by the use of machines?

184. A MACHINE can not, under any circumstances, create power, or increase the quantity of power, or force, applied to it.

A machine will enable us to concentrate, or divide, any quantity of force which we may possess, but they no more increase the quantity of force applied than a mill-pond increases the quantity of water flowing in the stream.*

Do not machines in reality diminish force?

Machines, in fact, do not increase an applied force, but they diminish it, or, in other words, no machine ever transmits the whole amount of force imparted to it by the moving power; since a part of the power is necessarily expended in overcoming the inertia of matter, the friction of the machinery, and the resistance of the atmosphere.

* "Power is always a product of nature. God has not vouchsafed to man the means of its primary creation. He finds it in the moving air and the rapid cataract; in the burning coal and the heaving tide. He transfers it from these to other bodies, and renders it the obedient servant of his will—the patient drudge which, in a thousand ways, administers to his wants, his convenience, and his luxuries, and enables him to reserve his own energy for the higher purposes of the development of his mind and the expression of his thoughts."—*Prof. Henry.*

Is Perpetual Motion in machinery possible?

185. PERPETUAL MOTION, or the construction of machines which shall produce power sufficient to keep themselves in motion continually, is, therefore, an impossibility, since no combination of machinery can create, or increase, the quantity of power applied, or even preserve it without diminution.

What example of continued motion have we in nature?

In nature we have an example of continued and undiminished motion in the revolution of the earth upon its axis, and of the planets around the sun. These bodies have been moving with undiminished velocity for ages past, and, unless prevented by the agency which created them, will continue so to do for ages to come.

How do we derive advantages from machines?

186. We derive advantages from machines in three different ways; 1st, from the additions they make to human power; 2d, from the economy they produce of human time; 3d, from the conversion of substances apparently worthless and common into valuable products.

How do machines make additions to human power?

187. Machines make additions to human power, because they enable us to use the power of natural agents, as wind, water, steam. They also enable us to use animal power with greater effect, as when we move an object easily with a lever, which we could not with the unaided hand.

How do machines produce economy of human time?

188. Machines produce economy of human time, because they accomplish with rapidity what would require the hand unaided much time to perform.

A machine turns a gun-stock in a few minutes; to shape it by hand would be the work of hours.

How do machines convert worthless objects into valuable products?

189. Machines convert objects apparently worthless into valuable products, because by their *great power, economy, and rapidity of action*, they make it profitable to use objects for manufacturing purposes which it would be unprofitable or impossible to use if they were to be manufactured by hand.

Without machines, iron could not be forged into shafts for gigantic engines; fibers could not be twisted into cables; granite, in large masses, could not be transported from the quarries.

Define Power, Weight, and Working Point, as applied to machinery.

190. In machinery, we designate the moving force as the POWER ; the resistance to be overcome, whatever may be its nature, as the WEIGHT ; and the part of the machine immediately applied to the resistance to be overcome, as the WORKING POINT.

What is the great general advantage of machinery?

191. The great general advantage that we obtain from machinery is, that it enables us to exchange time and space for power.

Thus, if a man could raise to a certain height two hundred pounds in one minute, with the utmost exertion of his strength, no arrangement of machinery could enable him unaided to raise 2,000 pounds in the same time. If he desired to elevate this weight, he would be obliged to divide it into ten equal parts, and raise each part separately, consuming ten times the time required for lifting 200 pounds. The application of machinery would enable him to raise the whole mass at once, but would not decrease the time occupied in doing it, which would still be ten minutes.

Again. A boy who can not exert a force of fifty pounds may, by means of a claw-hammer, draw out a nail which would support the weight of half a ton. It may seem that the use of the hammer in this case creates power, but it does not, since the hand of the boy is required to move through perhaps *one foot of space* to make the nail rise *one quarter of an inch.* But it has been already shown that the force of a small body moving with great velocity may equal the force of a large body with a slight velocity. On the same principle, the small weight, or power, exerted by the boy on the end of the hammer handle, moving through a large space with an increased velocity, acquires sufficient momentum to overcome the great resistance of the nail.

In both of these examples space and time are exchanged for power.

How is the mechanical effect of a power determined?

192. The mechanical force, or momentum, of *a body*, is ascertained by multiplying its weight by the space through which it moves in a given time, that is to say, by its velocity. The mechanical force, or momentum, of *a power* may also be found, by multiplying the power, or its equivalent weight, by its velocity.

What is the law of equilibrium of all machines?

193. The power, multiplied by the space through which it moves in a vertical direction, is equal to the weight multiplied by the space through which it moves in a vertical direction.

This is the general law which determines the equilibrium of all machines.

Under what conditions will motion take place in a machine?

194. The power will overcome the resistance of the weight, and motion will take place in a machine, when the product arising from the power multiplied by the space through which

it moves in a vertical direction, is greater than the product arising from the weight multiplied by the space through which it moves in a vertical direction.

What is meant by the expression power is gained at the expense of time?

Practical men express the principle of equilibrium in machinery by saying "that what is gained in power is lost in time." Thus, if a small power acts against a great resistance, the motion of the latter will be just as much slower than that of the power, as the resistance, or weight, is greater than the power; or if one pound be required to overcome the resistance of two pounds, the one pound must move over two feet in the same time that the resistance, two pounds, requires to move over one.

SECTION I.

THE ELEMENTS OF MACHINERY.

How many simple machines are there?

195. All machines, no matter how complex and intricate their construction, may be reduced to one or more of six simple machines, or elements, which we call the "MECHANICAL POWERS."

Enumerate the six elementary machines.

196. They are the LEVER, the WHEEL and AXLE, the PULLEY, the INCLINED PLANE, the WEDGE, and the SCREW.

These simple Machines may be further reduced to three—the lever, the pulley, and the inclined plane; since the wheel and axle, the screw and the wedge, may be regarded as modifications of them.

The name "mechanical powers" which has been applied to the six elementary machines, is unfortunate, since it serves to convey an idea that they are really powers, when in fact they possess no power in themselves, and are only instruments for the application of power.

What is a Lever?

197. A LEVER consists of a solid bar, straight or bent, turning upon a pivot, prop, or axis.

What are the Arms of a Lever?

198. The ARMS of the lever are those parts of the bar extending on each side of the axis.

What is the Fulcrum?

199. The FULCRUM, or prop, is the name applied to the axis, or point of support.

How many kinds of levers are there?

200. Levers are divided into three kinds, or classes, according to the position which the fulcrum has in relation to the power and the weight.

What are the relative positions of the power, fulcrum and weight in the three kinds of levers?

201. In the first class the fulcrum is between the power and the weight; in the second class, the fulcrum is at one end of the lever, and the weight is between the fulcrum and the power; in the third class, the fulcrum is at one end of the lever, and the power is between the fulcrum and the weight.

Fig. 61 represents the three classes of levers, numbered in their order, 1, 2, 3. P is the power, W the weight, and F the fulcrum.

FIG. 61.

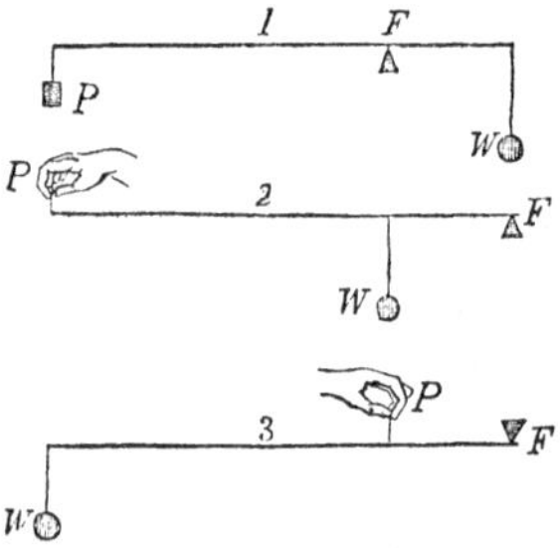

What are examples of levers of the first class?

A crowbar applied to elevate a stone, is an example of a lever of the first kind. In Fig. 62, which represents a lever of this class, *a* indicates the fulcrum which suppports the bar, *b* the power applied by the hand at the end of the longest arm, and *c* the weight, or stone, raised at the end of the short arm. A poker applied to stir up the fuel of a grate is a lever of the first class, the fulcrum being the bars of the grate; the break, or or handle of a pump, is also a familiar example. Scissors, pincers, etc., are composed of two levers of the first kind, the fulcrum being the joint, or pivot, and the weight the resistance of the substance to be cut, or seized. The power of the fingers is applied at the other end of the levers.

FIG. 62.

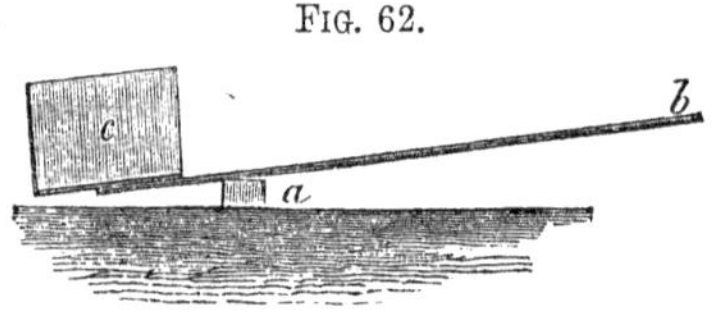

What is the law of equilibrium of the lever?

202. A lever will be in equilibrium, when the power and the weight are to each other inversely as their distances from the fulcrum.

Thus, if in a lever of the first class the power and the weight are equal, and are required to exactly balance each other, they must be placed at equal distances from the fulcrum. If the power is only half the weight, it must be at double the distance from the fulcrum; if one third of the weight, three times the distance. If we suppose, in Fig. 62, *c* to represent a weight of 300 pounds, placed two feet from the fulcrum *a*, and *b* a power of 100 pounds placed six feet from *a*, then *c* and *b* will be in equilibrium, for $(300\times 2)=(100\times 6)$.

203. When the weight and lengths of the two arms

Weight, and the length of the arms of a lever being given, how we find the equivalent power?

of a lever are given, the power requisite to balance the weight may be ascertained, by dividing the product of the weight multiplied into its distance from the fulcrum, by the distance of the power from the fulcrum.

What are examples of levers of the second class?

204. Cork, or lemon-squeezers, Fig. 63, are examples of the levers of the second class, which have the fulcrum at one end, and the weight, or resistance to be overcome, between the fulcrum and the power. An oar is a lever of the second class, in which the reaction of the water against the blade is the fulcrum, the boat the weight, and the hand of the boatman the power. A door moved on its hinges is another example. A wheel-barrow is a lever of the second class, the fulcrum being the point at which the wheel presses upon the ground, the barrow and its load the weight, and the hands the power. Nut-crackers are two levers of the second class, the hinge which unites them being the fulcrum, the resistance of the shell placed between them the weight, and the hand the power.

Fig. 63.

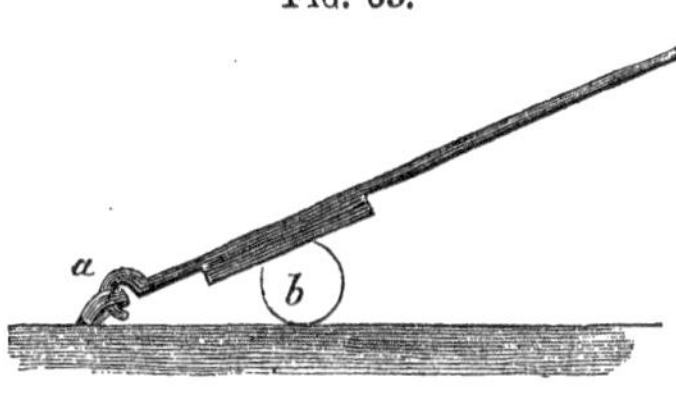

What are examples of levers of the third class?

205. A pair of sugar-tongs represents a lever of the third class, in which the power is applied between the fulcrum and the resistance, or weight. In Fig. 64, the fulcrum is at *a*, the resistance is the piece of sugar to be lifted at *b*, and the power is the fingers applied at *c*. When a man raises a ladder against a wall, he employs a lever of the third class; the fulcrum being the foot of the ladder resting upon the ground, the power being the hands applied to raise it, and the resistance being the weight of the ladder.

Fig. 64.

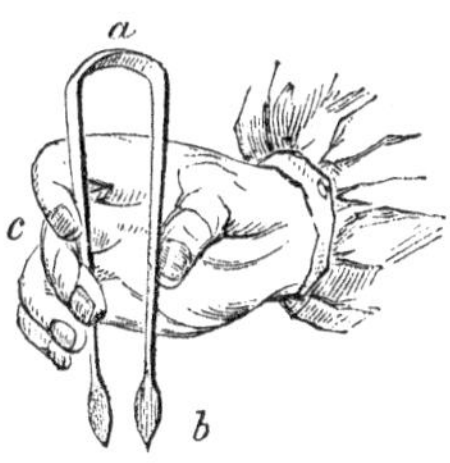

What is the relation between the power and the weight in levers of the third class?

206. In levers of the third class, the power, being between the fulcrum and the weight, will be at a less distance from the fulcrum than the weight; and, consequently, in this form of lever the power must be always greater than the weight.

Thus (in No. 3, Fig. 61), if the length from the point where the weight, W, is suspended to F be three times the length of P F, then a weight of 100 pounds suspended at W will require a power of 300 applied at P to sustain it.

Under what circumstances do we employ levers of the third class?

Owing to its mechanical disadvantages, this class of levers is rarely used, except where a quick motion is required, rather than great force. The most striking examples of levers of the third class are found in the animal kingdom. The limbs of animals are generally levers of this description. The socket of the bone, *a*, Fig. 65, is the fulcrum; a strong muscle attached to the bone near the socket, *c*, and extending to *d*, is the power; and the weight of the limb, together with whatever resistance, *w*, is opposed to its motion, is the weight. A very slight contraction of the muscle in this case gives considerable motion to the limb.

FIG. 65.

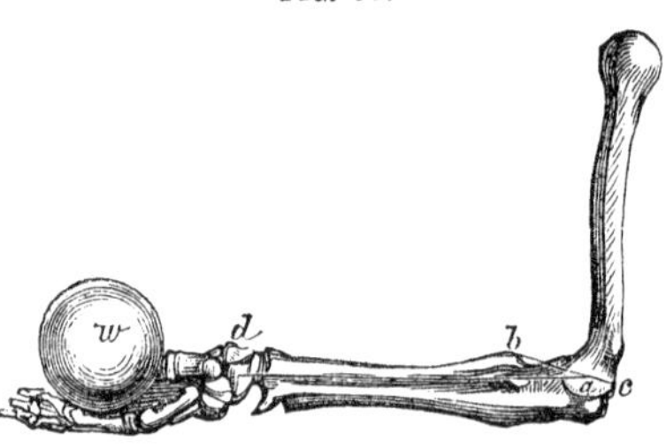

The leg and claws of a bird, are examples of the third class of levers, the whole arrangement being admirably adapted to the wants of the animal. When a bird rests upon a perch, its body constitutes the weight, the muscles of the leg the power, and the perch the fulcrum. Now, the greater the weight of the body, the more strain it exerts upon the muscles of the claws, which, in turn, grasp the perch more firmly: consequently, a bird sits upon its perch with the greatest ease, and never falls off in sleeping, since the weight of the body is instrumental in sustaining it.

What is a Compound Lever?

207. A COMPOUND LEVER is a combination of several simple levers, so arranged that the shorter arm of one may act upon the longer arm of another. In this way, the power of a small force in overcoming a large resistance is greatly multiplied.

FIG. 66.

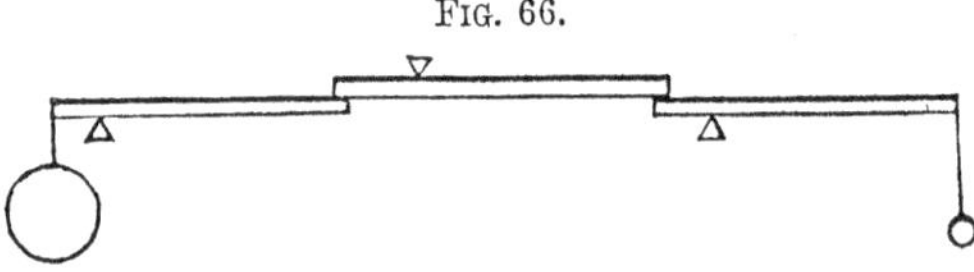

An arrangement of compound levers is shown in Fig. 66. Here, by means of three simple levers, 1 pound may be made to balance 1000; for if the long arm of each of the levers is ten times the length of the short one, 1 pound at the end of the first one will exert a force of 10 pounds upon the end of the second one, which will in turn exert ten times that amount, or 100 pounds, upon the end of the third one, which will balance ten times that amount, or 1000 pounds, at the other extremity.

What are the disadvantages of a compound lever?

208. The disadvantage of a compound lever is, that its exercise is limited to a very small space.

Describe the common steel-yard.

209. The different varieties of weighing machines are varieties or combinations of levers. The common steel-yard is a lever of unequal arms, belonging to the first class. It consists of a bar (Fig. 67) marked with notches to indicate pounds and ounces, and a weight which is movable along the notches. The bar is furnished with three hooks, or rings, on the largest of which the article to be weighed is always hung. The other hooks serve to support the instrument when it is in use, and the pivot by which they are attached to the bar serves as the fulcrum. The weight, Q, sliding upon the bar, balances the article, P, which is to be weighed, it being evident that a pound weight at D will balance as many pounds at P as the distance A C is contained in the space D C.

Fig. 67.

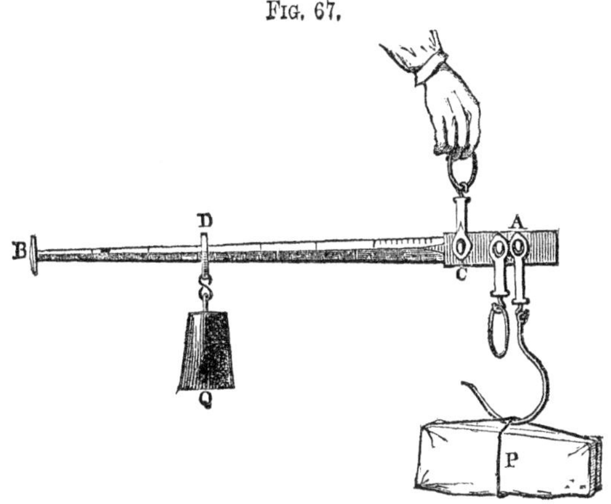

It may happen that when the weight Q is moved to the last notch upon the bar B C, that the article P will still preponderate. In this case, the steel-yard is held by the hook or ring nearer to A, which hangs down in the figure, and the steel-yard turned over, it being furnished with two sets of notches on opposite sides of the bar. By this means the distance of P, the article weighed, from the fulcrum is diminished, and the weight Q, at the given distance upon the opposite side of the fulcrum, will balance a proportionally greater resistance, or weight.

Describe the ordinary balance.

210. The ordinary balance is a lever of the first class, with equal arms, in which the power and the weight are necessarily equal. Fig. 68 shows the common form. The fulcrum or axis, is made wedge-like, with a sharp knife-like edge, and rests upon a

FIG. 68.

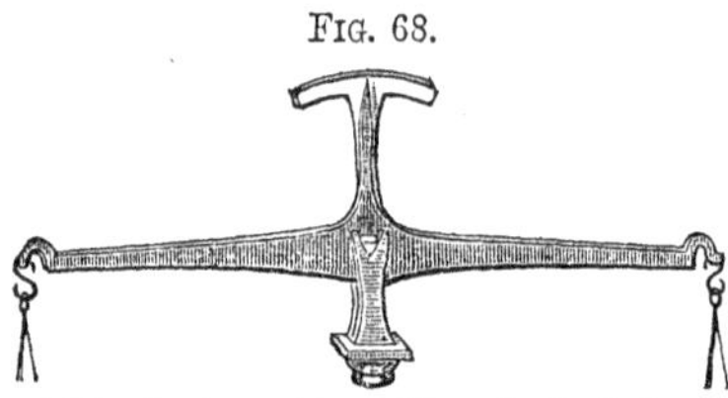

surface of hardened steel, or agate, in order that the beam may turn easily. The scale-pans are suspended by chains from points precisely at equal distances from the fulcrum, and being themselves adjusted so as to have precisely equal weights, the two sides will perfectly balance when the pans are empty.

Under what circumstances will a balance indicate false weights?

211. If the two arms of a scale-beam be not of perfectly equal length, a smaller weight at the end of the larger arm will balance a greater weight at the end of the shorter. An excess of half an inch in the length of the arm of the beam, to which merchandise is attached, where the arm should be eight inches long, would cheat the buyer exactly one ounce in every pound. This fraud, if suspected, might be detected instantly, by transposing the weight and the article balanced; the lightest would then be at the end of the short arm, and would appear lighter than it actually is.

FIG. 69.

What is the construction of platform scales?

212. Platform scales, and scales intended for weighing hay, etc., are usually compound levers, and are constructed in very various forms, but all depend on the principles above explained. Fig. 69 represents one of the varieties, and Fig. 70 a sec-

FIG. 70.

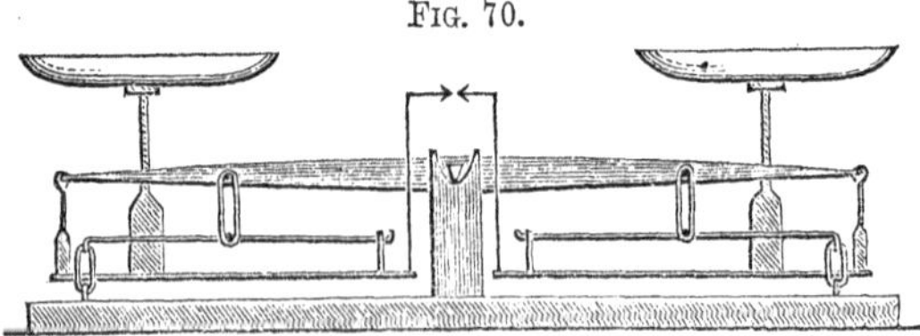

tion of the same, showing the arrangement and combination of the levers.

What circumstances limit the utility of the lever?

213. When a lever is applied to raise a weight or overcome a resistance, the space through which it acts at any one time is small, and the work must be accomplished by a succession of short and intermitting efforts. These circumstances, therefore, limit the utility of the common lever, and restrict its use to those cases only in which weights are required to be raised through small spaces.

How is continuous motion obtained?

214. When, however, a continuous motion is required, as in raising ore from a mine, or in lifting the anchor of a ship, in order to remove the intermitting action of the lever, and render it continual, we employ the simple machine known as the wheel and axle, which is only another form of the lever, in which the power is made to act without intermission.

What is a Wheel and Axle?

215. The form of the simple machine denominated the WHEEL and AXLE, consists of a cylinder, termed an axle, revolving on an axis, and having a wheel of larger diameter immovably attached to it, so that the two revolve with a common motion.

Describe the action of the wheel and axle.

FIG. 71.

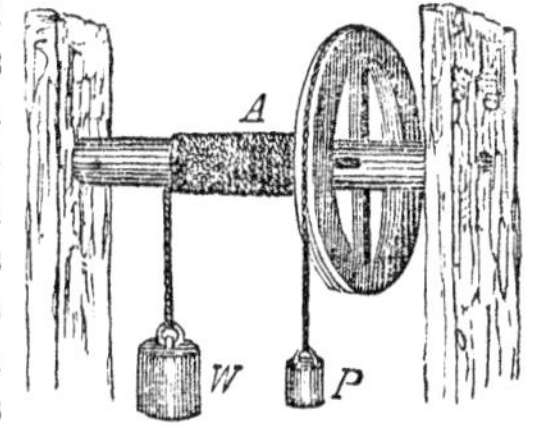

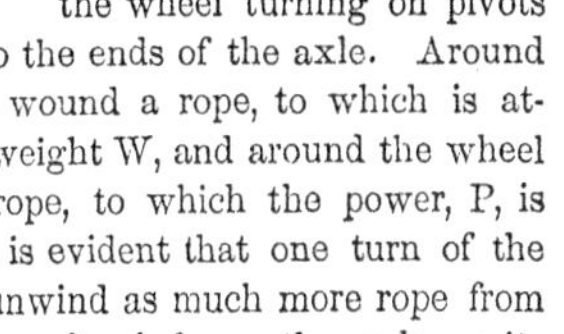

In Fig. 71, A represents the axle with a wheel immovably attached to it, and the wheel turning on pivots inserted into the ends of the axle. Around this axle is wound a rope, to which is attached the weight W, and around the wheel is another rope, to which the power, P, is applied. It is evident that one turn of the wheel will unwind as much more rope from the wheel than it winds on the axle, as its circumference is greater. The power, P, will therefore pass over a much greater space than the weight W. The weight on the axle, which may be considered as acting on the short arm of a lever which is the radius* of the axle, may be much heavier than the power which acts at the long arm of a lever, which is the radius of the wheel.

Hence the advantage gained in the wheel and axle is equal to the number of times that the radius of the axle is contained in the radius of the wheel, and to estimate the mechanical advantage gained by the wheel and axle, we have the following rule:

How do we estimate the advantage of the wheel and axle?

216. The power is to the weight, as the diameter of the wheel is to the diameter of the axle.

* The radius of a wheel, or cylinder, is its semi-diameter, or a line drawn from its center to its circumference. The spoke of a carriage wheel represents its radius.

Fig. 72 represents a section of the wheel and axle, showing the radius of the axle, *b c*, and the radius of the wheel, *a c*. The two being in a straight line, the weights hanging in opposition are always as if they were connected by a horizontal lever, *a c b*, turning on a fulcrum at *c*. If the radius of the wheel, or the length of the longer arm of the lever, *a c*, be 24 inches, and the radius of the axle, or the length of the shorter arm, *c b*, be 3 inches, then the advantage gained would be $24 \div 3 = 8$, and a power of 100 pounds applied to the wheel would balance a weight of 800 applied to the axle.

Fig. 72.

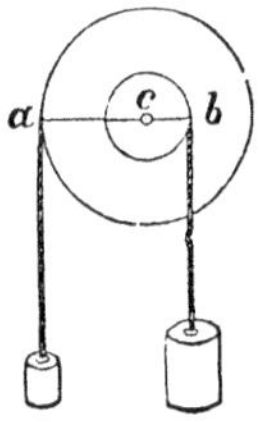

How do we apply power in the wheel and axle?

217. The methods of applying power in the wheel and axle are very various, it not being essential that the power should be applied by a rope. The axle is sometimes placed in a vertical or upright position, and the power applied by means of levers, or bars, inserted into holes in one end of the axle. A capstan of a ship, Fig. 73, is an example of this.

Fig. 73.

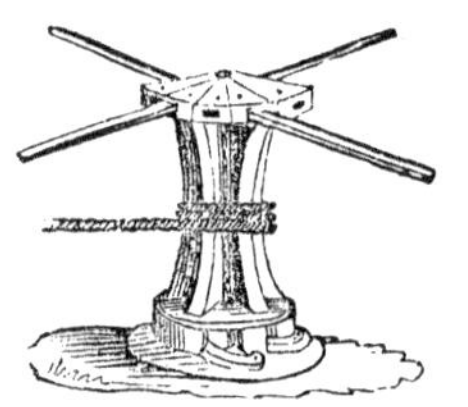

In the windlass, a handle, or winch, is substituted in the place of a wheel. (See Fig. 74.) In this case, the advantage gained is equal to the number of times that the length of handle is greater than the radius of the axle. Thus, if the handle is 20 inches and the radius of the axle is 2 inches, then the advantage would be 10, and a power of 50 pounds applied at the handle would just raise a weight of 10 times 50, or 500 pounds.

When a weight, or resistance, of comparatively great amount is to be raised by a very small power, by means of the simple wheel and axle, either of two inconveniences would ensue; either the diameter of the axle would become too small to support the weight, or the diameter of the wheel would become so great as to be unwieldy. This has been remedied by a very simple arrangement, called the double axle, Fig. 74.

Fig. 74.

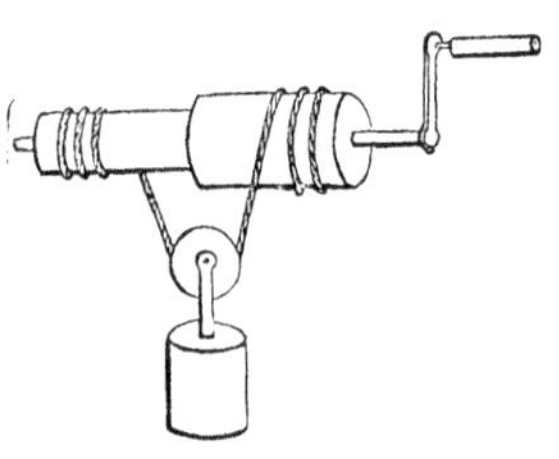

The axle of the windlass here consists of two parts of unequal diameters, and the rope winds around them in different directions; therefore, every turn of the windlass, or handle, winds up a portion equal to the circumference of the one, but unwinds a portion equal to the circumference of the other, and if the two be nearly equal, the weight moves very slow. If the weight rise 1 inch while the handle describes 100 inches, 1 pound at the handle will balance 100 attached to the rope.

In this arrangement space and time are exchanged for power in a most convenient manner.

When great power is required, wheels and axles may be combined together in a manner similar to that of the compound lever already explained (§ 207). By such a combination we gain the advantage of using a very large wheel with a small axle, without their inconveniences.

What is the most frequent method of transmitting motion through a combination of wheels?

218. The most frequent method of transmitting motion through a combination of wheels, is by the construction of teeth upon their circumference, so that the teeth of each wheel falling between those of the other, the one necessarily pushes forward the other. When teeth are thus affixed to the circumference of a wheel, they are termed *cogs;* upon an axle, they are termed *leaves*, while the axle itself is called a *pinion.*

FIG. 75.

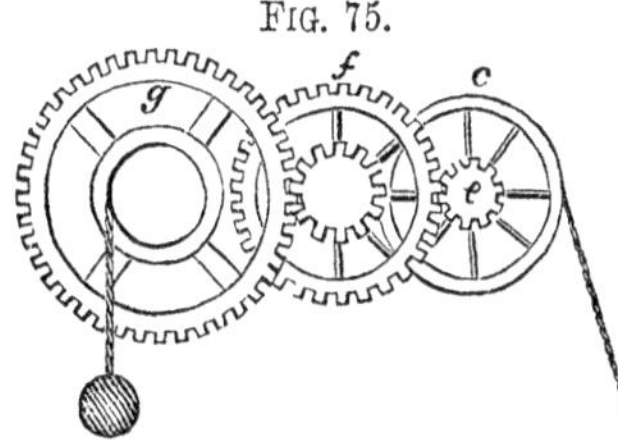

Fig. 75 represents a combination of wheels and axles for the transmission of power. If the teeth on the axle of the wheel *c* act on six times the number of teeth on the circumference of the second wheel, the second will turn only once for every six turns of the first. In the same manner the second wheel, by turning six times, turns the third wheel once; consequently, if the proportion between the wheels and their axles be preserved in all three, the third turns once, the second six times, and the first thirty-six times. Now, as the wheel and axle act in all respects like a simple lever, and a combination of wheels and axles as a combination of levers, there is no difficulty in understanding how a mechanical advantage is gained by this contrivance. The power is to the weight as the product of the diameter of all the axles is to the product of the diameter of all the wheels. Thus, if the diameter of all the axles be expressed by the numbers 2, 3, and 4, and the diameters of the wheels, *c*, *f*, and *g*, be expressed by the numbers 20, 25, and 30, then power will be to the weight as $2 \times 3 \times 4 = 24$, is to $20 \times 25 \times 30 = 15,000$;—or a power of 24 at the first wheel will balance 15,000 at the axle of the last wheel.

What are familiar illustrations of compound wheel-work?

219. One of the most familiar instances of combined wheel-work is exhibited in clocks and watches. One turn of the axle on which the watch-key is fixed, is rendered equivalent, by a train of wheel-work, to about 400 turns, or beats, of the balance-wheel; and thus the exertion, during a few seconds, of the hand which winds up, gives motion for twenty-four, or thirty hours. By increasing the number of wheels, time pieces are made which go for a year, or a greater length of time.

FIG. 76.

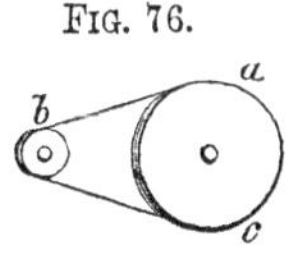

FIG. 77.

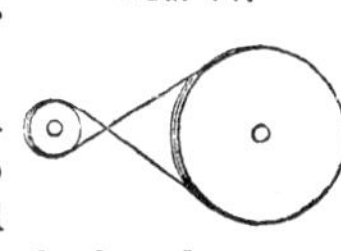

Wheels may be connected and motion communicated from one to the other, by bands, or belts, as well as by teeth. This principle is seen in the spinning-wheel and common turning-lathe. A spinning-wheel, as *a c*, Fig. 76, of thirty inches in circumference, turns by its band a smaller wheel, or spindle, *b*, of half an inch, sixty times for every revolution of *a c*.

When the wheel is intended to revolve in the same direction with the one from which it receives its motion, the band is attached as in Fig. 76; but when it is to revolve in a contrary direction, the band is crossed, as in Fig. 77.

In many wheels power is communicated by means of a weight applied to the circumference.

FIG. 78.

In the tread-mill (Fig. 78) a number of persons stepping upon the circumference of a wheel cause it to revolve. Similar machines are often adopted in ferry-boats, moved by horses, and called "horse-boats."

In most water-wheels, power is obtained by the action of water applied to the circumference of the wheel, which is caused to revolve, either through the weight, or pressure of the water, or by both conjointly.

What is a Pulley?

220. The PULLEY is a small wheel fixed in a block, and turning on an axis, by means of a cord, which runs in a groove formed on the edge of the wheel.

This simple machine is represented in Fig. 79.

FIG. 79.

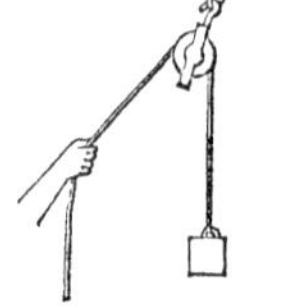

How many kinds of pulleys are there?

220. Pulleys are of two kinds; —fixed and movable.

What is a fixed pulley?

221. By a fixed pulley we mean one that merely revolves on its axis, but does not change its place.

Describe the working and advantage of the fixed pulley.

Figs. 79 and 80 are illustrations of fixed pulleys. In Fig. 80, C is a small wheel turning upon its axis, around which a cord passes, having at one end the power P, and at the other, the resistance, or weight, W. It is evident that by pulling the cord at P, the weight, W, must ascend as much and as fast as the cord is drawn down. As, therefore, the power and the weight move with the same velocity, it is clear that they balance one another; and that no mechanical advantage is gained.

FIG. 80.

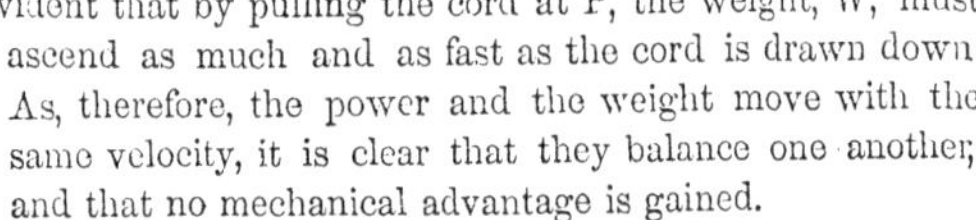

In all the applications of power there are always some directions in which it may be exerted to greater advantage and convenience than others; and in many cases the power is capable of acting in only one particular direction. Any arrangement of machinery, therefore, which will enable us to render power more available, by applying it in the most advantageous direction, is as convenient and valuable as one which enables a small power to balance or overcome a great weight. Thus, if we wish to apply the strength of a horse to lift a heavy weight to the top of a building, we should find it a difficult matter to accomplish directly, since the horse exerts his strength mainly, and to the best advantage, in drawing horizontally; but by changing the direction of the power of the horse, by an arrangement of fixed pulleys, as is represented in Fig. 81, the weight is lifted most readily, and the horse exerts his power to the best advantage.

FIG. 81.

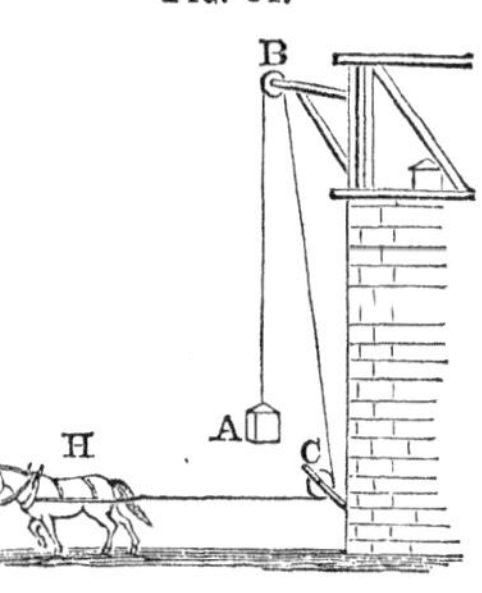

What are familiar applications of fixed pulleys?

223. A fixed pulley is most useful for changing the direction of power, and for applying power advantageously. By it a man standing on the ground can raise a weight to the top of a building. A curtain, a flag, or a sail, can be readily raised to an elevation by a fixed pulley, without ascending with it, by drawing down a cord running over the pulley.

What is a movable pulley?

224. A MOVABLE PULLEY differs from a fixed pulley in being attached to the weight; it therefore rises and falls with the weight.

Fig. 82 represents a movable pulley, B, associated, as it most commonly is, with a fixed pulley, C. The movable pulley, B, is often called a "Runner."

FIG. 82.

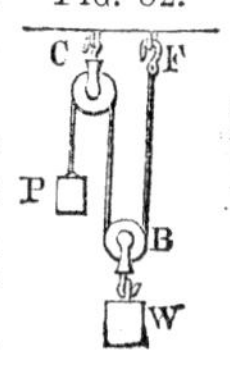

What is the advantage gained by the use of a movable pulley?

225. In the fixed pulley, Fig. 80, it will be readily seen that to move the weight, W, at one end of the cord, passing over the pulley, a greater weight must be applied at P, for if P is only equal to W, they will balance one another. If, however, we fasten one end of the cord to a fixed support, as at F, Fig. 82, and pass it under the groove in the movable pulley B, to which the weight, W, we desire to raise is attached, and then carry it over the fixed pulley C, we may lift a force of 100 pounds at W by an application of 50 pounds at P. To understand this, we must remember that the weight W is supported by the cords B F and B C on each side of the movable pulley B; and as each are equally stretched, the weight must be equally divided between them; or, in other words, the point of support, F, sustains half the weight, and the power, P, the other half. A person, therefore, pulling at P, will raise the weight by exerting a force equal to its half. But the cord at P must move through two feet to raise the weight W one foot.

When still greater power is required, pulleys are compounded into a system containing two more single pulleys, called BLOCKS, and these again are combined in a compound system of fixed and movable pulleys.

A single movable pulley may be so arranged that the power will sustain three times its own weight. Such an arrangement is represented in Fig. 83.

In this we have four cords, one employed in sustaining the power, P, and the other sustaining the weight; consequently the power will be to the weight as 1 to 3. In Fig. 84, we have two blocks, each containing two single pulleys. The rope is thus divided into five portions, each equally stretched; one is employed in supporting the power P, and four sustain the weight. With this system a power of 1 will balance a weight of 4.

FIG. 83.

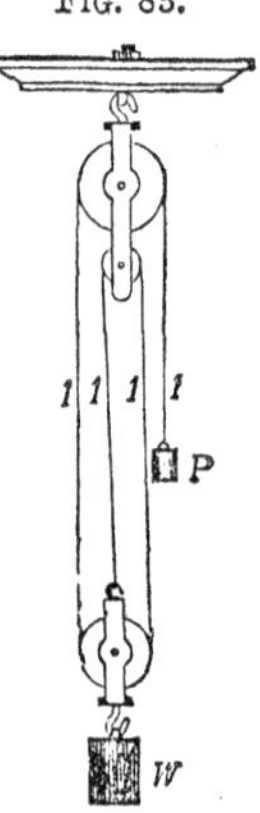

How is power gained at the expense of time in a system of pulleys?

226. In all these arrangements of pulleys, the increase of power has been gained at the expense of time, and the space passed over by the power must be double the space passed over by the weight, multiplied by the number of pulleys. That is, in the case of the single pulley, the power must pass over two feet to raise the weight one foot; and with two movable pulleys, as in Fig. 84, the power must fall four feet to raise the weight one foot.

FIG. 84.

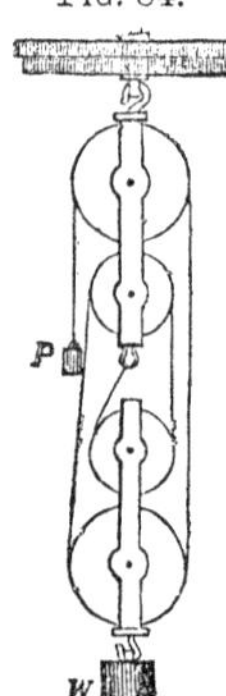

Instead of folding the string on the pulleys entire, it is sometimes doubled into separate portions, each pulley hanging by a separate cord, one end of which is attached to a fixed support. Here a very great mechanical advantage is gained, attended, however, with a corresponding loss of time. In an arrangement of such a character, represented in Fig. 85, the weight W, is supported by the two parts of the cord passing round the movable pulley, C; and as each of these parts is equally stretched, the fixed support will sustain one half the weight, and the next pulley in order above C, namely B, may be considered as sustaining the other half. But the two parts of the string which support the pulley B, again divide the weight, so that the pulley A, which is attached to one of them, only sustains one quarter of the first weight, W. The string which passes around A again divides this weight, so that each part of it sustains only one eighth of W. The fixed pulley serves merely to change the direction of the motion. In this system, therefore, a power of 1 will balance a weight of 8.

FIG. 85.

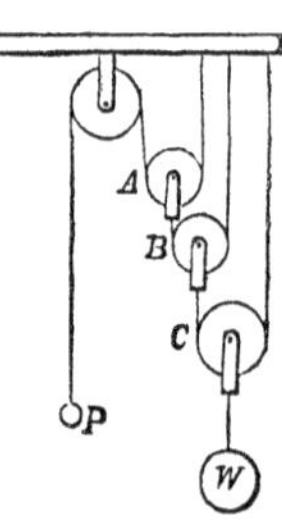

How may the advantage gained by pulleys be ascertained?

227. In general, the advantage gained by pulleys is found by multiplying the number of movable pulleys by two, or by multiplying the power by the number of folds in the rope which sustains the weight, where one rope runs through the whole.

Thus a weight of 72 pounds may be balanced by four movable pulleys by a weight or power of 9 pounds; with two pulleys, by a power of 18 pounds, with one movable pulley, by a power of 36 pounds.

These rules apply only to movable pulleys in the same block, when the parts of the rope which sustain the weight are parallel to each. The mechanical advantage which the pulley appears to possess in theory, is considerably diminished in practice, owing to the stiffness of the ropes, and the friction of the ropes and wheels. From these causes it is estimated that two thirds of the power is lost. When the parts of the cord are not parallel, the strength of the pulley is very greatly diminished.

What are Cranes and Derricks, Tackle and Fall?

228. Fixed and movable pulleys are arranged in a great variety of forms, but the principle upon which all are constructed is the same. What is called a "tackle and fall," or "block and tackle," is nothing but a pulley. Cranes and derricks are pieces of mechanism usually consisting of combinations of toothed wheels and pulleys, by means of which materials are lifted to different elevations—as goods from vessels to the wharves, building materials from the ground to the stage where the builders are engaged, and for similar purposes. One of the most simple forms of movable cranes is represented in Fig. 86. It consists of a strong triangular ladder, at the top of which is a fixed pulley, C, over which the rope attached to the object to be elevated passes, and is carried down to the cylindrical axle, T, upon which it is wound by means of bars inserted in holes, or by a crank. This ladder is inclined more or less from the upright position by means of a rope, C D, which is attached to some fixed point at a distance.

Fig. 86.

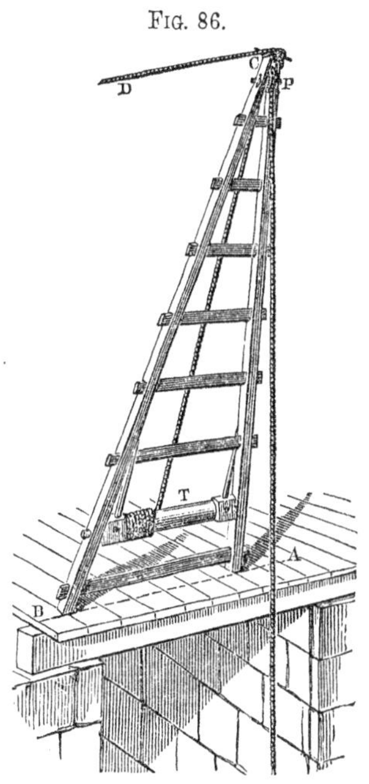

What is an Inclined Plane?

229. The Inclined Plane consists of a hard plane surface, inclined at an angle.

Illustrate the use of an Inclined Plane.

In Fig. 87, *a b c* represents an inclined plane.

230. If we attempt, for instance, to raise a cask, or any other heavy body into a wagon, we may find that our strength is unequal to lifting it

Fig. 87.

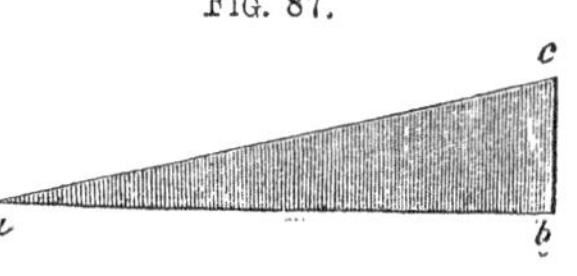

directly, while to haul it up by pulleys would be very inconvenient, if not impossible. We may, however, accomplish our object with comparative ease by rolling the cask up an inclined plank, and exerting our force in a direction parallel to the inclined surface of the plank.

How do we derive a mechanical advantage from an inclined plane?

The plank, in this instance, forms an inclined plane, and we gain a mechanical advantage, because it supports a part of the weight.

If we place a body upon a horizontal plane, or surface, it is evident that the surface will support its whole weight; if we incline the surface a little, it will support less of the weight, and as we elevate it more, it will continue to support less and less, until the surface becomes perpendicular, in which case no support will be afforded.

231. The advantage gained by the use of the inclined plane may be estimated by the following rule:

How can we estimate the advantage gained by the use of the inclined plane?

232. The power is to the weight as the perpendicular height of the plane is to its length.

From this it will appear that the less the height of the inclined plane, and the greater its length, the greater will be the mechanical advantage. Thus, in Fig. 88, if the plane, *c d*, be twice as long as the height, *e d*, one pound at *p*, acting over the pulley, would balance two pounds any where between *c* and *d*. If the plane, *c d*, were three times the length of *d e*, then one pound at *p* would balance three pounds any where on the plane, *c d*, and so of all other quantities and proportions.

FIG. 88.

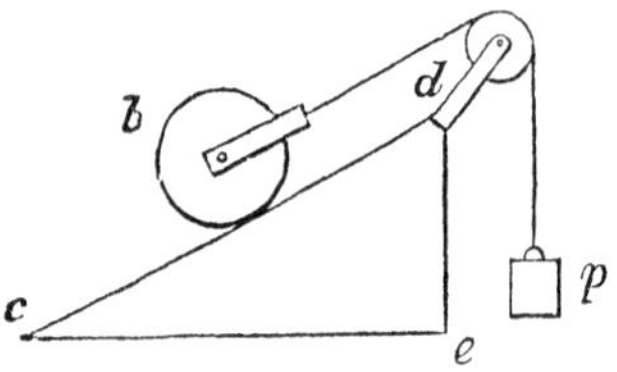

How do we estimate the inclination of roads?

233. Roads which are not level may be considered as inclined planes, and the inclination of a road is estimated by the height which corresponds to some proposed length. Thus, we say a road rises one foot in twenty, or one in fifty, meaning that if twenty or fifty feet of the road be taken, as the length of an inclined plane, the corresponding height of such a plane would be one foot, and the difference of level between the two extremities of such a length of road would be one foot.

How ought roads to be constructed?

According to this method of estimating the inclination of roads, the power required to sustain, or draw up a load, friction not considered, is always proportioned to the rate of elevation. On a level road, the carriage moves when the horse exerts a strength sufficient to overcome the friction and resistance of the atmosphere; but in going up a hill, where the road rises one foot in twenty, the horse, beside these impediments, is obliged to exert an extra force in the proportion of one to twenty, or, in other words, he is obliged to lift one twentieth of the load. It is, therefore, bad policy ever to construct a road directly over the summit of a hill, when it can be avoided, because, in addition to the force necessary to overcome the friction in drawing a heavy load up the steep incline, we must add additional force to overcome the gravity, which acts parallel with the inclined plane of the road, and tends constantly to make the load roll back to the bottom of the slope. This force increases most rapidly with the steepness, and consequently requires an immense expenditure of power. An equal power expended on a road gently winding round the hill, with an increase of speed, would gain the same elevation in much less time.

An intelligent driver, in ascending a steep hill on which there is a broad road, winds from side to side, since by so doing he diminishes the abruptness of the ascent (the plane being made longer in proportion to its height), and thus favors the horses.

Our common stairs are inclined planes, the steps being merely for the purpose of obtaining a good foot-hold.

How is power gained at the expense of time in the inclined plane?

234. In the inclined plane, as in all other simple machines, a gain in power is attended with a corresponding loss of time. A body, in ascending an inclined plane, has a greater space to pass over than if it should rise perpendicularly. The time, therefore, of its ascent will be greater, and it will thus oppose less resistance, and consequently require less power.

What is a Wedge?

235. The WEDGE is a movable inclined plane. It is also defined to be two inclined planes united at their bases, as A B, Fig. 89.

FIG. 89.

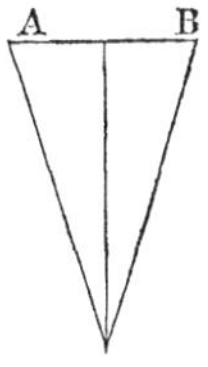

In the inclined plane, the weight moves upon the plane, which remains stationary; but in the wedge, the plane itself is moved under the weight.

In what cases are Wedges used in the arts?

236. The cases in which wedges are most generally used in the arts, are those in which an intense force is required to be exerted through a very small space. It is, therefore, used for splitting masses of wood, or stone, for blocking up buildings, raising vessels in docks, and pressing out the oil from seeds. In this last instance, the seeds are placed in bags, between two surfaces of hard wood, which are pressed together by wedges.

Upon what does the influence of the Wedge depend?

237. The usefulness of the wedge depends on friction; for if there were no friction, the wedge would fly back after each stroke of the driving force.

How does the power of the Wedge increase?

238. The power of the wedge increases as the length of its back, compared with that of its sides, is diminished. Hence, it follows that the power of the wedge is in proportion to its sharpness.

The power commonly used in the case of the wedge, is not pressure, but percussion. Its edge being inserted into a fissure, the wedge is driven in by blows upon its back. The tremor produced when the wedge is struck with a violent blow, causes it to insinuate itself much more rapidly than it otherwise would.

What are familiar examples of the use or application of the Wedge in the arts?

239. The edges of all cutting and piercing instruments, such as knives, razors, chisels, nails, pins, etc., are wedges. The angle of the wedge in all these cases is more or less acute, according to the purpose to which it is applied. Chisels intended to cut wood have their edges at an angle of about 30°; for cutting iron from 50° to 60°, and for brass about 80° to 90°. In general, tools which are urged by pressure admit of being sharper than those which are driven by percussion. The softer, or more yielding the substance to be divided is, the more acute the wedge may be constructed.

What is the Screw?

240. The Screw is an inclined plane winding round a cylinder.

This may be illustrated by cutting a strip of paper in such a way as to represent an inclined plane, and then winding it round a cylinder, or common lead pencil, as is represented in Fig. 90.

Fig. 90.

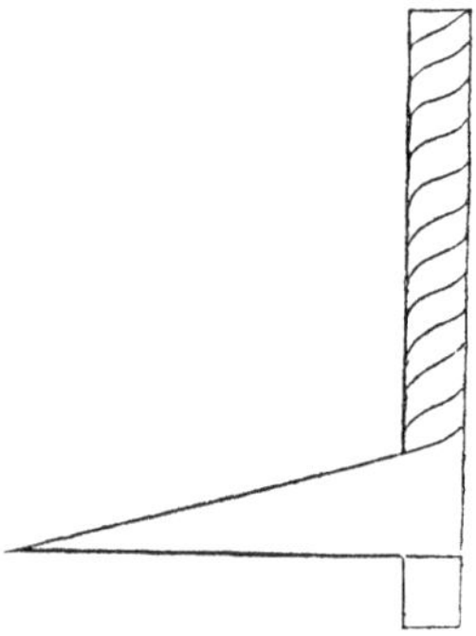

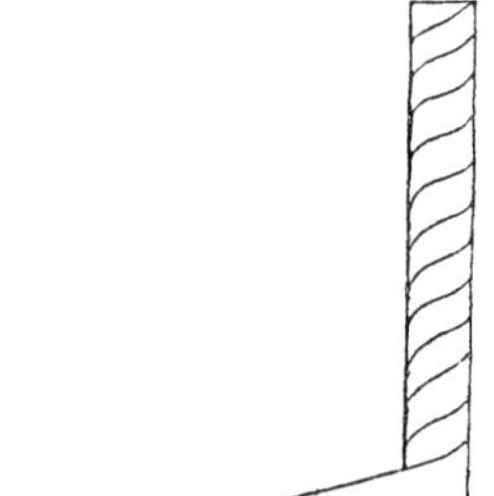

What is the Thread of a Screw?

241. The edge of the inclined plane winding about the cylinder, or the coil of the spiral line which it describes upon the cylinder, constitutes the Thread of the screw, and the distance between the successive coils is called the distance between the threads.

The screw, surrounded by its spiral line is represented in Fig. 91.

Fig. 91.

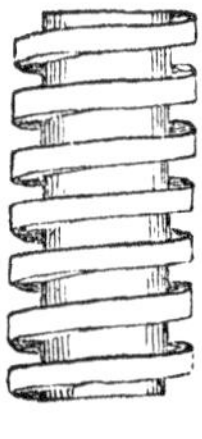

The screw is not applied directly to the resistance to be overcome, as in the case of the inclined plane and wedge, but the power is transmitted by means of what is called the Nut.

What is the Nut of a Screw?

242. The Nut of a screw is a block, with a cylindrical cavity, having a spiral groove cut round upon the surface of this cavity corresponding with the thread of the screw.

In this groove the thread of the screw will move by causing the screw to rotate. Each turn of the screw in the nut will cause it to advance or recede a distance just equal to the interval between the threads.

Is the Screw, or the Nut, movable?

Generally, the nut is stationary and the screw movable, but the nut may be movable, and the screw stationary.

How is power applied to the Screw?

243. Power is commonly applied to the screw by means of a lever, either attached to the nut, or to the head of the screw, as seen in Fig. 92. By varying the length of this, the power may be indefinitely increased at the point of resistance. The screw, therefore, acts with the combined power of the lever and the inclined plane.

Fig. 92.

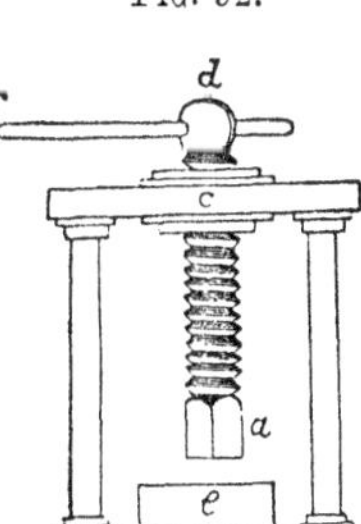

Thus, in Fig. 92, *f d* is the lever, *c* the nut, *a d* the screw, and *e* the block upon which the substance to be pressed is placed. As in all the other simple machines, the advantage in this is estimated by the relative distances passed over by the power and the weight. If the distance of the spiral threads of the screw is 1 inch, and the handle of the screw, that is the lever, is 2 feet in length, then the extremity of the lever will describe a circle of over 12 feet in turning once round, but the screw will only advance 1 inch. The ratio between the power and the weight will be, therefore, as 1 inch to 12 feet, or as 1 to 144. Consequently, if a man is capable of exerting a force of 60 pounds at the end of the lever, the screw will advance with a force of 8,640 pounds. If the distance of the threads had been ½ an inch, the power exerted by the screw would have been doubled. In this illustration friction has not been taken into account; this will diminish the total effect nearly one fourth.

How is the advantage gained by the Screw estimated?

244. The advantage gained by the screw is in proportion as the circumference of the circle described by the power (that is by the handle of the lever) exceeds the distance between the threads of the screw.

Hence the enormous mechanical force exerted by the screw is rendered evident. There is no limit to the smallness of the distance between the threads except the strength it is necessary to give them; and there is no limit to the magnitude of the circumference to be described by the power, except the necessary facility for moving it.

What are familiar applications of the Screw?

245. The screw is generally used where great pressure is to be exerted through small spaces; hence its application in presses of all kinds; for extracting the juices of seeds and fruits, in compressing cotton, hay, etc., as also for coining and punching. For the two latter operations it is caused to act with enor-

Fig. 93.

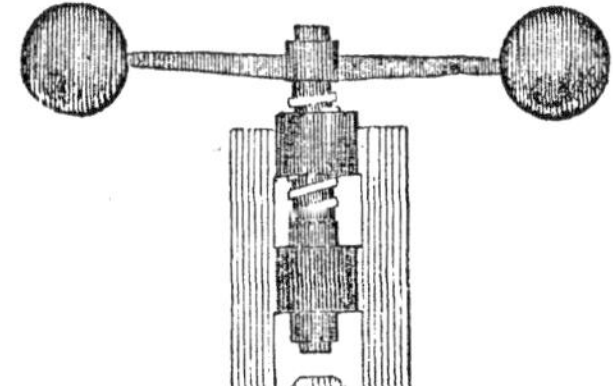

mous energy by means of the momentum of two heavy balls attached to the end of a long lever, or handle, as is represented in Fig. 93. A force of several tons may thus be applied at one effort.

What is an endless Screw?

When the thread of a screw works in the teeth of a wheel, as is shown in Fig. 94, it constitutes what is called an endless screw. Such a contrivance is oftentimes a very convenient method of applying power.

FIG. 94.

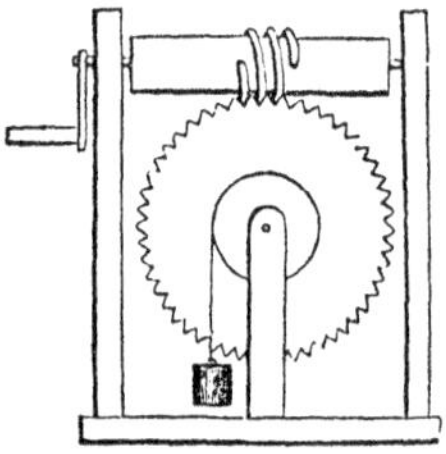

Describe the construction and advantage of Hunter's Screw.

246. The efficacy of a screw increases with the fineness of the thread; but a practical limit is soon attained, for if the thread be made too fine, it will become weak, and be liable to be torn off. To obtain an indefinite increase of the strength of the screw without diminishing the strength of the thread, we have a contrivance known as "Hunter's screw," represented in Fig. 95. It consists of a screw, A, working in a nut. To a movable bottom-board, D, a second screw, B, is affixed. This second screw works in the interior of A, which is hollow, and in which a corresponding thread is cut. When, therefore, A is screwed downward, the threads of B pass upward, and the movable piece, D, urged forward by the screw which has the greater thread, it is drawn back by that which has the less; so that during each revolution the screw instead of being advanced through a space equal to the breadth of either of the threads, moves through a space equal to their difference. Suppose the distance between the threads of A to be 1-20th of an inch, and of B 1-21st of an inch; then in turning the screw A once, the board D will descend a distance equal to the difference between 1-20th and 1-21st, or the 1-420th of an inch. Hence, if the circle described by the handle be 26 inches while the screw advances 1-420th of an inch, the power will be to the weight as 1 to 8,400.

FIG. 95.

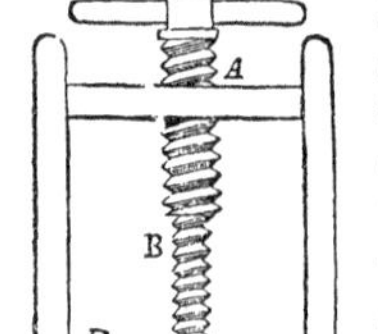

247. All machines, however complicated, are made up of combinations of the six simple machines. If we examine the construction of any complex machine, as a steam-engine, a loom, a spinning machine, or a time-piece, we shall find that they are composed of simple levers, wheels and axles, screws, etc., connected together in an endless variety of forms, to form a complete whole.

Is the moving force in machinery applied directly?

In the practical application of machinery, it rarely or never happens that the moving force is capable of producing directly, the particular kind of motion required by the machine to perform the work to which it is adapted. Expedients must therefore be resorted to, by means of which the motions which the moving power is capable of exerting directly can be converted into those which are necessary for the purposes to which the machine is applied.

How many kinds of motion are considered in machinery?

248. The varieties of motion which occur in machinery are divided into two classes, viz.: ROTARY and RECTILINEAR MOTION.

What is Rotary Motion?

249. In Rotary Motion, the several parts revolve round an axis, each performing a complete circle, or similar parts of a circle, in the same time.

What is Rectilinear Motion?

250. In Rectilinear Motion, the several parts of a moving body proceed in parallel straight lines with the same speed.

Examples of rotary motion are seen in all kinds of wheel work, and examples of rectilinear motion in the rod of a common pump, the piston of a steam-engine, the motion of a straight saw.

What is Reciprocating Motion?

In rotary and rectilinear motion, if the parts move constantly in the same direction, the motion is called continued rotary, or continued rectilinear motion. If the parts move alternately backward and forward in opposite directions, passing over the same spaces from end to end continually, the motion is called reciprocating motion.

How are rotary and reciprocatory motion converted into each other?

251. The method by which a power having one of these motions may be made to communicate the same or a different kind of motion, involves a lengthy description of a great variety of machinery; but the most simple and common plan of converting rotary motion into rectilinear, and rectilinear motion back again into rotary, is by means of what is called a *Crank*.

What is a Crank?

252. The CRANK is a double winch, or handle, and is formed by bending an axle so as to form four right angles, facing in opposite directions.

It is represented complete in Fig. 96. Attached to the middle of C D, by a joint, G, is a rod, H, which is the means of imparting power to the crank. This rod is driven by an alternate motion, like the brake of a pump. The bar C D is turned with a circular motion round the axle A F.*

FIG. 96.

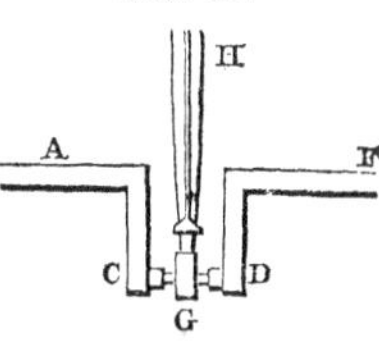

What disadvantages attend the use of the crank?

The disadvantage attending the use of the crank is, that it is incapable of transmitting a constant force to the resistance. This is illustrated in Fig. 97. In No. 1,

* The terms axis, axle, arbor, and shaft, in mechanics, are generally understood to mean the bar, or rod, which passes through the center of a wheel. A gudgeon is the pin, or support, on which a horizontal shaft turns; the pins upon which an upright shaft turns are called pivots.

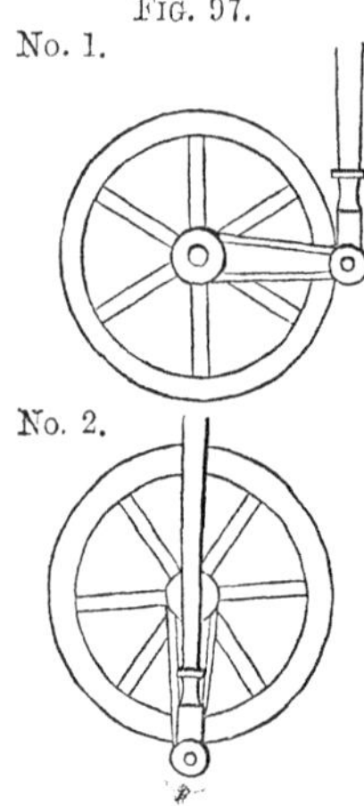

FIG. 97.

where the arm of the crank is horizontal, the power from the rod acts with the greatest advantage, as at the extremity of a lever. But when the rod which communicates motion stands perpendicular with the arm of the crank, as in No. 2, which is the case twice during every revolution, the power, however great, can exert no effect upon the resistance, the whole force being expended in producing pressure upon the axle and pivots of the crank. Such a situation of the rod and the arm of the crank is called the *dead point*, and when the mahinery stops, as is often the case, it is said to be "set," or "caught on its center." The difficulty is generally overcome by the employment of a fly-wheel (§ 21), which, by its inertia, keeps up the motion.

SECTION II.

FRICTION.

What proportion of power in machinery is lost by friction?

253. The most serious obstacle to the perfection of machinery is Friction; and it is usually considered to destroy one third of the power of a machine.

How many kinds of friction are there?

254. Friction is of two kinds: sliding and rolling. Sliding friction is produced by the sliding, or dragging of one surface over another; rolling friction is caused by the rolling of a circular body upon the surface of another.

How does friction increase?

Friction increases as the weight, or pressure increases, as the surfaces in contact are more extensive, and as the roughness of the surfaces increase. With surfaces of the same material, friction is nearly proportional to the pressure.

How does friction diminish?

Friction diminishes as the weight or pressure is less, as the polish or smoothness of the moving surfaces is more perfect, and as the surfaces in contact are smaller. It may also be diminished by applying to the surfaces some unguent, or greasy material: oils, tallow, black-lead, etc., are commonly used for this purpose; they diminish friction by filling up the minute cavities and smoothing the irregularities that exist upon the surface.* Oils are the best adapted for diminishing the friction of metals, and tallow the friction of wood.

* All bodies, however much they may be polished, appear rough and uneven when examined with a microscope.

What are the advantages of friction?

255. Friction, although an obstacle in the working of machinery generally, is not without some advantages. Without friction, the stones and bricks used in building would tend to fall apart from one another. When nails and screws are driven into bodies, with a view of holding them together, it is friction alone that maintains them in their places. The strength of cordage depends on the friction of the short fibers of the cotton, flax, or hemp, of which it is composed, which prevents them from untwisting. In walking, we are dependent on friction for our foothold upon the ground: the difficulty of walking upon smooth ice illustrates this most clearly. Without friction we could not hold any body in the hand; the difficulty of holding a lump of ice is an example of this. Without friction, the locomotive could not propel its load; for if the tire of the driving wheel and the rail were both perfectly smooth, one would slip upon the other without affording the requisite adhesion.

How does friction between the same and different substances compare?

256. Experiments seem to show that the friction of two surfaces of the same substance is generally greater than the friction of two unlike substances. The friction of polished steel against polished steel, is greater than that of polished steel upon copper, or on brass. So of wood and various other metals.

Why are large wheels used for transporting heavy weights?

257. For transporting very heavy timbers, or large castings, wheels of great size are used, as by their use the weight is moved with greater facility, and the roughness of the road more easily overcome than with small wheels. The reason of this is, that the large wheels bridge over the cavities of the road, instead of sinking into them; and in surmounting an obstacle, the large circumference of the wheel, causes the load to rise very gradually.

The resistance of sliding friction is much greater than that of rolling friction. In the wheel of a carriage there is rolling friction at the circumference of the wheel, but sliding friction at the axles. In a locomotive, the so-called driving wheels are turned by the force of the steam-engine; the whole carriage rolls on in consequence of this rotation; for if the locomotive were to remain at rest, the wheels could not revolve without sliding on the rails, and overcoming a great amount of sliding friction; but by rolling, the wheels have only the much smaller rolling friction to overcome. The machine, therefore, moves onward, this being the direction in which its motion will experience the least resistance.

The load which a locomotive is capable of drawing depends, not only upon the force of its steam power, but also upon the weight of the engine, or, in other words, upon the pressure of the driving wheels upon the rails, the friction increasing with the pressure. If we assume that two locomotives have equally strong engines, but that one is heavier than the other, a greater weight will be propelled by the heavier of the two.

Friction is generally resorted to as the most convenient method of retarding the motion of bodies, and bringing them to rest. The different modifications of machinery employed for this purpose are termed *Brakes*.

PRACTICAL PROBLEMS IN MECHANICS.

1. What must be the horse-power of a locomotive engine which moves at the constant speed of 25 miles per hour, on a level track, the weight of the train being 60 tons, and the resistance from friction being equal to 480 pounds?

2. If a lever, twelve feet long, have its fulcrum 4 feet from the weight at one end, and this weight be 12 pounds, what power at the other end will balance?

3. In a lever of the first class a power of 20 at one end balances a weight of 100 at the other: what is the comparative length of the two arms?

4. In a lever of the first class, 6 feet in length, the power is 75, and the weight 150 pounds: where must the fulcrum be placed in order that the two may balance?

5. Two persons carry a weight of 200 pounds suspended from a pole 10 feet long; one of them being weak can carry only 75 pounds, leaving the rest of the load to be carried by the other: how far from the end of the pole must the weight be suspended?

6. A lever of the second class is 20 feet long: at what distance from the fulcrum must a weight of 80 pounds be placed in order that it may be sustained by a power of 60 pounds?

7. In a lever of the third class, 8 feet long, what power will be required to balance a weight of 100 pounds, the power being applied at a distance of 2 feet from the fulcrum?

8. A power of 5 pounds is required to lift a weight of 20, by means of the wheel and axle: what must be the proportionate diameters of the wheel and axle?

9. A power of 60 acts on a wheel 8 feet in diameter: what weight suspended from a rope winding round an axle 10 inches in diameter will balance this power?

10. In a set of cog-wheels the diameters of wheel and axle are, first 7 and 2, second 8 and 1, third 9 and 1: a power of 25 being applied at the circumference of the first wheel, what weight will be sustained at the axle of the third?

11. What weight will a power of 3 sustain with a system of 4 movable pulleys, one cord passing round all of them?

12. Suppose a power of 100 pounds applied to a set of 2 movable pulleys, what weight will it sustain, allowing a deduction of two thirds for friction?

13. If a man is able to draw a weight of 200 pounds up a perpendicular wall 10 feet high, how much will he be able to draw up a plank 40 feet long, sloping from the top of the wall to the ground, no allowance being made for friction?

Solution.—In this the height (10) is to the length (40) as the weight (200) is to the required weight.

14. If a man has just strength enough to lift a cask weighing 196 pounds perpendicularly into a wagon 3 feet high, what weight could he raise by means of a plank 10 feet long, with one end resting upon the wagon, and the other on the ground?

15. The length of a plane is 12 feet, the height is 4 feet: what is the proportion of the power to the weight to be raised?

16. The distance between the threads of a screw being half an inch, and the circumference described by the power 10 feet, what proportion will exist between the power and the weight?

Solution.—The power will be to the weight as half an inch, the distance between the threads, is to 10 feet (240 half inches), the circumference described by the power =1 to 240.

17. A power of 20 pounds acting at the end of a lever attached to a screw describes a circle of 100 inches: what resistance will the power overcome, the distance between the threads of the screw being 2 inches?

CHAPTER VII.

ON THE STRENGTH OF MATERIALS USED IN THE ARTS, AND THEIR APPLICATION TO ARCHITECTURAL PURPOSES.

SECTION I.

ON THE STRENGTH OF MATERIALS.

Upon what does the strength of a material depend?

258. WHEN materials are employed for mechanical purposes, their power, or strength, for resisting external force, apart from the nature of the material, depends upon the shape of the material, its bearing, or manner of support, and the nature of the force applied to it.

Under what circumstances will a beam sustain the greatest force?

259. A beam, or bar, will sustain the greatest application of force, when the strain is in the direction of its length.

How does the strength of different substances vary?

260. The strongest of all metals for resisting tension, or a direct pull, is iron in the condition of tempered steel. The strength of metals is affected by their temperature, being diminished, in general, as their temperature is raised. Wood of the same kind is subjected to very great variations of strength. Trees that grow in mountainous or windy places, have greater strength than those which grow on plains; and the different parts of a tree, such as the root, trunk, and branches, possess different degrees of strength. Cords of equal thickness are strong in proportion to the fineness of their strands, and also to the fineness of the fibers of these strands. Ropes which are damp, are stronger than those which are dry; those which are tarred than the untarred, the twisted than the spun, the unbleached than the bleached. Other things being equal, a rope of silk is three times stronger than a rope of flax.

How does the size of a body affect its strength?

261. Of two bodies of similar shape, but of different sizes, the larger is proportionably the weaker.*

* A knowledge of the strength of various materials in resisting the action of forces exerted in different directions, is of great importance in the arts. In the following tables are collected the results of the most recent and extensive experiments upon this subject. The bodies subjected to experiment are supposed to be in the form of long rods, the cross-

That a large body may have the proportionate strength of a smaller, it must contain a greater proportionate amount of material; and beyond a certain limit, no proportions whatever will keep it together, but it will fall to pieces by its own weight. This fact limits the size, and modifies the shape of most productions of nature and art—of trees, of animals, of architectural or mechanical structures.

In what position is a rectangular beam the strongest?

262. The strength of a rectangular beam, or a beam in the form of a parallelogram, when its narrow side is horizontal, is greater than when its broad side is horizontal, in the same proportion that the width of its broad side is greater than the width of its narrow side.

Hence, in all parts of structures where beams are subjected to transverse strain, as in the rafters of roofs, floors, etc., they are always placed with their narrow sides horizontal, and their broad sides vertical.

section of which measures a square inch; in the second column is given the amount of breaking weights, which are the measure of their strength in resisting a direct pull.

Name.		lbs.	lbs.
1st. Metals;—			
Steel, tempered....	from	114794	to 153471
Iron, bar..........	"	53182	— 84611
— plate, rolled..	"	53920	
— wire.........	"	58736	—112905
— Swedish malleable.....	"	72064	
— English do..	"	55872	
— cast..........	"	16243	— 19464
Silver, cast........	"	40997	
Copper, do.........	"	20320	— 37380
— hammered.	"	37770	— 39968
Brass, cast.........	"	17947	— 19472
— wire........	"	47114	— 58931
— plate........	"	52240	
Gold..............	"	20490	— 65237
Tin................	"	3228	— 6666

Name.		lbs.	lbs.
Metals;—			
Tin, cast...........	from	4736	
Zinc...............	"	2820	
Lead, wire.........	"	2543	to 6823
2d. Woods;—			
Teak..............	"	12915	— 15405
Sycamore..........	"	9630	
Beech.............	"	12225	
Elm...............	"	9720	— 15040
Larch.............	"	10240	
Oak...............	"	10367	— 25851
Alder.............	"	11453	— 21730
Box...............	"	14210	— 24043
Ash...............	"	13480	— 23455
Pine..............	"	10038	— 14965
Fir................	"	6991	— 12876

The following table shows the average weights sustained by wires of different metals, each having a diameter of about one twelfth of an inch;

Lead.....................	27 pounds.
Tin.......................	34 "
Zinc......................	109 "
Gold......................	150 "
Silver.....................	187 pounds.
Platinum..................	274 "
Copper....................	303 "
Iron.......................	549 "

Cords of different materials, but of the same diameter, sustained the following weights:

Common flax.............	1175 pounds.
Hemp....................	1633 "
New Zealand flax...........	2380 pounds.
Silk.......................	3400 "

The following table shows the weights necessary to crush columns or pillars composed of different metals; the numbers expressed in the second column being the total crushing weight in lbs. per square inch:

Name.		lbs.	lbs.
1st. Metals:			
Cast iron..........	from	115813	to 177776
Brass, fine...... ..	"	164864	
Copper, molten....	"	117088	
— hammered.	"	103040	
Tin, molten.......	"	15456	
Lead, molten......	"	7728	

Name.		lbs.	lbs.
2d. Woods:—			
Oak..................	from	3860	to 5147
Pine..................	"	1928	
Elm...................	"	1284	
3d. Stones:—			
Granite..............	"	4970	
Sandstone............	"	2556	
Brick, well baked.....	"	1092	

The strength of a structure depends, in a very great degree, on the manner in which the several parts are joined together, and by a skillful combination, or interlocking, very weak and fragile materials may be made to resist the action of powerful forces. Examples of this occur in the manufacture of ropes, strings, thread, etc.; in the weaving of baskets, and especially in the structure of cloth; in this last instance, a series of parallel threads called the *woof*, is made to interlock with another series of threads called the *warp*, running transversely across, and passing alternately over and under the first series. Fig. 98 represents the appearance of a piece of plain cloth seen through the microscope; the alternate intersections of the threads are seen in the lower figure, the dots representing the ends of the warp threads, and the cross line the woof.

Fig. 98.

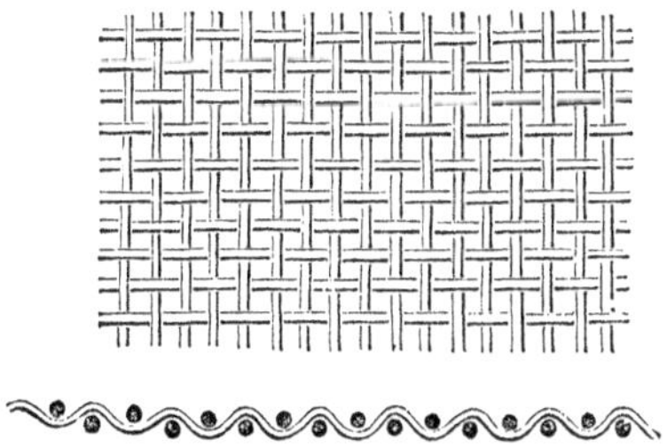

263. When a single beam can not be found deep enough to have the strength required in any particular case, several beams may be joined together, in a variety of ways, so that very great strength is obtained without a very great increase of bulk. Such methods of joining timber are known as scarfing and interlocking, tonguing, dovetailing, mortising, etc.

Fig. 99.

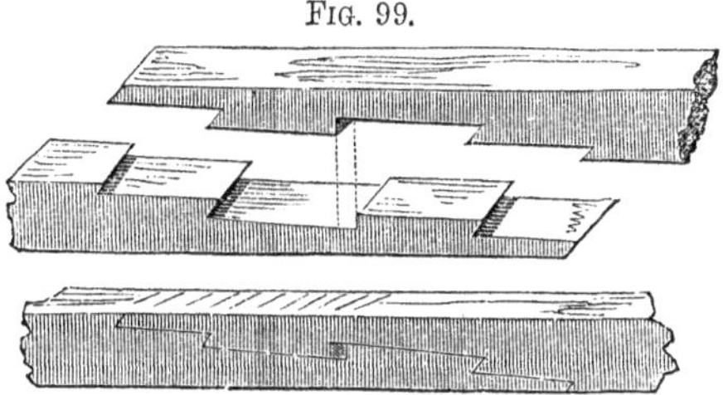

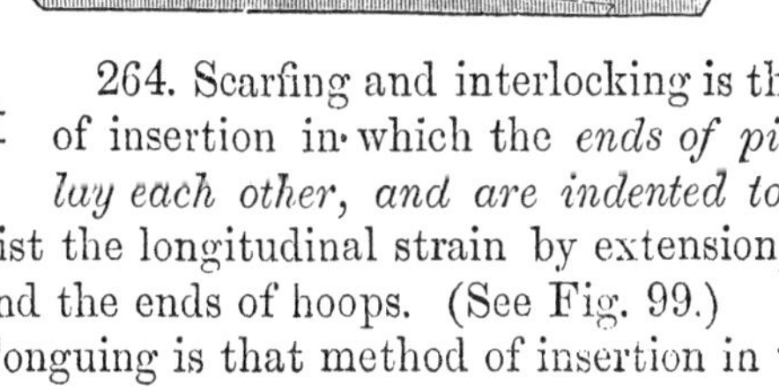

What is scarfing and interlocking?

264. Scarfing and interlocking is the method of insertion in which the *ends of pieces overlay each other, and are indented together*, so as to resist the longitudinal strain by extension, as in tie bearers and the ends of hoops. (See Fig. 99.)

265. Tonguing is that method of insertion in which the

What is tonguing?

edges of boards are wholly, or partially received by channels in each other.

What is dovetailing?

266. Dovetailing is a method of insertion in which the parts are connected *by wedge-shaped indentations* which permit them to be separated *only in one direction.* (See Fig. 100.)

Fig. 100.

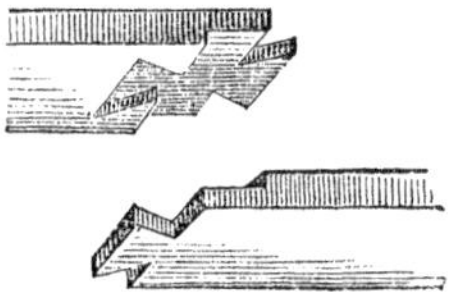

What is mortising?

267. Mortising is a method of insertion in which the projecting extremity of one timber is received into a perforation in another. (See Fig. 101.) The opening or hole cut in one piece of wood to receive or admit the projecting extremity of another piece, is called a mortise; and the end of the timber which is reduced in dimensions so as to be fitted into a mortise, for fastening two timbers together, is called a tenon.

Fig. 101.

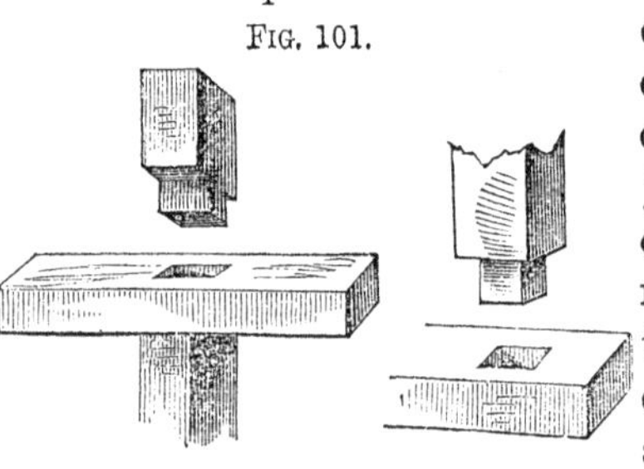

In what form can a given quantity of matter be arranged to oppose the greatest resistance?

268. The form in which a given quantity of matter can be arranged in order to oppose the greatest resistance to a bending force, is that of a hollow tube, or cylinder; and the strength of a tube is always greater than the strength of the same quantity of matter made into a solid rod.

What are illustrations of this principle?

The most beautiful and striking illustrations of this principle occur in nature. The bones of men and animals are hollow, and nearly cylindrical, because they can in this form, with the least weight of material, sustain the greatest force. The stalks of numerous species of vegetables, especially the grain-bearing plants, as wheat, rice, oats, etc., which are required to bear the weight of the ripened ear of grain, or seed, are hollow tubes, and their strength, compared with their lightness, is most remarkable. In this form they not only sustain the crushing weight of the ear which they bear at the summit, but also the force of the wind. In the construction of columns for architectural purposes, especially those made of metal, this principle is taken advantage of.*

* In that most gigantic work of modern engineering, the Britannia Tubular Bridge,

Why is a beam bent in the middle liable to break?

269. A beam, supported at its two ends, when bent by its weight in the middle, has its liability to break greatly increased, because the *destroying force* acts with the advantage of a *long lever*, reaching from the end of the beam to the center; and the *resisting force* or strength acts only with the force of a *short lever* from the side to the center; at the same time, a little only of the beam on the under side is allowed to resist at all.

This last circumstance is so remarkable, that the scratch of a pin on the under side of a beam, resting as here supposed, will sometimes suffice to begin the fracture.

SECTION II.

APPLICATION OF MATERIALS FOR ARCHITECTURAL OR STRUCTURAL PURPOSES.

What is Architecture?

270. Architecture, in its general sense, is the art of erecting buildings. In modern use, the name is often restricted to the external forms, or styles of buildings.

To what do the different varieties of architecture probably owe their origin?

The different varieties of architecture undoubtedly owe their origin to the rude structures which the climate or materials of any country obliged its early inhabitants to adopt for temporary shelter. These structures, with all their prominent features, have been afterward kept up by their refined and opulent posterity. Thus the Egyptian style of architecture had its origin in the cavern, or mound. The Chinese architecture is modeled from a tent; the Grecian is modeled from the wooden cabin; and the Gothic, it has been suggested, from the bower of trees.

On what does the strength of a building principally depend?

271. The strength of a building will principally depend on the walls being laid on a good and firm foundation, of sufficient thickness at the bottom, and standing perfectly perpendicular. Its usefulness will depend upon a proper arrangement of its parts.

crossing the Menai Straits, which separate the island of Anglesea from the mainland of Great Britain, advantage has been taken of the strength of matter arranged in the form of a tube or hollow cylinder. The entire bridge is formed of immense rectangular tubes of iron, 26 feet high in the center, 14 feet wide, and having an entire length of 1513 feet, with an elevation above the water of more than 100 feet. The sides of the tubes are also composed of smaller tubes, united together in a peculiar manner, so as to obtain the maximum of strength from the form of structure; and so great is this strength, that a train of loaded cars, weighing 280 tons, and impelled with great velocity, deflects the tubes in their centers less than three fourths of an inch. The entire weight of the tubes composing this bridge is upward of 10,500 tons, the length of two of the spans, or distances between the points of support, being 460 feet each. The same amount of iron in the form of a solid rod or beam, would not probably have sustained its own weight.

What is a pile?

272. A PILE, in architecture and engineering, is a cylinder of wood or metal pointed at one extremity, and driven forcibly into the earth, to serve as a support or foundation of some structure. It is generally used in marshy or wet places, where a stable foundation could not otherwise be obtained.

Why are columns supporting weights larger at the bottom than the top?

In constructing columns for the support of the various parts of a building, or of great weights, they are made smaller at the top than at the bottom, because the lower part of the column must sustain not only the weight of the superior part, but also the weight which presses equally on the whole column. Therefore the thickness of the column should gradually decrease from bottom to top.

What is an arch?

273. An ARCH is a concave or hollow structure, generally of stone or brick, supported by its own curve.

The base of an arch is supported by the support upon which it rests, while all the other parts constituting the curve are sustained in their positions by their mutual pressure, and by the adhesion of the cement interposed between their surfaces.

A continued arch is termed a vault.

Why is an arch stronger than a horizontal structure?

An arch is capable of resisting a much greater amount of pressure than a horizontal or rectangular structure constructed of the same materials, because the arrangement of the materials composing the arch is such, that the force which would break a horizontal beam or structure is made to compress all the particles of the arch alike, and they are therefore in no danger of being torn or overcome separately.

What is an abutment?

274. The vertical wall which sustains the base of an arch is termed an abutment: when there are two contiguous arches, the intermediate supporting wall is called a pier.

What are illustrations of the principles of the arch?

A beautiful application of the principles of the arch exists in the human skull, protecting the brain. The materials are here arranged in such a way as to afford the greatest strength with the least weight. The shell of an egg is constructed upon the principle of the arch; and it is almost impossible to break an egg with the hands, by pressing directly upon its ends. A thin watch-glass, for the same reason, sustains great pressure. A dished or arched wheel of a carriage is many times stronger to resist all kinds of shocks than a perfectly flat wheel. A full cask may fall without damage, when a strong square box would be dashed to pieces.

What is an order in architecture?

275. By an order in architecture we understand a certain mode of arranging and decor-

ating a column, and the adjacent parts of the structure which it supports or adorns.

How many orders in architecture are there?

276. Five orders are recognized in architecture—the Doric, Ionic, and Corinthian, derived from the Greeks; to these the Romans added two others, known as the Tuscan and Composite.

What is a Pilaster?

277. A Pilaster is a square column generally set within a wall, and not standing alone.

What is a Portico?

278. A Portico is a continued range of columns, covered at the top to shelter from the weather.

What are Balusters?

279. Balusters are small columns, or pillars of wood, stone, etc., used in terraces or tops of buildings for ornament; also to support a railing. When continued for some distance, they form a balustrade.

Into what two members is an order in architecture divided?

280. An order, in architecture, consists of two principal members—the column and the entablature—each of which is divided into three principal parts.

What is the Entablature?

281. The Entablature is the horizontal continuous portion which rests upon a row of columns.

Into how many parts is the Entablature divided?

It is divided into the architrave, which is the lower part of the Entablature; the frieze, which is the middle part; and the cornice, which is the upper, or projecting part.

Into how many parts is the column divided?

282. The column is divided into the base, the shaft, and the capital.

The base is the lower part, distinct from the shaft; the shaft is the middle, or longest part of the column; the capital is the upper, or ornamental part resting on the shaft.

The height of a column is always measured in diameters of the column itself, taken at the base of the shaft. Thus we say the height of the Doric column is six times its diameter, and the height of the Corinthian, ten diameters. Fig. 102 represents the various parts of an order of architecture.

What is the Façade of a Building?

283. The Façade of a building is its whole front.

Architecture ought to be considered as a useful, and not as a fine art. It is degrading the fine arts to make them entirely subservient to utility. It is out of taste to make a statue of Apollo hold a candle, or a fine painting stand as a fire-board. Our houses are for use, and architecture is, therefore, one of the useful arts. In building, we should plan the inside first, and then the outside to cover it. It is in bad taste to construct a dwelling-house in the form of a Grecian temple, because a Grecian temple was intended for external worship, not for a habitation, or a place of meeting.*

FIG. 102.

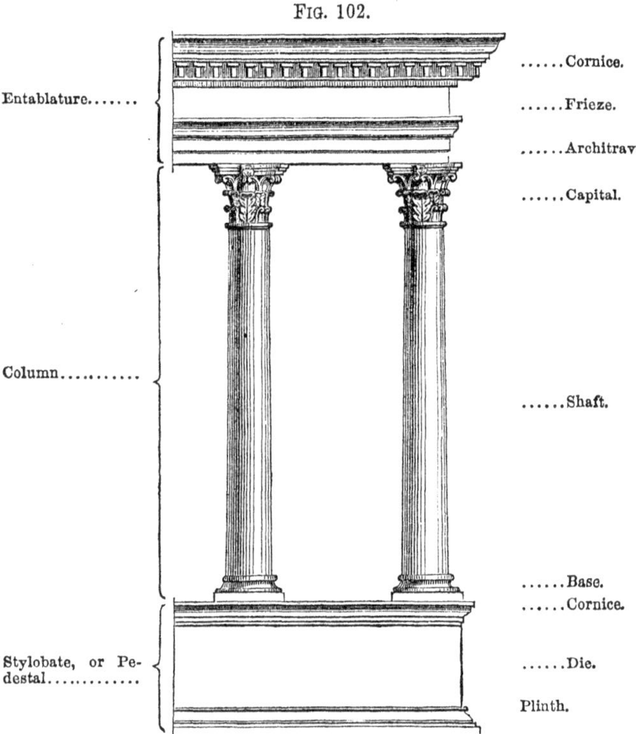

How may an estimate of the durability of stone for architectural purposes be made?

284. In selecting a stone for architectural purposes, we may be able to form an opinion respecting its durability and permanence. By visiting the locality from whence it was obtained, we may judge from the surfaces which have been long exposed to the weather if the rock is liable to yield to atmospheric influences, and the conditions under which it does so. For example, if the rock be a granite, and it be very uneven and rough, it may be inferred that it is not very durable; that the feldspar, which forms one of its compo-

* Prof. Henry.

nent parts, is more readily decomposed by the action of moisture and frost than the quartz, which is another ingredient; and therefore it is very unsuitable for building purposes. Moreover, if it possess an iron-brown or rusty appearance, it may be set down as highly perishable, owing to the attraction which this iron has for oxygen, causing the rock to increase in bulk, and so disintegrate.

Sandstones, termed freestones, are ill adapted for the external portions of exposed buildings, because they readily absorb moisture; and in countries where frosts occur, the freezing of the water on the wet surface continually peels off the external portions, and thus, in time, all ornamental work upon the stone will be defaced or destroyed.

CHAPTER VIII.

HYDROSTATICS.

What is the science of Hydrostatics?

285. HYDROSTATICS is that department of Physical Science which treats of the weight, pressure, and equilibrium of water,* and other liquids at rest.

* Water is a fluid composed of oxygen and hydrogen, in the proportion of 8 parts of oxygen to 1 of hydrogen. It is one of the most abundant of all substances, constituting three fourths of the weight of living animals and plants, and covering about three-fifths of the earth's surface, in the form of oceans, seas, lakes, and rivers.

In the northern hemisphere the proportion of land to water is as 419 to 1000; while in the southern hemisphere it is as 129 to 1000. The maximum depth of the ocean has never been ascertained. Soundings were obtained in the South Atlantic in 1853, between Rio Janeiro and the Cape of Good Hope, to the depth of 48,000 feet, or about 9 miles. Other soundings, made during the recent U. S. survey of the Gulf Stream, extended to the depth of 34,200 feet without finding bottom. The average depth of the ocean has been estimated at about 2000 fathoms.

Notwithstanding this apparent immensity of the ocean, yet, compared with the whole bulk of the earth, it is a mere film upon its surface; and if its depth were represented on an ordinary globe, it would hardly exceed the coating of varnish placed there by the manufacturer.

The source of all our terrestrial waters is the ocean. By the action of evaporation upon its surface, a portion of its water is constantly rising into the atmosphere in the form of vapor, which again descends in the form of rain, dew, fog, etc. These waters combine to form springs and rivers, which all at last discharge into the ocean, the point from which they originally came, thus forming a constant round and circulation. "All the rivers run into the sea, yet the sea is not full," because the quantity of water evaporated from the sea exactly equals the quantity poured into it by the rivers. In nature, water is never found perfectly pure; that which descends as rain is contaminated by the impurities it washes out of the air; that which rises in springs by the substances it meets with in the earth. Any water which contains less than fifteen grains of solid mineral matter in a gallon, is considered as comparatively pure. Some natural waters are known so pure that they contain only 1-20th of a grain of mineral matter to the gallon, but such instances are very rare. Water obtained from different sources may be classed, as regards com-

Are liquids compressible and elastic?

286. Liquids have but a slight degree of compressibility and elasticity, as compared with other bodies.

What are illustrations of the elasticity of water?

287. The elasticity of water may be shown in various ways. When a flat stone is thrown so as to strike the surface of water nearly horizontally, or at a slight angle, it rebounds with considerable force and frequency. Water also dashed against a hard surface shows its elasticity by flying off in drops in angular directions. Another familiar example of the elasticity of water is observed, when we attempt to separate a drop of water attached to some surface for which it has a strong attraction. The drop will elongate, or allow itself to be drawn out to a considerable degree, before the cohesion of its constituent particles is wholly overcome; and if the separating force is at any time relaxed, or discontinued, the elasticity of the water will restore the drop to very nearly its original form and position. Mercury is much more elastic than water, and rebounds from a reflecting surface with considerable velocity and violence. The exercise of both the elastic and compressive principle is, however, so extremely limited in liquids, that for all practical purposes this form of matter is regarded as inelastic and uncompressible; or, in other words, the elasticity and compressibility of water produce no appreciable effects.

To what extent has water been compressed?

The compressibility of water is not so easily demonstrated as is its elasticity, although the elasticity is a direct consequent of the compressibility. An experiment of Mr. Perkins showed that water, under a pressure of 15,000 pounds to the square inch, was reduced in bulk 1 part in 24.

In what manner do the particles of liquids move upon each other?

288. In liquid bodies, as has been already shown (§§ 34, 36), the attractive and repulsive forces existing between the particles are so nearly balanced, that the particles move upon each other with the greatest facility. The particles which make up a collection of fine sand, or dust, also move upon each other with great facility: but the particles of a liquid possess this additional quality, viz., that of moving upon themselves without friction. The particles of no solid substance, however fine they may be rendered, possess this property.

289. From this is derived a great fundamental principle lying at the basis of all the mechanical phenomena connected with liquid bodies, viz. :—

parative purity, as follows; Rain water must be considered as the purest natural water, especially that which falls in districts remote from towns or habitations; then comes river water; next, the water of lakes and ponds; next, spring waters; and then the waters of mineral springs. Succeeding these, are the waters of great arms of the ocean, into which immense rivers discharge their volumes, as the water of the Black Sea, which is only brackish; then the waters of the ocean itself; then those of the Mediterranean and other inland seas; and last of all, the waters of those lakes which have no outlet, as the Dead Sea, Caspian, Great Salt Lake of Utah, etc. etc.

All natural waters contain air, and sometimes other gaseous substances. Fishes and other marine animals are dependent upon the air which water contains for their respiration and existence. It is owing to the presence of air in water that it sparkles and bubbles.

What great law constitutes the basis of all the mechanical phenomena of liquids?

290. Liquids transmit pressure equally in all directions.

This remarkable property constitutes a very characteristic distinction between solids and liquids; since solids transmit pressure only in one direction, viz., in the line of the direction of the force acting upon them, while liquids press equally in all directions, upward, downward, and sideways.

Illustrate the equality of pressure in liquids.

In order to obtain a clear understanding of the principle of the equality of pressure in liquids, let us suppose a vessel, Fig. 103, of any form, in the sides of which are several tubular openings, A B C D E, each closed by a movable piston. If now we exert upon the top of the piston at A, a downward pressure of 20 pounds, this pressure will be communicated to the water, which will transmit it equally to the internal face of all the other pistons, each of which will be forced outward with a pressure equal to 20 pounds, provided their surfaces in contact with the water are each equal to that of the first piston. But the same pressure exerted on the pistons is equally exerted upon all parts of the sides of the vessel, and therefore a pressure of 20 pounds upon a square inch of the surface of the piston A, will produce a pressure of 20 pounds upon every square inch of the interior of the surface of the vessel containing the liquid.

FIG. 103.

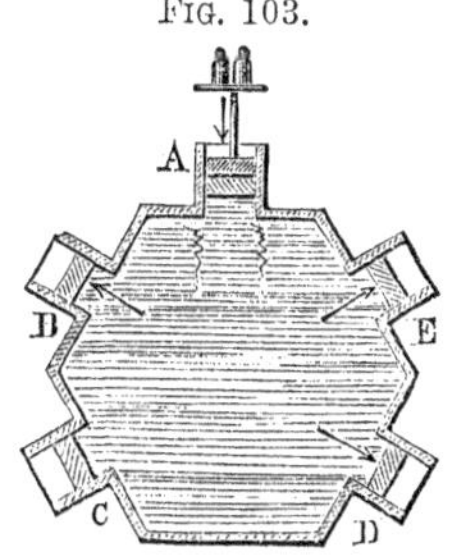

FIG. 104.

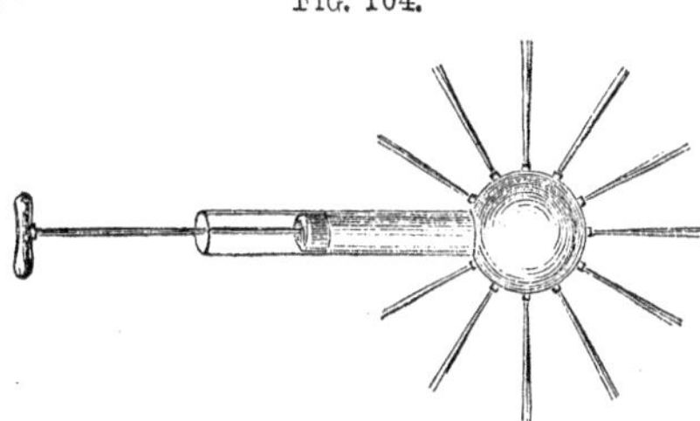

The same principle may also be shown by another experiment. Suppose a cylinder, Fig. 104, in which a piston is fitted, to terminate in a globe, upon the sides of which are little tubular openings. If the globe and the cylinder are filled with water, and the piston pressed down, the liquid will jet out equally from all the orifices, and not solely from the one which is in a direct line with, and opposite to the piston.

In what manner may a liquid act as a machine?

291. This property of transmitting pressure equally and freely in every direction, is one in virtue of which a liquid becomes a machine, and can be made to receive, distribute, and apply power. Thus, if water be confined in a vessel, and a mechanical force exerted on any portion of it, this force will be at once transmitted throughout the entire mass of liquid.

What is the Hydrostatic Paradox?

The effects of the practical application of this principle are so remarkable that it has been called the Hydrostatic Paradox, since the weight, or force, of one pound, applied through the medium of an extended surface of some liquid, may be made to produce a pressure of hundreds, or even thousands of pounds. Thus, in Fig. 105, A and *a* are two cylinders containing water connected by a pipe, each fitted with a piston in such a way as to render the whole a close vessel. Suppose the area of the base of the piston, *p*, to be one square inch, and the area of the base of the piston, P, to be 1,000 square inches. Now any pressure applied to the small piston will be transmitted by the water to the large piston; so that every portion of surface in the large piston will be pressed upward with the same force that an equal portion of the surface in the small piston is pressed downward. A pressure, therefore, of 1 pound acting on the base of the piston *p*, will exert an outward pressure of 1,000 pounds acting on the base of the piston P; so that a weight of 1 pound resting upon the piston *p*, would support a weight of 1,000 pounds resting upon the piston P.

Fig. 105.

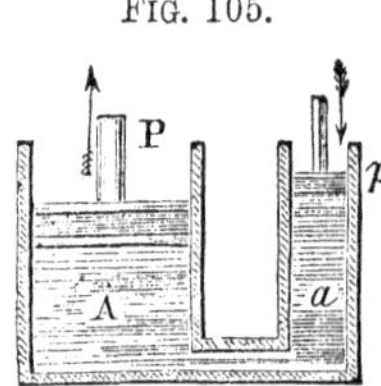

How do the forces acting in the Hydrostatic Paradox compare with the forces acting on the arms of a lever?

The action of the forces here supposed differs in nothing from that of like forces acting on a lever having unequal arms in the proportion of 1 to 1,000. A weight of 1 pound acting on the longer arm of such a lever, would support, or raise a weight of 1,000 pounds acting on the shorter arm. The liquid contained in the vessel, in the present case, acts as the lever, and the inner surface of the vessel containing it acts as the fulcrum. If the piston *p* descends one inch, a quantity of water which occupies one inch of the cylinder *a* will be expelled from it, and as the vessel A *a* is filled in every part, the piston P must be forced upward until space is obtained for the water which has been expelled from the cylinder *a*. But as the sectional area of A is 1,000 times greater than that of *a*, the height through which the piston P must be raised to give this space, will be 1,000 times less than that through which the piston *p* has descended. Therefore, while the weight of 1 pound on *p* has moved through 1 inch, the weight of 1,000 pounds on P will be raised through only 1-1,000th part of an inch. If this process were repeated a thousand times the weight of 1,000 pounds on P would be raised through 1 inch; but in accomplishing this, the weight of 1 pound acting on P would be moved successively through 1,000 inches. The mechanical action, therefore, of the power in this case, is expressed by the force of 1 pound acting successively through 1,000 inches, while the mechanical effect produced upon the resistance is expressed by 1,000 pounds raised through 1 inch.

What is a Hydraulic Press?

292. The Hydraulic, or Hydrostatic Press, is a machine arranged in such a manner, that the advantages derived from the principle that

liquids transmit pressure equally in all directions, may be practically applied.

The principle of the construction and action of the hydraulic press is explained in the preceding paragraph (§ 291), and Fig. 105, represents a section of its several parts.

Fig. 106.

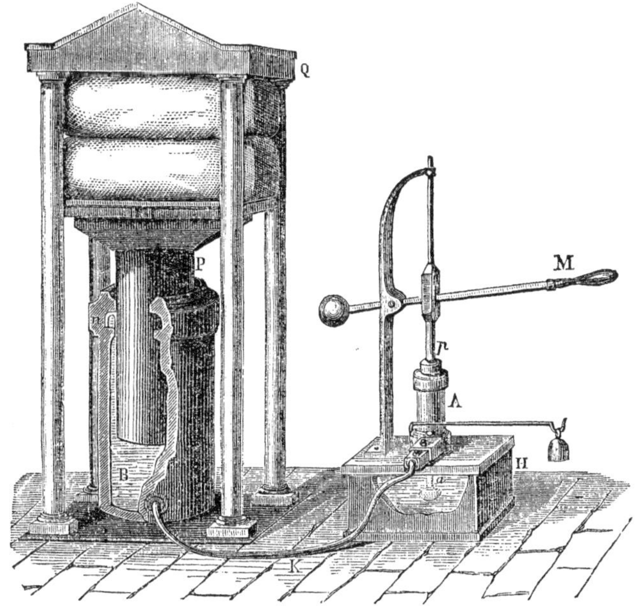

Fig. 106 represents the hydraulic press as constructed for practical purposes. In a small cylinder, A, the piston of a forcing-pump, P, works by means of the handle M. The cylinder of the forcing-pump, A, connects, by means of a tube, K, leading from its base, with a large cylinder, B. In this moves also a piston, P, having its upper extremity attached to a movable iron plate, which works freely up and down in a strong upright frame-work, Q. Between this plate and the top of the frame-work the substance to be pressed is placed. To operate the press, water is raised in the forcing-pump, A, by raising the handle M, from a small reservoir beneath it, *a;* by depressing the handle, the water filling the small cylinder A is forced through a valve, H, and the pipe K, into the larger cylinder B, where it acts to raise the larger piston, and causes it to exert its whole force upon the object confined between the iron plate and the top of the frame-work. If the area of the base of the piston *p* is a square inch in diameter, and the area of the base of the piston P 1,000 square inches, then a downward pressure of one pound on *p* will exert an upward pressure of 1,000 pounds on P.

As thus constructed, the hydraulic press constitutes the most powerful mechanical engine with which we are acquainted, the limits to its power being bounded only by the strength of the machinery and material. By means of this press, cotton is pressed into bales, ships are raised from the water for repair, chain-cables are tested, etc. etc.

Will liquids press upward as well as downward?

293. As liquids transmit pressure equally in all directions, it follows that any given portion of a liquid contained in a vessel will press upward upon the particles above it, as powerfully as it presses downward upon the particles below it.

How is the upward pressure of liquids shown by experiment?

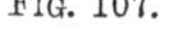
Fig. 107.

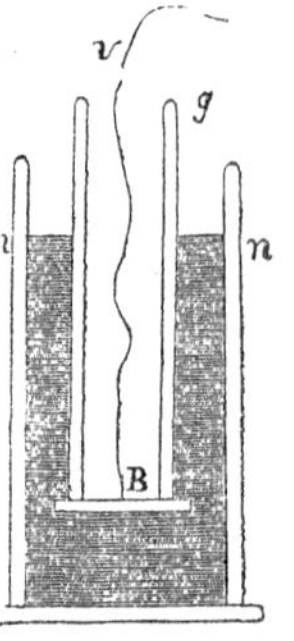

This fact may be illustrated by means of the apparatus represented in Fig. 107. If a plate of metal, B, be held against the bottom of a glass tube, *g*, by means of a string, *v*, and immersed in a vessel of water, the water being up to the level *n n*, the plate B will be sustained in its place by the upward pressure of the water; to show that this is the case, it is only necessary to pour water into the tube *g*, until it rises to the level *n n*, when the plate will immediately fall, the upward pressure below the plate B being neutralized by the downward pressure of the water in the tube *g*.

"Some persons find it difficult to understand why there should be an upward pressure in a mass of liquid, as well as a downward and lateral pressure. But if in a mass of liquid the particles below had not a tendency upward equal to the weight, or downward pressure of the particles of liquid above them, they could not support that part of the liquid which rests upon them. Their tendency upward is owing to the pressure around them from which they are trying to escape."*

To what is the pressure of a column of liquid proportional?

Fig. 108.

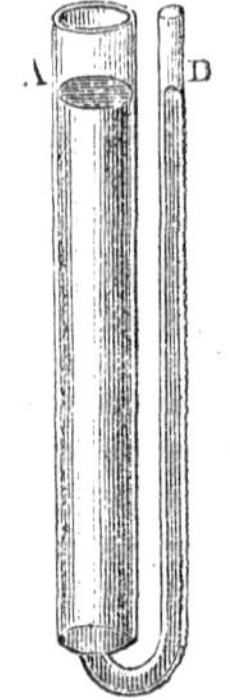

294. The pressure exerted by a column of liquid is proportioned to, or measured by the height of the column, and not by its bulk, or quantity.

If we take a tube in the form of the letter U, with one of its branches much smaller than the other, as in Fig. 108, and pour water into one of the branches, we shall find that the liquid will stand at the same height in both tubes. The great mass of liquid contained in the large tube, A, exerts no more pressure on the liquid contained in the small tube, D, than would a smaller mass contained in a tube of the same dimensions as D. And if A contained 10,000 times the quantity of water that D contained, the water would rise to no greater elevation in D than in A.

* Arnott.

What is the principle and action of the Hydrostatic Bellows?

The principle that the pressure exerted by a column of water is as its height, and not as its quantity, may be also illustrated by the Hydrostatic Bellows, Fig. 109. This consists of two boards, B C and D E, united together by means of cloth, or leather, A, as in a common bellows. A small vertical pipe, T, attached to the side communicates with the interior of the bellows. Heavy weights, W W, are placed upon the top of the bellows when empty. If water be poured into the vertical pipe, the top of the bellows, with the weights upon it, will be lifted up by the pressure of the water beneath; and as the height of the column of water increases, so in like proportion may the weights upon the top of the bellows be increased. It is a matter of no consequence what may be the diameter of the vertical tube, since the power of the apparatus depends upon the height of the column of water in the small tube, and the area of the board, B C; *that is, the weight of a small column of water in the vertical pipe, T, will be capable of supporting a weight upon the board, B C, greater than the weight of the water in the pipe, in the same proportion as the area of board B C is greater than the sectional area of the bore of the pipe.* Thus, if the area of the bore of the pipe be a quarter of an inch, and the area of the board forming the top of the bellows a square foot, then the proportion of the pipe to the board will be that of 576 to 1; and, consequently, the weight capable of being supported by the board will be 576 times the weight of the water contained in the pipe.

FIG. 109.

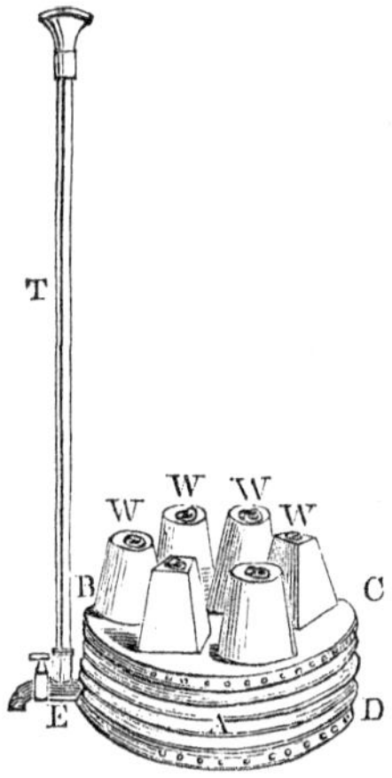

What are illustrative examples of the pressure of liquids?

In this manner a strong cask, *a*, Fig. 110, filled with liquid, may be burst by a few ounces of water poured into a long tube, *b c*, communicating with the interior of the cask.

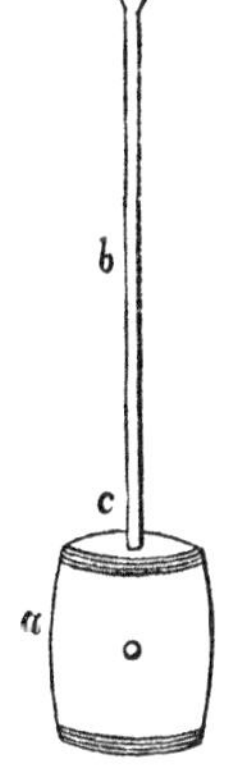

FIG. 110.

This law of pressure is sometimes exhibited on a great scale in nature, in the bursting of rocks, or mountains. Suppose a long vertical fissure, as in Fig. 111, to communicate with an internal cavity formed in a mountain, without any outlet. Now, when the fissure and cavity become filled, an enormous pressure is exerted, sufficient, it may be, to crack, or disrupture, the whole mass of the mountain.

The most striking effects of the pressure of the water at great depths are exhibited in the ocean. If a strong, square glass bottle, empty and firmly corked, be sunk in water, its sides are generally crushed in by the pressure, before it has reached a depth of 60 feet. Divers plunge with impunity to certain depths, but there is a limit beyond which they can not sustain the

6*

Fig. 111.

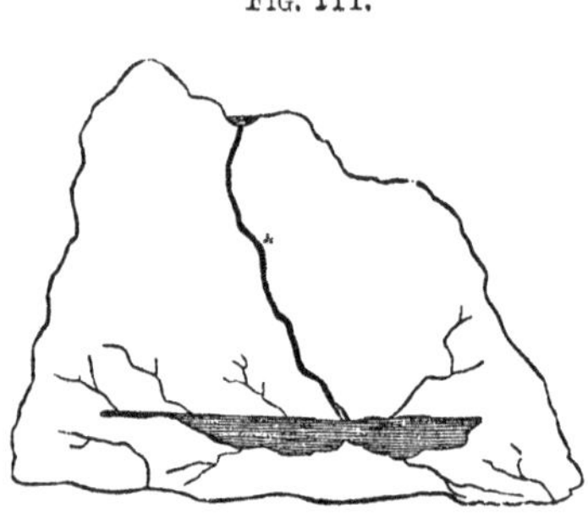

immense pressure on the body exerted by the water. It is probable, also, that there is a limit of depth beyond which each species of fish can not live. The principle of the equal transmission of pressure by liquids, however, enables fishes to sustain a very great pressure of water without being crushed by it; the fluids contained within them pressing outward with as great a force as the liquid which surrounds them presses inwards.

When a ship founders at sea, the great pressure at the bottom forces the water into the pores of the wood, and increases its weight to such an extent that no part can ever rise again.

Upon what does the pressure upon the bottom of a vessel containing liquid depend?

295. The pressure upon the bottom of a vessel containing a liquid, is not effected by the shape of the vessel, but depends solely upon the area of the base, and its depth below the surface.

This arises from the law of equal distribution of pressure in liquids. Fig. 112 represents two different vessels having equal bases, and the same perpendicular depth of water in them. Although the quantity of water contained in one is much greater than in the other, the pressure sustained by these bases will be thesame.

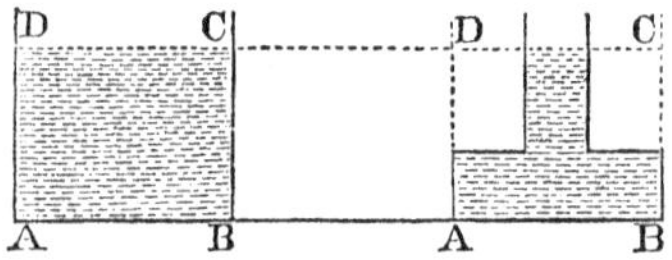

Fig. 112.

In a conical vessel, Fig. 113, the base, C D, sustains a pressure measured by the height of the column, A B C D; for all the rest of the liquid only presses on A B C D laterally, and resting

Fig. 113. Fig. 114.

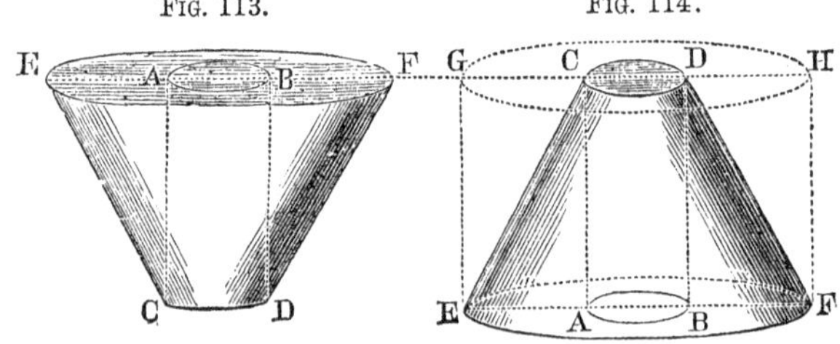

on the sides, E C and F D, can not contribute any thing to the pressure on the base, C D. But in a conical vessel, of the shape represented in Fig. 114, the pressure on A B a portion of the base, E F, is measured by the column A B C D as before; but the other portions of the liquid not resting on the sides also press upon the bottom, E F; and as the pressure of the column A B C D is transmitted equally, every portion of the base, E F, sustains an equal pressure as that portion of the base, A B, which is directly beneath the column, A B C D; therefore the whole pressure on the base, E F, is the same as if the vessel had been cylindrical, and filled throughout to the height indicated by the dotted lines, G H.

296. Hence, to find the pressure of water upon the bottom of any vessel, we have the following rule:

How can we calculate the pressure upon the bottom of a vessel containing water?

297. Multiply the area of the base by the perpendicular depth of the water, and this product by the weight of a cubic foot of water.*

Thus, suppose the area of the base of a vessel to be 2 square feet, and the perpendicular depth of the water to be 3 feet; required the pressure on the bottom of the vessel, the weight of a cubic foot of water being assumed to be 1,000 ounces (see § 82).

2×3=6 cubic feet.

6×1,000=6,000 oz.=pressure on the base of the vessel.

* "The actual pressure of water may also be calculated from the following data. It is ascertained that the weight of a cubic inch of water of the common temperature of 62° Fahrenheit, is a portion of a pound expressed by the decimal 0·036065. The pressure, therefore, of a column of water one foot high, having a square inch for its base, will be found by multiplying this by 12, and consequently will be 0·4328 lb.

"The pressure produced upon a square foot by a column one foot high, will be found by multiplying this last number by 144, the number of square inches forming a square foot; it will therefore be 62·3232 lbs.

Table showing the pressure in lbs. per square inch and square foot, produced by water at various depths.

Depth in Feet.	Pressure per Square Inch.	Pressure per Square Foot.	Depth in Feet.	Pressure per Square Inch.	Pressure per Square Foot.
	lbs.	lbs.		lbs.	lbs.
I.	0·4328	62·3232	VI.	2·5968	373·9392
II.	0·8656	124·6464	VII.	3·0296	436·2624
III.	1·2984	186·9696	VIII.	3·4624	498·5856
IV.	1·7312	249·2928	IX.	3·8952	560·9088
V.	2·1640	311.6160	X.	4·3280	623·2320

"By the aid of the above table, the actual pressure of water on each part of the surface of a vessel containing it can always be determined, the depth of such part being given. Thus, for example, if it be required to know the pressure upon a square foot of the bottom of a vessel where the depth of the water is 25 feet, we find, from the above table, that the pressure upon a square foot at the depth of 2 feet is 124·6464 lbs.; and, consequently, the pressure at the depth of 20 feet is 1246·464 lbs.; to this, let the pressure at the depth of 5 feet, as given in the table, be added: 1246·464+311·616=1558·080 lbs., which is, therefore, the required pressure.

"If the liquid contained in the vessel be not water, but any other whose relative weight compared with water is known, the calculation is made first for water, and the result being multiplied by the number expressing the proportion of the weight of the given liquid to that of water, the result will be the required pressure."—*Lardner*.

How is the pressure of a liquid exerted laterally?

298. As liquids transmit pressure equally in all directions, this pressure will act sideways as well as downward, and the pressure at any point upon the side of a vessel containing a liquid, will be in proportion to the perpendicular depth of that point below the surface.

Fig. 115.

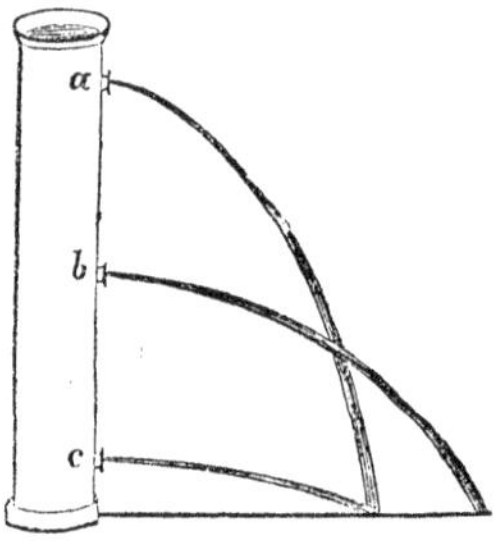

Fig. 115 represents a vessel of water with orifices at the side, at different distances from the surface. The water will flow out with a force proportionate to the pressure of the water at these several points, and this pressure is proportionate to the depth below the surface. Thus, at *a* the water will flow out with the least force, because the pressure is least at that point. At *b* and *c* the force and pressure will be greater, because they are situated at a greater depth below the surface.

How may the pressure upon the side of a vessel of water be calculated?

299. To find the pressure upon the side of a vessel containing water, multiply the area of the side by one half its whole depth below the surface, and this product again by the weight of a cubic foot of water.

Fig. 116.

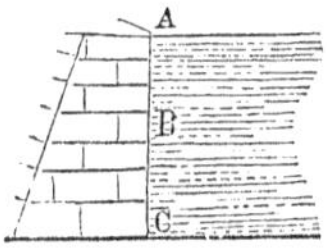

Suppose A C, Fig. 116, to represent the section of the side of a canal, or a vessel filled with water, and let the whole depth, A C, be 10 feet: then at the middle point, B, the depth, A B, will be 5 feet. Now the pressure at C is produced by a column of water whose depth is 10 feet, but the pressure at B is produced by a column whose depth is 5 feet, which is the average between the pressure at the surface and at the bottom, or the average of the entire pressure upon the side. Hence the total pressure upon the side of a vessel containing water will be equal to the weight of a column of water whose base is equal to the area of that side, and whose height is equal to one half the depth of the liquid in the vessel, or, in other words, to the depth of the middle point of the side below the surface.

Why should an embankment be made stronger at the bottom than at the top?

As the pressure upon the sides of a reservoir containing water increases with the depth, the walls of embankments, dams, canals, etc., are made broader or thicker at the bottom than at the top (as in Fig. 114). For the same reason, in order to render a cistern equally strong throughout, more hoops should be placed near the bottom than at the top.

If a surface equal to the side of a vessel containing liquid were laid upon the bottom, then the pressure upon the surface would be double the actual pressure on the side; for in this instance the surface sustains the weight of a column equal in height to the whole depth, while the column of pressure upon the side is only equivalent to one half the depth.

How does the pressure of a given quantity of liquid compare with its weight?

300. The actual pressure produced upon the bottom and sides of a vessel which contains a liquid, is always greater than the weight of the liquid.

In a cubical vessel, for example, the pressure upon the bottom will be equal to the weight of the liquid, and the pressure on each of the four sides will be equal to one half the weight; consequently the whole pressure on the bottom and sides will be equal to three times the weight of the liquid.

In what condition is the surface of a liquid at rest?

301. The surface of a liquid when at rest is always HORIZONTAL, or LEVEL.

Why is the surface of a liquid at rest level?

The particles of a liquid having perfect freedom of motion among themselves, and all being equally attracted by gravitation, the whole body of liquid will tend to arrange itself in such a manner that all the parts of its surface shall be equally distant from the earth's center, which is the center of attraction.

What is the true definition of a spherical surface?

A perfectly level surface really means one in which every part of the surface is equally near the center of the earth; it must be, therefore, in fact, a spherical surface. But so large is the sphere of which such a surface forms a part, that in reservoirs and receptacles of water of limited extent, its sphericity can not be noticed, and it may be considered as a perfect plane and level; but when the surface of water is of great extent, as in the case of the ocean, it exhibits this rounded form, conforming to the figure of the earth, most perfectly.* This sphericity of the surface of the ocean is illustrated by the fact, that the masts of a ship appproaching us at sea, are visible long before the hull of the vessel can be seen. In Fig. 117 only that part of the ship above the line A C can be seen by the spectator at A, because the rest of the vessel is hidden by the swell of the curve of the surface of the ocean, or rather of the earth, D E.

FIG. 117.

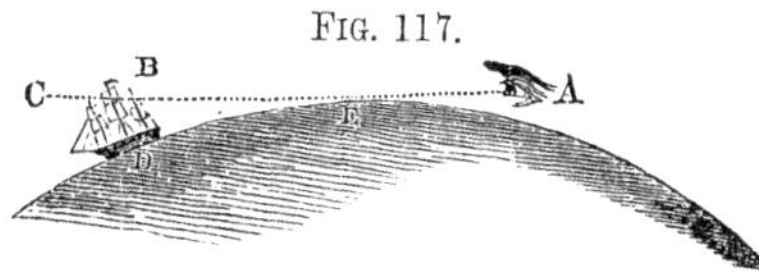

In what manner will a liquid rise in a series of tubes or vessels communicating with each other?

302. Water, or other liquids will always rise to an exact level in any series of different tubes, pipes, or other vessels communicating with each other.

* A hoop surrounding the earth would bend from a perfectly straight line eight inches in a mile. Consequently, if a segment of the surface of the earth, a mile long were cut off, and laid on a perfect plane, the center of the segment would be only four inches higher than the edges. A small portion of it, therefore, for all ordinary purposes, may be considered as a perfect plane.

This fact is sufficiently illustrated by reference to Fig. 118.

FIG. 118.

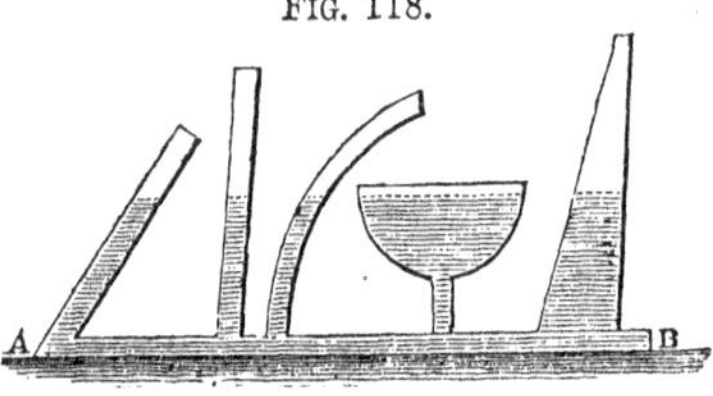

On what principle are we enabled to convey water in aqueducts over uneven surfaces?

303. It is upon the application of the principle that water in pipes will always rise to the height, or level of its source, that all arrangements for conveying water over uneven surfaces in aqueducts, or closed pipes depend. The water brought from any reservoir or source of supply, in or near a town or building, may be delivered by the effect of gravity alone to every location beneath the level of the reservoir; the result not being affected by the inequalities of the surface over which the water pipes may pass in their connection between the reservoir and the point of delivery. So long as they do not rise above the level of the source of supply, so long will the water continue to flow.

Fig. 119 represents the line of a modern aqueduct:—*a a a* represents the water level of a pond or reservoir upon elevated ground. From this pond a line of pipe is laid, passing over a bridge or viaduct at *d*, and under a river at *c*. The fountains at *b b*, show the stream rising to the level of its source in the pond *a*, at two points of very different elevation.

FIG. 119.

The ancients, in constructing aqueducts, do not seem to have ever practically applied this principle, that water in pipes rises to the level of its source. When, in conducting water from a distant source to supply a city, it became necessary to cross a ravine or valley, immense bridges, or arches of masonry were built across it, with great labor and at enormous expense, in order that the water-flow might be continued nearly horizontally. At the present day the same object is effected more perfectly by means of a simple iron pipe, bending in conformity with the inequalities of surface over which it passes.

In what manner should pipes for the conveyance of water be constructed?

In the construction of pipes for conveying water, it is necessary that those parts which are much below the level of the reservoir, should have a great degree of strength, since they sustain the bursting pressure of a column of water whose height is equal to the difference of level. A pipe with a diameter of 4 inches, 150 feet below the level of a reservoir, should have sufficient strength to bear with security a bursting pressure of nearly 5 tons for each foot of its length.

Upon the principle that water tends to rise to the level of its source, ornamental fountains may be constructed. Let water spout upward through a pipe communicating with the bottom of a deep vessel, and it will rise nearly to the height of the upper surface of the water in the vessel. The resistance of the air, and the falling drops, prevent it from rising to the exact level. Let A, Fig. 120, represent a cistern filled with water to a constant height, B. If four bent pipes be inserted in the side of the cistern at different distances below the surface, the water will jet upward from all the orifices to nearly the same level.

Fig. 120.

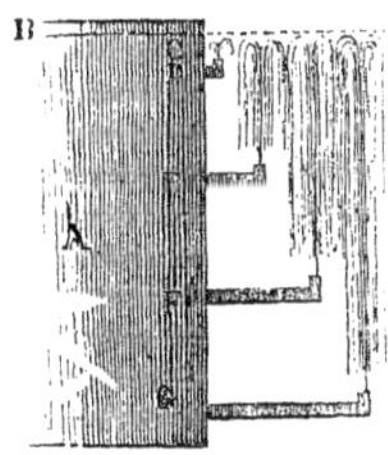

The phenomena of Artesian Wells, and the plan of boring for water, depend on the same principle.

What is an Artesian Well?

304. An Artesian Well is a cylindrical excavation formed by boring into the earth with a species of auger, until a sheet or vein of water is found, when the water rises through the excavation. Such excavations are called Artesian, because this method was employed for obtaining water at Artois in France.

Why does the water rise in an Artesian Well?

The reason that the water rises in Artesian, and sometimes in ordinary wells, to the surface, is as follows: The surface of the globe is formed of different layers, or strata, of different materials, such as sand, gravel, clay, stone, etc., placed one upon the other. In particular situations, these strata do not rest horizontally upon one another, but are inclined, the different strata being like cups, or basins placed one within the other, as in Fig. 121. Some of these strata are composed of materials, as sand or gravel, through which water will soak most readily; while other strata, like clay and rock, will not allow the water to pass through them. If, now, we suppose a stratum like sand, pervious to water, to be included as at *a a*, Fig. 121, between two other strata of clay or rock, the water falling upon the uncovered margin of the sandy stratum *a a*, will be absorbed, and penetrate through its whole depth. It will be prevented from rising to the surface by the impervious stratum above it, and from sinking lower, by the equally impervious stratum below it. It will, therefore, accumulate as in a reservoir. If, now, we

Fig. 121.

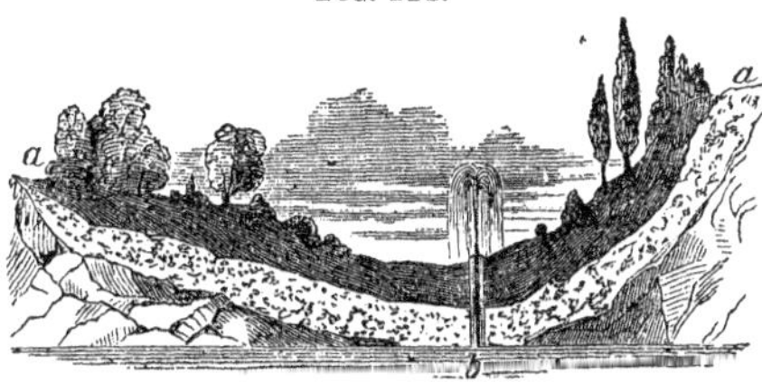

bore down through the upper stratum, as at *b*, until we reach the stratum containing the water, the water will rise in the excavation to a certain height, proportional to the height or level of the water accumulated in the reservoir *a a* from which it flows.*

What is the origin of springs?

305. The rain which falls upon the surface of the earth sinks downward through the sandy and porous soil, until a bed of clay or rock, through which the water can not penetrate, is reached. Here it accumulates, or running along the surface of the impervious stratum, bursts out in some lower situation, or at some point where the impervious bed or stratum comes to the surface in consequence of a valley, or some depression. Such a flow of water constitutes a spring. Suppose *a*, Fig. 122, to be a gravel hill, and *b* a stratum of clay or rock, impervious to water. The fluid percolating through the gravel would reach the impervious stratum, along which it would run until it found an outlet at *c*, at the foot of the hill, where a spring would be formed.

FIG. 122.

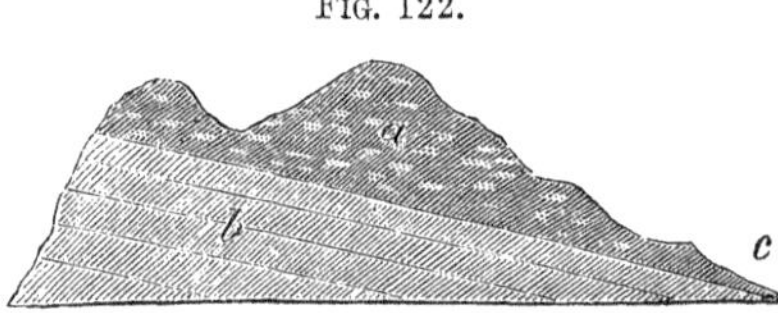

Why does water collect in an ordinary well?

306. If there are no irregularities in the surface, so situated as to allow a spring to burst forth, or if a spring issues out at some point of the porous earth considerably above the surface of the clay, or rock, upon which at some depth all such earth rests, the water soaking downward will not all be drained off, but will accumulate, and rise among the particles of soil, as it would among shot, or bullets, in a watertight vessel. If a hole, or pit, be dug into such earth, reaching below the level of the water accumulated in it, it will soon be filled up with water to this level, and will constitute a well. The reason why some wells are deeper than others, is, that the distance of the impervious stratum of clay below the surface is different in different localities.

From what source do all wells and springs derive their water?

307. All wells and springs, therefore, are merely the rainwater which has sunk into the earth, appearing again, and gradually accumulating, or escaping at a lower level.

What is a Water, or Spirit Level?

308. The property of liquids to assume a horizontal surface is practically taken advantage of in ascertaining whether a surface is perfectly horizontal, or level, and is accomplished by means of an instrument known as the "WATER" or "SPIRIT LEVEL." This consists of a small glass tube, *b c*, Fig. 123, filled with spirit, or water, except a small space occupied with air, and called

* In the great Artesian wells of Grenelle, near Paris, and of Kissingen, in Bavaria, the water rises from depths of 1,800 and 1,900 feet to a considerable height above the surface of the earth. The well of Paris is capable of supplying water at the rate of 14 millions of gallons per day. The region of country in which this water fell, from the curvature of the layers, or strata of material through which the excavation was made, must have been distant two hundred miles or more.

FIG. 123.

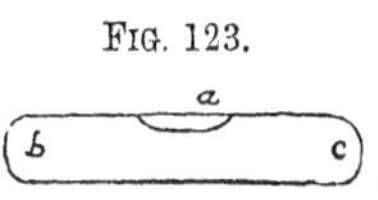

the air-bubble, *a*. In whatever position the tube may be placed, the bubble of air will rest at the highest point. If the two ends of the tube are level, or perfectly horizontal, the air-bubble will remain in the center of the tube; but if the tube inclines ever so little, the bubble rises to the higher end. For practical use the glass-tube is inclosed in a wood, or brass case, or box.

Upon what principle are canals constructed and operated?

309. The method of conducting a canal through a country, the surface of which is not perfectly horizontal, or level, depends upon this same property of liquids. In order that boats may sail with ease in both directions of the canal, it is necessary that the surface of the water should be level. If one end of a canal were higher than the other, the water would run toward the lower extremity, overflow the banks, and leave the other end dry. But a canal rarely, if ever, passes through a section of country of any great extent, which is not inclined, or irregular in its surface. By means, however, of expedients called LOCKS, a canal can be conducted along any declivity. In the formation of a canal, its course is divided into a series of levels corresponding with the inequalities of the surface of the country through which it passes. These levels communicate with each other by locks, by means of which boats passing in any direction can be elevated, or lowered with ease, rapidity, and safety.

FIG. 124.

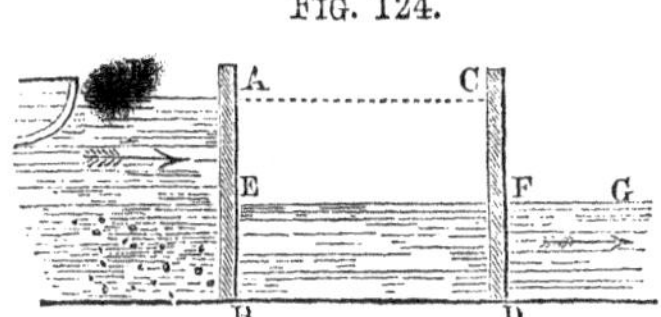

Fig. 124 represents a section of a lock, and Fig. 125 the construction of the LOCK GATES. The section of Fig. 125 represents a place where there is a sudden fall of the ground, along which the canal has to pass. A B and C D are two gates which completely intercept the course of the water, but at the same time admit of being opened and closed. A H is the level of the water in that part of the canal lying above the gate A B, and E F and F G the levels below the gate A B. The part of the canal included between two gates, as E F, is called a lock, because when a vessel is let into it, it can be shut by closing both pair of gates. If now it is required to let a boat down from the higher level, A H, to the lower level, E G, the gates C D are closed tightly, and an opening made in the gates A B (shown in Fig. 125), which allows the water to flow gradually from A H into the lock A E F C, until it attains a common level, H A C. The gate A B is then opened, and the boat floats into the lock A B C D. The gates A B are then closed, and an opening made in gates C D, which allows the water to flow from the space A E F C, until it comes to the common level, E F G. The gate C D is then opened, and the boat floats out of the locks into the continuation of the canal. To enable a boat to pass from the lower level, E F G, to the superior level, A H, the process here described is reversed.

FIG. 125.

With what force is a floating body pressed upward?

310. When a solid is immersed in a liquid it will be pressed upward with a force equal to the weight of the liquid it displaces.

How much water will a solid immersed in it displace?

311. A solid immersed in water will displace as much of the liquid as is equal in volume to the part immersed.

What is Buoyancy?

312. BUOYANCY is the name applied to the force by which a solid immersed in a liquid is heaved, or pressed upward.

The resistance offered when we attempt to sink a body lighter than water in that liquid, proves that the water presses with a force upward as well as downward. Upon this fact the laws of floating bodies depend; and for this reason the bottoms of large ships are constructed with a great degree of strength.

How is a body floating upon a liquid maintained in equilibrio?

313. A body floating upon a liquid is maintained in EQUILIBRIO by the operation of gravity drawing the mass downward, and by the pressure of the particles of the liquid upon which it rests, pressing it upward.

What is essential to the stability of a floating body?

314. In order that a body may float with stability, it is necessary that its center of gravity should be situated as low as possible.

What is the use of ballast in vessels?

For this reason, all vessels which are light in proportion to their bulk, require to be ballasted by depositing in the lowest portions of the vessel, immediately above the keel, a quantity of heavy matter, usually iron or stone. The center of gravity may thus be brought so low that no force of the wind striking the vessel sideways can capsize it. By raising the center of gravity, as when men in a boat stand upright, the equilibrium is rendered unstable.

When is a floating body in its most stable position?

A body floating is most stable when it floats upon its greatest surface: thus a plank floats with the greatest stability when placed flat upon the water; and its position is unstable when it is made to float edgewise.

When will a solid float, and when sink?

A solid can never float that is heavier, bulk for bulk, than the liquid in which it is immersed.

If the weight of a solid be exactly equal to the weight of an equal bulk of liquid, it will sink in it until it is entirely immersed; but when once it is entirely immersed, then, the upward and downward pressure being equal, the solid will neither sink or rise, but will remain suspended at any depth at which it may be placed.

Fig. 126.

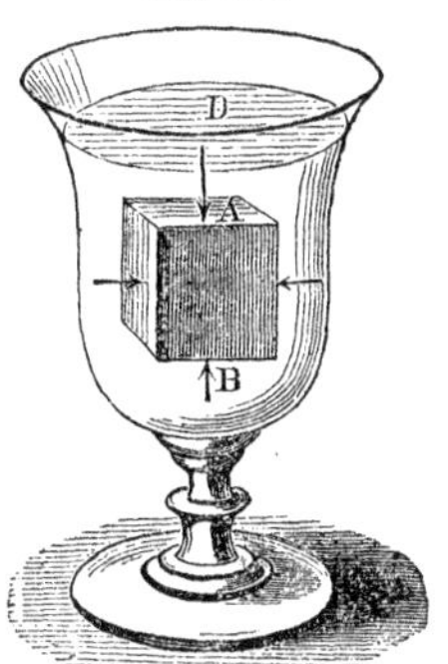

Let A B, Fig. 126, be a cube of wood floating in water; then the weight of the water displaced, or the weight of a volume of water equal to A B, is equal to the whole weight of the wood; since the upward pressure on the bottom of A B is the same as that which would support a portion of water equal in bulk to the displaced water, or to the cube A B; and as the downward pressure of the body is equal to the upward pressure of the liquid, it follows that the weight of the cube is equal to the weight of the water displaced. Hence A B will neither sink or rise.

A mass of stone, or any other heavy substance beneath the surface of water is more easily moved than upon the land because, when immersed in the water, it is lighter by the weight of its own bulk of water than it would be on land. A boy will often wonder why he can lift a stone of a certain weight to the surface of water, but can carry it no farther. The least force will lift a bucket immersed in water to the surface; but if it be lifted farther, its weight is felt just in proportion to the part of it which is above the surface.

The weight of the human body does not differ much from the weight of its own bulk of water; consequently, when bathers walk in water chin-deep, their feet scarcely press upon the bottom, and they have not sufficient hold upon the ground to give them stability; a current, therefore, will easily take them off their feet.

The facility with which different persons are able to float or swim, depends upon the physical constitution of the body. Corpulent people are lighter, bulk for bulk, than those of sparer habits: and as fat possesses a less specific gravity than water, a fat person will swim or float easier than a thin one.

315. It is not, however, necessary, in order that a body should float upon a liquid, that the materials of which it is composed should be specifically lighter than the liquid. If the entire mass of a solid is lighter than an equal volume of the liquid, it will float.

A thick piece of iron, weighing half an ounce, loses in water nearly one eighth of its weight; but if it is hammered into a plate or vessel, of such a form that it occupies eight times as much space as before, it will then weigh less than an equal bulk of water, and will consequently float, sinking just to the brim. If made twice as large, it will displace one ounce of water, consequently, twice its own weight; it will then sink to the middle, and can be loaded with half an ounce weight before sinking entirely.

How can a body heavier than an equal bulk of water be made to float?

316. A body composed of any material, however heavy, can be made to float on any liquid, however light, by giving it such a shape as will render its bulk or volume lighter than an equal bulk of water.

Iron ships and boats are illustrations of this principle. A ship carrying a thousand tons' weight will displace just as much water, or float to the same depth, whether her cargo be feathers, cotton, or iron. A ship made of iron floats just as high out of water as a ship of similar form and size made of wood, provided that the iron be proportionally thinner than the wood and therefore not heavier on the whole.

The buoyancy of hollow solids is frequently used for lifting or supporting heavy weights in water. Life-preservers, which are inflated bags of India-rubber, are an example. Hollow boxes, or tanks, are used for the purpose of raising sunken vessels. These boxes are sunk, filled with water, and attached to the side of the vessel to be raised. The water, by a connection of pipes, is then pumped out of them, when the upward pressure of the liquid becoming greater than the gravity or weight of the entire mass, the whole will rise and float.

To what is the buoyancy of liquids proportional?

317. The buoyancy of liquids is in proportion to their density or specific gravity, or, in other words, a solid is buoyant in a liquid, in proportion as it is light, and the liquid heavy.

Thus quicksilver, the heaviest, or most dense fluid known, supports iron upon its surface; and a man might float upon mercury as easily as a cork floats upon water. Many varieties of wood which will sink in oil, float readily upon water.

318. The principle that the buoyancy of liquids varies in proportion as their specific gravity varies, furnishes a very ready method of determining the specific gravity of a liquid. This is done by means of an instrument called the hydrometer.

What is a Hydrometer?

319. The HYDROMETER consists of a hollow glass tube, on the lower part of which a spaerical bulb is blown, the latter being filled with a suitable quantity of small shot, or quicksilver, in order to cause it to float, in a vertical position. The upper part of the tube contains a scale graduated into suitable divisions. (See Fig. 127.)

FIG. 127.

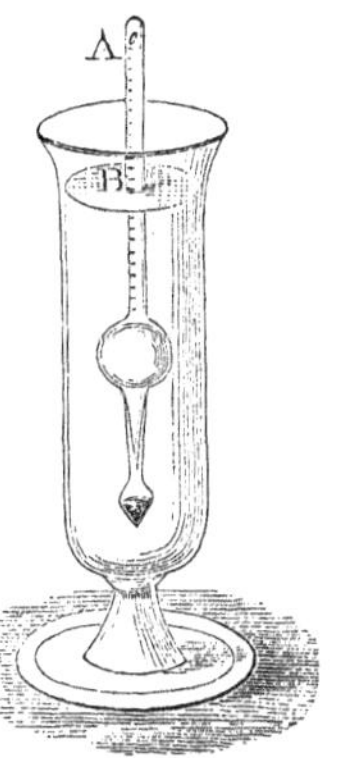

How may the specific gravity of a liquid be determined by the Hydrometer?

It is obvious that the hydrometer will sink to a greater or less depth in different liquids; deeper in the lighter ones, or those of small specific gravity, and not so deep in those which are denser, or which have great specific gravity. The specific gravity of a liquid may, therefore, be estimated by the number of divisions on the scale which remain above the surface of the liquid. Tables are constructed, so that, by their aid, when the number on the scale at which the hydrometer floats in a given liquid is determined by experiment, the specific gravity is expressed by figures in a column directly opposite that number in the table.

There are various forms of the hydrometer especially adapted for determining the density, or specific gravity, of spirits, oils, syrups, lye, etc. It affords a ready method of determining the purity of a liquid, as, for instance, alcohol. The addition of water to alcohol adds to its density, and therefore increases its buoyancy. The addition of water, therefore, will at once be shown by the less depth to which the hydrometer will sink in the liquid. The adulteration of sperm oil with whale, or other cheaper oils, may be shown in the same manner.

320. For the reason that the buoyancy of a liquid is proportioned to its density, a ship will draw less water, or sail lighter by one thirty-fifth in the heavy salt water of the ocean, than in the fresh water of a river; for the same reason it is easier to swim in salt than in fresh water.*

* "A floating body sinks to the same depth whether the mass of liquid supporting it be great or small, as is seen when an earthen cup is placed first in a pond, and then in a second cup only so much larger than itself, that a very small quantity of water will suffice to fill up the interval between them. An ounce of water in this way may be made to float substances of much greater weight. And if a large ship were received into a dock, or case, so exactly filling it that there were only half an inch of interval between it and the wall, or side of the containing space, it would float as completely when the few hogsheads of water required to fill this little interval up to its usual water-mark were poured in, as if it were on the high seas. In some canal locks, the boats just fit the place in which they have to rise and fall, and thus diminish the quantity of water necessary to supply the lock."—*Arnott.*

SECTION I.

CAPILLARY ATTRACTION.

Explain the phenomena observed when the hand is plunged into different liquids.

321. If we plunge the hand into a vessel of water, and withdraw it, it is said to be wet; that is, it is covered with a thin film, or coating of water, which adheres to it, in opposition to the tendency of the attraction of gravitation to make it fall off. There is, therefore, an attraction between the particles of the water and the hand, which, to a certain extent, is stronger than the influence of gravitation.

If now we plunge the hand into a vessel of quicksilver, no adhesion of the particles of the mercury to the hand will take place, and the hand, when withdrawn, will be perfectly dry.

If we plunge a plate of gold, however, into water and quicksilver, it will be wet equally by both, and will come out of the quicksilver covered with a white coating of that liquid.

It is, therefore, obvious that a certain molecular attraction exists between certain liquids and certain solids, which does not prevail to the same extent between others.

What is Capillary Attraction?

322. That variety of molecular force which manifests itself between the surfaces of solids and liquids is called CAPILLARY ATTRACTION.

What is the origin of the term?

This name originates from the circumstance, that this class of phenomena was first observed in small glass tubes, the bore of which was not thicker than a hair, and which were hence called *Capillary Tubes*, from the Latin word *capillus*, which signifies a hair.

How may Capillary Attraction be illustrated?

323. If we take a series of glass tubes of very fine bore, but of different diameters, and place them in a vessel of water, which has been colored in order to show the effect more strikingly, we shall see that the water will rise in the tubes to various heights, attaining the greatest degree of elevation in the smallest tube. The height at which the same liquid will rise in any given tube is always uniform, but it varies for different liquids.

FIG. 128.

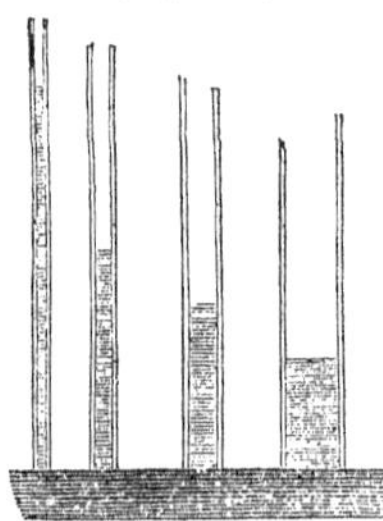

Fig. 128 is an enlarged representation of the manner in which water will rise in tubes of different diameters.

The simplest method of exhibiting capillary attraction is to immerse the end of a piece of thermometer tube in water (see Fig. 129) which has been tinted with ink. The liquid will be seen to ascend, and will remain elevated in the tube at a considerable height above the surface of the liquid in the vessel.

The ordinary definition of capillary attraction is, that form of attraction which causes liquids to ascend above their level in capillary tubes. It, however, is not strictly correct, as this force not only acts in elevating but in depressing liquids in tubes, and is at work wherever liquids are in connection with solid bodies.

FIG. 129.

What will be the condition of the surface of a liquid which wets the sides of the vessel containing it?

324. If a liquid be poured into a vessel, as water in glass, whose sides are of such a nature as to be wetted by it, the liquid will be elevated above the general level of its surface at the points where it touches the sides of the vessel. This is shown in Fig. 130.

When the liquid does not wet the sides of the vessel, what will be the condition of its surface?

If, however, the liquid is poured into a vessel whose sides are of such a nature that they are not wetted by it, as in the case of quicksilver in a glass vessel, then the liquid will be depressed below the general level of its surface at the points where it comes in contact with the sides of the vessel. This is shown in Fig. 131.

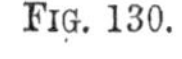

FIG. 130.

FIG. 131.

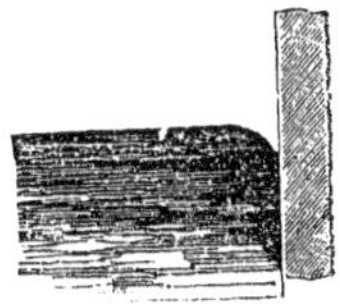

325. If two plates of glass, A and B, Fig. 132, be plunged into water at their lower extremities, with their faces vertical and parallel, and at a certain distance asunder, the water will rise at the points m and n, where it is in contact with the glass; but at all intermediate points, beyond a small distance from the plates, the general level of the surfaces E, C, and D, will correspond.

FIG. 132.

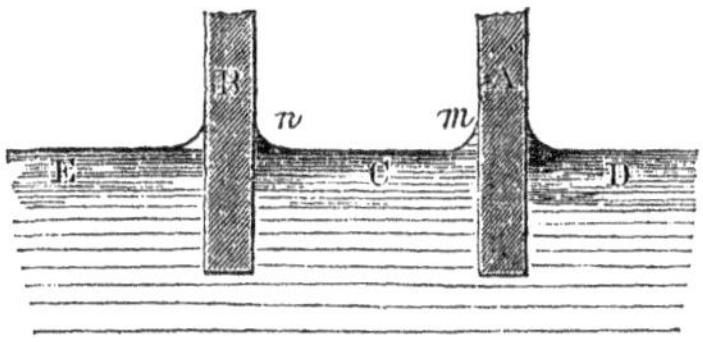

If the two plates, A and B, are brought near to each other, as in Fig. 133, the two curves, m and n, will unite, so as to form a concave surface, and the water at the same time between them will be raised above the general level, E and D, of the water in the vessel. If the plates be brought still nearer together, as in Fig. 134, the water between them will rise still higher, the force which sustains the column being increased as the distance between the plates is diminished

FIG. 133.

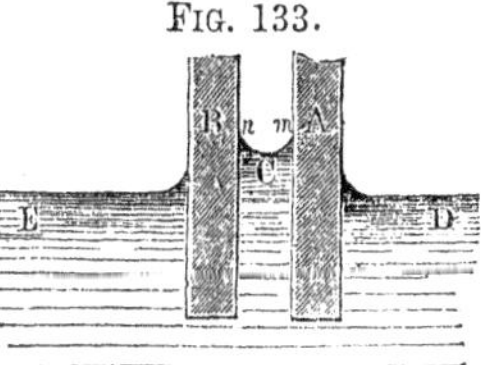

To what is the elevation of water in capillary tubes proportioned?

326. The height to which water will rise in capillary tubes is in proportion to the smallness of their diameters.

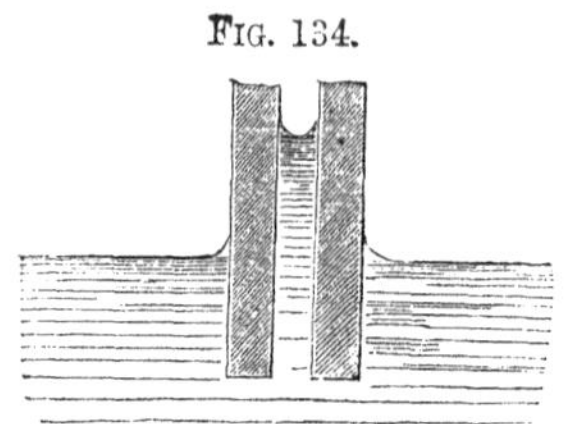
FIG. 134.

Thus in two tubes, one of which is double the diameter of the other, the fluid will rise to twice the height in the small tube that it will in the larger. The truth of this principle can be made evident by the following beautiful and simple experiment. Two square pieces of plate-glass, C and B, Fig. 135, are arranged so that their surfaces form a minute angle at A. This position may be easily given them by fastening with wax or cement. When the ends of the plates are placed in the water, as shown in the figure, the water rises in the space between them, forming the curve, which is called an hyperbola. The elevation of the water between the two surfaces will be the greatest at the points where the distance between the plates is the least.

FIG. 135.

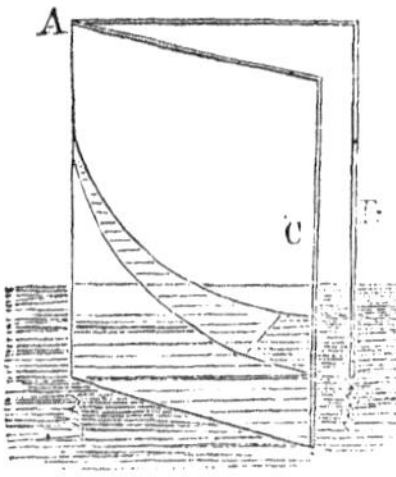

327. The figure of the surface which bounds a liquid in a capillary tube will depend upon the extent of the attraction which exists between the particles of the liquid and the surface of the tube. Thus, a column of water contained in a glass capillary tube will have a concave form of surface, as in Fig. 136, since the attraction of glass for water exceeds the attraction of the particles of water for each other; a surface of mercury, on the contrary, in a similar tube, will be convex, see Fig. 137, since the attraction of glass for mercury is less than the mutual attraction of the particles of mercury.

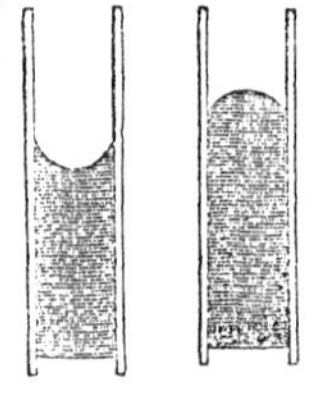
FIG. 136. FIG. 137.

When will a liquid be elevated and when depressed in a capillary tube?

328. In a capillary tube a liquid will ascend above its general level, when it wets the tube; and is depressed below its level when it does not wet it.

How may a needle be made to float upon water?

329. If the surface of a body repels a liquid, such a body, though heavier, bulk for bulk, than the liquid, may, under some circumstances, float upon it; and so present an apparent exception to the general hydrostatic law by which solids which are heavier than liquids, bulk for bulk, will sink in them. An example of this may be shown by slightly greasing a fine sewing-needle, and then placing it carefully in the direction of its length upon the surface of water. The needle, although heavier, bulk for bulk, than water, will float.

The power of certain insects to walk upon the surface of water without sinking, has been explained upon the same principle. The feet of these insects, like the greased needle, have a capillary repulsion for the water, and when they apply them to the surface of water, instead of sinking in it, they produce depressions upon it.

For a like reason, water will not flow through a fine sieve, the wires of which have been greased.

When will a liquid fail to wet a solid?

330. A liquid will not wet a solid when the force of adhesion developed between the particles of the liquid and the surface of the solid, is less than half the cohesive force which exists between the particles of the liquid.

What is a "Rope Pump?"

331. The fact of the strong adhesion which exists between water and the fibers of a rope, has been taken advantage of in the construction of a kind of pump, called the "Rope," or "Vera's" Pump, Fig. 138. It consists of a cord passing over two wheels, *a* and *b*, the lower one of which is immersed in water. A rapid motion is given to the wheels by means of the crank *d*, and the water, by adhering, follows the rope in its movements, and is discharged into a receptacle above.

FIG. 138.

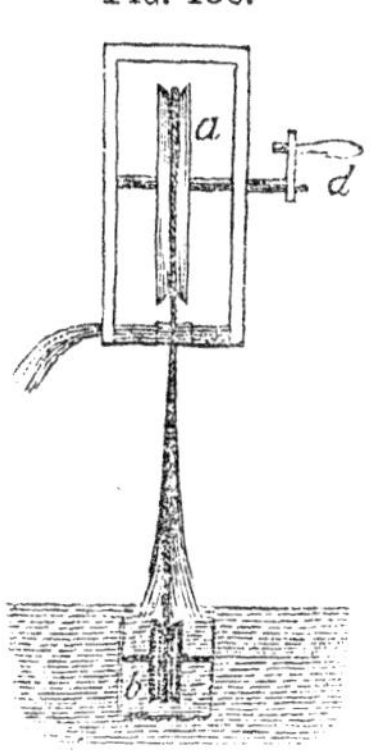

What are familiar illustrations of capillary attraction?

Illustrations of capillary attraction are most familiar in the experience of every-day life. The wick of a lamp, or candle, lifts the oil, or melted grease which supplies the flame, from a surface often two or three inches below the point of combustion. In a cotton-wick, which is the material best adapted for this purpose, the minute, separate fibers of the cotton themselves are capillary tubes, and the interstices between the filaments composing the wick are also capillary tubes; in these the oil ascends. The oil, however, can not be lifted freely beyond a certain height by capillary attraction: hence, when the surface of the oil is low in the lamp, the flame becomes feeble, or expires.

If the end of a towel, or a mass of cotton thread, be immersed in a basin of water, and the remainder allowed to hang over the edge of the basin, the water will rise through the pores and interstices of the cloth, and gradually wet the whole towel. In this way the basin may be entirely emptied.

If sand, a lump of sugar, or a sponge, have moisture beneath and slightly in contact with it, it will ascend through the pores by the agency of capillary attraction in opposition to gravity, and the entire mass will become wet.

The lower story of a house is sometimes damp, because the moisture of the ground ascends through the pores of the materials constituting the walls of the building. Wood imbibes moisture by the capillary attraction of its pores, and expands or swells in consequence. This fact has been taken advantage of for splitting stones; wedges of dry wood are driven into grooves cut in the stone, and on being moistened, swell with such irresistible force as to split the block in a direction regulated by the groove.

An immense weight suspended by a dry rope, may be raised a little way, by merely wetting the rope; the moisture imbibed by capillary attraction into the substance of the rope causes it to swell laterally and become shorter.

Capillary attraction is also instrumental in supplying trees and plants with moisture through the agency of the roots and underground fibers.

What are the phenomena of Exosmose and Endosmose?

332. The terms EXOSMOSE and ENDOSMOSE are applied to those currents in contrary directions which are established between two liquids of a different nature, when they are separated from each other by a partition composed of a membrane, or any porous substance.

FIG. 139.

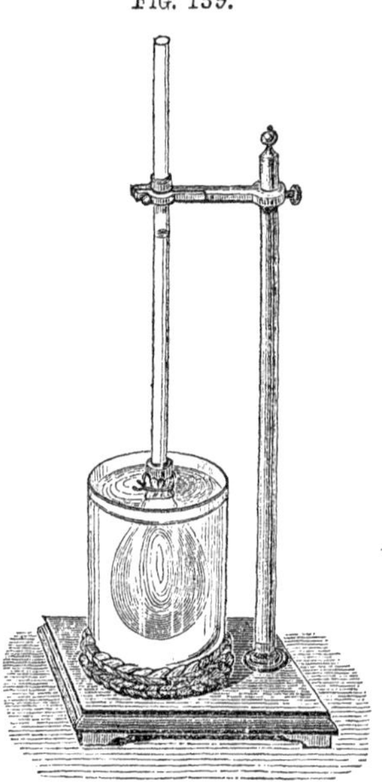

The name *Endosmose*, derived from a Greek word, signifies *going in*, and is applied to the stronger current; while the name *Exosmose*, signifying *going out*, is applied to the weaker current.

The phenomena of *Endosmose* and *Exosmose*, which are undoubtedly dependent on capillary attraction, may be illustrated by the following simple experiment:—If we take a small bladder, or any other membranous substance, and having fastened it on a tube open at both ends, as is represented in Fig. 139, fill the bladder with alcohol, and immerse it, connected with the tube, in a basin of water, to such an extent that the top of the bladder filled with alcohol corresponds with the level of the water in the vessel, in a short time it will be observed, that the liquid is rising in the tube connected with the bladder, and will ultimately reach the top and flow over. This rising of the alcohol in the tube is evidently due to the circumstance that the water permeates through the bladder, with a certain degree of force, producing the phenomena which we call *endosmose*, "*going in*;" the effect being to elevate the alcohol to a considerable height in the tube. At the same time, a certain quantity of the alcohol has passed out through the pores of the bladder, and mixed with the water in the external vessel. This outward passage of the alcohol we call *exosmose*, "going out." A less quantity of the alcohol will pass out of the bladder in a given time to mingle with the water, than of the water will pass in, and consequently the bladder containing the alcohol having more liquid in it than at first, becomes strained, and presses the liquid up in the tube.

If we have a box divided by a partition of porous clay, or any other substance of like nature, and place a quantity of syrup on one side, and water on the other, or any other two liquids of different densities which freely mix with one another, currents will be established between the two in opposite directions through the porous partition, until both are thoroughly mingled with each other.

333. If a liquid is placed in contact with a surface of the body, divested of its epidermis, or outer skin, or in contact with a mucous membrane, the liquid will be absorbed into the vessels of the body through the force of endosmose.

PRACTICAL QUESTIONS AND PROBLEMS IN HYDROSTATICS.

1. Why are stones, gravel, and sand so easily moved by waves and currents?

Because the moving water has only to overcome about half the weight of the stone.

2. Why can a stone which, on land, requires the strength of two men to lift it, be lifted and carried in water by one man?

Because the water holds up the stone with a force equal to the weight of the volume of water it displaces.

3. Why does cream rise upon milk?

Because it is composed of particles of oily, or fatty matter, which are lighter than the watery particles of the milk.

4. How are fishes able to ascend and descend quickly in water?

They are capable of changing their bulk by the voluntary distension, or contraction of a membraneous bag, or air bladder, included in their organization; when this bladder is distended, the fish increases in size, and being of less specific gravity, *i. e.*, lighter, it rises with facility; when the bladder is contracted, the size of the fish diminishes, and its tendency to sink is increased.

5. Why does the body of a drowned person generally rise and float upon the surface several days after death?

Because, from the accumulation of gas within the body (caused by incipient putrefaction), the body becomes specifically lighter than water, and rises and floats upon the surface.

6. How are life-boats prevented from sinking?

They contain in their sides air-tight cells, or boxes, filled with air, which by their buoyancy prevent the boat from sinking, even when it is filled with water.

7. Why does blotting-paper absorb ink?

The ink is drawn up between the minute fibers of the paper by capillary attraction.

8. Why will not writing, or sized paper, absorb ink?

Because the sizing, being a species of glue into which writing papers are dipped, fills up the little interstices, or spaces, between the fibers, and in this way prevents all capillary attraction.

9. Why is vegetation on the margin of a stream of water more luxuriant than in an open field?

Because the porous earth on the bank draws up water to the roots of the plants by capillary attraction.

10. Why do persons who water plants in pots frequently pour the water into the saucer in which the pot rests, and not over the plants?

Because the water in the saucer is drawn up by capillary attraction through the little interstices of the mold with which the pot is filled, and is thus presented to the roots of the plant.

11. Why does dry wood, immersed in water, swell?

Because the water enters the pores of wood by capillary attraction, and forces the particles further apart from each other.

12. Why will water, ink, or oil, coming in contact with the edge of a book, soak further in than if spilled upon the sides?

Because the space between the leaves acts in the same manner as a small capillary tube would—attracts the fluid, and causes it to penetrate far inward. The fluid penetrates with more difficulty upon the side of the leaf, because the pores in the paper are irregular, and not continuous from leaf to leaf.

13. In a hydrostatic press, the area of the base of the piston in the force-pump is one square inch, and the area of the base of the piston in the large cylinder is fourteen square inches; what will be the force exerted, supposing a power of eight hundred pounds applied to the piston of the force-pump?

14. A flood-gate is five feet in breadth, and sixteen feet in depth: what will be the pressure of water upon it in pounds?

15. What pressure will a vessel, having a superficial area of three feet, sustain when lowered into the sea to the depth of five hundred feet?

16. What pressure is exerted upon the body of a diver at the depth of sixty feet, supposing the superficial area of his body to be two and a half square yards?

17. What will be the pressure upon a dam, the area of the side of which is one hundred and fifty superficial feet, and the height of the side fifteen feet, the water rising even with the top?

CHAPTER IX.

HYDRAULICS.

What is the science of Hydraulics?

334. HYDRAULICS is that department of physical science which treats of the laws and phenomena of liquids in motion.*

Hydraulics considers the flow of liquids in pipes, through orifices in the sides of reservoirs, in rivers, canals, etc., and the construction and operation of all machines and engines which are concerned in the motion of liquids.

* From ὕδωρ (hudor), *water*, and αυλός (aulos), *a pipe*.

Upon what does the velocity of a flowing liquid depend?

335. When an opening is made in a reservoir containing a liquid, it will jet out with a velocity proportioned to the depth of the aperture below the surface.

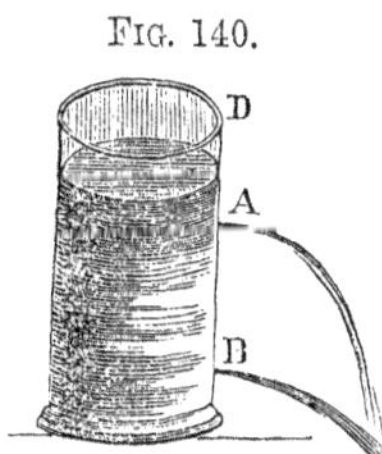

FIG. 140.

Supposing the surface of water in a vessel, D, Fig. 140, to be kept at a constant height by the water flowing into it, and that the water flows out through openings in the side of precisely the same size; then a quart measure would be filled from the jet issuing from B as soon as a pint measure from the upper opening, A.

As the flow of liquids is in consequence of the attraction of gravity, and as the pressure of a liquid is equal in all directions, we have the following principle established:—

What is the velocity of a liquid flowing from a reservoir equal to?

336. The velocity which the particles of a liquid acquire when issuing from an orifice, whether sideways, upward, or downward, is equal to that which they would have acquired in falling perpendicularly through a space equal to the depth of the aperture below the surface of the liquid.

Thus, if an aperture be made in the bottom, or side, of a vessel containing water, 16 feet below the surface, the velocity with which the water will jet out will be 32 feet per second, for this is the velocity which a body acquires in falling through a space of 16 feet.

As the velocity acquired by a falling body is as the square root of the space through which it falls, the velocity with which water will issue from an aperture may be calculated by the following rule:—

How may the velocity of a liquid flowing from a reservoir be calculated?

337. The velocity with which water spouts out from any aperture in a vessel is as the square root of the depth of the aperture below the surface of the water.

The water must, therefore, flow with ten times greater velocity from an opening 100 inches below the level of the liquid, than from a depth of only one inch below the same level.

What is the theoretical law for determining the quantity of water discharged from an aperture?

338. The theoretical law for determining the *quantity* of water discharged from an orifice is as follows:—

The quantity of water discharged from an orifice in each second may be calculated by multiplying the velocity by the area of the aperture.

The above rules for calculating the velocity and quantity of water flowing from orifices, are not found strictly to hold good in practice. The friction of water against the sides of vessels, pipes, and apertures, and the formation of what is called the "contracted vein," tend very much to diminish the motion and discharge of water.

Fig. 141.

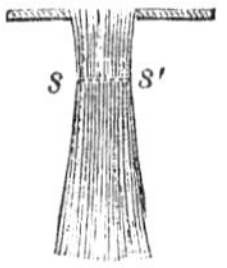

What is the "contracted vein" in a current of water?

When water flows through a circular aperture in a vessel, the diameter of the issuing stream is contracted, and attains its smallest dimensions at a distance from the orifice equal to the diameter of the orifice itself. The section of the jet at this point, Fig. 141, *s s'*, will be about two thirds of the magnitude of the orifice. This point of greatest contraction is called the *vena contracta*, or *contracted vein.*

What is the cause of this phenomenon?

This phenomenon arises from the circumstance that a liquid contained in a vessel rushes from all sides toward an orifice, so as to form a system of converging currents. These issuing out in oblique directions, cause the shape of the stream to change from the cylindrical form, and contract it in the manner described.

How may the effect of the contracted vein be avoided?

By the attachment of suitable tubes to the aperture, the effect of the contracted vein may be avoided, and the quantity of flowing water be very greatly increased. A short pipe will discharge one half more water in the same time, than a simple orifice of the same dimensions. The tube, however, must be entirely without the vessel, as at B, Fig. 142, for if continued inside, as at A, the quantity of liquid discharged will be diminished instead of augmented.

Fig. 142.

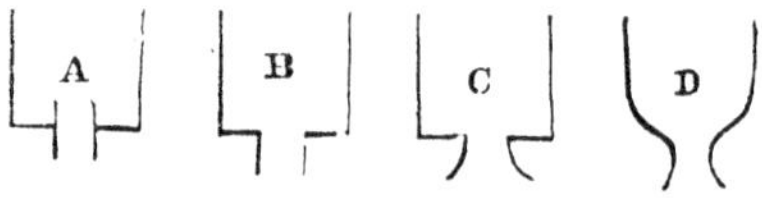

The rapidity of the discharge of the water will also depend much on the figure of the tube, and that of the bottom of the vessel, since more water will flow through a conical, or bell-shaped tube, as at C, Fig. 142, than through a cylindrical tube. A still further advantage may be gained by having the bottom of the vessel rounded, as at D, and the tube bell-shaped.

An inch tube of 200 feet in length, placed horizontally, will discharge only one fourth as much water as a tube of the same dimensions an inch in length; hence, in all cases where it is proposed to convey water to a distance in pipes, there will be a great disappointment in respect to the quantity actually delivered, unless the engineer takes into account the friction, and the turnings of the pipes, and makes large allowances for these circumstances. If the quantity to be actually delivered ought to fill a two inch pipe, one of three inches will not be too great an allowance, if the water is to be conveyed to any considerable distance.

In practice, it will be found that a pipe of two inches in diameter, one hundred feet long, will discharge about five times as much water as one of one inch in diameter of the same length, and under the same pressure. This difference is accounted for, by supposing that both tubes retard the motion of the fluid, by friction, at equal distances from their inner surfaces, and consequently, the effect of this cause is much greater in proportion, in the small tube, than in the large one.

As the velocity with which a stream issues depends upon the height of the column of fluid, it follows that when a liquid flows from a reservoir which is not replenished, but the level of which constantly descends, its velocity will be uniformly retarded. The following principle has been established:—

What will be the difference in the flow of a liquid when the vessel is kept full and when it is allowed to empty itself?

339. If a vessel be filled with a liquid and allowed to discharge itself, the quantity issuing from an orifice in a given time, will be just one half what would be discharged from the same orifice in the same time, if the vessel was kept constantly full.

What is the principle and construction of the water-clock?

340. Before the invention of clocks and watches, the flow of water through small orifices was applied by the ancients for the measurement of time, and an arrangement for this purpose was called a *Clepsydra*, or water-clock. One form of this instrument consisted of a cylindrical vessel filled with water, and furnished with an orifice which would discharge the whole in twelve hours. If the whole depth through which the water in the vessel would sink in this time be divided into 144 parts, it will sink through 23 in the first hour, 21 in the second, 19 in the third, and so on, according to a series of odd numbers: this diminishing rate depending on the constantly decreasing height and pressure of the column above the point of discharge. The spaces indicated upon a scale attached to the side of the vessel, and compared with the position of the descending column, marks the time. Fig. 143 represents the form of the water clock.

Fig. 143.

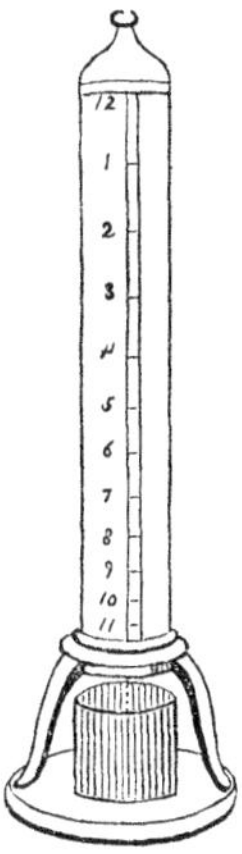

How is the velocity of water in pipes and rivers retarded?

341. The force of currents, whether in pipes, canals, or rivers, is more or less resisted, and their velocity retarded, by the friction which takes place between those surfaces of the liquid and the solid which are in contact.

At what part of a stream is the velocity greatest?

This explains a fact which may be observed in all rivers: that the velocity of a stream is always greater at the center than near the bank, and the velocity at the surface is greater than the velocity at the bottom.

In a channel of unequal section, how will the velocity of a current be affected?

342. If a given quantity of liquid must pass through pipes or channels of unequal section in the same time, its velocity will increase as the transverse section diminishes, and diminish as the area of the section increases.

This fact is familiar to every one who observes the course of brooks or rivers: wherever the bed contracts, the current becomes rapid, and on the contrary if it widens, the stream becomes more sluggish.

What inclination is sufficient to give motion to running water?

343. A very slight declivity is sufficient to give motion to running water. Three inches to a mile in a smooth, straight channel, gives a velocity of about three miles per hour.

The river Ganges, at a distance of 1,800 miles from its mouth, is only 800 feet above the level of the sea. The average rate of inclination of the surface of the Mississippi is 1.80 for the first hundred miles from the Gulf of Mexico, 2 inches for the second hundred, 2.30 for the third, and only 2.57 for the fourth.

What is the average velocity of rivers?

The velocity of rivers is extremely variable; the slower class moving from two to three miles per hour, or three or four feet per second, and the more rapid as much as six feet per second. The mean velocity of the Mississippi, near its mouth, is 2.26 miles per hour, or 2.95 feet per second.*

The quantity of water which passes over the beds of rivers in a given time is very various. In the smaller class of streams it amounts to from 300 to 350 cubic feet per second. In the smaller class of navigable rivers, it amounts to from 1,000 to 1,200 cubic feet; and in the larger class to 14,000 cubic feet and upward. It is estimated that the Mississippi discharges 12 billions of cubic feet of water per minute.†

* In the construction of water-channels for drainage, the regulation of inclination necessary to produce free flowage of the water, is a matter of great importance. This inclination varies greatly with the size of the stream of water to be conducted off. Large and deep rivers run sufficiently swift with a fall of a few inches per mile; smaller rivers and brooks require a fall of two feet per mile, or 1 foot in 2,500. Small brooks hardly keep an open course under 4 feet per mile, or 1 in 1,200; while ditches and covered drains require at least 8 feet per mile, or 1 in 600. Furrows of ridges, and drains partially filled with loose materials, require a much greater inclination.

† A question of some interest relative to the course and flow of rivers, may, perhaps, be appropriately considered in this connection. The question is as follows: Do the Mississippi, and other rivers whose courses are northerly and southerly, flow up hill or down hill? The Mississippi runs from north to south. If its source were at the pole and its mouth at the equator, the elevation of the mouth would be thirteen miles higher than its source, as this is the difference between the equatorial and the polar radii of the earth. On this principle, the mouth of the Mississippi is two and a half miles more elevated than its source. Does it run up hill, and if so, how has its course and motion originated? The problem, although apparently one of difficulty, admits of an easy solution.

The centrifugal force, caused by the rotation of the earth, has changed the form of our planet from that of a perfect sphere to that of an ellipsoid, or a sphere flattened at the poles, in which the length of the largest radius, exceeds the shorter by thirteen miles, the present form being the figure of equilibrium under the present conditions. The cohesion of the solid particles of the earth has resisted, and does resist, to a limited extent, the influence of the centrifugal force which has changed the original figure; but the particles of liquid on the earth's surface, being perfectly free to move, yield to the influence, and are at rest only so long as the condition of equilibrium is undisturbed, and always move in such a way as to restore it when it is disturbed. Water, consequently, always flows from places which are above the figure of equilibrium, to those which are below it. Now the mouth of the Mississippi is two and a half miles more distant from the center of

How are waves upon liquid surfaces formed?

344. When one portion of a liquid is disturbed, the disturbance (in consequence of the freedom with which the particles of a liquid move upon each other) is communicated to all the other portions, and a wave is formed. This wave propagates itself into the unmoved spaces adjoining, continually enlarging as it goes, and forming a series of undulations.

What is the origin of sea waves?

345. Ordinary sea waves are caused by the wind pressing unequally upon the surface of the water, depressing one part more than another: every depression causes a corresponding elevation.

Does the substance of the wave actually advance, or is it stationary?

Where the water is of sufficient depth, waves have only a vertical motion, *i. e.*, up and down. Any floating body, as a buoy, floating on a wave, is merely elevated and depressed alternately; it does not otherwise change its place. The apparent advance of waves in deep water is an ocular deception: the same as when a corkscrew is turned round, the thread, or spiral, appears to move forward.

Why do waves always break against the shore?

346. A wave is a form, not a thing; the form advances, but not the substance of the wave. When, however, a rock rises to the surface, or the shore by its shallowness prevents or retards the oscillations of the water, the waves forming in deep water are not balanced by the shorter undulations in shoal water, and they consequently move forward and form breakers. Thus it is that waves always break against the shore, no matter in what direction the wind blows.

When the shore runs out very shallow for a great extent, the breakers are distinguished by the name of surf.

On the Atlantic, during a storm, the waves have been observed to rise to a height of about forty-three feet above the hollow occupied by a ship; the total distance between the crests of two large waves being 559 feet, which distance was passed by the wave in about seventeen seconds of time.

the earth (*i. e.*, the center of figure) than the source is. But if it had not been for the restraining influence of the cohesive force prevailing among the solid particles, it would have been, through the action of the centrifugal force, three miles higher, instead of two and a half. It is therefore below the surface of equilibrium, and the water flows south to fill up the proper level.

The question as to whether the river flows up, or down, depends on the meaning we attach to the words used. If by DOWN we mean toward the earth's center of figure, or toward that part of the earth's surface where the attraction of gravity is the greatest, as at the poles, then the Mississippi runs up hill. If, on the contrary, DOWN means below the surface of equilibrium, and UP means above the surface of equilibrium, then the Missisipi flows downward. If the earth were a perfect sphere, and without rotation, the river would flow northward. A more complete explanation of this subject will be found in a paper read before the American Academy by Prof. Lovering in 1856, and in the "Annual of Scientific Discovery" for 1857, pp. 179—182.

How does the resistance of a liquid to a solid moving through it vary?

347. The resistance which a liquid opposes to a solid body moving through it, varies with the form of the body.

The resistance which a plane surface meets with while it moves in a liquid, in a direction perpendicular to its plane, is in general, proportioned to the square of its velocity.

What advantage has an oblique surface in moving against a liquid?

If the surface of a solid moved against a liquid be presented obliquely with respect to the direction of its motion, instead of perpendicularly, the resistance will be modified and diminished; the quantity of liquid displaced will be less, and the surface, acting as a wedge, or inclined plane, will possess a mechanical advantage, since in displacing the liquid it pushes it aside, instead of driving it forward.

The determination of the particular form which should be given to a mass of matter in order that it may move through a liquid with the least resistance, is a problem of great complexity and celebrity in the history of mathematics, inasmuch as it is connected with nearly all improvements in navigation and naval architecture. The principles involved in this problem require that the length of a vessel should coincide with the direction of the motion imparted to it; and they also determine the shape of the prow and of the surfaces beneath the water. Boats which navigate still waters, and are not intended to carry a great amount of freight, are so constructed that the part of the bottom immersed moves against the liquid at a very oblique angle.

Vessels built for speed should have the greatest possible length, with merely the breadth necessary to stow the requisite cargo.

The form and structure of the bodies of fishes in general, are such as to enable them to move through the water with the least resistance.

When are the paddles of a steamboat most effective?

348. In the paddles of steamboats, that one is only completely effectual in propelling the vessel which is vertical in the water, because upon that one alone does the resistance of the water act at right angles, or to the best advantage. In the propulsion of steamboats, it is found that paddle-wheels of a given diameter act with the greatest effect when their immersion does not exceed the width, or depth, of the lowest paddle-board; their effect also increases with the diameter of the wheel.

Is the paddle-wheel an advantageous method of applying power for propelling vessels?

The amount of power lost by the use of the paddle-wheel as a means of propelling vessels is very great, since, in addition to the fact that only the paddle which is vertical in the water is fully effective, the series of paddles in descending into the water, are obliged to exert a downward pressure, which is not available for propulsion, and in ascending, to lift a considerable weight of water that opposes the ascent, and adheres to the paddles. The rolling of the vessel, also, renders it impossible to maintain the paddles at the requisite degree of immersion necessary to give them their greatest efficiency; one wheel on one side being occasionally immersed too deeply, while the other wheel, on the other side may be lifted entirely out of water.

Describe the construction and action of the screw-propeller.

349. To remedy in some degree these causes of inefficiency and waste, the submerged propelling-wheel, known as the *screw-propeller*, has been introduced within the last few years. The screw-propeller consists of a wheel resembling in its form the threads of a screw, and rotating on an axle. It is placed in the stern of the vessel, below the water-line, immediately in front of the rudder. Fig. 144 represents one form of the screw-propeller, and its location in reference to the other parts of the vessel.

FIG. 144.

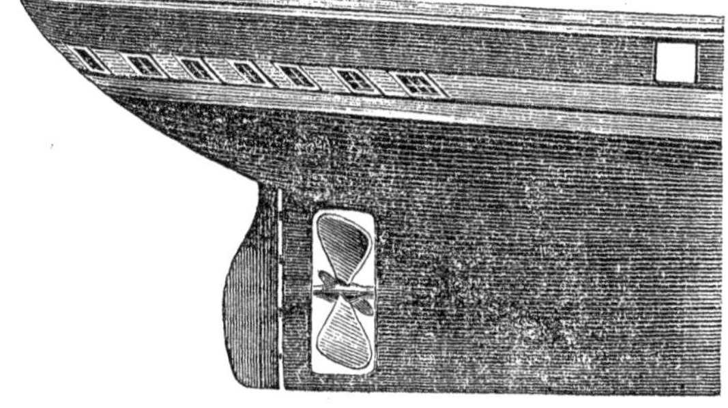

The manner in which the screw-propeller acts in impelling the vessel forward, may be understood by supposing the wheel to be an ordinary screw, and the water surrounding it a solid substance. By turning the screw in one direction or the other, it would move through the water, carrying the vessel with it, and the space through which it would move in each revolution would be equal to the distance between two contiguous threads of the screw. In fact, the water would act as a fixed nut, in which the screw would turn. But the water, although not fixed in its position as a solid nut, yet offers a considerable resistance to the motion of the screw-wheel; and as the wheel turns, driving the water backward, the reaction of the water gives a propulsion to the vessel in a contrary direction, or forward.

What is the great advantage of the screw-propeller over the paddle-wheel?

The great advantage of the screw-propeller is, that its action on the water will be the same, no matter to what degree it may be immersed in it, or how the position of the vessel on the surface of the water may be changed.

What is the simplest method of using water as a motive power?

350. The application of the force of water in motion for impelling machinery, is most extensive and familiar. The simplest method of applying this force as a mechanical agent, is by means of wheels, which are caused to revolve by the

weight, or pressure, of the water applied to their circumferences. These wheels are mounted upon shafts, or axles, which are in turn connected with the machinery to which motion is to be imparted.

Into how many classes are water-wheels divided?

351. The water-wheels at present most generally used may be divided into four classes—the UNDERSHOT, the OVERSHOT, the BREAST WHEEL, and the TOURBINE WHEEL.

Describe the construction of an Undershot Wheel.

352. The Undershot Wheel consists of a wheel, on the circumference of which are fixed a number of flat boards called "*float-boards*," at equal distances from each other. It is placed in such a position that its lower floats are immersed in a running stream, and is set in motion by the impact of the water on the boards as they successively dip into it. A wheel of this kind will revolve in any stream which furnishes a current of sufficient power. Fig. 145 represents the construction of the undershot wheel.

FIG. 145.

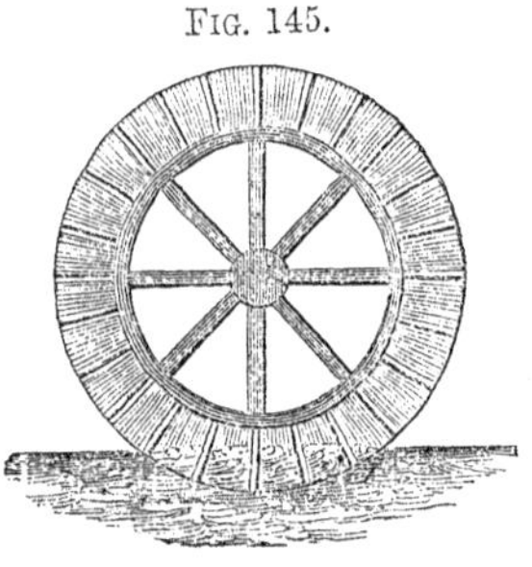

This form of wheel is usually placed in a "race-way," or narrow passage, in such a manner as to receive the full force of a current issuing from the bottom of a dam, and striking against the float-boards. And it is important to remember, that the moving power is the same, whether water falls downward from the top of a dam to a lower level, or whether it issues from an opening made directly at the lower level. This will be obvious, if it is considered that the force with which water issues from an opening made at any point in the dam will be equal to that which it would acquire in falling from the surface or level of the water in the dam down to the same point.

What proportion of power is lost by the undershot wheel?

The undershot wheel is a most disadvantageous method of applying the power of water, not more than 25 per cent. of the moving power of the water being rendered available by it.

Describe the construction of the Overshot Wheel.

353. In the Overshot Wheel, the water is received into cavities or cells, called "buckets," formed in the circumference of the wheel, and so shaped as to retain as much of the water as possible, until they arrive at the lowest part of the wheel, where they empty themselves. The buckets then ascend empty on the other side of the wheel to be filled as before. The wheel is moved by the weight of the water contained in the buckets on the descending side. Fig. 146 represents an overshot wheel.

FIG. 146.

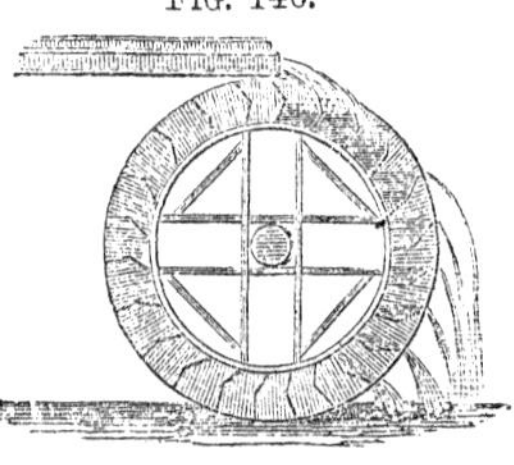

What proportion of the moving power is utilized by the overshot wheel?

The overshot wheel is one of the most effective varieties of water-wheels, and receives its name from the circumstance that the water shoots over it. It requires a fall in the stream, rather higher than its own diameter. Wheels of this kind, when well constructed, utilize nearly three fourths of the moving force of the water.

Describe the construction of the Breast-Wheel.

354. The Breast Wheel may be considered as a variety intermediate between the overshot and the undershot wheels. In this, the water, instead of falling on the wheel from above, or passing entirely beneath it, is delivered just below the level of the axis. The race-way, or passage for the water to descend upon the side of the wheel, is built in a circular form, to fit the circumference of the wheel, and the water thus inclosed acts partially by its weight, and partially by its impulse, or momentum. Fig. 147 represents a breast-wheel, with its circular race-way.

FIG. 147.

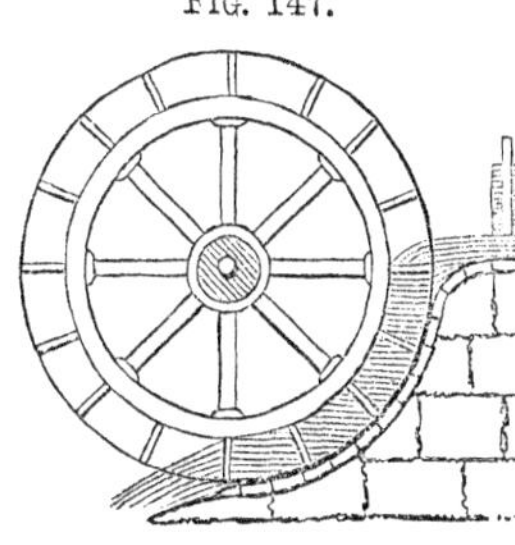

The breast-wheel, when well constructed, will utilize about 65 per cent. of the moving power of the water. It is more efficient than the undershot wheel, but less than the overshot. It is therefore only used where the fall happens to be particularly adapted for it.

355. The fourth class of water-wheels, the "Tourbine," or "Turbine," is a wheel of modern invention, and is the most powerful and economical of all water-engines.

FIG. 148.

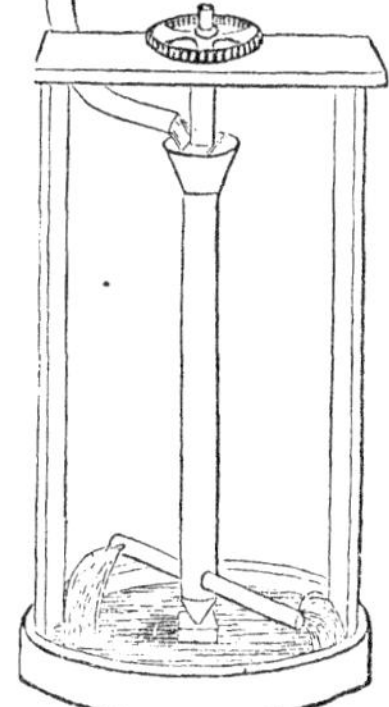

The principles of the construction and action of the Tourbine wheel may be best understood by a previous examination of the construction of another water-engine known as "Barker's Mill." (See Fig. 148.)

Describe the construction of Barker's Mill.

This consists of an upright tube or cylinder, furnished with a smaller cross-tube at the bottom, and enlarged into a funnel at the top. The whole cylinder is so supported upon pivots at the top and bottom, that it revolves freely about a vertical axis. It is evident if there are no openings in the ends of the cross-tubes, and the whole is filled with water, that the entire arrangement will be simply that of a close vessel filled with water, without any tendency to motion. If, however, the ends of the arms, or cross-tube, have openings on the sides, opposite to one another, as is represented in the figure, the sides of the tube on which the openings are, will be relieved from the pressure of the column of water in the upright tube by the water flowing out, while the pressure on the sides opposite to them, which have no openings, will remain the same. The machine, therefore, will revolve in the direction of the greater pressure, that is, in a direction contrary to that of the jets of water. A supply of water poured into the funnel-head, keeps the cylinder full, and the pressure of the column of water constant.

The action of this machine may also be explained according to another view: the pressure of the column of water in the upright tube, will cause the water to be projected in jets from the openings at the ends of the arms in opposite directions; when the recoil, or reaction of these jets upon the extremities of the cross-tubes, gives a rotary motion to the whole machine upon its vertical axis.

Describe the construction and action of the Tourbine Wheel.

The Tourbine wheel derives its motion, like the Barker's mill, from the action of the pressure of a column of water. It consists of a fixed, horizontal cylinder, A B, Fig. 149, in the center of which the water enters from an upright tube or cylinder, corresponding in position to the upright cylinder of a Barker's mill. The water descending in the tube diverges from the center in every direction, through curved water-channels, or compartments, A and B, formed in the horizontal cylinder, and escapes at the circumference. Around the fixed horizontal cylinder, a horizontal wheel, D, in the form of a ring or circle, is fitted, with its rim formed into compartments exactly similar to the compartments of the fixed cylinder, with the exception that their sides curve in an opposite direction. The water issuing from the guide-curves A B, strikes against the curved compartments of the wheel C B, and causes it to revolve. The wheel, by attachments beneath the fixed cylinder A B, is connected with a shaft, E, which passes up through the fixed and upright cylinder, and by which motion is imparted to machinery.

Fig. 149.

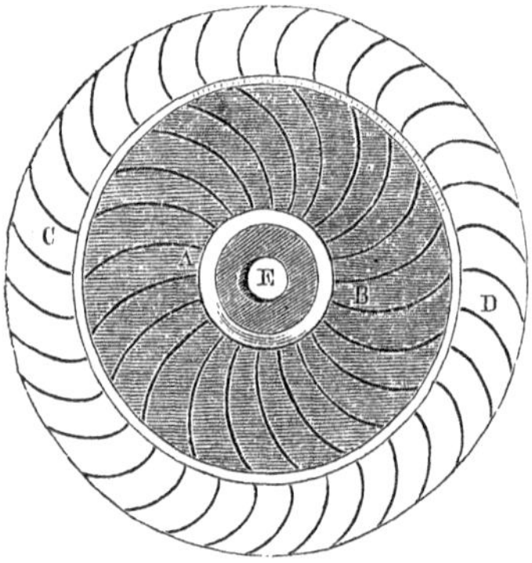

What is the efficiency of the Tourbine wheel?

The Tourbine wheel may be used to advantage with a fall of water of any height, and will utilize more of the force of the moving power than any other wheel—amounting, in some instances, as at the cotton factories at Lowell, Mass., to upward of 95 per cent. of the whole force of the water.

Is it possible to construct a water-wheel which will render the whole power available?

356. It may appear strange to those unacquainted with the action of hydraulic engines, that so much of the power existing in the agent we use for producing motion, as running water, should be lost, amounting in the undershot wheel to 75 per cent. of the whole power. This is due partially to the

friction of the water against the surfaces upon which it flows, and to the friction of the wheel which receives the force of the current. Force is also lost by changing the direction of the water in order to convey it to the machinery; in the sudden change of velocity which the water undergoes when it first strikes the wheels; and more than all, from the fact that a considerable amount of force is left unemployed in the water which escapes with a greater or less velocity from every variety of wheel. It may be considered as practically impossible to construct any form of water-engine which will utilize the whole force of a current of water.

357. Water, although one of the most abundant substances in nature, and a universal necessity of life, is not always found in the location in which it is desirable to use it. Mechanical arrangements, therefore, adapted to raise water from a lower to a higher level, have been among the earliest inventions of every country.

What were the earliest arrangements for raising water?

358. The application of the lever, in the form of the old-fashioned well-sweep (still used in many parts of this country, and throughout Eastern Asia), of the pulley and rope, and the wheel and axle in the form of the windlass, were undoubtedly the earliest mechanical contrivances for raising water.

Describe the Archimedes screw.

The screw of Archimedes, invented by the philosopher whose name it bears, is a contrivance for raising water, of great antiquity.

Fig. 150.

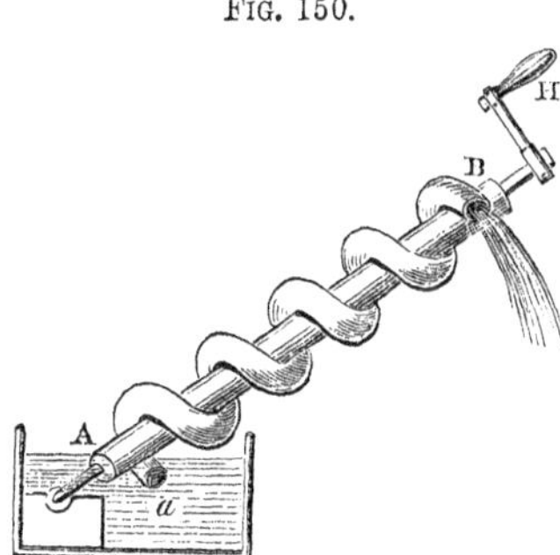

This machine, represented in Fig. 150, consists of a tube wound in a spiral form about a solid cylinder, A B, which is made to revolve by turning the handle H. This cylinder is placed at a certain inclination, with its lower extremity resting in the water. As the cylinder is made to revolve, the end of the tube dips into the water, and a certain portion enters the orifice *a*. By continuing the revolution of the cylinder, the water flows down a series of inclined planes, or to the under side of the tube, and if the inclination of the tube be not too great, the water will finally flow out at the upper orifice into a proper receptacle.

The following diagram, Fig. 151, representing the curved tube in two opposite positions, will illustrate the action of the Archimedes screw. Suppose a marble dropped into the tube at *a*, fig. 1,: if it was kept stationary in the

Fig. 151.

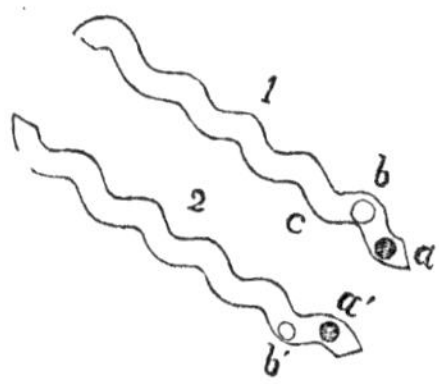

tube until it was turned half round, as in the position, fig. 2, the marble would be at a'; now, if at liberty to move, it would roll down to b'; but this effect, which we have supposed accomplished all at once, is really, gradually performed, and a rolls down toward b' by the gradual turning of the tube, and reaches b' as soon as the screw comes into the position marked in fig. 2; another half turn of the screw would bring it into its first position, and the marble would gradually roll forward to c.

When was the common pump invented?

359. The common suction-pump is a later discovery than the screw of Archimedes, and is supposed to have been invented by Ctesibius, an Athenian engineer who lived at Alexandria, in Egypt, about the middle of the second century before the Christian era.*

Describe the construction of the chain-pump.

360. The chain-pump consists of a tube, or cylinder, the lower part of which is immersed in a well or reservoir, and the upper part enters the bottom of a cistern into which the water is to be raised. An endless chain is carried round a wheel at the top, and is furnished at equal distances with flat discs, or plates, which fit tightly in the tube. As the wheel revolves, they successively enter the tube, and carry the water up before them, which is discharged into the cistern at the top of the tube. The machine may be set in motion by a crank attached to the upper wheel.

Fig. 152.

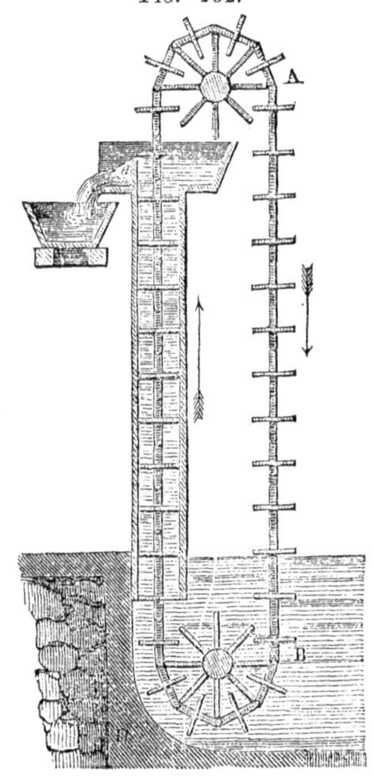

Fig. 152 represents the construction and arrangement of the chain-pump.

In what situations is this chain-pump generally used?

The chain-pump will act with its greatest effect, when the cylinder in which the plates and chain move, can be placed in an inclined position, instead of vertically. It is used generally on board of ships and in situations where the height through which the water is to be elevated is not very great, as in cases where the foundations of docks, etc., are to be drained.

* The suction-pump, and other machines for raising water which depend upon the pressure of the atmosphere, are described under the head of Pneumatics.

For what other purposes than raising water is the chain-pump used?

This machine is not, however, used exclusively for raising water. Its application, in principle, may be seen in any grist-mill, where it conveys the flour discharged from the stones, to an upper part of the building, where it is bolted. Dredging machines for elevating mud from the bottom of rivers, are also constructed on the same principle.

What is an Hydraulic Ram?

361. The HYDRAULIC RAM is a machine constructed to raise water by taking advantage of the impulse, or momentum, of a current of water suddenly stopped in its course, and made to act in another direction.

Describe the construction of the Hydraulic Ram.

The simplest construction of the hydraulic ram is represented in Fig. 153, and its operation is as follows:—At the end of a pipe, B, connected with a spring, or reservoir, A, somewhat elevated, from which a supply of water is derived, is a valve, E, of such weight as just to fall when the water is quiet, or still, within the pipe; this pipe is connected with an air-chamber, D, from which the main pipe, F, leads; this air-chamber is provided with a valve opening upward, as shown in the cut. Suppose now, the water being still within the tube, the valve E to open by its own weight; immediately the stream begins to run, and the water flowing through B soon acquires a momentum, or force, sufficient to raise the valve E up against its seat. The water, being thus suddenly arrested in its passage, would by its momentum burst the pipe, were it not for the other valve in the air-chamber, D, which is pressed upward, and allows the water to escape into the air-chamber, D. The air contained in the chamber D is condensed by the sudden influx of the water, but immediately reacting by means of its elasticity, forces a portion of the water up into the tube F.

FIG. 153.

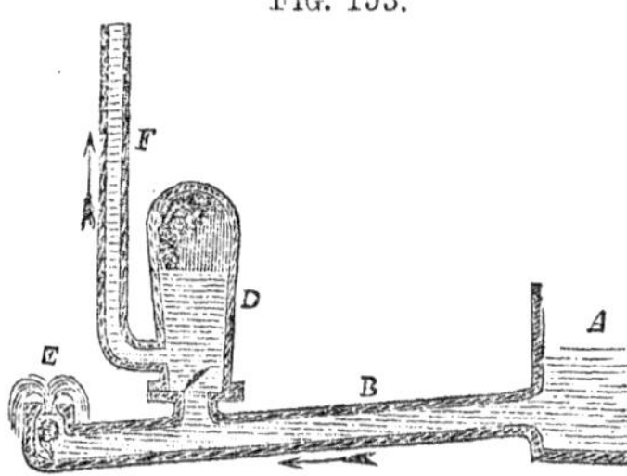

As soon as the water in the pipe B is brought to a state of rest, the valve of the air-chamber closes, and the valve E falls down or opens; again the stream commences running, and soon acquires sufficient force to shut the valve E; a new portion is then, by the momentum of the stream, urged into the air-chamber and up the pipe F; and by a continuance of this action, water will be continually elevated in the pipe F.

Fig. 154 represents a more improved construction of the ram, in which by the use of two air-chambers, C and F, the force of the machine is greatly increased. A represents the main pipe, B the valve from whence the water escapes, G the pipe in which it is elevated.

Fig. 154.

As this machine produces a kind of intermitting motion from the alternate flux and reflux of the stream, accompanied with a noise arising from the shock, its action has been compared to the butting of a ram; and hence the name of the machine.

It will be seen from these details, that a very insignificant pressing column of water, running in the supply pipe, is capable of forcing a stream of water to a very great height, so that a sufficient fall of water may be obtained in any running brook, by damming up its upper end to produce a reservoir, and then carrying the pipe down the channel of the stream until a sufficient fall is obtained. A considerable length of descending pipe is desirable to insure the action of the stream, otherwise the water, instead of entering the air-vessel, may be thrown back, when the valve is closed, into the reservoir.

CHAPTER X.

PNEUMATICS.

What is the science of Pneumatics?

362. PNEUMATICS is that department of physical science which treats of the motion and pressure of air,* and other aeriform, or gaseous substances.

Into what two classes may all aeriform substances be divided?

363. Aeriform, or gaseous bodies, may be divided into two classes, viz., the permanent gases, or those which under all ordinary circumstances of temperature and pressure are always in the gaseous state, as common air; and the vapors, which may readily be condensed by pressure, or the diminution of temperature, into liquids, as steam, or the vapor of water.

364. Atmospheric air is taken as the type, or representative, of all permanent gases, and steam as the type of all vapors, because these substances possess the general properties of gases and vapors in the utmost perfection.

What is the atmosphere?

365. The atmosphere is a thin, transparent fluid, or aeriform substance, surrounding the earth to a considerable height above its surface, and which by its peculiar constitution supports and nourishes all forms of animal or vegetable life.

* Atmospheric air is composed of oxygen and nitrogen mixed together in the proportion of seventy-nine parts of nitrogen and twenty-one of oxygen, or about four-fifths nitrogen to one-fifth oxygen. These two gases existing in the atmosphere are not chemically combined with each other, but merely mixed.

Beside these two ingredients there is always in the air, at all places, carbonic acid gas and watery vapor, in variable proportions, and sometimes also the odoriferous matter of flowers, and other volatile substances.

The air in all regions of the earth, and at all elevations, never varies in composition, so far as regards the proportions of oxygen and nitrogen which it contains, no matter whether it be collected on the top of high mountains, over marshes, or over deserts.

It is a wonderful principle, or law of nature, that when two gases of different weights, or specific gravities, are mixed together, they can not remain separate, as fluids of different densities do, but diffuse themselves uniformly throughout the whole space which both occupy. It is, therefore, by this law that a vapor, arising by its own elasticity from a volatile substance, is caused to extend its influence and mingle with the surrounding atmosphere, until its effects become so enfeebled by dilution as to be imperceptible to the senses. Thus we are enabled to enjoy and perceive at a distance the odor of a flower-garden, or a perfume which has been exposed in an apartment.

Is the atmosphere visible? The atmosphere is not, as is generally regarded, invisible. When seen through a great extent, as when we look upward in the sky on a clear day, the vault appears of an azure, or deep blue color. Distant mountains also appear blue. In both these instances the color is due to the great mass of air through which we direct our vision.

Why does not a small quantity of air exhibit color? The reason that we do not observe this color in a small quantity of air is, that the portion of colored light reflected to the eye by a limited quantity is insufficient to produce the requisite sensation upon the eye, and in this way excite in the mind a perception of the color. Almost all slightly transparent bodies are examples of this fact.

If a glass tube of small bore be filled with sherry wine, or wine of a similar color, and looked at through the tube, it will be found to have all the appearance of water, and be colorless. If viewed from above, downward, in the direction of its length, it will be found to possess its original color. In the first instance, there can be no doubt that the wine has the same color as the liquid of which it originally formed a part; but in the case of small quantities, the color is transmitted to the eye so faintly, as to be inadequate to produce perception. For the same reason, the great mass of the ocean appears green, while a small quantity of the same water contained in a glass is perfectly colorless.

Does air possess all the essential qualities of matter? 366. Air, in common with other material substances, possesses all the essential qualities of matter, as impenetrability, inertia, and weight.

What are proofs of the impenetrability of air? 367. The impenetrability of air may be shown by taking a hollow vessel, as a glass tumbler, and immersing it in water with its mouth downward; it will be found that the water will not fill the tumbler. If a cork is placed upon the water under the mouth of the tumbler, it will be seen that as the tumbler is pressed down, the air in it will depress the surface of the water on which the cork floats. The diving-bell is constructed on the same principle.

What are proofs of the inertia of air? 368. The inertia of the air is shown by the resistance which it opposes to the motion of a body passing through it. Thus, if we open an umbrella, and endeavor to carry it rapidly with the concave side forward, a considerable force will be required to overcome the resistance it encounters. A bird could not fly in a space devoid of air, even if it could exist without respiration, since it is the inertia, or resistance of the particles of the atmosphere to the beating of the wings, which enables it to rise. The wings of birds are larger, in proportion to their bodies, than the fins of fishes, because the fluid on which they act is less dense, and has proportionally less inertia, than the water upon which the fins of fishes act.

To what extent is air compressible? 369. Air is highly compressible and perfectly elastic.

By these two qualities air and all other gaseous substances

are particularly distinguished from liquids, which resist compression, and possess but a small degree of elasticity. Illustrations of the compressibility of air are most familiar. A quantity of air contained in a bladder, or India-rubber bag, may be easily forced by the pressure of the hand, to occupy less space. There is, indeed, no theoretical limit to the compression of air, for with every additional degree of force, an additional degree of compression may be obtained.

Does air possess any constant size or volume?

The elasticity, or expansibility of air, also manifests itself in an unlimited degree. Air cannot be said to have any original size or volume, for it always strives to occupy a larger space.

What are illustrations of the expansibility of air?

When a part of the air inclosed in any vessel is withdrawn, that which remains, expanding by its elastic property, always fills the dimensions of the vessel as completely as before. If nine tenths were withdrawn, the remaining one tenth would occupy the same space that the whole did formerly.

This tendency of air to occupy a larger space, or in other words, to increase its volume, causes it when confined in a vessel, to continually press against the inner surface. If no corresponding pressure acts from the outer surface, the air will burst it, unless the vessel is of considerable strength. This fact may be shown by the experiment of placing a bladder partially filled with air beneath the receiver of an air-pump, and by exhausting the air in the receiver the pressure of the external air upon the outer surface of the bladder is removed. The elasticity of the air contained in the bladder being then unresisted by any external pressure, will dilate the bladder to its fullest extent, and oftentimes burst it.

Has air weight?

369. Air, as well as all other gases and vapors, possesses weight.

The weight of air may be shown by first weighing a suitable vessel filled with air; then exhausting the air from it by means of an air-pump, and weighing again. The difference between the two weights will be the weight of the air contained in the vessel.

The weight of 100 cubic inches of air is about 31 grains.

To what is the elasticity of air due?

370. The elasticity, or expansion of air is due to the peculiar action of the molecular forces among its particles, which manifest themselves in a very different manner from what they do in solid and liquid bodies.

In solid bodies, these forces hold the molecules, or particles together so closely, that they can not change their respective positions; they also hold together the particles of liquid bodies, but to such a limited extent only, as to enable the particles to move upon each other with perfect freedom. But in gases, or æriform substances, the molecular forces act repulsively, and give to the particles a tendency to move away from each other; and this to so great an extent, that nothing but external impediments can hinder their further expansion.

What limits the atmosphere to the earth?

The question, therefore, naturally occurs in this connection, viz.: If air expands unlimitedly, when unrestricted, why does not our atmosphere leave the earth and diffuse itself throughout space indefinitely? This it would do were it not for the action of gravitation. The particles of air, it must be remembered, possess weight, and by gravity are attracted toward the center of the earth. This tendency of gravity to condense the air upon the earth's surface, is opposed by the mutual repulsion existing between the particles of air. These two forces counterbalance each other: the atmosphere will therefore expand, that is, its particles will separate from one another, until the repulsive force is diminished to such an extent as to render it equal to the weight of the particles, or what is the same thing, to the force of the attraction of gravitation, when no further expansion can take place. We may therefore conceive the particles of air at the upper surface of the atmosphere resting in equilibrium, under the influence of two opposite forces, viz., their own weight, tending to carry them downward, and the mutual repulsion of the particles, which constitutes the elasticity of air, tending to drive them upward.

What law regulates the density of the atmosphere?

371. The density of the air, or the quantity contained in a given bulk, decreases with the altitude, or height above the surface of the earth.

Fig. 155.

This is owing to the diminished pressure of the air, and the decreasing force of gravity. Those portions directly incumbent upon the earth are most dense, because they bear the weight of the superincumbent portions; thus, the hay at the lower part of the stack bears the weight of that above, and is therefore more compact and dense. (See Fig. 155.) This idea may be conveyed by the gradual shading of the figure, which indicates the gradual diminution in the density of the atmosphere in proportion to its altitude.

When is air said to be rarefied?

372. Air is said to be rarefied when it is caused to expand and occupy a greater space.

Generally, when we speak of rarefied air, we mean air that is expanded to a greater degree, or is thinner, than the air at the immediate surface of the earth.

373. The great law governing the compressibility of air, which is known from its discoverer as "Mariotte's Law," may be stated as follows:

What is Mariotte's Law?

The volume of space which air occupies is inversely as the pressure upon it.

If the compressing force be doubled, the air which is compressed will occupy one half of the space: if the compressing force be increased in a threefold proportion, it will occupy one third the space; if fourfold, one fourth the space, and so on.

The relation between the compressibility of air, and its elasticity and density, also obeys a certain law which may thus be expressed:—

What relation exists between the compressibility of air and its elasticity and density?

374. The density and elasticity of air are directly as the force of compression.

This relation is clearly exhibited by the following table:—

With the same amount of air, occupying the space of

$$1, \tfrac{1}{2}, \tfrac{1}{3}, \tfrac{1}{4}, \tfrac{1}{5}, \tfrac{1}{6}, \tfrac{1}{100},$$

the elasticity and density will be 1, 2, 3, 4, 5, 6, 100.

What are illustrations of the elastic force of air?

Hence by compressing air into a very small space, by means of a proper apparatus, we can increase its elastic force to such an extent as to apply it for the production of very powerful effects. The well-known toy, the pop-gun, is an example of the application of this power. The space A of a hollow cylinder, Fig. 156, is inclosed by the stopper, *p*, at one end, and by the end of the rod, S, at the other end. This rod being pushed further into the cylinder, the air contained in the space, A, is compressed until its elastic force becomes so great as to drive out the stopper, *p*, at the other end of the cylinder with great force,

Fig. 156.

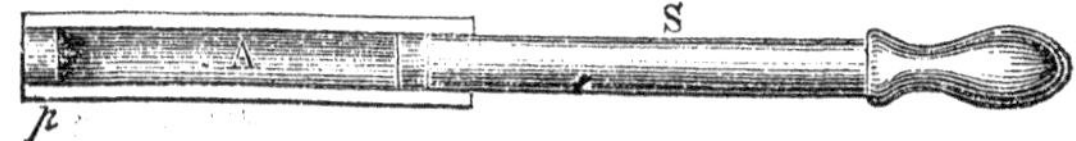

accompanied with a report. The air-gun is constructed and operated on a similar principle.

Prove and illustrate the laws of Mariotte.

375. The laws of Mariotte may be illustrated and proved by the following experiment: let A B C D be a long, bent glass tube, open at its longer extremity, and furnished with a stop-cock at the shorter. The stop-cock being open so as to allow free communication with the air, a quantity of mercury is poured into the open end. The surfaces of the mercury will, of course, stand at the same level, E F, in both legs of the tube, and will both sustain the weight of a column of air reaching from E and F to the top of the atmosphere. If we now close the stop-cock, D, the effect of the weight of the whole atmosphere above that point is cut off, so that the surface, F, can sustain no pressure arising from the weight of the atmosphere. Still, the level of the mercury in both legs of the tube remains the same, because the elasticity of the air inclosed in F D is precisely equal, and sufficient to balance the weight of the whole column

Fig. 157.

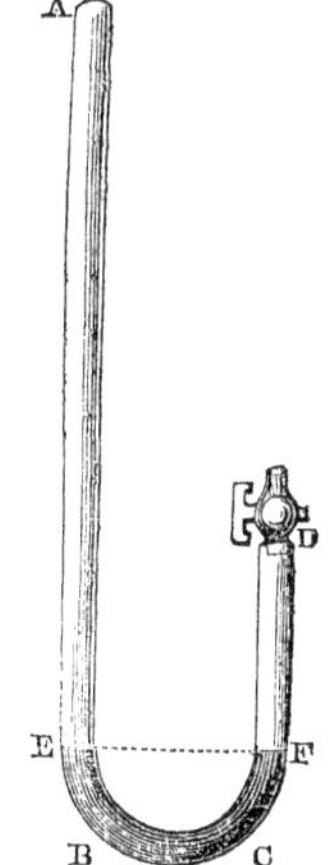

of atmosphere pressing upon the surface, E. If this were not the case, or if there were no air in F D, then the weight of the atmosphere pressing upon the surface E would force the mercury, E B C F, up into the space, F D. *The elasticity of air is, therefore, directly proportionate to the force, or compression, exerted upon it.*

It is evident that the pressure exerted upon the surface, E, Fig. 157, whatever may be its amount, is that of a column of air reaching from E to the top of the atmosphere, or, as we express it, the weight of one atmosphere. The amount of this pressure, accurately determined, is equal to the weight, or pressure, which a column of mercury 30 inches high would exert on the same surface. If then, we pour into the tube, A E, Fig. 157, as much mercury as will raise the surface in the leg A B 30 inches above the surface of the mercury in the leg D C, we shall have a pressure on the surface of E equal to two atmospheres; and since liquids transmit pressure equally in all directions, the same pressure will be exerted on the air included in the leg D F. This will reduce it in volume one half, or compress it into half the space, and the mercury will rise in the leg D F from F to F'. This weight of two atmospheres reduces a given quantity of air into one half its volume. In the same manner, if mercury be again poured into the tube A E until the surface of the column in A E is 60 inches above the level of the mercury in D F, then the air in D F will be compressed into one third of its original volume. In the same manner it could be shown, by continuing these experiments, that the diminution of the volume of air will always be in the exact proportion of the *increase* of the compressing force, and its volume can also be increased in exact proportion to the *diminution* of the compressing force. In fact this law has been verified by actual experiment, until the air has been condensed 27 times and rarefied 112 times.

Air has been allowed to expand into more than 2,000 times its bulk, and it would have expanded still more if greater space had been allowed. Air has also been compressed into less than a thousandth of its usual bulk, so as to become denser than water. In this state it still preserved its gaseous form and condition.

Was the weight of air known to the ancients?

376. The fact that air possesses weight, and consequently exerts pressure, was not known until about two hundred years ago. The ancient philosophers recognized the fact, that air was a substance, or a material thing, and they also noticed that when a solid, or a liquid, was removed, that the air rushed in and filled up the space that had been thus deserted. But when called to give a reason for this phenomenon, they said "that nature abhorred an empty space," or a "vacuum," and therefore filled it up with air, or some liquid, or solid body.

What is a vacuum?

377. A vacuum is a space devoid of matter; in general, we mean by a vacuum a space devoid of air.

No perfect vacuum can be produced artificially; but confined spaces can be deprived of air sufficiently for all experimental and practical purposes.

We do not know, moreover, that any vacuum exists in nature, although there is no conclusive evidence that the spaces between the planets are filled with any material substance.

If we dip a pail into a pond, and fill it with water, a hole (or vacuum) is made in the pond as big as the pail; but the moment the pail is drawn out, the hole is filled up by the water around it. In the same manner air rushes in, or rather is pressed in by its weight, to fill up an empty space.

How does water rise in a straw by suction?

When we place one end of a straw, or tube in the mouth, and the other end in a liquid, we can cause the liquid to rise in the straw, or tube by sucking it up, as it is called. We, however, do no such thing; we merely draw into the mouth the portion of air confined in the tube, and the pressure of the external air which is exerted on the surface of the liquid into which the tube dips, being no longer balanced by the elasticity of the air in the tube, forces the liquid up into the mouth. If, however, the straw were gradually increased in length, we should find that above a certain length we should not be able to raise water into the mouth at all, no matter how small the tube might be in diameter; or, in other words, if we made the tube 34 feet long, we should find that no power of suction, even by the most powerful machinery instead of the mouth, could raise the water to that height. The water rises in the common pump in the same way that it does in the straw; but not above a height of 33 or 34 feet above the level of the reservoir.

How was the ascent of water in tubes by suction first explained and demonstrated?

378. The reason why water thus rises in a straw, or pump, remained a mystery until explained and demonstrated by Torricelli, a pupil of Galileo. It is clear that the water is sustained in the tube by some force, and Torricelli argued that whatever it might be, the weight of the column of water sustained must be the measure of the power thus manifested; consequently, if another liquid be used, heavier or lighter, bulk for bulk, than water, *then the same force must sustain a lesser or greater column of such liquid.* By using a much heavier liquid, the column sustained would necessarily be much shorter, and the experiment in every way more manageable.

Torricelli verified his conclusions in the following manner:—He selected for his experiment mercury, the heaviest known liquid. As this is 13½ times heavier than water, bulk for bulk, it followed that if the force imputed to a vacuum could sustain 33 feet of water, it would necessarily sustain 13½ times less, or about 30 inches of mercury. Torricelli therefore made the following experiment, which has since become memorable in the history of science:—

He procured a glass tube (Fig. 158) more than 30 inches long, open at one end, and closed at the other. Filling this tube with mercury, and applying his finger to the open end, so as to prevent its escape, he inverted it, plunging the end into mercury contained in a cistern. On removing the finger, he observed that the mercury in the tube fell, but did not fall altogether into the cistern; it only subsided until its surface was at a height of about 30 inches above the surface of the mercury in the cistern. The result was what Torricelli expected, and he soon perceived the true cause of the phenomenon. The weight of the atmosphere acting upon the surface of the mercury in the vessel, supports the liquid in the tube, this last being protected from the pressure of the atmosphere by the closed end of the tube.

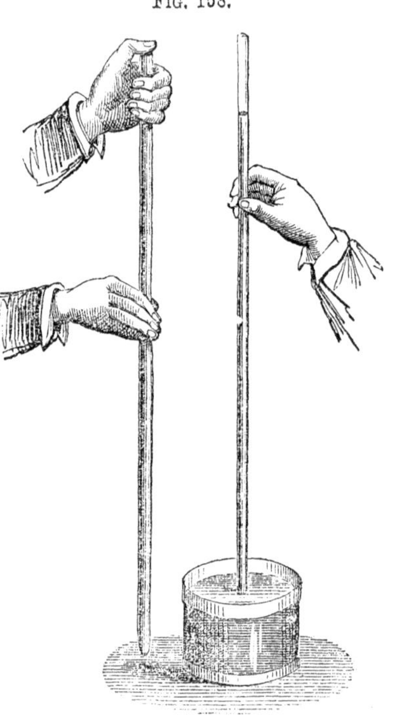

Fig. 158.

How was the conclusion of Torricelli further verified?

379. The fact that the column of mercury in the tube was sustained by the pressure of the atmosphere, was further verified by an experiment made by Pascal in France. He argued, that if the cause which sustained the column in the tube was the weight of the atmosphere acting on the external surface of the mercury in the cistern, then, if the tube was transported to the top of a high mountain, where a less quantity of atmosphere was above it, the pressure would be less, and the length of the column less. The experiment was tried by carrying the tube to the top of a mountain in the interior of France, and correctly noting the height of the column during the ascent. It was noticed that the height of the column gradually diminished as the elevation to which the instrument was carried increased.

The most simple way of proving that the column of mercury contained in the tube, as in Fig. 158, is only balanced against the equal weight of a column of air, is to take a tube of sufficient length, and having tied over one end a bladder, to fill it up with mercury, and invert it in a cup of the same liquid; the mercury will now stand at the height of about 30 inches; but if with a needle we make a hole in the bladder closing the top of the tube, the mercury in the tube immediately falls to the level of that in the cup.

How did the experiment of Torricelli lead to the invention of the Barometer?

These experiments by Torricelli led to the invention of the Barometer. It was noticed that a column of mercury sustained in a tube by the pressure of the atmosphere, the tube being kept in a fixed position, as in Fig. 159, fluctuated from day to day, within certain small limits. This effect was naturally attributed to the variation in the weight or pressure of the incumbent atmosphere, arising from various meteorological causes.

Fig. 159.

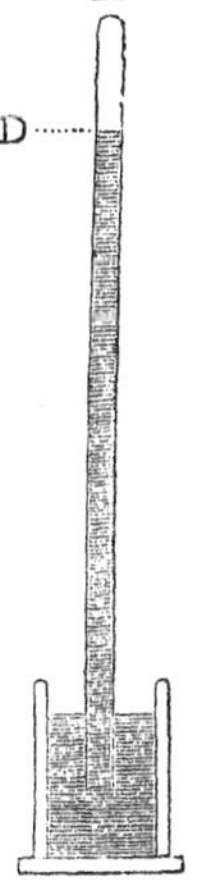

Thus, when the air is moist or filled with vapors, it is lighter than usual, and the column of mercury stands low in the tube; but when the air is dry and free from vapor, it is heavier, and supports a longer column of mercury.

Why should the presence of condensed vapor of water in the atmosphere affect its pressure?

So long as the vapor of water exists in the atmosphere, as a constituent part of it, it contributes to the atmospheric pressure, and thus a portion of the column of mercury in the barometer tube is sustained by the weight of the vapor; but when the vapor is condensed, and takes on a visible form, as clouds, etc., then it no longer forms a constituent part of the atmosphere, any more than dust, smoke, or a balloon floating in it does, and the atmospheric pressure being diminished, the mercury in the tube falls. In this way the barometer, by showing variations in the weight of the air, indicates also the changes in the weather.

What is the most perfect vacuum with which we are acquainted?

380. The space above the mercury in the barometer tube, A D, Fig. 159, is called the *Torricellian vacuum*, and is the nearest approach to a perfect vacuum that can be procured by art; for upon pressing the lower end deeper in the mercury, the whole tube becomes completely filled; the fluid again falling upon elevating the tube, it is therefore a perfect vacuum, with the exception of a small portion of mercurial vapor.

Fig. 160.

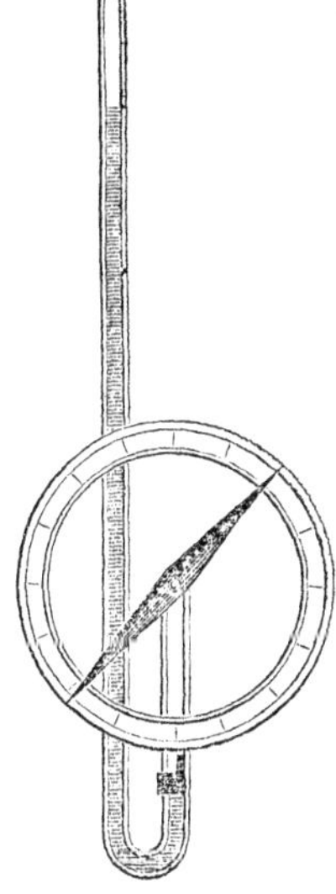

381. Barometers are constructed in very different forms—the principle remaining the same, of course, in all. The first barometer constructed was simply a tube closed at one end, filled with mercury, and inverted in a vessel containing mercury, as in Fig. 159.

What is the construction of the Wheel-Barometer?

A very common form of barometer, called the "Wheel-Barometer," consists of a glass tube, bent at the bottom, and filled with mercury. (See Fig. 160.) The column of mercury in the long arm of the tube is sustained by the pressure of the atmosphere upon the surface of the mercury in the shorter arm, the end of which is open. A small float of iron or glass rests upon the mercury in the shorter arm of the tube, and is suspended by a slender thread, which is passed round a wheel carrying an index, or pointer. As the level of the mercury is altered by a variation of the pressure of the atmosphere, the float resting upon the open surface, is raised or lowered in the tube, moving the index over a dial-plate, upon which the various changes of the weather are lettered.

Fig. 161.

Fig. 160 represents the internal structure of the wheel-barometer, and Fig. 161 its external appearance, or casing, with a thermometer attached.

Describe the Aneroid Barometer.

A very curious barometer, called the "*Aneroid Barometer,*" has been invented and brought into use within the last few years. Fig. 162 respresents its appearance and construction. Its action is dependent on the effect produced by atmospheric pressure on a metal box, from which the air has been exhausted. In the interior of the box is a circular spring of metal, fastened at one extremity to the sides of the box, and attached at the other extremity by a suitable arrangement to a pointer, which moves over a dial-plate, or scale. The interior of the box being deprived of air, the atmospheric pressure upon the external surfaces of the metal sides is very great, and as the pressure varies, these surfaces will be elevated and depressed to a slight degree. This motion is communicated to the spring in the interior, and from thence to the pointer, which, moving upon the dial, thus indicates the changes in the weather, or the variation in the pressure of the atmosphere.

Fig. 162.

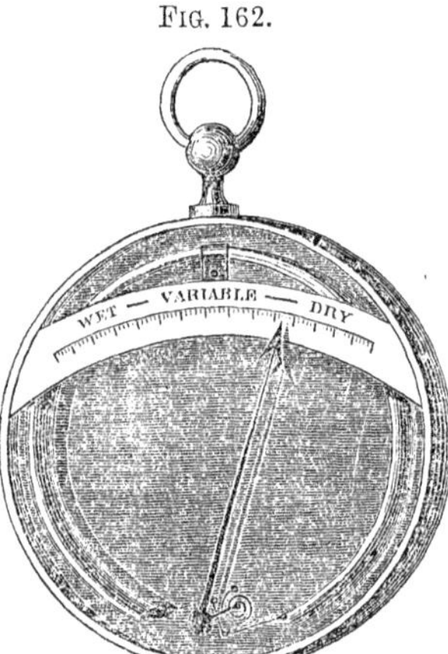

What are the peculiarities of the water-barometer?

Water, or some other liquid than mercury, may be used for filling the tube of a barometer. But as water is 13½ times lighter than mercury, the height of the column in the water-barometer supported by atmospheric pressure, will be 13½ times greater than that of mercury, or about 34 feet high; and a change which would produce a variation of a tenth of an inch in a column of mercury, would produce a variation of an inch and a third in the column of water. The water-barometer is rarely used, for various reasons, one of which is, that a barometer 34 feet high is unwieldy and difficult to transport.

What is the value of the barometer as a weather indicator?

382. The ordinary use of the barometer on land as a weather indicator is extremely limited and uncertain. It has been already stated that the weight of 100 cubic inches of air is about 30 grains. To obtain this result, it is necessary that the experiment should be performed at the level of the sea, and it is also requisite that the temperature of the air should be about 60° Fahrenheit's thermometer, and that the height of the column of mercury in the barometer tube should be 30 inches. As these conditions vary, the weight, or pressure of the atmosphere, and consequently the height of the mercury in the barometer tube must also vary. Especially will the height of the mercurial column vary with every change in the position of the instrument as regards its elevation above the level of the sea. A barometer at the base of a lofty tower will be higher at the same moment than one at the top of the tower, and consequently two such barometers would indicate different coming changes in the weather, though absolutely situated in the same place. No correct judgment, therefore, can be formed relative to the density of the atmosphere as affecting the state of the weather, without reference to the situation of the instrument at the time of making the observation. Consequently, no attention ought to be paid to the words "*fair, rain, changeable,*" etc., frequently engraved on the plate of a barometer, as they will be found no *certain* indication of the correspondence between the heights marked, and the state of the weather.

To what extent may the barometer be relied on for foretelling changes in the weather?

The barometer, however, may be generally relied on for furnishing an indication of the state of the weather to this extent;—that a fall of the mercury in the tube shows the approach of foul weather, or a storm; while a rise indicates the approach of fair weather.

At sea, the indications of the barometer respecting the weather, are generally considered, from various circumstances, more reliable than on land: the great hurricanes which frequent the tropics, are almost always indicated, some time before the storm occurs, by a rapid fall of the mercury.

How may the barometer be used for determining the height of mountains?

383. If a barometer be taken to a point elevated above the surface of the earth, the mercury in the tube will fall; because as we ascend above the level of the sea, the pressure of the atmosphere becomes less and less. In this way the barometer may be used to determine the heights of mountains, and tables have been prepared showing the degrees of elevation corresponding to the amount of depression in the column of mercury.

What is the supposed height of the atmosphere?

384. The absolute height to which the atmosphere extends above the surface of the earth is not certainly known. There are good reasons, however, for believing that its height does not exceed fifty miles.

This envelope of air is about as thick, in proportion to the whole globe, as the liquid layer adhering to an orange after it has been dipped in water, is to the entire mass of the orange. Of the whole bulk of the atmosphere, the zone, or layer which surrounds the earth to the height of nearly 2 3-4 miles from its surface, is supposed to contain one half. The remaining half being relieved of all superincumbent pressure, expands into another zone, or belt, of unknown thickness. Fig. 163 will convey an idea of the proportion which the highest mountains bear to the curvature of the earth, and the thickness of the atmosphere. The concentric lines divide the atmosphere into six layers, containing equal quantities of air, showing the great compression of the lower layers by the weight of those above them.

FIG. 163.

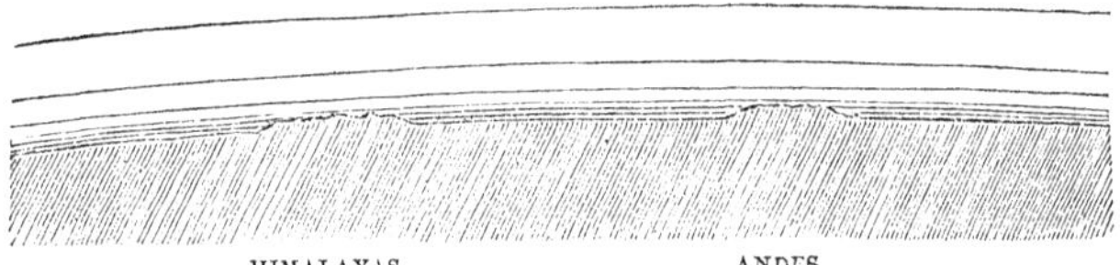

What is the comparative weight of the atmosphere?

Water is about 840 times the weight of air, taken bulk for bulk, and the weight of the whole atmosphere enveloping our globe has been estimated to be equal to the weight of a globe of lead sixty miles in diameter.

If the whole air were condensed, so as to occupy no more space than the same weight of water, it would extend above the earth to an elevation of thirty-four feet.

How is the pressure of aeriform substances exerted?

385. All aeriform, or gaseous substances, like liquids, transmit pressure in every direction equally; therefore, the atmosphere presses upward, downward, laterally, and obliquely, with the same force.

What is the amount of pressure exerted by the atmosphere?

386. The amount of pressure which the atmosphere exerts at the level of the ocean is equal to a force of 15 pounds for every square inch of surface.

What pressure is sustained by the human body?

The surface of a human body, of average size, measures about 2,000 square inches. Such a body, therefore, sustains a pressure from the atmosphere amounting to 30,000 pounds, or about 15 tons.

Why are we not crushed by the pressure of the atmosphere?

The reason we are not crushed beneath so enormous a load, is because the atmosphere presses equally in all directions, and our bodies are filled with liquids capable of sustaining pressure, or with air of the same density as the external air;

so that the external pressure is met and counterbalanced by the internal resistance.

If a man, or animal were at once relieved of all atmospheric pressure, all the blood and fluids of the body would be forced by expansion to the surface, and the vessels would burst.

What effect is experienced in rising to great elevations?

Persons who ascend to the summits of very high mountains, or who rise to a great elevation in a balloon, have experienced the most intense suffering from a diminution of the atmospheric pressure. The air contained in the vessels of the body, being relieved in a degree of the external pressure, expands, causing intense pain in the eyes and ears, and the minute veins of the body to swell and open. Travelers, in ascending the high mountains of South America, have noticed the blood to gush from the pores of the body, and the skin in many places to crack and burst.

What is the principle of "cupping?"

We become painfully sensible of the effect of withdrawing the external pressure of the atmosphere from a portion of the skin of the body in the operation of cupping. This is effected in the following manner: a vessel with an open mouth is connected with a pump, or apparatus for exhausting the air. The mouth of the vessel is applied in air-tight contact with the skin; and by working the pump a part of the air is withdrawn from the vessel, and consequently the skin within the vessel is relieved from its pressure. All other parts of the body being still subjected to the atmospheric pressure, and the elastic force of the fluids contained in the body having an equal degree of tension, that part of the skin which is thus relieved from the pressure swells out, and will have the appearance of being sucked into the cupping-glass.

If the lips be applied to the back of the hand, and the breath drawn in so as to produce a partial vacuum in the mouth, the skin will be drawn, or sucked in—not from any force resident in the lips or the mouth drawing the skin in, but from the fact that the usual external pressure of air is removed, and the pressure from within the skin is allowed to prevail.

Why do we often feel oppressed before a storm?

The sense of oppression and lassitude experienced in summer previous to a storm, is caused by a diminished pressure of the atmosphere. The external air, in such instances, becomes greatly rarefied by extreme heat and by the condensation of vapor, and the air inside us (seeking to become of the same rarity) produces an oppressive and suffocating feeling.

FIG. 164.

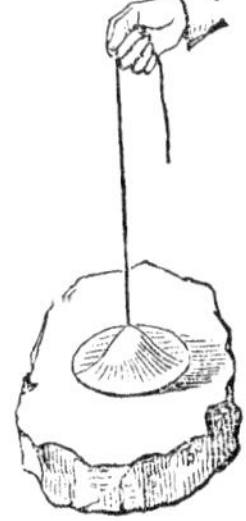

Describe the common sucker.

387. The direct effects of atmospheric pressure may be illustrated by many practical experiments. If a piece of moist leather, called a sucker, Fig. 164, be placed in close contact with any heavy body, such as a stone, or a piece of metal, it will adhere to it, and if a cord be attached to the leather, the stone, or metal, may be raised

by it. The effect of the sucker arises from the exclusion of the air between the leather and the surface of the stone. The weight of the atmosphere presses their surfaces together with a force amounting to 15 pounds on every square inch of the surface of contact. If the sucker could act with full effect, a disc an inch square would support a weight of 15 pounds; two square inches, 30 pounds, etc. The practical effect, however, of the sucker is much less.

Upon what principle are flies enabled to walk upon the ceiling, etc.?

388. The power of flies and other small insects to walk on ceilings, and surfaces presented downward, or upon smooth panes of glass, in opposition to the gravity of their bodies, is generally refered to a sucker-like action of the palms of their feet. Recent investigations have, however, proved, that the effect is rather due to the mechanical action of certain minute hairs growing upon the feet, which are tubular and excrete a sticky liquid.

Explain the principle and construction of the exhausting syringe and air-pump?

389. For the purpose of exhibiting the effects produced by the atmosphere in different conditions, and for various practical purposes, instruments have been contrived by which air may be removed from the interior of a vessel, or condensed into a small space to any extent, within certain limits. The first of these requirements may be obtained by the use of the instruments known as the exhausting syringe and the air-pump.

Fig. 165.

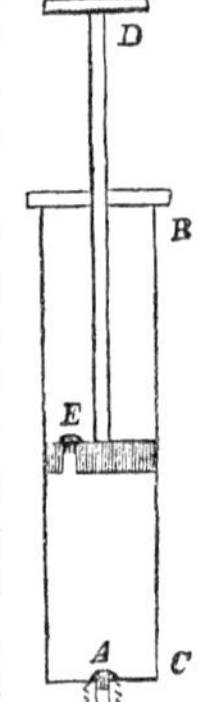

The exhausting syringe consists of a hollow cylinder, generally of metal, B C, Fig. 165, very truly and smoothly bored upon the inside, and having a piston moving in it air-tight. This cylinder communicates by a screw and pipe at the bottom, with any vessel, generally called a receiver, from which it is desirable to withdraw the air. The piston has a valve at E, opening upward, and at the bottom of the cylinder another valve precisely similar is placed, which also opens upward, shown at A. Suppose now the piston to be at the bottom of the cylinder and the receiver to be in proper connection—upon raising the piston by the handle, D, a vacuum is made in the cylinder; immediately the air in the receiver expands, passes through the valve A at the bottom of the cylinder, and fills its interior; upon depressing the piston, the valve E opening upward permits the air to pass through, and the valve A at the bottom of the cylinder closing, prevents it from passing back into the receiver. Upon again raising the piston, a further portion of air expanding from the receiver, enters the interior of the syringe, and upon depressing the piston, passes out through its valve. It is evident that this operation may be continued as long as the air within the receiver has elasticity sufficient to force open the valves.

The process of removing air from a vessel, or receiver, by means of the exhausting syringe is slow and tedious, and more powerful instruments, known as air-pumps, are generally employed for this purpose. The modern form of constructing the air-pump is represented by Fig. 166. The principle of its construction is the same as that of the exhausting syringe, the piston being worked by a lever or handle, as in the common pump, the valves opening and closing with great nicety and perfection.

Fig. 166.

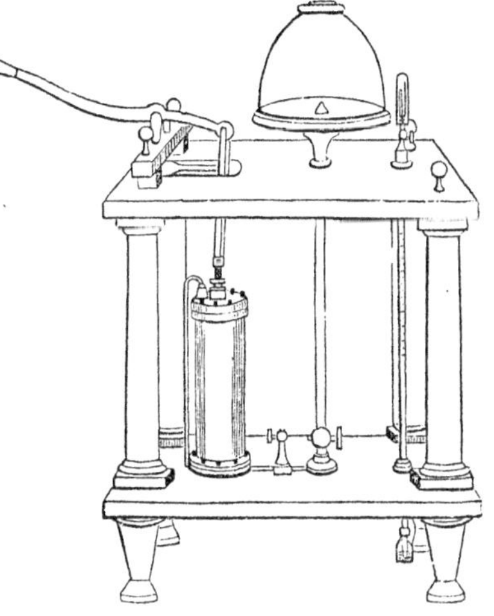

What is the construction of the condensing syringe?

381. When the density of the air is required to be increased, the condensing syringe, the converse of the exhausting syringe, is employed. It consists merely of an exhausting syringe, or air-pump, reversed, its valves being so arranged as to force air into a chamber, instead of drawing it out. For this purpose, the valves open inward in respect to the interior of the cylinder, while in the exhausting syringe and air-pump, they open outward.

What is an experimental proof of the crushing force of the atmosphere?

Fig. 167.

382. That the air in the inside of vessels is the force which resists and counterbalances the great pressure of the external atmosphere, may be proved by the following experiment: A strong glass vessel, Fig. 167, is provided, open both at top and bottom, and having a diameter of four or five inches. Upon one end is tied a bladder, so as to be completely air-tight, while the other end is placed upon the plate of an air-pump. Upon exhausting the air from beneath the bladder, it will be forced inward by the pressure of the air outside, and when the exhaustion has been carried to such an extent that the strength of the bladder is less than this pressure, it will burst with a loud report.

Fig. 168.

What is the experiment of the Magdeburg Hemispheres?

383. The air pump was invented, in the year 1654, by Otto Guericke, a German, and at a great public exhibition of its powers, made in the presence of the emperor of Germany, the celebrated experiment known as the "Magdeburg Hemispheres," was first shown. The Magdeburg Hemispheres, so called from the city where Guericke resided, consist of two hollow hemispheres of

brass, Fig. 168, which fit together air-tight. By exhausting the air in their interior, by means of the air-pump, and a stop-cock arrangement affixed to one of the hemispheres, it will be found that they can not be pulled apart without the exertion of a very great force, since they will be pressed together with a force of 15 pounds for every square inch of their surface. In the exhibition above referred to, given of these hemispheres by Guericke, the surfaces of a pair constructed by him were so large, that thirty horses, fifteen upon a side, were unable to pull them apart. By admitting the air again to their interior, the Magdeburg hemispheres fall apart by their own weight.

FIG. 169.

Another interesting example of atmospheric pressure is, to fill a wine-glass, or tumbler with water to the brim, and, having placed a card over the mouth, to invert it cautiously. If the card be kept in a horizontal position, the water will be supported in the glass by the pressure of the air against the surface of the card. (See Fig. 169.)

Describe the principle and construction of the gasometer.

384. In a like manner, if we take a jar, and having filled it with water, invert it in a reservoir or trough, as is represented in Fig. 170, it will continue to be completely filled with water, the liquid being sustained in it by the pressure of the atmosphere upon the water in the vessel. Such an arrangement enables the chemist to collect and preserve the various gases without admixture with air; for if a pipe or tube through which a gas is passing be depressed beneath the mouth of the jar, so that the bubbles may rise into it, they will displace the water, and be collected in the upper part of the jar, free of all admixture.

FIG. 170.

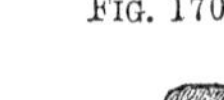

The gasometers, or large cylindrical vessels in which gas is collected in gas-works for general distribution, are constructed on this principle. They consist, as is shown in Fig. 171, of a large cylindrical reservoir suspended with its mouth downward, and plunged in a cistern of water of somewhat greater diameter. A pipe which leads from the gas-works is carried through the water, and turned upward, so as to enter the mouth of the gasometer. The gas, flowing through the pipe, rises into the gasometer, filling the upper part of it, and pressing down the water. Another pipe, descending from the gasometer through the water, is continued to the service pipes, which supply the gas. The gasometer is balanced by counter weights supported by chains, which pass over pulleys, and just such a preponderance is allowed to it as is sufficient to give the gas contained in it the compression necessary to drive it through the pipes to the remotest part of the district to be illuminated.

FIG. 171.

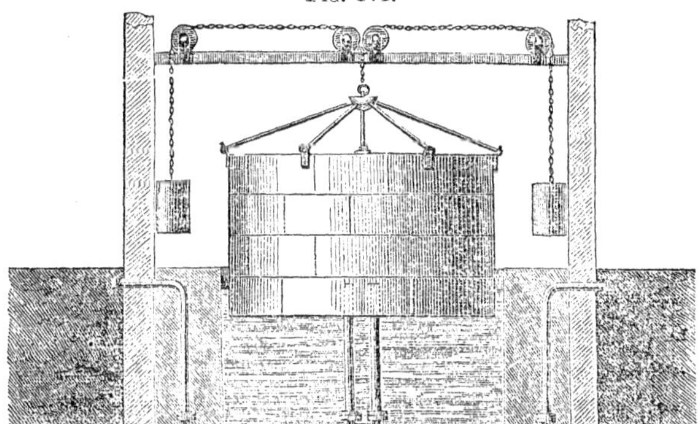

Why will not a liquid flow from a tight cask with only one opening?

385. A liquid will not flow continuously from a tight cask after it has been tapped or pierced, unless another opening is made as a vent-hole, in the upper part of the cask. The cask being air-tight, with the exception of a single opening the surface of the liquid in the vessel will be excluded from the atmospheric pressure, and it can only flow out in virtue of its own weight. But if the weight of the liquid be less than the force of the air pressing upon the mouth of the opening, the liquid can not flow from the cask; the moment, however, that the air is enabled to act through the vent-hole in the upper part of the cask, the pressure below is counterbalanced, and the liquid descends and runs freely through the opening by its own weight,.

If the lid of a tea-pot or kettle be air-tight, the liquid will not flow freely from the spout, on account of the atmospheric pressure. This is remedied by making a small hole in the lid, which allows the air to enter from without.

What is the principle and construction of the Pneumatic Ink-stand?

The Pneumatic Ink-stand, designed to prevent the ink from thickening, by the exposure of a small surface only to the air, is constructed upon the principles of atmospheric pressure. It consists of a close glass vessel, represented in Fig. 172, from the bottom of which a short tube proceeeds, the depth of which is sufficient for the immersion of the pen. By filling the ink-stand in an inclined position, we exclude the

FIG. 172.

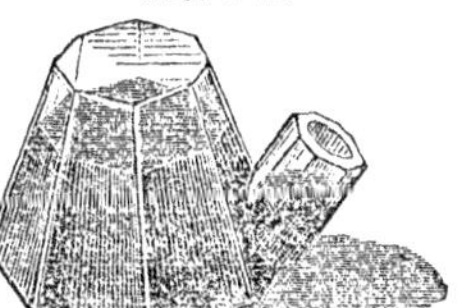

air in great part from the interior, and on replacing it in an upright position, the ink will be prevented from rising in the small tube and flowing over, on account of the atmospheric pressure upon the exposed surface of the ink in the small tube, which is much greater than the pressure of the column of liquid in the interior of the vessel. As the ink in the small tube is consumed by use, its surface will gradually fall; a small bubble of air will enter and rise to the top of the bottle, where it will exert an elastic pressure, which causes the surface of the ink in the short tube to rise a little higher, and this effect will be repeated until all the ink in the bottle has been used.

Why does a bottle gurgle when a liquid is poured freely out of it?

386. The peculiar gurgling noise produced when liquid is freely poured from a bottle, is produced by the pressure of the atmosphere forcing air into the interior of the bottle. In the first instance, the neck of the bottle is filled with liquid, so as to stop the admission of air. When a part has flowed out, and an empty space is formed within the bottle, the atmospheric pressure forces in a bubble of air through the liquid in the neck, which by rushing suddenly into the interior of the bottle, produces the sound. The bottle will continue to gurgle so long as the neck continues to be choked with liquid. But as the contents of the bottle are discharged, the liquid, in flowing out, only partially fills the neck; and, while a stream passes out through the lower half of the neck, a stream of air passes in through the upper part. The flow being now continued and uninterrupted, no sound takes place.

Does air exist in water?

387. Water, and most liquids exposed to the air, absorb a greater or less quantity of it, which is maintained in them by the pressure of the atmosphere acting on their surfaces.

Boiled water is flat and insipid, because the agency of heat expels the air which the water previously contained. Fishes and other marine animals could not live in water deprived of air.

How may the presence of air in water be shown?

The presence of air in water may be shown by placing a tumbler containing this liquid under the receiver of an air-pump, and exhausting the air. The pressure of the air being removed from the surface of the water, minute bubbles will make their appearance in the whole mass of the water, and rising to the surface, escape.

Why do some bottled liquids froth and sparkle?

The reason that certain bottled liquors froth and sparkle when uncorked and poured into an open vessel is, that when they are bottled, the air confined under the cork is condensed, and exerts upon the surface a pressure greater than that of the atmosphere. This has the effect of holding, in combination with the liquor, air or gas, which, under the atmospheric pressure only, would escape. If any air or gas rise from the liquor after being bottled, it causes a still greater condensation, and an increased pressure above its surface. When the cork is drawn from a bottle containing liquor of this kind, the air fixed in the liquid, being released from the pressure of the air which was condensed under the cork, instantly makes its escape, and rising in bubbles, produces effervescence and froth.

It sometimes happens that the united force of the air and gases thus confined in the bottle, becomes greater than the cohesive strength of the particles of matter composing the bottle; the sides of the bottle in such cases give way or burst.

Those liquors only froth which are viscid, glutinous, or thick, like ale, porter, etc., because they retain the little bubbles of air as they rise; while a thin liquor, like champagne, which suffers the bubbles to escape readily, sparkles.

How is the pressure of the atmosphere connected with the act of breathing?

388. The pressure of the atmosphere is connected with the action of breathing. The air enters the lungs, not because they draw it in, but by the weight of the atmosphere forcing it into the empty spaces formed by the expansion of the air-cells of the lungs. The air in turn escapes from the lungs by means of its elasticity; the lungs, by muscular action, compress the air contained in them, and give to it by compression a greater elasticity than the air without. By this excess of elasticity it is propelled, and escapes by the mouth and nose.

What is the proposed construction of the atmospheric telegraph?

389. It has been proposed to take advantage of the pressure of the atmosphere for the construction of an atmospheric telegraph, or apparatus for conveying the mails and other matter over great distances with great rapidity. The plan proposed is as follows;—a long metal tube is laid down, the interior surface of which is perfectly smooth and even. A piston is fitted to the tube in such a manner as to move freely in it and yet be air-tight. To one side of this piston the matter to be moved, made up in the form of a cylindrical bundle, is attached. A partial vacuum is then made in the tube before the piston, by means of large air-pumps, worked by steam-power, located at the further end of the tube, when the pressure of the atmosphere on the other side of the piston impels it forward through the whole length of the exhausted tube. It has been estimated that a piston, drawing after it a considerable weight of matter, could in this way be forced through a tube at the rate of 600 miles per hour.

390. The pressure of the atmosphere is taken advantage of in the construction of a great variety of machines for raising water; the most important and familiar of which is the common, or suction pump.

Describe the construction of the common pump.

The common, or suction pump, consists of a hollow cylinder, or barrel, open at both ends, in which is worked a movable piston, which fits the bore of the cylinder exactly, and is air-tight. The pump is further provided with two valves, one of which is placed in the piston, and moves with it, while the other is fixed in the lower part of the pump-barrel. These valves are termed *boxes.*

Fig. 173 represents the construction of the common pump. The body consists of a cylinder, or barrel, *b*, the lower part of which, called the suction-pipe, descends into the water which it is designed to raise. In the barrel works a piston containing a valve, *p*, opening upward. A similar valve, *g*, is fixed in the body of the pump, at the top of the suction-pipe. S is a spout from which the water raised by the working of the piston is discharged.

FIG. 173.

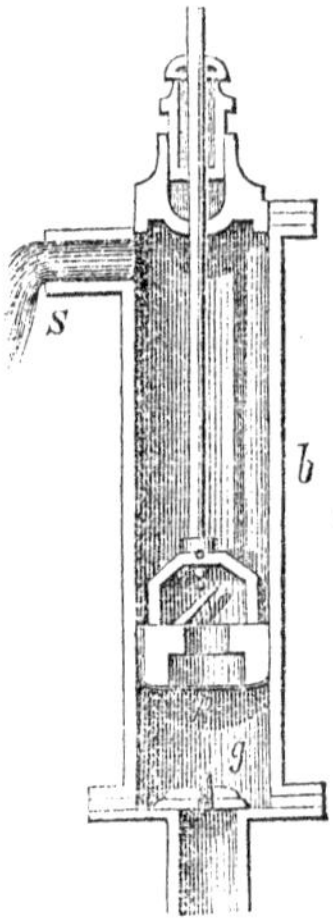

The operation of the pump in raising water is as follows;—when the piston is raised from the bottom of the cylinder, the air above it is drawn up, leaving a vacuum below the piston; the water in the well then rushes up through the valve *g*, and fills the cylinder; the piston is then forced down, shutting the valve, *g*, and causing the water to rise through the piston-valve, *p*; the piston is then raised, closing its valve, and raising the water above it, which flows out of the spout, S.

Why does water rise in a common pump?

391. Water rises in a pump simply and entirely by the pressure of the atmosphere (15 pounds on every square inch), which pushes it up into the void, or vacuum left by the up-drawn piston.

To what height will water rise in the common pump?

392. The common, or suction pump, can not raise water beyond the point of height at which the column of water in the pump tube is exactly balanced by the weight of the atmosphere. The utmost limit of this does not exceed 34 feet.

The height to which water is thus forced up in a pump is simply a question of balance; 15 pounds' pressure of the atmosphere can support only 15 pounds' weight of water; and a column of water, one inch square and 34 feet high, will weigh 15 pounds. As the pressure of the atmosphere is subject to variations, and as the mechanism of the pump is never absolutely perfect, the length of the pipe through which water is to be elevated ought never to exceed in practice 30 feet above the level of the water in the well, or reservoir.

What is a Valve?

393. A valve, in general, is a contrivance by which water or other fluid, flowing through a tube or aperture, is allowed free passage in one direction, but is stopped in the other. Its structure is such, that, while the pressure of fluid on one side has a tendency to close it, the pressure on the other side has a tendency to open it.

Figs. 174, 175, and 176, represent the various forms of valves used in pumps, water-engines, etc.

FIG. 174.

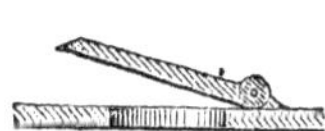

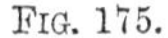

FIG. 175.

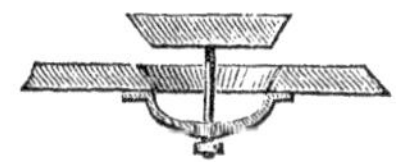

FIG. 176.

394. When it is desired to raise water to a greater height than 34 feet, a modification of the pump, called the forcing-pump, is employed.

What is a Forcing-Pump? The Forcing-Pump is an apparatus which raises water from a reservoir, on the principle of the suction-pump, and then, by the pressure of the piston on the water, elevates it to any required height.

FIG. 177.

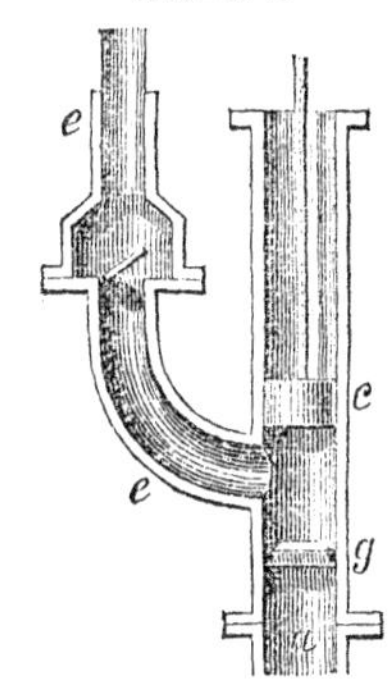

Fig. 177 represents the principle of the construction of the forcing-pump. There is no valve in the piston *c* (Fig. 177), but the water raised through the suction-pipe *a*, and the valve *g*, by the elevation of the piston, is forced by each depression of the piston up through the pipe *e e*, which is furnished with a valve to prevent the return of the liquid.

The forcing-pump, as constructed in Fig. 177, ejects the water only at each stroke of the piston, in the manner of a syringe. When it is desired to make the flow of the water continuous as in a fire-engine, an air, chamber is added to the force-pump, as is represented at A, Fig. 178. The water then, instead of immediately passing off through the discharging-pipe, partially fills the air vessel, and by the action of the piston in the pump, compresses the air contained in it. The elasticity of the air, thus compressed, being increased, it reacts upon the water, and forces its ascent in the discharge, or force-pipe. When the air in the chamber is condensed into half its original bulk, it will act upon the surface of the water with double the atmospheric pressure, while the water in the force-pipe, being subject to only one atmospheric pressure, there will be an unrestricted force, press-

FIG. 178.

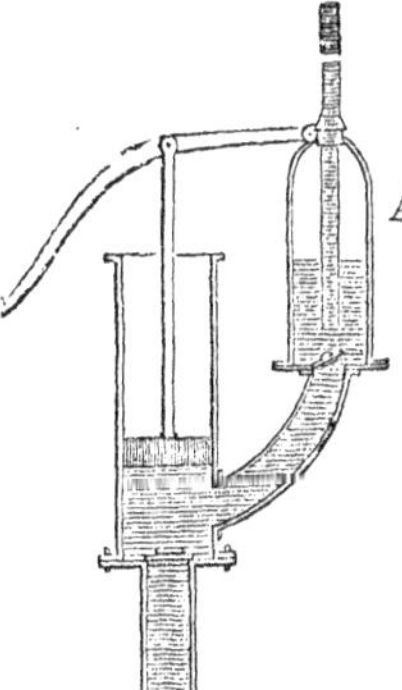

ing the water up, equal to one atmosphere: consequently, a column of water will be sustained, or projected to a height of 34 feet. When the air is condensed into one third of its bulk, its elastic force will be increased three-fold, and it will then not only counterbalance the ordinary atmospheric pressure, but will force the water upward with a pressure equal to two atmospheres, or 64 feet, and so on. The ordinary fire-engine is simply a convenient arrangement of two forcing-pumps, furnished with a strong air-chamber, and which are worked successively by the elevation and depression of two long levers called *brakes*.

What is a Syphon?

395. The Syphon is an apparatus by which a liquid can be transferred from one vessel to another without inverting, or otherwise disturbing the position of the vessel from which the liquid is to be removed.

In its simplest form, the syphon consists of a bent tube, A B C, Fig. 179, having one of its branches longer than the other. If we immerse the short arm in a vessel of water, and by applying the mouth to the long arm, as at C, exhaust the air in the tube, the water will be pressed over by atmospheric pressure, and continue to flow so long as the end of the lower arm is below the level of the water in the vessel.

FIG. 179.

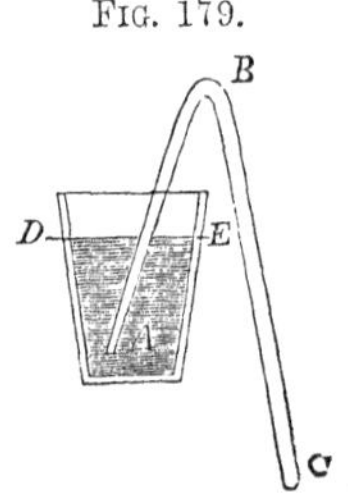

Upon what principle does the syphon act?

The action of the syphon is readily explained: the column of liquid in the longer arm, and that reaching in the shorter arm from the top of the curve or bend to the surface of the liquid in the vessel, have both a tendency to obey the attraction of gravity and fall out of the tube. This tendency is opposed, however, on both sides, by atmospheric pressure, acting on one side at the opening C, and upon the other upon the surface of the liquid in the vessel, thus preventing, in the interior of the tube, the formation of a vacuum, which would take place at the curve, if the two columns ran down on both sides. But the column on one side being longer than upon the other, the weight of the long column overbalances the short one, and determines the direction of the flow; and in proportion as the liquid escapes from the long arm, a fresh portion is forced into the short arm on the other side by the pressure of the air. The syphon is, therefore, kept full by the pressure of the atmosphere, and kept running by the irregularity of the lengths of the columns in its branches.

A suction-tube is sometimes attached to the syphon to make it more useful and efficient, as is represented in Fig. 180. By this means we may fill the whole syphon without the liquid entering the mouth, by sucking at the end of the suction-tube, and temporarily closing the end of the longer arm.

In order that the discharge of a liquid by means of the syphon should be perfectly constant, it is necessary that the difference of lengths of the columns of liquid in both branches should be immovable. This may be effected by connecting the syphon with a float and pulley, as is represented in Fig. 181.

Fig. 180.

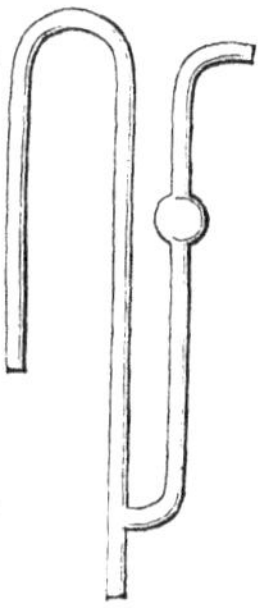

Fig. 181.

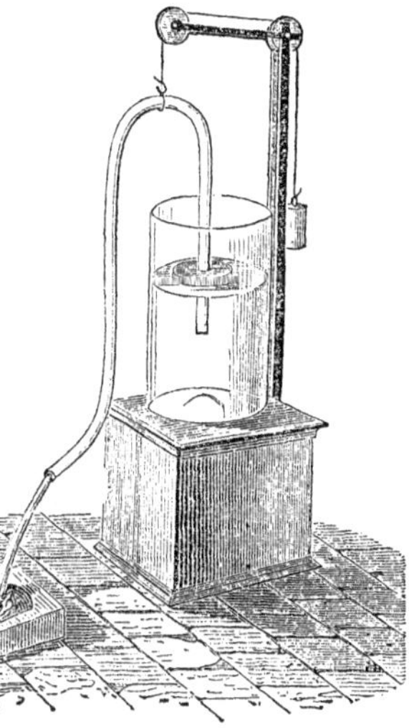

Explain the phenomenon of intermitting springs.

The curious phenomenon of intermitting springs may be explained upon the principle of the syphon. These springs run for a time and then stop altogether, and after a time run again, and then stop. If we suppose a reservoir in the interior of a hill or mountain, with a syphon-like channel running from it, as in Fig. 182, then as soon as the water collecting in the reservoir rises to the height shown by the dotted line, the stream will begin to flow, and continue flowing till the reservoir is nearly emptied. Again, after an interval long enough to fill the reservoir to the required height, it will again flow, and so on.

Fig. 182.

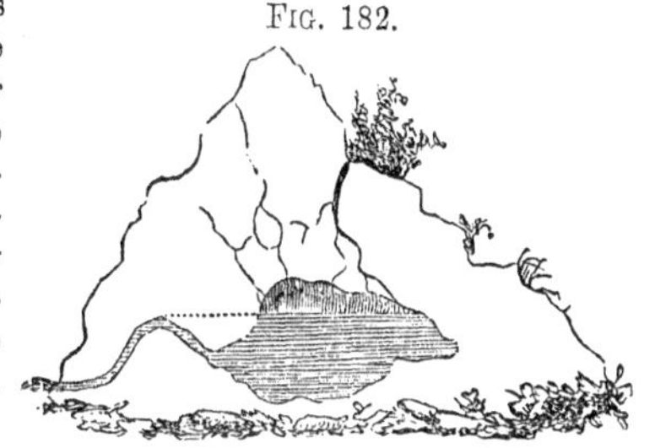

When will a body remain suspended in the air?

396. If a solid substance have the same density as atmospheric air, it will, when immersed in air, lose its entire weight, and will remain suspended in it in any position in which it may be placed.

When will a body rise in the air?

397. If a solid body, bulk for bulk, be lighter than atmospheric air, it is pressed upward by the surrounding particles of air, and rises, upon the same principle as a cork rises from the bottom of a vessel of water. (See § 85.)

At what point will an ascending body remain stationary?

As the density of the air continually diminishes as we ascend from the surface of the earth, it is evident that such a body, as it goes up, will finally attain a height where the air will have the same density as itself, and at such a point the body will remain stationary. Upon this principle clouds, at different times, float at different degrees of elevation.

It is also upon these principles that aerostation, or the art of navigating the air, depends.

What are Balloons?

398. Balloons are machines which ascend through the atmosphere, and float at a certain height, in virtue of being filled with a gas or air lighter than the same bulk of atmospheric air.

What are the two varieties of balloons?

Balloons are of two kinds. MONTGOLFIER, or rarefied air balloons, and HYDROGEN GAS balloons. The first are filled with common air rarefied by heat, and thus made lighter than the surrounding atmosphere; while the second are filled with hydrogen, a gas about fourteen times lighter than air.

Describe the Montgolfier, or rarefied air balloon.

The rarefied air-balloon was invented by Montgolfier, a French gentleman, in 1782, who first filled a paper bag with heated air, and allowed it to pass up a chimney. He afterward constructed balloons of silk, of a spherical shape, with an aperture formed in the lower surface. Beneath this opening a light wire basket was suspended, containing burning material. The hot air arising from the burning substances, enters the aperture, and rendering the balloon specifically lighter than the air, causes it to ascend with considerable velocity. Small balloons of a similar character are frequently made at the present day of paper, the air within them being rarefied by means of a sponge soaked in alcohol, suspended by a wire beneath the mouth, and ignited.

Describe the hydrogen gas balloon.

The hydrogen gas balloon consists of a light silken bag, filled either with hydrogen, or common illuminating gas. The difference between the specific weight of either of these gases and common air is so great, that a large balloon filled with them possesses ascensional power sufficient to rise to great heights, carrying with it considerable additional weight. The aeronaut can descend by allowing the gas to escape by means of a valve, thereby diminishing the bulk of the balloon. To enable him to rise again, ballast is provided, generally consisting of bags of sand, by throwing out which, the balloon is lightened, and accordingly rises.

By means of one of these machines Gay Lussac, an eminent French chemist, ascended in 1804, for the purpose of making meteorological observations, to the great height of 23,000 feet.

Do the laws of motion apply to air?

399. Air obeys the laws of motion which are common to all other material and ponderable substances.

How is the momentum of air calculated?

400. The momentum of air, or the amount of force which it is capable of exerting upon bodies opposed to it, is estimated in the same way as in the case of solids, viz., by multiplying its weight by its velocity.

What are illustrations of the momentum of air?

The momentum of air is usefully employed as a mechanical agent in imparting motion to wind-mills and to ships. Its most striking effects are seen in the force of wind, which occasionally, in hurricanes and tornadoes, acts with fearful power, prostrating trees and buildings. Such results are caused by the momentum of the air being greater than the force by which a building, or a tree is fastened to the earth.

What causes the rings of smoke observed in smoking and in the discharge of cannon?

401. Any force acting suddenly upon the air from a center, imparts to it a rotary movement. A very beautiful illustration of this is seen in the rings of smoke which are produced by the mouth of a skilful tobacco-smoker, and frequently also upon a much larger scale by the discharge of cannon, on a still day. In these cases a portion of air acted upon suddenly from a center is caused to rotate, and the particles of smoke render the motion visible. The whole circumference of each circle is in a state of rapid rotation, as is shown by the arrows in Fig. 183. The rapid rotation in short, confines the smoke within the narrow limits of a circle, and causes the rings to be well defined.

FIG. 183.

PRACTICAL PROBLEMS IN PNEUMATICS.

1. If 100 cubic inches of air weigh 31 grains, what will be the weight of one cubic foot?

2. If the pressure of the atmosphere be 15 pounds upon a square inch, what pressure will the body of an animal sustain, whose superficial surface is forty square feet?

3. When the elevation of the mercury in the barometer is 28 inches, what will be the height of a column of water supported by the pressure of the atmosphere?

Solution: Column of mercury supported by the atmosphere = 28 inches. Mercury being $13\frac{1}{2}$ times heavier than water, the column of water supported by the atmosphere = $13\frac{1}{2}\times 28=31$ feet.

4. When the elevation of the mercury in the barometer is 30 inches, what will be the height of a column of water supported by the atmosphere?

5. To what height may water be raised by a common pump, at a place where the barometer stands at 24 inches?

6. If a cubic inch of air weighs .30 of a grain, what weight of air will a vessel whose capacity is 60 cubic inches, contain?

CHAPTER XI.

ACOUSTICS.

What is the science of Acoustics?

402. Acoustics is that department of physical science which treats of the nature, phenomena, and laws of sound. It also includes the theory of musical concord or harmony.

What is Sound?

403. Sound is the sensation produced on the organs of hearing, when any sudden shock or impulse, causing vibrations, is given to the air, or any other body, which is in contact, directly or indirectly, with the ear.

Under what circumstances do vibratory movements arise?

404. When an elastic body is disturbed at any point, its particles execute a series of vibratory movements, and gradually return to a position of rest.

Thus when a glass tumbler is struck by a hard body, a tremulous agitation is transmitted to its entire mass, which movement gradually diminishes in force until it finally ceases. Such movements in matter are termed vibrations, and when communicated to the ear produce a sensation of sound.

The nature of these vibratory movements may be illustrated by noticing the visible motions which occur on striking or twitching a tightly extended cord, or wire. Suppose such a cord, represented by the central line in Fig. 184 to be forcibly drawn out to A, and let go; it would immediately recover its original position by virtue of its elasticity; but when it reached the central point, it would have acquired so much momentum as would cause it to pass onward to *a*; thence it would vibrate back in the same manner to B, and back again to *b*, the extent of its vibration being gradually diminished by the resistance of the air, so that it would at length return to a state of rest.

Fig. 184.

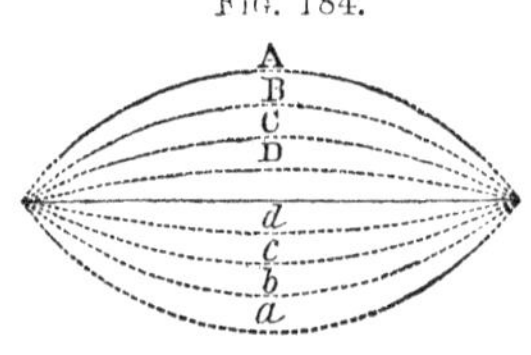

Describe the nature of a stationary vibration.

In vibratory movements of this kind all the separate particles come into motion at the same time, simultaneously pass the point of equilibrium, or rest, simultaneously reach the maximum of their vibration, and simultaneously begin their retrograde motion. Such vibrations are therefore called stationary, or fixed vibrations.

Describe the nature of a progressive vibration.

If, however, the motions of the vibrating body are of such a character that the agitation proceeds from one particle to another, so that each makes the same vibration, or oscillation, as the preceding one, with the sole exception of the motion beginning later, we have what is called progressive vibrations. Thus if we fasten a cord at one end, and move the other end up and down, a wave, or progressive vibration, is produced.

As the clearest conception can be formed of vibrations by comparing them to the waves produced by throwing a stone into smooth water, the term undulatory, or wave movement, has been adopted in general to express the phenomena of vibrations.

405. Daily experience teaches us that almost every motion of bodies in our vicinity is accompanied by a noise perceptible to our ears. All such sounds are the result of the vibrations of a portion of matter, and the nature of the tone, or sound, depends only on the manner in which these vibrations originate.

How may the sound-vibrations in solid bodies be rendered visible?

406. Sound-vibrations in solid bodies may be rendered visible by many simple contrivances. If we attach a ball by means of a string to a bell, and strike the bell, the ball will vibrate so long as the bell continues to sound. When a bell is sounding, also, the tremulous motion of its particles may be perceived by gently touching it with the finger. If the finger is pressed firmly against the bell, the sound is stopped, because the vibrations are interrupted. When sounds are produced by drawing the wet finger around the edge of a glass containing water, waves will be seen undulating from the sides toward the center of the glass.

FIG. 185.

When a tuning-fork is struck and made to sound, its vibrations are clearly visible, both branches alternately approaching and receding from each other, as is represented in Fig. 185.

If we strike a tuning-fork, and then touch the surface of mercury with one of its extremities, the surface of the mercury will exhibit little undulations or waves.

How are the so-called acoustic figures produced?

The most interesting method of exhibiting the character of sound is by means of the so-called "acoustic figures," which may be produced in the following manner:—Sprinkle some fine sand over a square or round piece of thin glass or metal, and holding the plate firmly by means of a pair of pincers, draw a violin bow down the edge; the sand is put in motion, and finally arranges itself along those parts of the surface which have the least vibratory motion. By changing the point by which the plate is held, or by varying the parts to which the violin bow is applied, the sand may be made to assume various interesting figures, as is represented in Fig. 186.

FIG. 186.

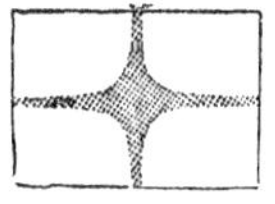

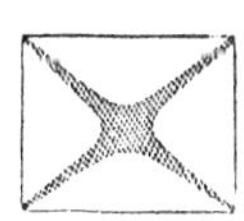

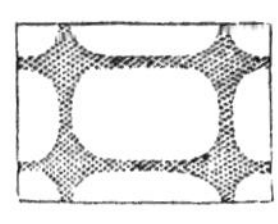

What is the usual medium through which sound is propagated?

407. Air is the usual medium through which sound is conveyed to the ear. The vibrating body imparts to the air in contact with it an undulatory, or wave-like movement, which, propagating itself in every direction, reaches the ear, and produces the sensation of sound.

What are sonorous bodies?

408. Vibrating bodies which are capable of thus imparting undulations to the air, are termed sounding, or sonorous bodies.

The aerial vibrations, or undulations thus caused, propagate themselves from the center of disturbance in concentric circles, in the same way that waves spread out upon the smooth surface of water. If such waves of water, propagated from a center, encounter any obstruction, as a floating body, they will bend their course round the sides of the obstacle, and spread out obliquely beyond it. So the undulations of air, if interrupted in their progress by a high wall or other similar impediment, will be continued over its summit and propagated on the opposite side of it.

In a sound-wave or undulation of the air, as in a wave of water, there is no permanent change of place among the particles, but simply an agitation, or tremor, communicating from one particle to another, so that each particle, like a pendulum which has been made to oscillate, recovers at length its original position.

This motion may be best illustrated by comparing it to the motion produced by the wind in a field of grain. The grassy waves travel visibly over the field in the direction in which the wind blows; but this appearance of an object moving is only delusive. The only real motion is that of the heads of the grain, each of which goes and returns as the stalk stoops or recovers itself. This motion affects successively a line of ears in the direction of the wind, and affects simultaneously all the ears of which the elevation or depression forms one visible wave. The elevations and depressions are propagated in a constant direction, while the parts with which the space is filled only vibrate to and fro. Of exactly such a nature is the propagation of sound through air.

Under what circumstances should we be unable to hear a sound?

409. If no substance intervenes between the vibrating body and the organs of hearing, no sensation of sound can be produced.

This is readily proved by placing a bell, rung by the action of clock-work, beneath the receiver of an air-pump, and exhausting the air. No sound will then be heard, although the striking of the tongue upon the bell, and the vibration of the bell itself, are visible. Now, if a little air be admitted into the receiver, a faint sound will begin to be heard, and this sound will become gradually louder in proportion as the air is gradually readmitted, until the air within the receiver is in the same condition as that without.

Sound, therefore, cannot be propagated through a vacuum.

"The loudest sound on earth, therefore, cannot penetrate beyond the limits of our atmosphere; and in the same manner, not the faintest sound can reach our earth from any of the other planets. Thus the most fearful explosions might take place in the moon, without our hearing anything of them."

How does the transmission of sound in air vary?

410. The power of air to transmit sound varies with its uniformity, its density, and its humidity.

Whatever tends to agitate or disturb the condition of the atmosphere, affects the transmission of sounds. When a strong wind blows from the hearer toward a sounding body, a sound often ceases to be heard which would be audible in a calm. Falling rain, or snow, interferes with the undulations of sound-waves, and obstructs the transmission of sound.

Why do we hear sounds more distinctly by night than by day?

The fact that we hear sounds with greater distinctness by night than by day, may be, in part, accounted for by the circumstance, that the different layers or strata of the atmosphere are less liable to variations in density and to currents, caused by changes of temperature, at night than by day. The air at night is also more still, from the suspension of business and hum of men. Many sounds become perceptible during the night, which during the day are completely stifled, before they reach the ear, by the din and discordant noises of labor, business, and pleasure.

Sound of any kind is transmitted to a greater distance in cold and clear weather than in warm weather, the density of air being increased by cold and diminished by heat.

What are illustrations of the variation of sound in air?

On the top of high mountains, where the air is greatly rarefied, the sound of the human voice can be heard for a short distance only; and on the top of Mont Blanc, the explosion of a pistol appears no louder than that of a small cracker. When persons descend to any considerable depth in a diving-bell, the air around them is compressed by the weight of a considerable column of water above them. In such circumstances, a whisper is almost as loud as a shout in the open air; and when one speaks with ordinary force it produces an effect so loud as to be painful.

Is air necessary for the production of sound?

411. Air is not necessary to the production of sound, although most sounds are transmitted by its vibrations. Sound can be produced under water, and all bodies are more or less fitted, not only to produce, but also to transmit sounds.

What substances communicate sound most readily?

412. Sound is communicated more rapidly and more distinctly through solid bodies than through either liquids or gases. It is trans-

mitted by water near four times more rapidly than by air, and by solids about twice as rapidly as by water.

If we strike two stones together under water, the sound will be as loud as if they had been struck in the air.

When a stick is held between the teeth at one extremity, and the other is placed in contact with a table, the scratch of a pin on the table may be heard with great distinctness, though both ears be stopped.

The earth often conducts sound, so as to render it sensible to the ear, when the air fails to do so. It is well known that the approach of a troop of horse can be heard at a distance by putting the ear to the ground, and savages practice this method of ascertaining the approach of persons from a great distance.

The principle that solids transmit sounds more perfectly than air, has been applied to the construction of an instrument called the "stethoscope."

Describe the Stethoscope.

The stethoscope consists of a hollow cylinder of wood, somewhat resembling in form a small trumpet. The wide mouth is applied firmly to the breast, and the other is held to the ear of the medical examiner, who is thus enabled to hear distinctly the action of the organs of respiration, and judge whether they are in a healthy condition, or the reverse.

How is the intensity of sound affected by distance?

413. Sound decreases in intensity from the center where it originates, according to the same law by which the attraction of gravitation varies, viz., inversely as the square of the distance. That is to say, at double the distance it is only one fourth part as strong; at three times the distance, one ninth, and so on.

This law applies with its full force only when no opposing currents of air, or other obstacles, interfere with the wave movements, or undulations. By confining the sound undulations in tubes, which prevent their spreading, the force of sound diminishes much less rapidly. It will, therefore, under such circumstances, extend to much greater distances. This principle is taken advantage of in the construction of speaking-trumpets.

Why are sounds heard more distinctly upon water than on land?

Sounds can generally be heard, especially on a calm day, at a greater distance upon water than upon land. The plane surface of water, as a smooth wall, prevents the lateral spreading and dispersion of the sound-waves, although on only one side. The air over water, owing to the presence of moisture, is also generally more dense, and the density more uniform than over the land. Water, in addition, is a better conductor of sound than the earth.

The transmission of sound from one apartment to another may be prevented by filling up the spaces between the partition walls with shavings, or any porous substances. The number of *media* through which the sound must pass is thus greatly increased, and every change of medium diminishes the strength of sound-waves.

What law governs the velocity of sound?

414. The velocity of the sound undulations is uniform, passing over equal intervals in equal times.

The softest whisper, therefore, flies as fast as the loudest thunder.

With what velocity does sound travel?

415. Sounds of every kind travel, when the temperature is at 62° Fahrenheit's thermometer, at a rate of 1,120 feet per second, or about 13 miles per minute, or 765 miles per hour. The velocity of sound increases or diminishes at the rate of 13 inches for every variation of a degree in temperature above or below the temperature of 62° Fahrenheit.

Why do we see the flash of a gun before we hear the report?

When a gun is fired at some distance, we see the flash a considerable time before we hear the report, for the reason that light travels much faster than sound. Light would go round the earth 480 times while sound was traveling 13 miles.

A knowledge of these circumstances is taken advantage of for the measurement of distances.

How may a knowledge of the velocity of sound be applied for the measurement of distances?

Thus, suppose a flash of lightning to be perceived, and on counting the seconds that elapse before the thunder is heard, we find them to amount to 20; then as sound moves 1,120 feet in a second, it will follow that the thunder-cloud must be distant $1,120 \times 20 = 22,400$ feet.

When a long column of soldiers are marching to a measure beaten on the drums which precede them, we may observe an undulatory motion transmitted from the drummers through the whole column, those in the rear stepping a little later than those which precede them. The reason of this is, that each rank steps, not when the sound is actually made, but when in its progress down the column at the rate of 1,120 feet in a second of time, it reaches their ears. Those who are near the music hear it first, while those at the end of the column must wait until it has traveled to their ears at the above rate.

Explain the phenomena of the interference of sound.

416. If two waves of water, advancing from opposite directions, meet in such a way that their points of elevation coincide, a wave of double the height of the single one will be formed at the point of interception; or if two wave depressions on the surface of water meet, a depression of double depth will be produced. If, however, the two waves come into contact in such a manner that an elevation of one wave coincides with the depression of another, both will be destroyed. Such a result is termed an interference of waves. In the same manner when two series of sound undulations, propagated from different sounding bodies, intersect each other, a like phenomena of interference is produced—the two undulations destroy each other, and silence is produced.

Fig. 187.

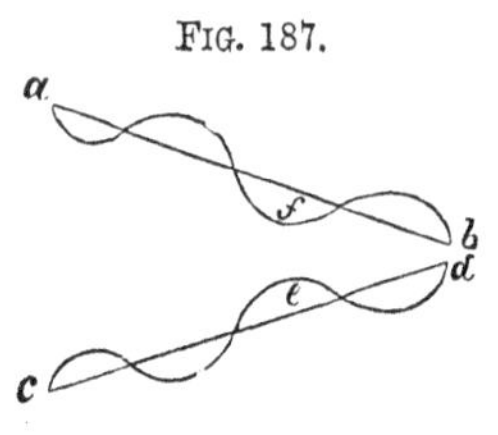

Let *a b* and *c d*, Fig. 187, represent two series of sound undulations, advancing in such a manner as to cause the elevation of one at *e* to correspond with the depression of the other at *f*; then if both are equal in intensity, they will neutralize each other, and an instant of silence will be produced. This fact may be very prettily illustrated by holding a common tuning-fork, after it has been put in vibration, over the mouth of a cylindrical glass vessel, as A, Fig. 188. The air contained within the vessel will assume sonorous vibrations, and a tone will be produced. If now a second glass cylinder be held in the position B, at right angles to A, the musical tone previously heard will cease; but if either cylinder be removed, the sound will be renewed again in the other. In this curious experiment, the silence arises from the interference of the two sounds.

Fig. 188.

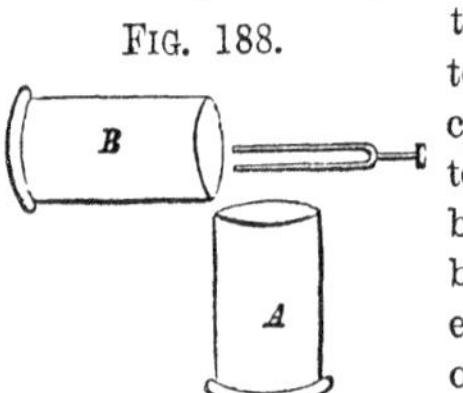

Another example of this phenomena may be produced by the tuning-fork alone. If this instrument, after being put into vibration, be held at a great distance from the ear, and slowly turned round its axis, a position of the two branches will be found at which the sound will become inaudible. This position will correspond to the points of interference of the two systems of undulations propagated from the two branches, or prongs of the fork.

Upon what does the loudness of a sound depend?

417. The loudness of a sound, or its degree of intensity, depends on the force with which the vibrations of a sounding body are made.

SECTION I.

MUSICAL SOUNDS.

What are musical sounds?

418. All vibrations of sonorous bodies which are uniform, regular, and sufficiently rapid, produce agreeable, or musical sounds.

What is the difference between a musical sound and a noise?

419. What constitutes the particular difference between a noise and a musical sound is not certainly known. A noise, however, is supposed to be occasioned by impulses communicated irregularly to the ear; but in a musical sound the vibrations of the sonorous body, and consequently the undulations of the air, must be all exactly similar in duration

and intensity, and must recur after exactly equal intervals of time.

What is meant by tone, or pitch in sound?

420. If the sound impulses be repeated at very short intervals, the ear is unable to attend to them individually, but hears them as a continued sound, which is uniform, or has what is called a tone or pitch, if the impulses be similar and at equal intervals.

What experiment illustrates the nature of a musical sound?

421. The nature of musical sounds, and indeed of all sounds, may be illustrated by the following experiment: If we take a thin elastic plate of metal, a few inches in length, firmly fixed at one end, and free at the other, and cause it to vibrate, it will be found to emit a clear, musical sound, having a certain tone.

If the plate be gradually lengthened, it yields tones, or notes, of different characters, until finally the vibrations become so slow that the eye can follow them without difficulty, and all sound ceases.

When is a tone grave or sharp?

422. When the impulses, or vibrations, are few in number in a given time, the tone is said to be grave; when they are many, the tone is said to be sharp. Musical sounds are spoken of as notes, or as high and low. Of two notes, the higher is that which arises from more rapid, and the lower from slower vibrations.

Beside this, sounds differ in their quality. The same musical note, produced with the same degree of loudness, and by the same number of vibrations in the flute, the clarionet, the piano, and the human voice, is in each instance peculiar and wholly different. Why this is we are unable to say. The French call this property, by which one sound is distinguished from another, the *timbre*.

Is there any limit to the number of vibrations requisite to produce sound?

To produce any sound whatever it is necessary that a certain number of vibrations should be made in a certain time. If the number produced in a second falls below a certain rate, no sound sensation will be made upon the ear. It is believed that the ear can distinguish a sound caused by fifteen vibrations in a second, and can also continue to hear though the number reaches 48,000 per second. Trained and sensitive ears are said to be able to exceed these limits.

When are two musical notes in unison?

423. Two musical notes are said to be in *unison* when the vibrations which cause them are performed in equal times.

What is an octave?

424. When one note makes twice the number of vibrations in a given time that another makes, it is said to be its octave. The relation, or interval which

exists between two sounds, is the proportion between their respective numbers of vibrations.

What is a chord, etc.?

425. A combination of harmonious sounds is termed a musical chord; a succession of harmonious notes, a melody; and a succession of chords, harmony.

A melody can be performed, or executed by a single voice; a harmony requires two or more voices at the same time.

Define concord and discord.

426. When two tones, or notes, sounded together produce an agreeable effect on the ear, their combination is called a musical concord; when the effect is disagreeable, it is called a discord.

Explain what is meant by the gamut, or scale of music.

427. Suppose we have a stretched string, as a wire or a piece of catgut, such as is used for stringed instruments: now the number of vibrations which such a cord will make in a given time, are inversely as its length; that is, if the whole cord makes a given number of vibrations in one second, as 100, on shortening it one half it will make twice as many, or 200, and this will yield a note exactly an octave higher than the former one. If we reduce its length three-fourths, it will make four times as many vibrations as at first, and yield a note two octaves higher.

Suppose the stretched string, or wire, to be 32 inches in length. When this is struck it will vibrate a certain number of times in a second, and give what is called a key-note. Reduce the string one half, and we have the octave of that note. But between the key-note and its octave there is a natural gradation by intervals in the pitch of the tone, which heard in succession are harmonious, the octave, as its name implies, being the eighth pitch of tone, or eighth successive note ascending from the key-note.

These eight notes, or intervals in the pitch of tone between the key-note and its octave, constitute what is called the gamut, or diatonic scale of music, because they are the steps by which the tone naturally ascends from any note to the corresponding tone above, produced by vibrations twice as rapid. These several notes are distinguished both by letters and names. They are:

C, D, E, F, G, A, B, C;

Or—do, re, me, fa, sol, la, si, do.

How are the notes of the scale indicated?

They may also be distinguished by numbers indicating the length of the strings and the number of vibrations required to produce them. Thus, the length of the string producing the primary, or key-note, being 32 inches, the lengths of the strings to produce the tones in the entire scale are—

32, 30, 27, 24, 21, 20, 18, 16;

or, supposing that whatever be the number of vibrations per second necessary to produce the first note in the scale, C, we agree to represent it by unity, or 1; then the numbers necessary to produce the other seven notes of the octave will be as follows:

Name of note	C,	D,	E,	F,	G,	A,	B,	C.
Number of vibrations .	1,	$\frac{9}{8}$,	$\frac{5}{4}$,	$\frac{4}{3}$,	$\frac{3}{2}$,	$\frac{5}{3}$,	$\frac{15}{8}$,	2.

However far this musical scale may be extended, it will still be found but a repetition of similar octaves. The vibrations of a column of air in a pipe may be regarded as obeying the same general laws; notes are naturally higher in proportion to the shortness of the pipes.

Is the same note in any instrument produced in the same manner?

The same note produced on any musical instrument is due to the same number of vibrations per second. Thus, a note produced by a string of a piano vibrating 256 times in a second, is also produced in the flute by a column of air vibrating the same number of times in a second, and also in the human voice by two chords contained in the upper part of the wind-pipe, also vibrating the same number of times in a second.

It has been already stated that the number of vibrations of a cord are inversely as its length; the number also increases as the square root of the force which stretches it. Thus an octave is given by the same length of string when stretched four times as strongly.

SECTION II.

REFLECTION OF SOUND.

What is meant by the reflection of sound?

428. When waves of sound strike against any fixed surface tolerably smooth, they are reflected, or rebound from that surface, and the angle of reflection is equal to the angle of incidence.

This law governing the reflection of sound is the same as that which governs the reflection of all elastic bodies, and also, as will be shown hereafter, the imponderable agents, heat and light.

What is an Echo?

429. An ECHO is a repetition of sound caused by the reflection of the sound waves, or undulations, from a surface fitted for the purpose, as the side of a house, a wall, hill, etc.; the sound, after its first production, returning to the ear at distinct intervals of time.

Thus if a body placed at a certain distance from a hearer produces a sound, this sound would be heard first by means of the sonorous undulations which produced it, proceeding directly and uninterruptedly from the sonorous body to the hearer, and afterward by sonorous undulations which, after striking on reflecting surfaces, return to the ear. These last constitute an echo.

In order to produce an echo, it is requisite that the reflecting body should be situated at such a distance from the source of sound, that the interval between the perception of the original and reflected sounds may be sufficient to prevent them from being blended together.

When the original and reflected sounds are blended together, the effect produced is called a resonance, and not an echo.

Thus, the walls of a room of ordinary size do not produce an echo, because the reflecting surface is so near the source of sound that the echo is blended with the original sound; and the two produce but one impression on the ear.

Large halls, spacious churches, etc., on the contrary, often reverberate or repeat the voice of a speaker, because the walls are so far off from the speaker, that the echo does not get back in time to blend with the original sound; and therefore each is heard separately.

The shortest interval sufficient to render sounds distinctly appreciable by the ear, is about 1-9th of a second; therefore when sounds follow at shorter intervals, they will form a resonance instead of an echo; so that no reflecting surface will produce a distinct echo, unless its distance from the spot where the sound proceeds is at least 62½ feet; as the sound will in its progress in passing to and from the reflecting surface, at the rate of 1,125 feet per second, occupy 1-9th part of a second, passing over $62\frac{1}{2} \times 2 = 125$ feet.

When is an echo multiplied?

430. Where separate surfaces are so situated that they receive and reflect the sound from one to the other in succession, multiplied echoes are heard.

Fig. 189.

An echo in a building near Milan, Italy, repeats a loud sound 30 times audibly. A river bounded by perpendicular walls of rock, where the sound is reflected backward and forward over the surface of still water, is a favorable situation for the production of repeated echoes. Fig. 189 represents the manner in which the sounds rebound, in such situations, as at 1, 2, 3, 4, from side to side.

What conditions of surface are requisite to produce a perfect echo?

It is not necessary that the surface producing an echo should be either hard or polished. It is often observed at sea that an echo proceeds from the surface of clouds. An echo at sea, however, or on an extensive plane, is heard but rarely, there being no surfaces to reflect sound. To insure a perfect echo, the reflecting surface must be tolerably smooth, and of some regular form. An irregular surface must break the echo; and if the irregularity be very considerable, there can be no distinct or audible reflection at all. For this reason an echo is much less perfect from the front of a house which has windows and doors, than from the plane end, or any plane wall of the same magnitude.

How is sound reflected from curved surfaces?

431. If the surface upon which the sound-waves strike be concave, it may concentrate sound, and reflect all that falls upon it to a point at some distance from the surface, called the focus.

FIG. 190.

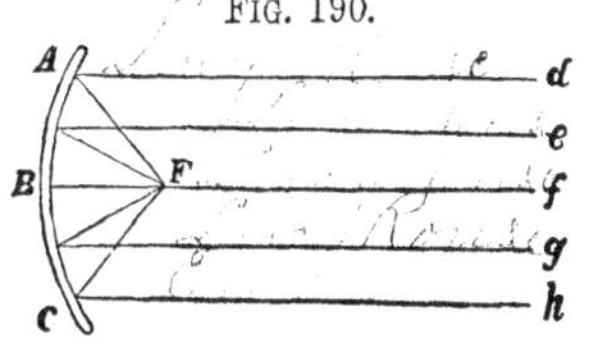

Thus, in Fig. 190, if the sound waves proceeding in right lines from the points *d*, *e*, *f*, *g*, *h*, strike upon the concave surface, A B C, they will all be reflected to the focus, F, and there concentrated in such a way as to produce a most powerful effect.

It is upon this principle that whispering galleries are constructed, and domes and vaulted ceilings often exhibit the same curious phenomena. In these instances a whisper uttered at one point is reflected from the curved surface to a focus at a distant point, at which situation it may be distinctly heard, while in all other positions it will be inaudible.

What occasions the noise heard in a sea-shell?

All are familiar with the resonance produced by placing a sea-shell to the ear—an effect which fancy has likened to the "roar of the sea." This is caused by the hollow form of the shell and its polished surface enabling it to receive and return the beatings of all sounds that chance to be trembling in the air around the shell.

432. Speaking-tubes and speaking-trumpets depend on the principles of the reflection of sound.

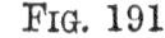
FIG. 191.

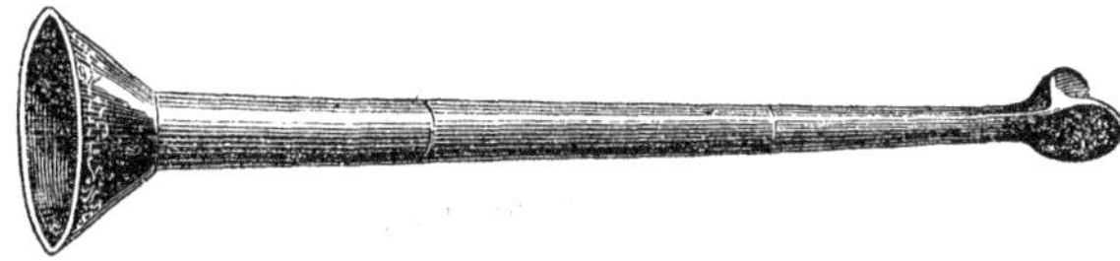

What is a Speaking-Trumpet?

433. A SPEAKING-TRUMPET (Fig. 191) is a hollow tube so constructed that the rays of sound (proceeding from the mouth when applied to it), instead of diverging, and being scattered through the surrounding atmosphere, are reflected from the sides, and conducted forward in straight lines, thus giving great additional strength to the voice.

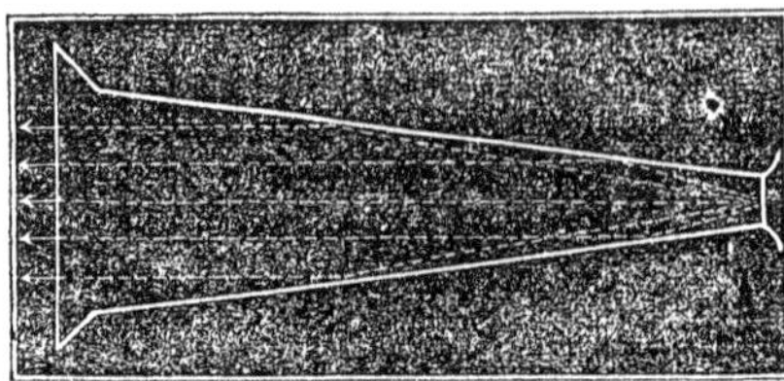

Fig. 192.

The course of the rays of sound proceeding from the mouth through this instrument, may be shown by Fig. 192. The trumpet being directed to any point, a collection of parallel rays of sound moves toward such point, and they reach the ear in much greater number than would the diverging rays which would proceed from a speaker without such an instrument.

What is an Ear-Trumpet?

434. An Ear-Trumpet is, in form and application, the reverse of a speaking-trumpet, but in principle the same. The rays of sound proceeding from a speaker, more or less distant, enter the hearing-trumpet and are reflected in such a manner as to concentrate the sound upon the opening of the ear.

Fig. 193.

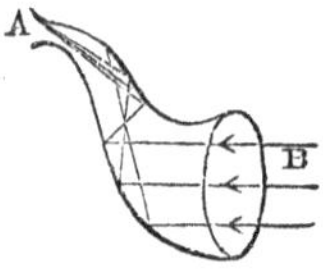

Fig. 193 represents the form of the ear-trumpet generally used by deaf persons. The aperture A is placed within the ear, and the sound which enters at B is, by a series of reflections from the interior of the instrument, concentrated at A.

In the same manner persons hold the hand concave behind the ear, in order to hear more distinctly. The concave hand acts, in some respects, as an ear-trumpet, and reflects the sound into the ear.

Most of the stories in respect to the so-called "haunted houses" can be all satisfactorily explained by reference to the principles which govern the reflection of sounds. Owing to a peculiar arrangement of reflecting walls and partitions, sounds produced by ordinary causes are often heard in certain localities at remote distances, in apparently the most unaccountable manner. Ignorant persons become alarmed, and their imagination connects the phenomenon with some supernatural cause.

435. A right understanding of the principles which govern the reflection of sound is often of the utmost importance in the construction of buildings intended for public speaking, as halls, churches, etc.

Experience shows that the human voice is capable of filling a larger space than was ever probably inclosed within the walls of a single room.

What circumstances are necessary to insure the utmost distinctness in hearing?

The circumstances which seem necessary in order that the human voice should be heard to the greatest possible distance, and with the greatest distinctness, seem to be, a perfectly tranquil and uniformly dense atmosphere, the absence of all extraneous sounds, the absence of echoes and reverberations, and the proper arrangement of the reflecting surfaces.

How does a pure atmosphere in a room favor hearing?

A pure atmosphere in a room for speaking, being favorable to the speaker's health and strength, will give him greater power of voice and more endurance, thus indirectly improving the hearing by strengthening the source of sound, and also by enabling the hearer to give his attention for a longer period undisturbed.

How should a room for public speaking be constructed?

In constructing a room for public speaking, the ceiling ought not to exceed 30 to 35 feet in height.

What is the reason of this?

The reason of this may be explained as follows:—If we advance toward a wall on a calm day, producing at each step some sound, we will find a point at which the echo ceases to be distinguishable from the original sound. The distance from the wall, or the corresponding interval of time, has been called the limit of perceptibility. This limit is about 30 to 35 feet; and if the ceiling of a building for speaking be arranged at this limit, the sound of the voice and the echo will blend together, and thus strengthen the voice of the speaker.

If the ceiling be constructed higher than this limit of perceptibility, or higher than 30 or 35 feet, the direct sound and the echo will be heard separately, and will produce indistinctness.

How may echoes in apartments to some extent be avoided?

Echoes from walls and ceilings may, to a certain extent, be avoided by covering their surfaces with thick drapery, which absorbs sound, and does not reflect it.

If the room is not very large, a curtain behind the speaker impedes rather than assists his voice.

What is meant by the key-note of an apartment?

436. In every apartment, owing to the peculiar arrangement of the reflecting surfaces, some notes or tones can be heard with greater distinctness than others; or, in other words, every apartment is fitted to reproduce a certain note, called the key-note, better than any other. If a speaker, therefore, will adapt the tones of his voice to coincide with this key-note, which may readily be determined by a little practice, he will be enabled to speak with greater ease and distinctness than under any other circumstances.

In a large room nearly square, the best place to speak from is near one corner, with the voice directed diagonally to the opposite corner. In most cases, the lowest pitch of voice that will reach across the room will be the most audible. In all rooms of ordinary form, it is better to speak along the length of a room than across it. It is better, generally, to speak from pretty near a wall or pillar, than far away from it.

SECTION III.

ORGANS OF HEARING AND OF THE VOICE.

Describe the construction of the human ear.

437. The Ear consists, in the first instance, of a funnel-shaped mouth, placed upon the external surface of the head. In many animals this is movable, so that they can direct it to the place from whence the sound comes. It is represented at *a*, Fig. 191.

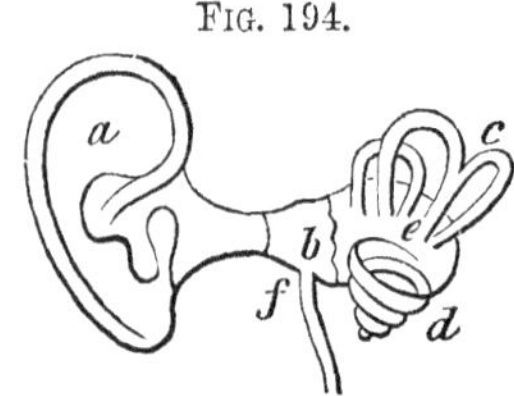

FIG. 194.

Proceeding inward from this external portion of the ear, is a tube, something more than an inch long, terminating in an oval-shaped opening, *b*, across which is stretched an elastic membrane, like the parchment on the head of a drum. This oval-shaped opening has received the name of the tympanum, or drum of the ear, and the membrane stretched across it is called the "membrane of the tympanum, or drum of the ear."

The sound concentrated at the bottom of the ear-tube falls upon the membrane of the drum, and causes it to vibrate. That its motion may be free, the air contained within and behind the drum has free communication with the external air by an open passage, *f*, called the *eustachian tube*, leading to the back of the mouth. A degree of deafness ensues when this tube is obstructed, as in a cold; and a crack, or sudden noise, with immediate return of natural hearing, is generally experienced when, in the effort of sneezing or otherwise, the obstruction is removed.

The vibrations of the membrane of the drum are conveyed further inward, through the cavity of the drum, by a chain of four bones (not represented in the figure on account of their minuteness), reaching from the center of the membrane to the commencement of an inner compartment which contains the nerves of hearing. This compartment, from its curious and most intricate structure is called the *Labyrinth*. Fig. 194, *c e d*.

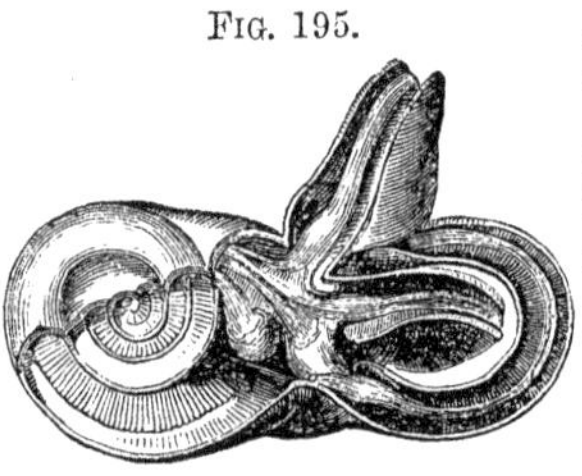
FIG. 195.

The Labyrinth is the true ear, all the other portions being merely accessories by which the sonorous undulations are propagated to the nerves of hearing contained in the labyrinth, which is excavated in the hardest mass of bone found in the whole body. Fig. 195 represents the labyrinth on an enlarged scale, and partially open.

The labyrinth is filled with a liquid substance, through which the nerves of hearing are distributed. When the membrane of the drum of the ear is made to vibrate by the undulations of sound striking against it, the vibrations are communicated to the little chain of bones, which, in turn, striking against a membrane which covers the external opening of the labyrinth, compresses the liquid contained in it. This action, by the law of fluid pressure, is communicated to the whole interior of the labyrinth, and consequently to all portions of the auditory nerve distributed throughout it: the nerve thus acted upon conveys an impression to the brain.

The several parts of the labyrinth consist of what is called the vestibule, *e*, Fig. 194, three semicircular canals, *c*, imbedded in the hard bone, and a winding cavity, called the *cochlea*, *d*, like that of a snail-shell, in which fibres, stretched across like harp-strings, constitute the *lyra*. The separate uses of these various parts are not yet fully known. The membrane of the tympanum may be pierced, and the chain of bones may be broken, without entire loss of hearing.

What are peculiarities of the hearing apparatus in the lower animals?

438. In the hearing apparatus of the lower orders of animals, all the parts belonging to the human ear do not exist. In fishes, the ear consists only of the labyrinth; and in lower animals the ear is simply a little membranous cavity filled with fluid in which the fibres of the nerves of hearing float.

Can all persons hear sound alike?

439. All persons can not hear sounds alike. In different individuals the sensibility of the auditory nerves varies greatly.

What is the range of human hearing?

440. The whole range of human hearing, from the lowest note of the organ to the highest known cry of insects, as of the cricket, includes about nine octaves.

What are the organs of voice?

441. In the human system, the parts concerned in the production of speech and music, are three: the wind-pipe, the larynx, and the glottis.

What is the Wind-pipe?

442. The Wind-pipe is a tube extending from one extremity of the throat to the other, which terminates in the lungs, through which the air passes to and from these organs of respiration.

What is the Larynx?

443. The Larynx, which is essentially the organ of speech, is an enlargement of the upper part of the wind-pipe. The Larynx terminates in two lateral membranes which approach near to each other, having a little narrow opening between them called the glottis. The edges of these membranes form what is called the vocal chords.

How is voice produced?

444. In order to produce sound, the air expired from the lungs passes through the wind-pipe and out at the larynx, through the opening between the membranes, the glottis: the vibration of the edges of these membranes, caused by the passage of air, produces sound.

How can the tones of the voice be rendered grave or acute?

By the action of muscles we can vary the tension of these membranes, and make the opening between them large or small, and thus render the tones of the voice grave or acute.*

Upon what does the loudness of the voice depend?

445. The loudness of the voice depends mainly upon the force with which the air is expelled from the lungs.

The force which a healthy chest can exert in blowing is about one pound per inch of its surface; that is to say, the chest can condense its contained air with that force, and can blow through a tube the mouth of which is ten feet under the surface of water.

What is the vocal action of coughing?

446. In coughing, the top of the windpipe, or the glottis, is closed for an instant, during which the chest is compressing and condensing its contained air; and on the glottis being opened, a slight explosion, as it were, of the compressed air takes place, and blows out any irritating matter that may be in the air-passages.

Does sound generally accompany the liberation of compressed air?

447. Sound, to some extent, appears to always accompany the liberation of compressed air. An example of this is seen in the report which a pop-gun makes when a paper-bullet is discharged from it. The air confined between the paper bullet and the discharging-rod is suddenly liberated, and strikes against the surrounding air, thus causing a report in the same manner as when two solids come into collision. In like manner an inflated bladder, when burst open with force, produces a sound like the report of a pistol.

To what is the sound of falling water due?

448. The sound of falling water appears in a great measure to be owing to the formation and bursting of bubbles. When the distance which water falls is so limited that the end of

* The power which the will possesses of determining with the most perfect precision the exact degree of tension which these membranes of the glottis, or vocal chords shall receive, is extremely remarkable. Their average length in man is estimated at 73-100ths of an inch in a state of repose, while in the state of greatest tension it is about 93-100ths of an inch. The average length of the membranes in the female is somewhat less. Each interval, or variation of tone which the human voice is capable of producing is occasioned by a different degree of tension of these membranes; and as the least estimated number of variations belonging to the voice is 240, there must be 240 different states of tension of the vocal chords, or membranes, every one of which can be at once determined by the will. Their whole variation in length in man being not more than one fifth of an inch, the variation required to pass from one interval of tone to another will not be more than 1-1200th of an inch.

It is on account of the greater length of the vocal chords, or membranes of the glottis, that the pitch of the voice is much lower in man than in woman: but the difference does not arise until the end of the period of childhood, the size of the larynx in both sexes being about the same up to the age of 14 or 15 years, but then changing rapidly in the male sex, and remaining nearly stationary in the female. Hence it is that boys, as well as girls and women, sing treble; while men sing tenor, which is about an octave lower than treble, or bass which is lower still.—*Dr. Carpenter.*

the stream does not become broken into bubbles and drops, neither sound or air-bubbles will be produced; but as soon as the distance becomes increased to a sufficient extent to break the end of the column into drops, both air-bubbles and sounds will be produced.

What is sneezing? 449. Sneezing is a phenomenon resembling cough; only the chest empties itself at one effort, and chiefly through the nose, instead of through the mouth, as in coughing.

What is laughing? 450. Laughing consists of quickly-repeated expulsions of air from the chest, the glottis being at the time in a condition to produce voice; but there is not between the expirations, as in coughing, a complete closure of the glottis.

What is crying? 451. Crying differs from laughing almost solely in the circumstance of the intervals between the gusts or expirations of air from the lungs being longer. Children laugh and cry in the same breath.

Insects generally excite sonorous vibrations by the fluttering of their wings, or other membraneous parts of their structure.

PRACTICAL QUESTIONS IN ACOUSTICS.

1. The flash of a cannon was seen, and ten seconds afterward the report was heard: how far off was the cannon?

2. At what distance was a flash of lightning, when the flash was seen seven seconds before the thunder was heard?

3. How long after a sudden shout will an echo be returned from a high wall 1,120 feet distant?

4. A stone being dropped into the mouth of a mine, was heard to strike the bottom in two seconds; how deep was the mine?

5. A certain musical string vibrates 100 times in a second: how many times must it vibrate in a second to produce the octave?

CHAPTER XII.

HEAT.

What is heat? 452. Heat is a physical agent, known only by its effects upon matter. In ordinary language we use the term heat to express the sensation of warmth.

What is caloric? 453. Caloric is the general name given to the physical agent which produces the sensation of warmth, and the various effects of heat observed in matter.

How is heat measured? 454. The quantity of heat observed in different substances is measured, and its effects on matter estimated, only by the change in bulk, or appearance, which different bodies assume, according as heat is added or subtracted.

What is temperature? 455. The degree of heat by which a body is affected, or the sensible heat a body contains, is called its TEMPERATURE.

What is cold? 456. Cold is a relative term expressing only the absence of heat in a degree; not its total absence, for heat exists always in all bodies.

What distinguishing characteristic does heat possess? 457. Heat possesses a distinguishing characteristic of passing through and existing in all kinds of matter at all times. So far as we know, heat is everywhere present, and every body that exists contains it without known limits.

Ice contains heat in large quantities. Sir Humphrey Davy, by friction, extracted heat from two pieces of ice, and quickly melted them, in a room cooled below the freezing-point, by rubbing them against each other.

In what manner does heat diffuse, or spread itself? 458. The tendency of heat is to diffuse, or spread itself among all neighboring substances, until all have acquired the same, or a uniform temperature.

A piece of iron thrust into burning coals becomes hot among them, because the heat passes from the coals into the iron, until the metal has acquired an equal temperature.

When do we call a body hot? 459. When the hand touches a body having a higher temperature than itself, we call it hot, because on account of the law that heat diffuses itself among neighboring bodies until all have acquired the same temperature, heat passes from the body of higher temperature to the hand, and causes a peculiar sensation, which we call warmth.

460. When we touch a body having a temperature

When do we call a body cold? lower than that of the hand, heat, in accordance with the same law, passes out from the hand to the body touched, and occasions the sensation which we call cold.*

461. Sensations of heat and cold are, therefore, merely degrees of temperature, contrasted by name in reference to the peculiar temperature of the individual speaking of them.

Under what circumstances may a body feel hot and cold to the same person at the same time? A body may feel hot and cold to the same person at the same time, since the sensation of heat is produced by a body colder than the hand, provided it be less cold than the body touched immediately before; and the sensation of cold is produced under the opposite circumstances, of touching a comparatively warm body, but which is less warm than some other body touched previously. Thus, if a person transfer one hand to common spring water immediately after touching ice, to that hand the water would feel very warm; while the other hand transferred from warm water to the spring water, would feel a sensation of cold.

Has heat weight? 462. Heat is imponderable, or does not possess any perceptible weight.

If we balance a quantity of ice in a delicate scale, and then leave it to melt, the equilibrium will not be in the slightest degree disturbed. If we substitute for the ice boiling water or red-hot iron, and leave this to cool, there will be no difference in the result. Count Rumford, having suspended a bottle containing water, and another containing alcohol to the arms of a balance and adjusted them so as to be exactly in equilibrium found that the balance remained undisturbed when the water was completely frozen, though the heat the water had lost must have been more than sufficient to have made an equal weight of gold red hot.

What do we know of the nature of heat? 463. The nature, or cause of heat is not clearly understood. Two explanations, or theories have been proposed to account for the various phenomena of heat, which are known as the mechanical and vibratory theories.

Explain the mechanical theory. 464. The mechanical theory supposes heat to be an extremely subtile fluid, or etherial

* There can not be a more fallacious means of estimating heat than by the touch. Thus, in the ordinary state of an apartment, at any season of the year, the objects which are in it have all the same temperature; and yet to the touch they will feel warm and cold in different degrees. The metallic objects will be the coldest; stone and marble less so; wood still less; and carpeting and woolen objects will feel warm. Now all these objects are at exactly the same temperature, as ascertained by the thermometer.

kind of matter pervading all space, and entering into combination in various proportions and quantities, with all bodies, and producing by this combination all the various effects noticed.

Explain the vibratory theory.

465. The vibratory theory, on the contrary, supposes heat to be merely the effect of a species of motion, like a vibration or undulation, produced either in the constituent particles of bodies, or in a subtle, imponderable fluid which pervades them.

When one end of a bar of iron is thrust into the fire and heated, the other end soon becomes hot also. According to the mechanical theory, a subtile fluid coming out of the fire enters into the iron, and passes from particle to particle until it has spread through the whole. When the hand is applied to the bar it passes into it also, and occasions the sensation of warmth. According to the vibratory theory, the heat of the fire communicates to the particles of the iron themselves, or to a subtile fluid pervading them, certain vibratory motions, which motions are gradually transmitted in every direction, and produce the sensation of heat, in the same way that the undulations or vibrations of air, produce the sensation of sound.

How are these two theories generally regarded?

There seems to be but little doubt at the present time among scientific men, that the theory which ascribes the phenomena of heat to a series of vibrations, or undulations, either in matter, or a fluid pervading it, is substantially correct. At the same time it is not wholly satisfactory, and neither theory will perfectly explain *all* the facts in relation to heat with which we are acquainted. For the purpose of describing and explaining the phenomena and effects of heat, it is convenient, in many cases, to retain the idea that heat is a substance.

What are evidences in favor of the respective theories of heat?

The fact that nature nowhere presents us, neither has art ever succeeded in showing us, heat alone in a separate state, is a strong ground for believing that heat has no separate material existence. Heat, moreover, can be produced without limit by friction, and intense heat is also produced by the explosion of gunpowder. On the contrary, as arguments in favor of the material existence of heat, we have the fact, that heat can be communicated very readily through a vacuum; that it becomes instantly sensible on the condensation of any material mass, as if it were squeezed out of it: as when, on reducing the bulk of a piece of metal by hammering, we render it very hot (the greatest amount of heat being emitted with the blows that most change its bulk); and, finally, that the laws of the spreading of heat do not resemble those of the spreading of sound, or of any other motion known to us.

What relation is there between heat and light?

466. The relation between heat and light is a very intimate one. Heat exists without light, but all the ordinary sources of light are also sources of heat; and by whatever artificial means natural light is condensed, so as to increase its splen-

dor, the heat which it produces is also, at the same time, rendered more intense.

When is a body incandescent or ignited?

467. When a body, naturally incapable of emitting light, is heated to a sufficient extent to become luminous, it is said to be incandescent, or ignited.

What is flame and fire?

468. Flame is ignited gas issuing from a burning body. Fire is the appearance of heat and light in conjunction, produced by the combustion of inflammable substances.

The ancient philosophers used the term fire as a characteristic of the nature of heat, and regarded it as one of the four elements of nature; air, earth, and water being the other three.

Heat and the attraction of cohesion act constantly in opposition to each other; hence, the more a body is heated, the less will be the attractive force between the particles of which it is composed.

SECTION I.

SOURCES OF HEAT.

What are the principal sources of heat?

469. Six great sources of heat are recognized. They are—1. The sun; 2. The interior of the earth; 3. Electricity; 4. Mechanical action; 5. Chemical action; 6. Vital action.

What is the greatest natural source of heat?

470. The greatest natural source of heat is the sun, as it is also the greatest natural source of light.

Although the quantity of heat sent forth from the sun is immense, its rays, falling naturally, are never hot enough, even in the torrid zone, to kindle combustible substances. By means, however, of a burning-glass, the heat of the sun's rays can be concentrated, or bent toward one point, called a focus, in sufficient quantity to set fire to substances submitted to their action.

FIG. 196.

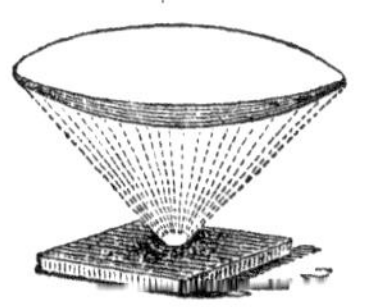

Fig. 196 represents the manner in which a burning-glass concentrates or bends down the rays of heat until they meet in a focus.

Two opinions, or theories, have been entertained in order to account for the production of heat and light by the sun; one supposes that the sun is an intensely-heated mass, which throws off its light and heat like an intensely-heated mass of iron: the other, based on the ground that heat is occasioned by the vibrations of an ethereal fluid occupying all space, supposes that the sun may produce the phenomena of light and heat without waste of its temperature or substance, as a bell may constantly produce the phenomena of sound.

Whatever may be the true theory, a series of experiments, made some years since by Arago, the eminent French astronomer, seem to prove that the temperature at the surface of the sun is much more elevated than any artificial heat we are able to produce. The experimental reasons which lead to this opinion are as follows:—

There are two states in which light is capable of existing—the ordinary state, and the state of polarization.* It has been proved that all bodies, in a solid or liquid state, which are rendered incandescent by heat, emit a polarized light, while bodies that are gaseous, when rendered incandescent, invariably emit light in its ordinary state. Thus the physical condition of a body may be distinguished when it is incandescent by examining the light which it affords. On applying the test to the direct light of the sun, it was found to be in the unpolarized or ordinary condition of light. Hence it has been inferred by Arago that the matter from which this light proceeds must be in the gaseous state, or, in other words, in a state of flame. From other experiments and observations, Arago was led to the conclusion that the sun was a solid, opaque, non-luminous body, invested with an ocean of flame.

Why is the relative heat of the sun always greater in some portions of the earth than at others?

471. Owing to the position of the earth's axis, the relative amount of heat received from the sun is always greater in some portions of the earth than at others, since the rays of the sun always fall more directly upon the central portions of the earth than they do at the poles, or extremities ; and the greatest amount of heat is experienced from the rays of the sun when they fall most perpendicularly.

Why is the heat of the sun greatest at noon?

472. The heat of the sun is greatest at noon, because for the day the sun has reached the highest point in the heavens, and its rays fall more perpendicularly than at any other time.

What occasions the difference in temperature in summer and winter?

For a like reason we experience the extremes of temperature, distinguished as summer and winter. In summer the position of the sun in relation to the earth is such, that although more remote from the earth than in winter, its rays fall more perpendicularly than at any other season, and impart the greatest amount of heat; while in winter the position of the sun is such that its rays fall more obliquely than at any other time, and impart the smallest amount of heat. The sun, moreover, is longer above the horizon in summer than in winter, which also produces a corresponding effect.

The reason why a difference in the inclination of the sun's rays falling upon

* For explanation of the term polarization, see chapter on Light.

the surface of the earth occasions a difference in their heating effect is, that the more the rays are inclined, the more they are diffused, or, in other words, the larger the space they cover. This may be rendered apparent by reference to Fig. 197.

FIG. 197.

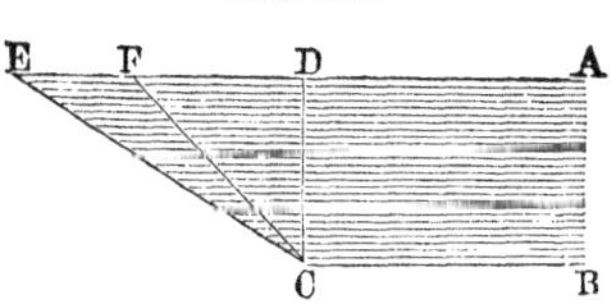

E F D A C B

Let us suppose A B C D to represent a portion of the sun's rays, and C D a portion of the earth's surface upon which the rays fall perpendicularly, and C E portions of the surface upon which they fall obliquely. The same number of rays will strike upon the surfaces C D and C E, but the surface C E being greater than C D, the rays will necessarily fall more densely upon the latter; and as the heating power must be in proportion to the density of the rays, it is obvious that C D will be heated more than C E, in just the same proportion as the surface C E is more extended. But if we would compare two surfaces upon neither of which the sun's rays fall perpendicularly, let us take C E and C F. They fall on C E with more obliquity than on C F; but C E is evidently greater than C F, and therefore the rays being diffused over a larger surface are less dense, and therefore less effective in heating.

What is the greatest natural temperature ever observed?

473. The greatest natural temperature ever authentically recorded was at Bagdad, in 1819, when the thermometer (Fahrenheit's) rose to 120° in the shade. On the west coast of Africa the thermometer has been observed as high as 108° F. in the shade. Burckhardt in Egypt, and Humboldt in South America, observed it at 117° F. in the shade.

What is the lowest temperature observed?

474. About 70° below the zero of Fahrenheit's thermometer is the lowest atmospheric temperature ever experienced by the Arctic navigators.

To what extent has artificial cold been produced?

475. The greatest artificial cold ever produced was 220° F. below zero.

This temperature was obtained some years since by M. Natterer, a German chemist. Professor Faraday also produced a cold equal to 166° F. below zero. At neither of these temperatures were pure alcohol or ether frozen.

The temperature of the space above the earth's atmosphere has been estimated at 58° below zero, Fahrenheit's thermometer.

To what depth in the earth does the heat of the sun extend?

476. The depth to which the influence of the heat of the sun extends into the earth varies from 50 to 100 feet; never, however, exceeding the latter distance.

How do we know that the earth is a source of heat?

Independently of the sun, however, the earth is a source of heat. The proof of this is to be found in the fact, that as we descend into the earth, and pass beyond the limits of the influence of the solar heat, the temperature constantly rises.

At what rate does the temperature of the earth increase?

477. The increase of temperature observed as we descend into the earth, is about one degree of the thermometerfor every fifty feet of descent.

Supposing the temperature to increase according to this ratio, at the depth of two miles water would be converted into steam; at four miles, tin would be melted; at five miles, lead; and at thirty miles, almost every earthy substance would be reduced to a fluid state.

The internal heat of the earth does not appear to have any sensible effect upon the temperature at the surface, being estimated at less than 1-30th of a degree. The reason why such an amount of heat as is supposed to exist in the interior of the earth does not more sensibly affect the surface is because the materials of which the exterior strata or crust of the earth is composed, do not conduct it to the surface from the interior.

Under what circumstances is electricity a source of heat?

478. When electricity passes from one substance to another, the medium which serves to conduct it is very frequently heated; but in what manner the heat is produced we have no positive information.

The greatest known heat with which we are acquainted, is thus produced by the agency of the electric or galvanic current. All known substances can be melted or volatilized by it.

Heat so developed has not been employed for practical or economical purposes to any great extent; but for philosophical experiments and investigations it has been made quite useful.

How is chemical action a source of heat?

479. Many bodies, when their original constitution is altered, either by the abstraction of some of their component parts, or by the addition of other substances not before in combination with them, evolve heat while the change is taking place.

In such cases, the heat is said to be due to chemical action.

What is chemical action?

480. We apply the term chemical action to those operations, whatever they may be, by which the form, solidity, color, taste, smell, and action of substances become changed; so that new bodies, with quite different properties, are formed from the old.

A familiar illustration of the manner in which heat is evolved by chemical action is to be found in the experiment of pouring cold water upon quicklime. The water and the lime combine together, and in so doing liberate a great amount of heat, sufficient to set fire to combustible substances.

How is heat affected by the change of form in matter?

481. Heat is always evolved when a fluid is transformed into a solid, and is always absorbed when a solid is made to assume a fluid condition. As water is changed from its liquid form when it is taken up by quicklime, therefore heat is given off.

The heat produced by the various forms of combustion, is the result of chemical action.

In what two conditions does heat exist?

482. Heat exists in two very different conditions, as FREE, or SENSIBLE HEAT, and as LATENT HEAT.*

What is sensible heat?

483. When the heat retained or lost by a body is attended with a sense of increased or diminished warmth, it is called sensible heat.

What is latent heat?

484. When the heat retained or lost by a body is not perceptible to our sense, it is called latent heat.

How do we know heat to exist in a body, if we can not perceive it?

Every substance contains more or less of latent heat. Although our senses give us no direct information of its presence, we may, by a variety of experiments, prove that it exists. Thus, the temperature of ice is 32° by the thermometer, but if ice be melted over a fire and converted into water, the water will be no hotter than the ice was before, although in the operation 140 degrees of heat have been absorbed by the water. When, on the contrary, water passes into ice, a large amount of heat which was before latent in the water, passes out of it, and becomes sensible.†

How is mechanical action a source of heat?

485. Another important source of heat is mechanical action, heat being produced by friction and by the condensation, or compression of matter.

What are illustrations of the production of heat by friction?

Savage nations kindle a fire by the friction of two pieces of dry wood; the axles of wheels revolving rapidly frequently become ignited; and in the boring and turning of metal the chisels often become intensely hot. In all these cases the friction of the surfaces of wood or of metal in contact, disturbs the latent heat of these substances, and renders it sensible.

The following interesting experiment was made by Count Rumford, to il-

* *Latent*, from the Latin word *lateo*, to be hid.

† The phenomena of latent heat are further considered under the head of liquefaction.

lustrate the effect of friction in producing heat:—A borer was made to revolve in a cylinder of brass, partially bored, thirty-two times in a minute. The cylinder was inclosed in a box containing 18 pounds of water, the temperature of which was at first 60°, but rose in an hour to 107°; and in two hours and a half the water boiled.

Is air necessary for the production of heat by friction?

Air does not appear to be necessary to the production of heat by the friction of solid bodies; since heat is produced by friction within a vacuum.

To whatever extent the operation may be carried, a body never ceases to give out heat by friction, and this fact is considered as a strong argument in favor of the theory that heat is not a substance, but merely a property of matter.

It was formerly supposed that solids alone could develop heat by friction, but recent experiments have proved, beyond a doubt, that heat is also generated by the friction of fluids.

The heat excited by friction is not in proportion to the hardness or elasticity of the bodies employed; on the contrary, a piece of brass rubbed with a piece of cedar-wood produced more heat than when rubbed with another piece of metal; and the heat was still greater when two pieces of wood were employed.

What are illustrations of the production of heat by condensation?

The reduction of matter into a smaller compass by an external or mechanical force, is generally attended with an evolution of heat. To such an act of compression we apply the term condensation.

Heat may be evolved from air by condensation. This may be shown by placing a piece of tinder in a tube, and suddenly compressing the air contained in it by means of a piston. The air being thus condensed, parts with its latent heat in sufficient quantity to set fire to the tinder at the bottom of the tube. Another familiar experiment of the evolution of heat by condensation, is the rendering of iron hot by striking it with a hammer. The particles of the iron being compressed by the hammer, can no longer contain so much heat in a latent state as they did before: some of it therefore becomes sensible, and increases the temperature of the metal, and the striking may be continued to such an extent as to render the iron red-hot.

When a match is drawn over sand-paper or other rough substance, certain phosphoric particles are rubbed off, and being compressed between the match and the paper, their heat is raised sufficiently high to ignite them and fire the match. If the match be drawn over a smooth surface, the compression must be increased, for the temperature of the whole phosphoric mass must be raised in order to cause ignition.

The fulminating substance of a percussion-cap explodes when struck by a hammer, because the blow occasions a compression of the particles, by which a sufficient amount of latent heat is liberated to produce ignition.

What is meant by vital heat?

486. Most living animals possess the property of maintaining in their system an equable tem-

perature, whether surrounded by bodies that are hotter or colder than they are themselves. The cause of this is due to the action of vital heat, or the heat generated or excited by the organs of a living structure.

The following facts illustrate this principle:—The explorers of the Arctic regions, during the polar winter, while breathing air that froze mercury, still had in them the natural warmth of 98° Fahrenheit above zero; and the inhabitants of India, where the same thermometer sometimes stands at 115° in the shade, have their blood at no higher temperature. Again, the temperature of birds is not that of the atmosphere, nor of fishes that of the sea.

What is the cause of vital heat?

487. The cause of animal heat is undoubtedly due to chemical action;—the result of respiration and nervous excitation.

Do plants possess this property?

Growing vegetables and plants also possess, in a degree, the property of maintaining a constant temperature within their structure. The sap of trees remains unfrozen when the temperature of the surrounding atmosphere is many degrees below the freezing point of water.

This power of preserving a constant temperature in the animal structure is limited. Intense cold suddenly coming upon a man who has not sufficient protection, first causes a sensation of pain, and then brings on an almost irresistible sleepiness, which if indulged in proves fatal. A great excess of heat also can not long be sustained by the human system.

Each species of animal and vegetable appears to have a temperatare natural and peculiar to itself, and from this diversity different races are fitted for different portions of the earth's surface. Thus, the orange-tree and the bird of Paradise are confined to warm latitudes; the pine-tree and the Arctic bear, to those which are colder

When animals and plants are removed from their peculiar and natural districts to one entirely different, they cease to exist, or change their character in such a way as to adapt themselves to the climate. As illustrations of this, we find that the wool of the northern sheep changes in the tropics to a species of hair. The dog of the torrid zone is nearly destitute of hair. Bees transported from the north to the region of perpetual summer, cease to lay up stores of honey, and lose in a great measure their habits of industry.

Man alone is capable of living in all climates, and of migrating freely to all portions of the earth.

Of all animals, birds have the highest temperature; mammalia, or those which suckle their young, come next; then amphibious animals, fishes, and certain insects. Shell-fish, worms, and the like, stand lowest in the scale of temperature. The common mud-wasp, in its chrysalis state, remains unfrozen during the most severe cold of a northern winter; the fluids of the body instantly congeal, however, in a freezing temperature, the moment the case, or shell which incloses it, is crushed.

SECTION II.

COMMUNICATION OF HEAT.

How may heat be communicated?

488. Heat may be communicated in three ways: by CONDUCTION, by CONVECTION, and by RADIATION.

How is heat communicated by conduction?

489. Heat is communicated by conduction when it travels from particle to particle of the substance, as from the end of the iron bar placed in the fire to that part of the bar most remote from the fire.

What is convection?

490. When heat is communicated by being carried by the natural motion of a substance containing it to another substance or place, as when hot water resting upon the bottom of a kettle rises and carries heat to a mass of water through which it ascends, the heat is said to be communicated by convection.

What is radiation of heat?

491. Heat is communicated by radiation when it leaps, as it were, from a hot to a cold body through an appreciable interval of space; as when a body is warmed by placing it before a fire removed to a little distance from it.

How does a heated body cool itself?

492. A heated body cools itself, first by giving off heat from its surface, either by conduction or radiation, or both conjointly; and secondly, by the heat in its interior passing from particle to particle by conduction, through its substance to the surface. A cold body, on the contrary, becomes heated by a process directly the reverse of this.

Do all bodies conduct heat equally well?

493. Different bodies exhibit a very great degree of difference in the facility with which they conduct heat: some substances oppose very little impediment to its passage, while through others it is transmitted slowly.

What are conductors and non-conductors of heat?

494. All bodies are divided into two classes in respect to their conduction of heat, viz., into conductors and non-conductors. The for-

mer are such as allow heat to pass freely through them; the latter comprise those which do not give an easy passage to it.

Dense solid bodies, like the metals, are the best conductors of heat;* light, porous substances, more especially those of a fibrous nature, are the worst conductors of heat.

The different conducting power of various solid substances may be strikingly shown by taking a series of rods of equal length and thickness, coating one of their extremities with wax, and placing the other extremities equally in a source of heat. The wax will be found to entirely melt off from some of the rods before it has hardly softened upon others.

What is the conducting power of liquids?

495. Liquids are almost absolute non-conductors of heat.

FIG. 198.

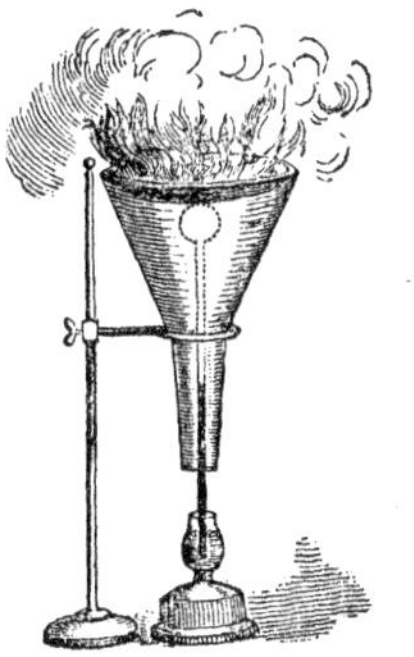

If a small quantity of alcohol be poured on the surface of water and inflamed, it will continue to burn for some time. (See Fig. 198.) A thermometer, immersed at a small depth below the common surface of the spirit and the water, will fail to show any increase in temperature.

Another and more simple experiment proves the same fact; as when a blacksmith immerses his red-hot iron in a tank of water, the water which surrounds the iron is made boiling hot, while the water not immediately in contact with it remains quite cold.

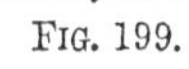

FIG. 199.

If a tube nearly filled with water is held over a spirit lamp, as in Fig. 199, in such a manner as to direct the flame against the upper layers of the water, the water will be observed to boil at the top, but remain cool below. If quicksilver, on the contrary, be so treated, its lower layers will speedily become heated. The particles of mercury will communicate the heat to each other, but the particles of water will not do so.

Why does a stone, or marble hearth feel colder than a carpet?

A stone, or marble hearth in an apartment, feels colder to the feet than a woolen carpet, or hearth-rug, not because the one is hotter than the other, for both are really of the same temperature, but because the stone and marble are good

* The following table exhibits the relative conducting power of different substances, the ratio expressing the conducting power of gold being taken at 100:

Substance	Power	Substance	Power
Gold	100·00	Tin	30·38
Platinum	98·10	Lead	17·96
Silver	97·30	Marble	2·34
Copper	89·82	Porcelain	1·22
Iron	37·41	Brick earth	1·13
Zinc	36·37		

conductors, and the woolen carpet and hearth-rug very bad conductors. The action of the two substances is as follows:—As soon as the hearth-stone has absorbed a portion of heat from the feet, it instantly disposes of it by conducting it away, and calls for a fresh supply; and this action will continue until the stone and the foot have established an equilibrium of temperature between them. The carpet and the rug also absorb heat from the feet in like manner, but they conduct or convey it away so slowly, that its loss is hardly perceptible.

Most varieties of wood are bad conductors of heat; hence, though one end of a stick is blazing, the other end may be quite cold. Cooking vessels, for this reason, are often furnished with wooden handles, which conduct the heat of the vessel too slowly to render its influx into our hands painful. For the same reason we use paper or woolen kettle-holders.

To what extent do aeriform bodies conduct heat?

496. Bodies in the gaseous, or aeriform condition are more imperfect conductors of heat than liquids. Common air, especially, is one of the worst conductors of heat with which we are acquainted.

How is air heated?

497. Air is, however, readily heated by convection. Thus, when a portion of air by coming in contact with a heated body has heat imparted to it, it expands, and becoming relatively lighter than the other portions around it, rises upward in a current, carrying the heat with it; other colder air succeeds, and (being heated in a similar way) ascends also. A series of currents are thus formed, which are called "convective currents."

In this way air, which is a bad conductor, rapidly reduces the temperature of a heated substance. If the air which encases the heated substance were to remain perfectly motionless, it would soon become, by contact, of the same temperature as the body itself, and the withdrawal of heat would be checked: but as the external air is never perfectly at rest, fresh and colder portions continually replace and succeed those which have become in any degree heated, and thus the abstraction of heat goes on.

For this reason a windy day always feels colder than a calm day of the same temperature, because in the former case the particles of air pass over us more rapidly, and every fresh particle takes some portion of heat.

How may the conducting power of bodies be diminished?

498. The conducting power of all bodies is diminished by pulverizing them, or dividing them into fine filaments.

Thus saw-dust, when not too much compressed, is one of the most perfect non-conductors of heat. Wool, fur, hair and feathers, are also among the worst conductors of heat.

Why are furs and woolens used for clothing?

Woolens and furs are used for clothing in cold weather, not because they impart any heat to the body, but because they are very bad conductors of heat; and therefore prevent the warmth of the body from being drawn off by the cold air. The heat generated in the animal system by vital action has constantly a tendency to escape, and be dissipated at the surface of the body, and the rate at which it is dissipated depends on the difference between the temperature of the surface of the body and the temperature of the surrounding medium. By interposing, however, a non-conducting substance between the surface of the body and the external atmosphere, we prevent the loss of heat which would otherwise take place to a greater or less degree.

To what are the non-conducting properties of fibrous substances due?

The non-conducting properties of fibrous and porous substances are due almost altogether to the air contained in their interstices, or between their fibers. These are so disposed as to receive and retain a large quantity of air without permitting it to circulate.

The warmest clothing is that which fits the body rather loosely, because more hot air will be confined by a moderately loose garment than by one which fits the body tightly.

Blankets and warm woolen goods are always made with a nap or projection of fibers upon the outside, in order to take advantage of this principle. The nap or fibers retain air among them, which, from its non-conducting properties, serves to increase the warmth of the material.

What influence has the fineness of the fibers upon the warmth of a material?

The finer the fibers of hair, or wool, the more closely they retain the air enveloped within them, and the more impermeable they become to heat. In accordance with this principle, the external coverings of animals vary not only with the climate which the species inhabit, but also in the same individual they change with the season. In warm climates the furs are generally coarse and thin, while in cold countries they are fine, close, light, and of uniform texture, almost perfect non-conductors of heat.

We have illustrations of this principle also in the vegetable kingdom. The bark of trees, instead of being compact and hard like the wood it envelops, is porous and formed of fibers, or layers, which, by including more or less of air between their surfaces, are rendered non-conductors, and prevent the escape of heat from the body of the tree.

An apartment is rendered much warmer for being furnished with double doors and windows, because the air confined between the two surfaces opposes both the escape of heat from within, and the admission of cold from without.

As a non-conducting substance prevents the escape of heat from within a body, so it is equally efficacious in preventing the access of heat from without. In an atmosphere hotter than our bodies, the effect of clothing would be to keep the body cool. Flannel is one of the warmest articles of dress, yet we can not preserve ice more effectually in summer than by enveloping it in its folds. Firemen exposed to the intense heat of furnaces and steam-boilers, invariably protect themselves with flannel garments.

Cargoes of ice shipped to the tropics, are generally packed for preservation in sawdust: a casing of sawdust is also one of the most effectual means of preventing the escape of heat from the surfaces of steam-boilers and steam-pipes. Straw, from its fibrous character, is an excellent non-conductor of heat, and is for this reason extensively used by gardeners for incasing plants and trees which are exposed to the extreme cold of winter.

How does snow protect the earth from cold?

Snow protects the soil in winter from the effects of cold in the same way that fur and wool protect animals, and clothing man. Snow is made up of an infinite number of little crystals, which retain among their interstices a large amount of air, and thus contribute to render it a non-conductor of heat. A covering of snow also prevents the earth from throwing off its heat by radiation. The temperature of the earth, therefore, when covered with snow, rarely descends much below the freezing-point, even when the air is fifteen or twenty degrees colder.* Thus roots and fibers of trees and plants, are protected from a destructive cold.

Under what circumstances is clothing considered warm and cold?

499. Clothing is considered warm or cool according as it impedes or facilitates the passage of heat to or from the surface of our bodies. The finer the cloth, the more slowly it conducts heat. Fine cloths, therefore, are warmer than coarse ones.

Woolen substances are worse conductors of heat than cotton, cotton than silk, and silk than linen. A flannel shirt more effectually intercepts heat than cotton, and a cotton than a linen one.

The sheets of a bed feel colder than the blankets, because they are better conductors of heat, and carry off the heat more rapidly from the body, the actual temperature of both, however, is the same. For the same reason, a linen handkerchief is cooler and more agreeable to the face than a cotton one.

Cellars feel cool in summer, and warm in winter, because the external air

* "Few can realize the protecting value of the warm coverlet of snow. No eider-down in the cradle of an infant is tucked in more kindly than the sleeping-dress of winter about the feeble flower-life of the Arctic regions. The first warm snows of August and September, falling on a thickly-blended carpet of grasses, heaths and willows, enshrine the flowery growths which nestle around them in a non-conducting air-chamber; and as each successive snow increases the thickness of the cover, we have, before the intense cold of winter sets in, a light cellular bed, covered by drift, six, eight, or ten feet deep, in which the plant retains its vitality. The frozen sub-soil does not encroach upon this narrow cover of vegetation. I have found, in mid-winter, in the high latitude of 78°, the surface so nearly moist as to be friable to the touch; and on the ice-floes commencing with a surface temperature of 30° below zero, I found, at two feet deep, a temperature of 8° below zero, at four feet 2° above zero, and at eight feet 26° above zero. My experiments prove that the conducting power of snow is proportioned to its compression by winds, rains, drifts, and congelation."—DR. KANE'S *Second Arctic Expedition.*

has not free access into them; in consequence of which they remain almost at an even temperature, which in summer is about 10 degrees colder, and in winter about 10 degrees warmer than the external air.

Upon what principle are refrigerators and fire-proof safes constructed?

500. Refrigerators, used for the preservation of animal and vegetable substances in warm weather, are double-walled boxes, with the spaces between the sides filled with powdered charcoal, or some other porous, non-conducting substance.

The so-called "fire-proof" safes are also constructed of double or treble walls of iron, with intervening spaces between them filled with gypsum, or "Plaster of Paris." This lining, which is a most perfect non-conductor, prevents the heat from passing from the exterior of the safe to the books and papers within. The idea of applying "Plaster of Paris" in this way for the construction of safes, originated, in the first instance, from a workman attempting to heat water in a tin basin, the bottom and sides of which were thinly coated with this substance. The non-conducting properties of the plaster were so great as to almost entirely intercept the passage of the heat, and the man, to his surprise, found that the water, although directly over the fire, did not get hot.

501. It has been already stated that liquids and gases are non-conductors of heat, and can not well be heated, like a mass of metal, or any solid, by the communication of heat from particle to particle.

Why can not liquids and gases be heated in the same manner as solids?

This peculiarity is owing to the mobility which subsists among the particles of all fluids, and to the change in the size of the particles, which is invariably produced by a change in their temperature.

The constituent particles of solid bodies being incapable of changing their relative position and arrangement, the heat can only pass through them, from particle to particle, by a slow process; but when the particles forming any stratum of liquid are heated, their mass, expanding, becomes lighter, bulk for bulk, than the colder stratum immediately above it, and ascends, allowing the superior strata to descend.

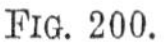
FIG. 200.

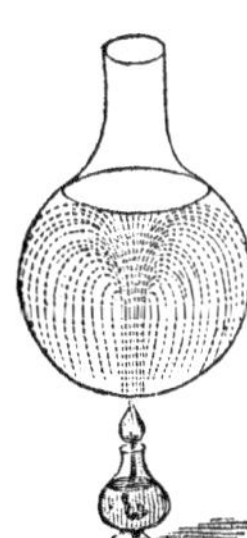

How is water made hot?

502. When the heat enters at the bottom of a vessel containing water, a double set of currents is immediately established—one of hot particles rising toward the surface, and the other of colder particles descending to the bottom. The portion of liquid which receives heat from below is thus continually diffused through the other

parts, and the heat is communicated by the motion of the particles among each other.

These currents take place so rapidly, that if a thermometer be placed at the bottom and another at the top of a long jar, the fire being applied below, the upper one will begin to rise almost as soon as the lower one. The movement of the particles of water in boiling will be understood by reference to Fig. 200. They may be rendered visible by adding to a flask of boiling water a few small particles of bituminous coal, or flowers of sulphur.

Air and other gases are heated in precisely the same manner as water, and this method of communicating heat is termed convection.

Heat, however, passes by conduction between the particles of both liquids and gases, but to such a slight extent, that they were for a long time regarded as entirely incapable of conducting heat.

In what manner is a liquid cooled?

503. The process of cooling in a liquid is directly the reverse of that of heating. The particles at the surface, by contact with the air, readily lose their heat, become heavier, and sink, while the warmer particles below in turn rise to the surface.

To heat a liquid, therefore, the heat should be applied at the bottom of the mass; to cool it, the cold should be applied at the top, or surface.

The facility with which a liquid may be heated or cooled depends in a great degree on the mobility of its particles. Water may be made to retain its heat for a long time by adding to it a small quantity of starch, the particles of which, by their viscidity or tenacity, prevent the free circulation of the heated particles of water. For the same reason soup retains its heat longer than water, and all thick liquids, like oil, molasses, tar, etc., require a considerable time for cooling.

Explain the phenomena of radiation.

504. When the hand is placed near a hot body suspended in the air, a sensation of warmth is perceived, even for a considerable distance. If the hand be held beneath the body, the sensation will be as great as upon the sides, although the heat has to shoot down through an opposing current of air approaching it. This effect does not arise from the heat being conveyed by means of a hot current, since all the heated particles have a uniform tendency to rise; neither can it depend upon the conducting power of the air, because aeriform substances possess that power in a very low degree, while the sensation in the present case is excited almost on the instant. This method of distributing heat, to distinguish it from heat passing by conduction, or convection, is called radiation, and heat thus distributed is termed radiant, or radiated heat.

Do all bodies radiate heat equally well?

505. All bodies radiate heat in some measure, but not all equally well; radiation being in proportion to the roughness of the radiating surface. All dull and dark substances are, for the most

part, good radiators of heat; but bright and polished substances are generally bad radiators.* Color, however, alone, has no effect on the radiation of heat.

If a metal surface be scratched, its radiating power is increased. A liquid contained in a bright, highly-polished metal pot, will retain its heat much longer than in a dull and blackened one. This is not due to the polish or brightness of the surface, but to the fact that, by polishing, the surface is rendered dense and smooth, and such surfaces do not allow the heat to escape readily. If we cover the polished metal surface with a thin cotton or linen cloth, so as to render the surface less dense, the radiation of heat, and consequent cooling, will proceed rapidly.

Black lead is one of the best known radiators of heat, and on this account is generally employed for the blackening of stoves and hot-air flues. As a high polish is unfavorable to radiation, stoves should not be too highly polished with this substance.

Heat radiated from the sun is all radiant heat.

How is heat propagated by radiation?

506. Heat is propagated through space by radiation in straight lines, and its intensity varies according to the same law which governs the attraction of gravitation, that is, inversely as to the square of the distance.†

Thus the heating effects of any hot body is nine times less at three feet than at one; sixteen times less at four feet; and twenty-five times less at five.

With what velocity does radiant heat move?

The velocity with which radiant heat moves through space is, in all probability, the same as the velocity of light. Some authorities, however, consider it to be only four fifths of that of light, or about 164,000 miles in a second of time.

Does radiation proceed constantly from all bodies?

507. The radiation of heat goes on at all times, and from all surfaces, whether their temperature be the same or different from that of surrounding objects: therefore the temperature of a body falls when it radiates more heat than it absorbs; its temperature is stationary when the quantities emitted and received are equal; and it grows warm when the absorption exceeds the radiation.

* The action of a blackened surface of tin being assumed as 100, it has been found that that of a steel plate was 15; of clean tin, 12; of tin scraped bright, 16; when scraped with the edge of a fine file in one direction, 26; when scraped again across, about 13; a surface of clean lead, 19; covered with a gray crust, 45; a thin crust of isinglass, 80; rosin 96; writing-paper, 98; ice, 85.

† It is an exceedingly curious fact that this law applies to all physical influences that spread from a center, such as gravitation, heat, light, electrical forces, magnetism and sound; and to all central forces, when not weakened by any resistance or opposing force.

If a body, at any temperature, be placed among other bodies, it will affect their condition of temperature, or as we express it, *thermally;* just as a candle brought into a room illuminates all bodies in its presence; with this difference, however, that if the candle be extinguished, no more light is diffused by it; but no body can be thermally extinguished. All bodies, however low be their temperature, contain heat, and therefore radiate it.

Why does a thermometer sink when brought near ice?

If a piece of ice be held before a thermometer, it will cause the mercury in its tube to fall, and hence it has been supposed that the ice emitted rays of cold. This supposition is erroneous. The ice and the thermometer both radiate heat, and each absorbs more or less of what the other radiates toward it. But the ice, being at a lower temperature than the thermometer, radiates less than the thermometer, and therefore the thermometer absorbs less than the ice, and consequently falls. If the thermometer placed in the presence of the ice had been at a lower temperature than the ice, it would, for like reasons, have risen. The ice in that case would have warmed the thermometer.

What do we mean by rays of heat or light?

509. Radiations, or effects which are propagated in straight lines only (such as light and radiant heat), are most conveniently considered by dividing them into innumerable straight lines, or *rays;* not that there are any such divisions in nature, but they enable us more readily to comprehend the nature of the phenomena with which these principles are concerned.

When radiant heat falls upon the surface of a body, how may it be disposed of?

510. When rays of heat radiated from one body fall upon the surface of another body, they may be disposed of in three ways: 1. They may rebound from its surface, or be reflected; 2. They may be received into its surface, or be absorbed; 3. They may pass directly through the substance of the body, or be transmitted.

In what manner is heat reflected?

511. A ray of heat radiated from the surface of a body proceeds in a straight line until it meets a reflecting surface, from which it rebounds in another straight line, the direction of which is determined by the law that the angle of incidence is equal to the angle of reflection.

The manner in which heat is reflected is strikingly shown by taking two concave mirrors, M and N, Fig. 201, of bright metal, about one foot in diameter, and placing them exactly opposite to each other at a distance of about ten feet. In the focus of one mirror, as at A, is placed a heated body, as a mass of red hot iron, and in the focus of the other mirror, as at B, a small quantity of gunpowder, or a piece of phosphorus. The rays of heat, radiated in diverging lines from the hot metal, strike upon the surface of the mirror M, and are reflected by it in parallel lines to the surface of the opposite mirror, N, where they will be caused to converge to its focus, B, and ignite the powder or phosphorus at that point.

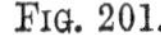

FIG. 201.

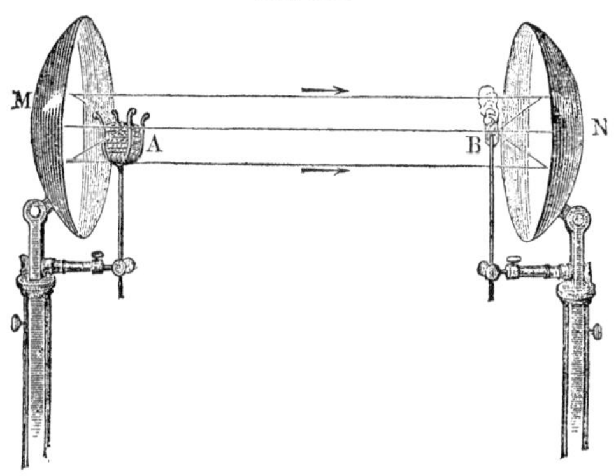

What are good reflectors of heat?

512. Polished metallic surfaces constitute the best reflectors of heat; but all bright and light colored surfaces are adapted for this purpose to a greater or less degree.*

Water requires a longer time to become hot in a *bright* tin vessel than in a dark colored one, because the heat is reflected from the bright surface, and does not enter the vessel.

How does the power of absorbing heat vary?

513. The power of absorbing heat varies with almost every form of matter. Surfaces are good absorbers of heat in proportion as they are poor reflectors. The best radiators of heat also are the most powerful absorbers, and the most imperfect reflectors.

Dark colors absorb heat from the sun more abundantly than light ones. This may be proved by placing a piece of black and a piece of white cloth upon the snow exposed to the sun; in a few hours the black cloth will have melted the snow beneath it, while the white cloth will have produced little or no effect upon it.

The darker any color is, the warmer it is, because it is a better absorbent of heat. The order may be thus arranged. 1, black (warmest of all); 2, violet; 3, indigo; 4, blue; 5, green; 6, red; 7, yellow; and 8, white (coldest of all).

* Of 100 rays falling at an angle of 60° from the perpendicular, polished gold will reflect 76; silver 62; brass, 62; brass without polish, 52; polished brass varnished, 41; looking-glass, 20; glass plate blackened on the back, 12; and metal plate blackened, 6.

10*

A piece of brown paper submitted to the action of a burning-glass, ignites much more quickly than a piece of white paper. The reason of this is, that the white paper reflects the rays of the sun, and though but slightly heated appears highly luminous; while the brown paper, which absorbs the rays, readily becomes heated to ignition. For the same reason a kettle whose bottom and sides are covered with soot, heats water more readily than a kettle whose sides are bright and clean.

Light-colored fabrics are most suitable for dresses in summer, since they reflect the direct heat of the sun, and do not absorb it; black outside garments, on the contrary, are most suitable for winter, as they absorb heat readily, but do not reflect it.

Hoar-frost, in the spring and autumn, may be observed to remain longer in the presence of the morning sun, on light-colored substances than upon the dark-colored soil, etc.; the former do not absorb the heat, as the dark-colored bodies do, but reflect it, and in consequence of this they remain too cold to thaw the frost deposited upon their surfaces.

How is the atmosphere heated?

514. Air absorbs heat very slowly, and does not readily part with it. Air rarely radiates heat, and is not heated to great extent by the direct rays of the sun. The sun, however, heats the surface of the earth, and the air resting upon it is heated by contact with it, and ascends, its place being supplied by colder portions, which in turn are heated also.

This reluctance of air to part with its heat occasions some very curious differences between its burning temperature and that of other bodies. Metals, which are generally the best conductors, and therefore communicate heat most readily, can not be handled with impunity when raised to a temperature of more than 120° F.; water becomes scalding hot at 150° F.; but air applied to the skin occasions no very painful sensation when its heat is far beyond that of boiling water.

Some curious experiments have been made in reference to the power of the human body to withstand the influence of heated air. Sir Joseph Banks entered an oven heated 52° above the boiling point, and remained there some time without inconvenience. During the time, eggs, placed on a metal frame, were roasted hard, and a beefsteak was overdone. But though he could thus bear the contact of the heated air, he could not bear to touch any metallic substance, as a watch-chain, money, etc. Workmen, also, enter ovens, in the manufacture of molds of plaster of Paris, in which the thermometer stands 100° above the temperature of boiling water, and sustain no injury.

In what manner is heat transmitted through different substances?

515. Heat, in passing through most substances, or media, is retained, or intercepted in its passage in a greater or less degree. The capacity of solids and liquids for transmitting heat is not always in proportion to their transparency, or capacity for transmitting light.

516. The heat of the sun passes through transparent bodies without loss; but heat from terrestrial sources is in great part arrested by many substances which allow light to pass freely,—such as water, alum, glass, etc.

Thus, a plate of glass held between one's face and the sun will not protect it, but held between the face and a fire, it will intercept a large proportion of the heat.

517. Those substances which allow heat to pass freely through them, are called *diathermanous*, and those which retain nearly all the heat they receive, are called *athermanous*.

Rock-salt allows heat to pass through it more readily than any other known substance; while a thin plate of alum, which is nearly transparent, almost entirely intercepts terrestrial heat. Heat, indeed, will pass more readily through a black glass, so dark that the sun at noon is scarcely discernible through it, than through a thin plate of clear alum. Water is one of the least diathermanous substances, although its transparency is nearly perfect. If, therefore, it is desired to transmit light without heat, or with greatly diminished heat, it is only necessary to let the rays pass through water, by which they will be strained of a great part of their heat.

How does the temperature of a body radiating heat affect its transmission?

It has been found that the power of heat to penetrate a dense, transparent substance, is increased in proportion as the temperature of the body from which it is radiated is increased. Heat, also, accompanied by light, is transmitted more readily than heat without light.

Is a ray of solar heat simple or compound in its nature?

518. Heat and light come to us conjointly from the sun. When a ray of light is caused to pass through a prism it is analyzed or separated into seven brilliant colors, or elementary parts. If the heat ray which accompanies the light is treated in a similar manner, our organs of sight are so constituted that we do not discover any separation to have taken place in it. It is, however, established beyond a doubt, that in the same manner as a ray of white light can be modified and divided, so a ray of radiant heat can be separated into parts possessing qualities corresponding to the various colors.

SECTION III.

THE EFFECTS OF HEAT.

What effect does heat produce upon all bodies?

519. The general and most obvious effect of heat upon material substances, is to expand them, or increase their dimensions.

Is the form of all bodies dependent upon heat?

520. The form of all bodies appears to be entirely dependent on heat; by its increase solids are converted into liquids, and liquids into vapor; by its diminution vapors are condensed into liquids, and these in turn become solids.

If matter ceased to be influenced by heat, all liquids, vapors, and doubtless even gases, would become permanently solid, and all motion on the surface of the earth would be arrested.

What are the three most apparent effects of heat?

521. The three most apparent effects of heat, so far as relate to the form and dimensions of bodies, are Expansion, Liquefaction, and Vaporization.

Heat operates to produce expansion by introducing a repulsive force among the particles of the body it pervades. This repulsive force, in the case of solids, weakens or overcomes the attraction of cohesion, and gives to the particles of all matter a tendency to separate, or increase their distance from one another. Hence the general mass of the body is made to occupy a larger space, or expand.

In what bodies does heat produce the greatest expansion?

522. The expansion occasioned by heat is greatest in those bodies which are the least influenced by the attraction of cohesion. Thus the expansion of solids is comparatively trifling, that of liquids much greater, and that of gases very considerable.

Do bodies continue to expand as long as heat enters them?

523. The expansion of the same body will continue to increase with the quantity of heat that enters it, so long as the form and chemical constitution of the body is preserved.

524. Among solids the metals expand the most; but an iron wire increases only 1-282 in bulk when heated from 32° of the thermometer up to 212.

Solids appear to expand uniformly from the freezing point of water up to 212°, the boiling point of water;—that is to say, the increase of volume which attends each degree of temperature which the body receives is equal. When solids are elevated, however, to temperatures above 212°, they do not dilate uniformly, but expand in an increasing ratio.

The expansion of solids by heat is clearly shown by the following experiment, Fig. 202: *m* represents a ring of metal, through which, at the ordinary temperature, a small iron or copper ball, *a*, will pass freely, this ball being a little less than the diameter of the ring. If this ball be now heated by the flame of an alcohol lamp, it will become so far expanded by heat as no longer to pass through the ring.

Fig. 202.

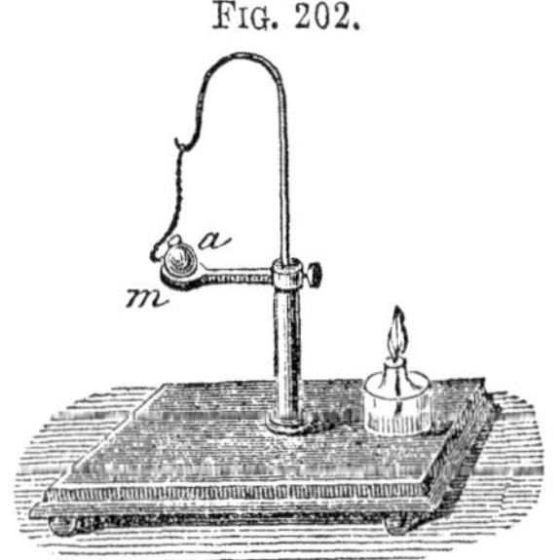

What applications of the expansion of solids by heat are made in the arts?

The expansion of solids by heat is made applicable for many useful purposes in the arts. The tires of wheels, and hoops surrounding water-vats, barrels, etc., are made in the first instance somewhat smaller than the frame-work they are intended to surround. They are then heated red hot and put on in an expanded condition; on cooling, they contract and bind together the several parts with a greater force than could be conveniently applied by any mechanical means. In like manner, in constructing steam-boilers, the rivets are fastened while hot, in order that they may, by subsequent contraction, fasten the plates together more firmly.

With what degree of force do bodies expand and contract?

525. The force with which bodies expand and contract under the influence of the increase or diminution of heat, is apparently irresistible, and is recognized as one of the greatest forces in nature.

The amount of force with which a solid body will expand or contract is equal to that which would be required to compress it through a space equal to its expansion, and to that which would be required to stretch it through a space equal to its contraction. Thus, if a pillar of metal one hundred inches in height, being raised in temperature, is augmented in height by a quarter of an inch, the force with which such increase of height is produced is equal to a weight which being placed upon the top of the pillar would compress it so as to diminish its height by a quarter of an inch.

In the same manner, if a rod of metal, one hundred inches in length, be contracted by diminished temperature, so as to render its length a quarter of an inch less, the force with which this contraction takes place is equal to that which being applied to stretch it would cause its length to be increased by a quarter of an inch.

This principle is sometimes practically applied when great mechanical force is required to be exerted through small spaces. Thus walls of buildings which, from a subsidence of the foundation, or an unequal pressure, have been thrown out of their perpendicular position, and are in danger of falling, may be restored in the following manner: A series of iron rods are carried across the building, passing through holes in the walls, and secured by nuts on the outside. The rods are then heated by lamps until they expand, thereby causing their ends to project beyond the building. The nuts with which these extremities are provided are then screwed up until they are in close contact with the outside wall, the lamps are then withdrawn and the rods allowed to cool. In cooling they gradually contract, and by their contraction draw up the walls.

On account of the expansion of metal by heat, the successive rails which compose a line of railway can not be placed end to end, but a small space is left between their extremities for expansion.

A stove snaps and crackles when a fire is first kindled in it, and also when the fire in it is extinguished. This noise is occasioned by the expansion and contraction of the several parts consequent on the increase and diminution of heat.

A glass or earthen vessel is liable to break when hot water is poured into it, on account of the unequal expansion of the inner and outer surfaces. Glass and earthen ware being poor conductors of heat, the inner surfaces in contact with the hot water become heated and expand before the outer are affected; the tendency of this is to warp or bend the sides unequally, and as the brittle material can not bend, it breaks.

Nails in old houses are often loose and easily drawn out; the iron expands in summer and contracts in winter more than the wood into which it has been driven, and thus in time the opening is enlarged.

When the stopper of a decanter or smelling-bottle sticks, a cloth dipped in hot water, and applied to the neck of the bottle will frequently loosen it, since by the heat of the cloth its dimensions are expanded and enlarged.

The tone of a piano is higher in a cold than in a warm room, for the reason that the strings, being contracted by cold, are drawn tighter.

To what extent do liquids expand by heat?

526. Liquids expand through the agency of heat more unequally, and to a much greater degree than solids.

A column of water contained in a cylindrical glass vessel will expand $\frac{1}{23}$ in length on being heated from the freezing to the boiling point, while a column of iron, with the same increase of temperature, will expand only $\frac{1}{846}$.

A familiar illustration of the expansion of water by heat is seen in the overflow of full vessels before boiling commences. Different liquids expand very unequally with an equal increase in temperature. Spirits of wine, on being heated from 32° to 212°, increase one ninth in bulk; oil expands about one twelfth; water, as before stated, about one twenty-third. A person buying oil, molasses and spirits in winter, will obtain a greater weight of the same material in the same measure than in summer. Twenty gallons of alcohol bought in January, will, with the ordinary increase of temperature, become, by expansion, twenty-one gallons in July.

What peculiarities of expansion does water exhibit?

527. Water, as it decreases in temperature toward the freezing point, exhibits phenomena which are wholly at variance with the general

law that bodies expand by heat and contract by cold, or by a withdrawal of heat.*

As the temperature of water is lowered, it continues to contract until it arrives at a temperature of 39° F., when all further contraction ceases. The volume or bulk is observed to remain stationary for a time, but on lowering the temperature still more, instead of contraction, expansion is produced, and this expansion continues at an increasing rate until the water is congealed. At the moment also of its conversion into ice, it undergoes a still further expansion.

When is water of the greatest density?

528. Water attains its greatest density, or the greatest quantity is contained in a given bulk, at a temperature of 39° F.

As the temperature of water continues to decrease below 39°, the point of its greatest density, its particles, from their expansion, necessarily occupy a larger space than those which possess a temperature somewhat more elevated. The coldest water, therefore, being lighter, rises and floats upon the surface of the warmer water. On the approach of winter this phenomenon actually takes place in our lakes, ponds and rivers. When the surface-water becomes sufficiently chilled to assume the form of ice, it becomes still lighter, and continues to float. By this arrangement, water and ice being almost perfect non-conductors of heat, the great mass of the water is protected from the influence of cold, and prevented from becoming chilled throughout.

If water constantly grew heavier as its temperature diminished (as is the case with most liquids), the colder particles at the surface would constantly sink, until the whole body of water was reduced to the freezing point. Again, if ice was not lighter than water, it would sink to the bottom, and by the continuance of this operation, a river or lake would soon become an immense solid mass of ice, which the heat of summer would be insufficient to dissolve. The temperate regions of the earth would thus be rendered uninhabitable. Among all the phenomena of the natural world, there is no more striking illustration of the wisdom of the Creator, and of the evidences of design, than in this wonderful exception to a great general law.

Why does water expand in freezing?

The expansion of water at the moment of freezing is attributed to a new and peculiar arrangement of its particles. Ice is, in reality, crystallized water, and during its formation the particles arrange themselves in ranks and lines which cross each other at angles of 60° and 120°, and consequently occupy more space than when liquid. This may be seen by examining the surface of water in a saucer while freezing.

A beautiful illustration of this crystallization of water in freezing is seen in the frost-work upon windows in winter, caused by the congelation of the vapor of the room when it comes in contact with the cold surface of the glass.

* A few other liquids besides water expand with a reduction of temperature. Fused iron, antimony, zinc, and bismuth, are examples of such expansion. Mercury is a remarkable instance of the reverse, for when it freezes, it suffers a very great contraction.

All these frost-work figures are limited by the laws of crystallization, and the lines which bound them, form among themselves no angles but those of 30°, 60°, and 120°. If we fracture thin ice, by allowing a pole or weight to fall upon it, the fracture will have more or less of regularity, being generally in the form of a star, with six equi-distant radii, or angles of 60°.

With what force does water expand in freezing?

529. The force exerted by the expansion of water in the act of freezing is very great. As an illustration, the following experiment may be quoted:—Cast-iron bomb-shells, thirteen inches in diameter and two inches thick, were filled with water, and their apertures or fuse-holes firmly plugged with iron bolts. Thus prepared, they were exposed to the severe cold of a Canadian winter, at a temperature of about 19° below zero. At the moment the water froze, the iron plugs were violently thrust out, and the ice protruded, and in some instances the shells burst asunder, thus demonstrating the enormous interior pressure to which they were subjected by water assuming a solid state.

The rounded and weather-worn appearance of rocks is mainly due to the expansion of freezing water, which penetrates into their fissures, and is absorbed into their pores by capillary attraction. In freezing, it expands and detaches successive fragments, so that the original sharp and abrupt outline is gradually rounded and softened down.

The bursting of earthen water vessels, and of water pipes, by the freezing of water contained in them, are familiar illustrations of the same principle.

By allowing the water to run in a service-pipe, we prevent its freezing, because the motion of the current prevents the crystals from forming and attaching themselves to the sides of the pipe.

At what temperature does water freeze?

530. The ordinary temperature at which water freezes is 32°, Fahrenheit's thermometer. This rule applies only to fresh water; salt water never freezes until the surface is cooled down to 27°, or five degrees lower than the freezing point of winter.

Under some circumstances pure water may be cooled down to a temperature much below 32° without freezing. Thus, if pure, recently-boiled water, be cooled very slowly and kept very tranquil, its temperature may be lowered to 21° without the formation of ice; but the least motion causes it to congeal suddenly, and its temperature rises to 32°.

Why is the ice produced by the freezing of salt water free from salt?

531. The ice produced by the freezing of sea or salt water is generally fresh and free from salt, since water in freezing, if sufficient freedom of motion be allowed to its particles, expels all impurities and coloring matters. The ice formed in the congelation of a solution of indigo is colorless, since the water in which the indigo was dissolved expels the blue coloring matter in freezing.

What is the origin of the minute bubbles seen in ice?

Blocks of ice are generally filled with minute air-bubbles; this is owing to the fact that the water in freezing expels the air contained in it, and many of the liberated bubbles become lodged and imbedded in the thickening fluid.

In what manner do gases expand by heat?

532. Gases and aeriform substances expand 1-490th of the bulk which they possess at 32° for every degree of heat which they receive above that point, and contract in the same proportion for every degree of heat withdrawn from them.

Thus, 490 cubic inches of air at 32° would so expand as to occupy an inch more space at 33°, and by the addition of another degree of heat, raising its temperature to 34°, it would occupy an additional inch, and so on. In a like manner, by the withdrawal of heat, 490 cubic inches of air would occupy an inch less space at 31° than at 32°; two inches less at 30°, and so on. The same law holds good for all other gases, and for vapors and steam.

Illustrations of the expansion of air by heat are most familiar. If a bladder partially filled with confined air be laid before the fire, the air contained in it may be expanded to a degree sufficient to burst the bladder. Chestnuts laid upon a heated surface, burst with a loud report on account of the expansion of the air within their shells. The process of warming and ventilating buildings depends entirely upon the application of this principle of the expansion and contraction of air by the increase and diminution of heat.

How may the expansion and contraction of bodies be applied to the measurement of heat?

533. As the magnitude of every body changes with the heat to which it is exposed, and as the same body, when subjected to calorific influences under the same circumstances has always the same magnitude, the expansions and contractions which are the constant effects of heat, may be taken as the measure of the cause which produced them.

What are the instruments for measuring heat called?

534. The instruments for measuring heat are Thermometers and Pyrometers. The former are used for measuring moderate temperatures; the latter for determining the more elevated degrees of heat.

Liquids are better adapted than either solids or gases for measuring the effects of heat by expansion and contraction; since in solids the direct expansion by heat is so small as to be seen and recognized with difficulty, and in air or gases it is too extensive, and too liable to be affected by variations in the atmospheric pressure. From both of these disadvantages liquids are free.

The liquid generally used in the construction of thermometers is mercury, or quicksilver.

Why is mercury especially adapted for the construction of thermometers?

Mercury possesses greater advantages for this purpose than any other liquid. It is, in the first place, eminently distinguished for its fluidity at all ordinary temperatures; it is, in addition, the only body in a liquid state whose variations in volume, or magnitude, through a considerable range of temperature are exactly uniform and proportional with every increase and diminution of heat. Mercury, moreover, boils at a higher temperature than any other liquid, except certain oils; and, on the other hand, it freezes at a lower temperature than all other liquids, except some of the most volatile, such as ether and alcohol. Thus a mercurial thermometer will have a wider range than any other liquid thermometer. It is also attended with this convenience, that the extent of temperature included between melting ice and boiling water stands at a considerable distance from the limits of its range, or its freezing and boiling points.

Describe the mercurial thermometer.

535. The mercurial thermometer consists essentially of a glass tube with a bulb at one end, partially filled with mercury. The mercury introduced through an opening in the end of the tube is afterward boiled, so as to expel all air and moisture, and fill the tube with its own vapor. The open end of the tube is then closed, by fusing the glass, and as the mercury cools it contracts, and collects in the bulb and lower part of the tube, leaving a vacuum above, through which it may again expand and rise on the application of heat. In this condition the thermometer is complete, with the exception of graduation.

How are thermometers graduated?

536. As thermometers are constructed of different dimensions and capacities, it is necessary to have some fixed rules for graduating them, in order that they may always indicate the same temperature under the same circumstances, as the freezing-point, for example. To accomplish this end the following plan has been adopted:—The thermometers are first immersed in melting snow or ice. The mercury will be observed to stop in each thermometer-tube at a certain height; these heights are then marked upon the tubes. Now it has been ascertained that at whatever time and place the instruments may be afterward immersed in melting snow or ice, the mercury contained in them will always fix itself at the point thus marked. This point is called the freezing point of water.

Another fixed point is determined by immersing the instruments in boiling water. It has been found that at whatever time or place the instruments are immersed in pure water, when boiling, provided the barometer stands at the height of thirty inches, the mercury will always rise in each to a certain height. This, therefore, forms another fixed point on the scale, and is called the boiling point.

How is the thermometer of Fahrenheit graduated?

Thus far all thermometers are constructed alike. In the thermometer most generally used, and which is known as Fahrenheit's, the intervals on the scale, between the freezing and boiling points, are divided into 180 equal parts. This division is similarly continued below the freezing point to the place 0, called zero, and each division upward from that is marked with the successive numbers 1, 2, 3, etc. The freezing point will now be the 32d division, and the boiling point will be the 212th division. These divisions are called degrees, and the boiling point will therefore be 212°, and the freezing temperature, 32°. Fig. 203 represents the usual form of thermometer, with its graduated scale.

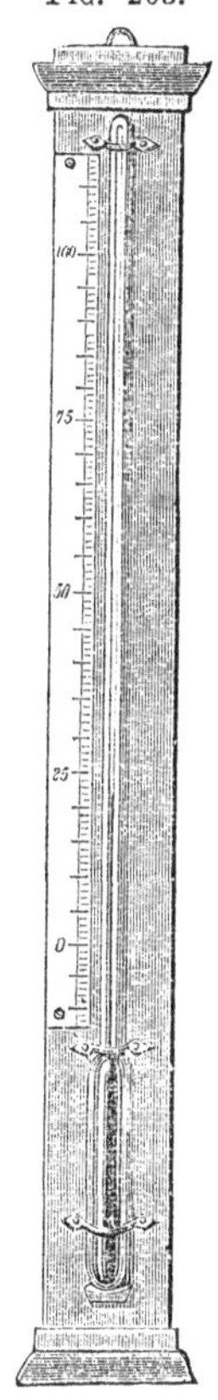

Fig. 203.

Thermometers of this character are called Fahrenheit's, from a Dutch philosophical instrument-maker who first introduced this method of graduation in the year 1724.

What other thermometers beside Fahrenheit's are used?

537. In addition to Fahrenheit's thermometer, two others are extensively used, which are known as Reaumur's, and the Centigrade thermometer, or thermometer of Celsius.

What constitutes the difference between the different varieties of the thermometer?

The only difference between these three kinds of thermometers is the difference in graduating the interval between the freezing and boiling points of water. Reaumur's is divided into eighty degrees, the Centigrade into one hundred, and Fahrenheit's into one hundred and eighty. According to Reaumur, water freezes at 0°, and boils at 80°; according to Centigrade, it freezes at 0°, and boils at 100°; and according to Fahrenheit, it freezes at 32°, and boils at 212°; the last, very singularly, commences counting, not at the freezing point, but 32° below it.

Fig. 204.

R C F

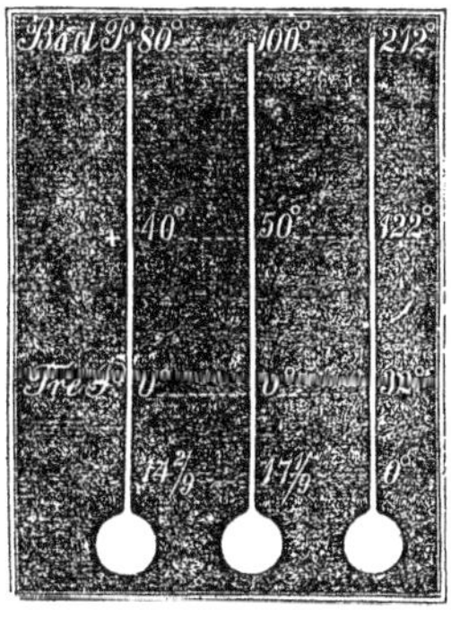

The difference between these instruments can be easily seen by reference to Fig. 204.

In England, Holland, and the United States, the thermometer most generally used is Fahrenheit's. Reaumur's scale is used in Germany, and the Centigrade in France, Sweden, and some other parts of Europe. The scale of the Centigrade is by far the simplest and most rational method of graduation, and at the present it is almost universally adopted for scientific purposes.

538. The thermometer was invented about the year 1600; but, like many other inventions, the merit of its discovery is not to be ascribed to one person, but to be distributed among many.

How is cold of great intensity indicated?

539. As the temperature is lowered, the mercury in Fahrenheit's thermometer gradually sinks, until it reaches a point 39° below zero, where it freezes. Mercury, therefore, can not be made available for measuring cold of a greater intensity. This difficulty is, however, obviated by using a thermometer filled with alcohol colored red, as this fluid, when pure, never freezes, but will continue to sink lower and lower in the tube as the cold increases. Such a thermometer is called a spirit thermometer.

How is heat of great intensity measured?

540. If a Fahrenheit's thermometer be heated, the mercury contained in it will rise in the tube until it reaches 660°, at which temperature it begins to boil. A slight additional heat forms vapor sufficient to burst the tube. Mercury, therefore, can not be used to measure degrees of heat of greater intensity than 660° F. Temperatures greater than this are determined by means of the expansion of solids; and instruments founded upon this principle are commonly called pyrometers.

FIG. 205.

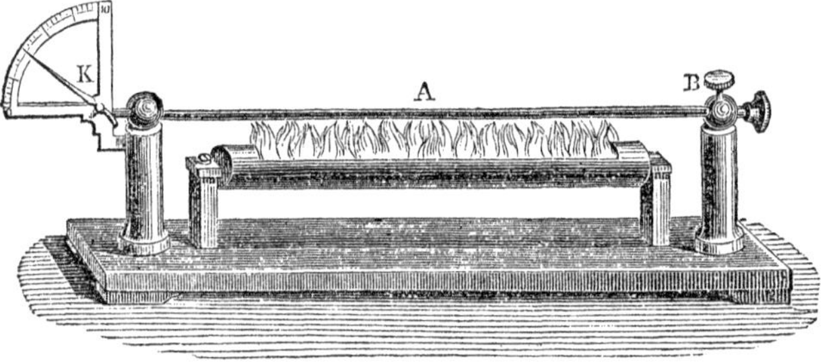

Explain the construction of the pyrometer.

The construction of the pyrometer is represented in Fig. 205. A represents a metallic bar, fixed at one end, B, but left free at the other, and in contact with the end of a pointer K, moving freely over a graduated scale. If the bar be heated by the flame of alcohol, the metal expands, and pressing upon the end of the pointer, moves it, in a greater or less degree. In this manner, the effect of heat, applied for a given length of time, to bars of different metals, having the same length and diameter, may be determined.

What is an air-thermometer?

541. The first thermometer used consisted of a column of air confined in a glass tube over colored water. Heat expands the air and increases the length of the column downward, pushing the water before it: cold produces a contrary effect. The temperature is thus indicated by the height at which the water is elevated in the tube. Fig. 206 represents the principle of the construction of the air-thermometer.

FIG. 206.

Does a thermometer inform us how much heat a substance contains?

A thermometer does not inform us how much heat any substance contains, but it merely points out the difference in the temperature of two or more substances. All we learn by the thermometer is whether the temperature of one body is greater or less than that of another; and if there is a difference, it is expressed numerically—namely, by the degrees of the thermometer. It must be remembered that these degrees are part of an arbitrary scale, selected for convenience, without any reference whatever to the actual quantity of heat present in bodies.

After the expansion of solids by heat, what other effect is next observed?

542. The first effect produced by heat upon solids is expansion. If the heat be augmented, they change their aggregate state and melt, or become liquid. Many solids become soft before melting, so that they may be kneaded; for instance, wax, glass, and iron. In this position, glass can be bent and molded with facility, and iron can be forged or welded.

What is Liquefaction?

543. By Liquefaction we understand the conversion of a solid into a liquid by the agency of heat, as solid ice is converted into water by the heat of the sun.

Heat is supposed to convert a solid into a liquid, by forcing its constituent particles asunder to such an extent that the force of cohesion is overcome or destroyed.

What is Solution?

544. When a solid is immersed in a liquid, and gradually disappears in it, the process is termed solution, and not liquefaction. A solution is the result of an attraction or affinity between a solid and a fluid; and when a solid disappears in a liquid, if the compound exhibits perfect transparency, we have an example of a perfect solution.

When is a solution said to be saturated?

When a fluid has dissolved as much of a solid as it is capable of doing, it is said to be saturated; or, in other words, the affinity or attraction of the fluid for the solid continues to operate to a certain point, where it is overbalanced by the cohesion of the solid; it then ceases, and the fluid is said to be saturated.

How does a solution differ from a mixture?

A solution is a complete union; a mixture is a mere mechanical union of bodies.

In most cases, the addition of heat to a liquid greatly increases its solvent properties. Hot water will dissolve much more sugar than cold water; and hot water will also dissolve many things which cold water is unable to affect.

What is Vaporization?

545. If heat be imparted in sufficient quantity to a body in a liquid state, it will pass into a state of vapor. Thus, water being heated sufficiently will pass into the form of steam. This change is called VAPORIZATION.

What is Condensation?

546. If a body in a state of vapor lose heat in sufficient quantity, it will pass into a liquid state. Thus, if a certain quantity of heat be abstracted from steam, it will become water. This change is called CONDENSATION.

The change from a state of vapor to a liquid is termed condensation, because, in so doing, the body always undergoes a very considerable diminution of volume, and therefore becomes condensed. Most solids become liquefied before they vaporize; but some pass at once, on the application of heat, from the state of a solid to that of a vapor, without assuming the liquid condition.

Is any particular temperature requisite for the formation of vapors?

547. The melting of a solid, or its conversion into a liquid, only occurs when the solid is heated up to a certain fixed point; but the conversion of a liquid into a vapor takes place at all temperatures.

If in a hot day we expose a vessel filled with cold water to the open air, we find that the quantity of water rapidly diminishes, that is, it evaporates, which means that it is converted into vapor and diffused through the air.

What is the appearance of vapor?

548. The vapor of water, and all other vapors, are invisible and transparent. The water which has become diffused through the air by evaporation only becomes visible when, on returning to its fluid condition, it forms mist, cloud, dew, or frost.

Steam, which is the vapor of boiling water, is invisible, but when it comes in contact with air, which is cooler, it becomes condensed into small drops, and is thus rendered visible.

The proof of this may be found in examining the steam as it issues from an orifice, or the spout of a boiling kettle: for a short space next to the opening no steam can be seen, since the air is not able to condense it; but as it spreads and comes in contact with a larger volume of air, the invisible vapor becomes condensed into drops, and is thus rendered visible.

The visible matter popularly called steam, should be, therefore, distinguished from steam proper, or the aeriform state of water. The cloud, or smoke-like matter observed, is really not an air or vapor at all, but a collection of minute bubbles of water, wafted by a current either of true steam, or, more frequently, of mere moist air.

The myriads of minute globules of water into which the steam is condensed are separately invisible to the naked eye, but each, nevertheless, reflects a minute ray of white light. The multitude of these reflecting points, therefore, make the space through which they are diffused appear like a cloudy body, more or less white, according to their abundance.

Is a boiling temperature requisite for the production of steam?

The surface of any watery liquid, whose temperature is about 20° warmer than any superincumbent air, rapidly gives off true steam. It is not necessary, therefore, for the production of steam that water should be raised to the boiling temperature.

Is vapor always present in air?

549. Air without vapor (theoretically called dry air) is not known to exist in nature, and is probably not producible by art.

What is the relative space occupied by liquids and vapors?

550. Liquids in passing into vapors occupy a much greater space than the substances from which they are produced. Water, in passing from its point of greatest density into steam, expands to nearly 1700 times its volume.

Fig. 207 represents the comparative volume of water and steam.

Fig. 207.

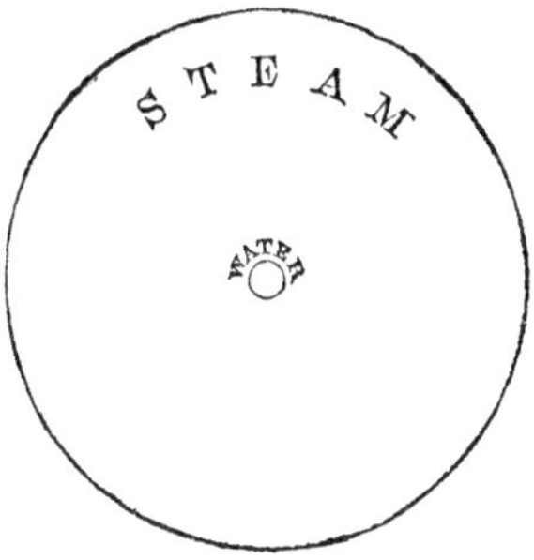

Is the density of vapors uniform?

551. Vapors are of all degrees of density. The vapor of water may be as thin as air, or almost as dense as water.

The opinion formerly prevailed that vapors could not exist by themselves as such, but that they were dissolved in the air in the same way as salt is dissolved in water. The fallacy of this idea is proved by the fact that evaporation goes on more rapidly in a vacuum, where no substance whatever is present, than in the air.

What circumstances influence evaporation?

552. Evaporation takes place from the surfaces of bodies only, and is influenced in a great degree by the temperature, dryness, stillness, and density of the atmosphere.

How does temperature influence evaporation?

The effect of temperature in promoting evaporation may be illustrated by placing an equal quantity of water in two saucers, one of which is placed in a warm and dry, and the other in a cold and damp, situation. The former will be quite dry before the latter has suffered an appreciable diminution.

How does the state of the air influence evaporation?

When water is covered by a stratum of dry air, the evaporation is rapid, even when its temperature is low; whereas it goes on very slowly if the atmosphere contains much vapor, even though the air be very warm.

Evaporation is far slower in still air than in a current. The air immediately in contact with the water soon becomes moist, and thus a check is put to evaporation. But if the air be removed by wind from the surface of the water as soon as it has become charged with vapor, and its place supplied with fresh air, then the evaporation continues on without interruption.

Evaporation is by no means confined to the surface of liquids; but takes place from the surface of the soil, and from all animal and vegetable productions. Evaporation takes place to a very considerable extent from the surface of snow and ice, even when the temperature of the air is far below the freezing point.

What singular circumstance is connected with the diffusion of vapors?

553. A very singular circumstance is connected with the diffusion of vapors throughout the atmosphere, viz.: that the vapors of all bodies arise into any space filled with air, in the same manner as if air were not present, the two fluids seeming to be independent of each other.

Thus as much vapor of water can be forced into a vessel filled with air as into one from which the air has been exhausted.

Explain the phenomena of the "spheroidal state" of liquids.

554. When a drop of water falls upon a surface highly heated, as of metal, it will be observed to roll along the surface without adhering, or immediately passing into vapor. The explanation of this is, that the drop of water does not in reality touch the heated surface, but is buoyed up and supported on a layer of vapor which intervenes between the bottom of the drop and the hot surface. This vapor is produced by the heat which is radiated from the hot substance, before the liquid can come in contact with it, and being constantly renewed, continues to support the drop. The drop generally rolls because the current of air which is always passing over a heated surface drives it forward. The drop evaporates slowly, because the layer of vapor between the hot surface and the liquid prevents the rapid transmission of heat. The liquid resting upon a cushion of steam continually evolved from its lower surface by heat, assumes a rounded, or globular shape, as the result of the gravity of its particles toward its own center.

The designation which has been given to the condition which water and other liquids assume when dropped upon very hot surfaces, is that of the "spheroidal state."

If the surface upon which the liquid rests is cooled down to such an extent that vapor is not generated rapidly, and in sufficient quantity to support the drop, it will come in contact with the surface, and heat being communicated by conduction, will transform it instantly into steam.

This is the explanation of the practice adopted by laundresses of touching a flat-iron with moisture to ascertain whether the surface is sufficiently hot.

If the temperature of the iron is not elevated sufficiently, the moisture wets the surface, and is evaporated; but at a higher degree of temperature, the moisture is repelled.

The phenomenon of the spheroidal condition of water furnishes an explanation of the feats often performed by jugglers, of plunging the hands with impunity into molten lead, or iron. The hand is moistened, and when passed into the liquid metal the moisture is vaporized, and interposes between the metal and the skin a sheath of vapor. In its conversion into vapor, the moisture absorbs heat, and thus still further protects the skin.

What is ebullition?

555. When a liquid is heated sufficiently to form steam, the production of vapor takes place principally at that part where the heat enters; and when the heating takes place not from above, but from the bottom and sides, the steam as it is produced rises in bubbles through the liquid, and produces the phenomenon of boiling, or ebullition.

What is the boiling point?

556. The temperature at which vapor rises with sufficient freedom to cause the phenomenon of ebullition, is called the boiling point.

Is the boiling point in different liquids the same?

557. Different liquids boil at different temperatures. The boiling point of a liquid is, therefore, one of its distinctive characters.

Thus water, under ordinary circumstances, begins to boil when it is heated up to 212° F.; alcohol at 173°; ether at 96°; syrup at 221°; linseed oil at 640°.

What is simmering?

The gentle tremor, or undulation, on the surface of water which precedes boiling, and which is termed "simmering," is owing to the collapse of the bubbles of steam as they shoot upward and are condensed by the colder water. The first bubbles which form are not steam, but air which the heat expels from the water. As the temperature of the whole mass of the water increases, the bubbles are no longer condensed and collapsed, but rise through to the surface; and the moment that this takes place boiling commences. The singing of a tea-kettle before boiling is occasioned by the irregular escape of the air and steam expelled from the water through the spout of the tea-kettle, which acts in the manner of a wind-instrument in producing a sound.

How does the pressure of the atmosphere affect the boiling of liquids?

558. Liquids, in general, being boiled in open vessels, are subjected to the pressure of the atmosphere. The tendency of this pressure is to prevent and retard the particles of water from expanding to a sufficient extent to form steam. Hence if the pressure of the atmosphere varies, as it does at different times and places, or if it be increased or diminished by artificial means, the boiling point of a liquid will undergo a corresponding change.

How may the temperature at which water boils be used for determining elevations?

559. As we ascend into the atmosphere the pressure is diminished, because there is less of it above us; it therefore follows, that water at different heights in the atmosphere will boil at different temperatures, and it has been found by observation, that an elevation of 550 feet above the level of the sea causes a difference of one degree in its boiling point. Hence the boiling point of water becomes an indication of the height of any station above the sea-level, or in other words, an indication of the atmospheric pressure; and thus by means of a kettle of boiling water and a thermometer, the height of the summit of any mountain may be ascertained with a great degree of accuracy. If the water boils at 211° by the thermometer, the height of the place is 550 feet; if at 210°, the height is 1100 feet, and so on, it being only necessary to multiply 550 by the number of degrees on the thermometer between the actual boiling point and 212°, to ascertain the elevation. In the city of Quito, in South America, water boils at 194° 2″ F.; its height above the sea-level is, therefore, 9,541 feet.

As we descend into mines, the pressure of the atmosphere is increased, there being more of it above us than at the surface of the earth. Water, therefore, must be heated to a higher temperature before it will boil, and it has been found that a descent of 550 feet, as before, makes a difference of one degree.

How can the boiling point of liquids be changed artificially?

560. In a like manner, if by artificial means we increase or diminish the pressure of the atmosphere on the surface of a liquid, we change its boiling point. If water be heated in a vacuum, ebullition will commence at a point 140° lower than in the open air. If a vessel of ether be placed under the receiver of an air-pump, and the atmospheric pressure removed from its surface, the vapor rises so abundantly that ebullition is produced without any increase of temperature.

How is sugar boiled in the process of refining?

Several beautiful applications in the arts have been made of the principle that liquids boil at a lower temperature when freed from the pressure of the atmosphere than in the open air.

In the refining of sugar, if the syrup is boiled in the open air, the temperature of the boiling point is so high that portions of the sugar become decomposed by the excess of heat, and lost or injured; the syrup is therefore boiled in close vessels from which the air has been previously exhausted, and in this way the water of the syrup may be evaporated at a temperature so low as to prevent all injury from heat.

For cooking, this application could not be carried out. The water might, indeed, be made to boil at a temperature much less than 212°, but owing to its diminished heat would not produce the desired effect.

What is distillation?

561. Distillation is a process by which one body is separated from another by means of heat, in cases where one of the bodies assumes the form of vapor at a lower temperature than the other; this first

rises in the form of vapor, and is received and condensed in a separate vessel.

By this means very volatile bodies can be easily separated from less volatile ones; as brandy and alcohol from the less volatile water which may be mixed with them. Water of extreme purity can also be obtained by distillation, because the non-volatile and earthy substances contained in all spring waters do not ascend with the vapor, but remain behind in the vessel.

Distillation upon a small scale is effected by means of a peculiar-shaped vessel, called a retort, Fig. 208, which is half filled with a volatile liquid and heated; the steam, as it forms, passes through the neck of the retort into a glass receiver set into a vessel filled with cold water, and is then condensed.

FIG. 208.

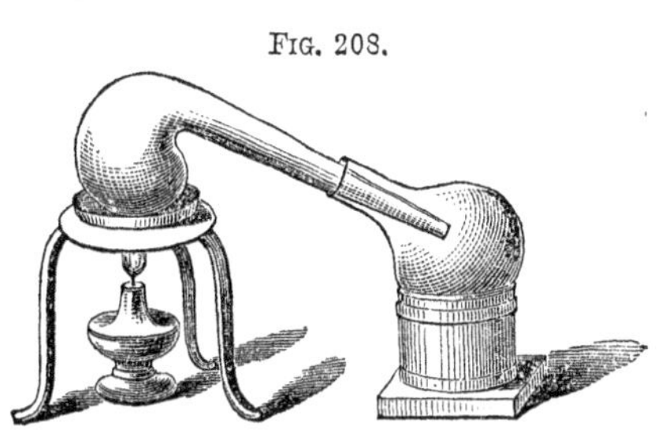

When the operation of distillation is conducted on an extensive scale, a large vessel called a "*still*" is used, and, for condensing the vapor, vats are constructed, holding serpentine pipes, or "worms," which present a greater condensing surface than if the pipe had passed directly through the vat. To keep the coil of pipe cool, the vats are kept filled with cold water. In Fig. 209, *a* is a furnace, in which is fixed a copper vessel, or still, to contain the liquid. Heat being applied, the steam rises in the head, *b*, and passes through the worm, *d*, which is placed in a vessel of water, the refrigerator. The vapor thus generated is condensed in its passage, and passes out as a liquid by the external pipe into a receiver.

FIG. 209.

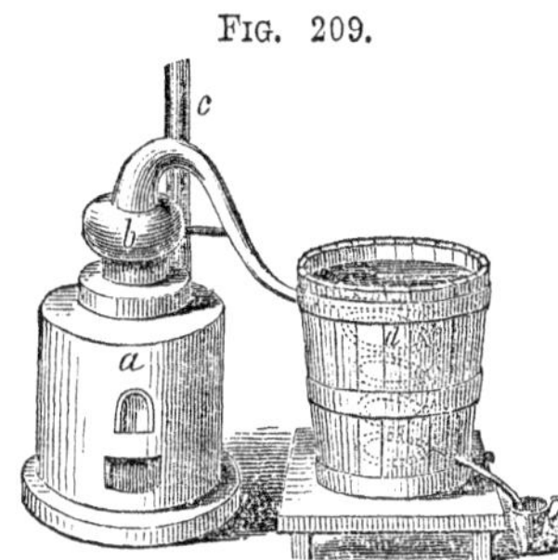

What is the difference between drying by heat and distillation?

The difference between drying by heat and distillation is, that in one case, the substance vaporized, being of no use, is allowed to escape or become dissipated in the atmosphere; while in the other, being the valuable part, it is caught and condensed into the liquid form. The vapor arising from damp linen, if caught and condensed would be distilled water; the vapor given out by bread while baking, would, if collected, be a spirit like that obtained in the distillation of grain.

What is sublimation?

562. As some substances, by the application of heat, pass directly from the solid condition to the state of vapor, so some substances, as camphor, sulphur, arsenic, etc., when vaporized

by heat, deposit their condensed vapors in a solid form. This process is termed sublimation.

What remarkable circumstance attends liquefaction and vaporization?

563. One of the most remarkable circumstances which accompany the phenomena, both of liquefaction and vaporization, is the disappearance of the heat which has effected the change.

How may this principle be illustrated?

Thus, if a thermometer be applied to a mass of snow, or ice just upon the point of melting, it will be found to stand at 32° F. If the ice be placed in a vessel over a fire, and the temperature tested at the moment it has entirely melted, the water produced will have only the temperature of 32°, the same as that of the original ice. Heat, however, during the whole process of melting, has been passing rapidly into the vessel from the fire, and if a quantity of mercury, or a solid of the same size, had been exposed to the same amount of heat, it would have constantly increased in temperature. It is clear, therefore, that the conversion of ice, a solid, into water, a liquid, has been attended with a disappearance of heat.

Again: if one pound of water, having a temperature of 174°, be mixed with one pound of snow at 32°, we shall obtain two pounds of water, having a temperature of 32°. All the heat, therefore, which was contained in the hot water is no longer to be detected by the thermometer, it having been entirely used up, or disposed of in converting snow at 32° into water at 32°. Such disappearances always occur whenever a solid is converted into a liquid.

If, however, a pound of water at 32°, instead of ice at the same temperature, had been mixed with a pound of water at 174°, we shall obtain two pounds at 103°, a temperature exactly intermediate between the temperatures of the components But if the pound at 32° had been solid instead of liquid, then the mixture, as before explained, would have had a temperature of 32°. It is evident, therefore, that it is the process of liquefaction, and it alone, which renders latent or insensible all that heat which is sensible when the pound of water at 32° is liquid.

How may the absorption of heat in evaporation be rendered evident?

In the same manner heat disappears when a liquid is converted into a vapor. The absorption of heat, in this instance, may be easily rendered perceptible to the feelings by pouring a few drops of some liquid which readily evaporates, such as ether, alcohol, etc., upon the hand. A sensation of cold is immediately experienced, because the hand is deprived of heat, which is drawn away to effect the evaporation of the liquid. On the same principle, inflammation and feverish heat in the head may be allayed by bathing the temples with Cologne water, alcohol, vinegar, etc.

If we surround the bulb of a thermometer loosely with cotton, and then moisten the latter with ether, the thermometer will speedily fall several degrees.

Why can not water impart additional heat after boiling?

Water when placed in a vessel over a fire, gradually attains the boiling temperature, or 212°; but afterward, however much we may increase the fire, it becomes no hotter, all the heat which is added serving only to convert the water at 212° from a liquid condition into steam, or vapor, at 212°.

How do we know that steam at 212° is hotter than water at the same temperature?

564. If we immerse a thermometer in boiling water, it stands at 212°; if we place it in steam immediately above it, it indicates the same temperature. We know, however, that steam contains more heat than boiling water, because if we mix an ounce of water at 212° with five and a half ounces of water at 32°, we obtain six and a half ounces of water at a temperature of about 60°; but if we mix an ounce of steam at 212° with five and a half ounces of water at 32°, we obtain six and a half ounces of water at 212°. The steam, from which the increased heat is all derived, contains as much more heat than the ounce of water at the same temperature, as would be necessary to raise six and a half ounces of water from the temperature of 60° to 212°, or six and a half times as much heat as would be requisite to raise one ounce of water through about 152° of temperature. This quantity of heat will, therefore, be found by multiplying 152° by six and a half, which will give a product of 983°—the excess of heat contained in an ounce of steam at 212° over that contained in an ounce of boiling water at the same temperature.

What becomes of the heat which disappears in liquefaction and vaporization?

565. In the conversion of solids into liquids, and liquids into vapors by heat, we may suppose the heat, the solid, and the liquid to have respectively combined together;—forming a liquid in the one case, and a vapor in the other. A liquid, therefore, may be regarded as a compound of a solid and heat, and a vapor as a compound of heat and the liquid from which it was formed. The heat which disappears in these combinations is called LATENT, or COMPOUND HEAT.

What are freezing mixtures?

The absorption of heat consequent on the conversion of solids into liquids, has been taken advantage of in the arts for the production of artificial cold; and the compounds of different substances which are made for this purpose, are called freezing mixtures.

Why does the mixture of snow and salt produce intense cold?

The most simple freezing mixture is snow and salt. Salt dissolved in water would occasion a reduction of temperature, but when the chemical relations of two solids are such, that on mixing, both are rendered liquid, a still greater degree of cold is produced. Such a relation exists between salt and snow, or ice, and therefore the latter substances are used in preference to water. When the two are mixed, the salt causes the snow to melt by reason of its attraction for water, and the water formed dissolves the salt: so that both pass from the solid to the liquid condition. If the operation is so conducted that no heat is supplied from any external source, it follows that the heat absorbed in liquefaction must be obtained from the salt and snow which comprise the mixture, and they must therefore suffer a depression of temperature proportional to the heat which is rendered latent.

In this way a degree of cold equal to 40° below the freezing point of

How great a degree of cold can be obtained by freezing mixtures?

water may be obtained. The application of this experiment to the freezing of ice-creams is familiar to all.

By mixing snow and sulphuric acid together in proper proportions, a temperature of 90° below zero can be obtained without difficulty.

Why is the air in spring cold and chilly?

The air in the spring of the year, when the ice and snow are thawing, is always peculiarly cold and chilly. This is due to the constant absorption of heat from the air by the ice and snow in their transition from a solid to a liquid state.

Why does a shower in summer cool the air?

A shower of rain cools the air in summer, because the earth and the air both part with their heat to promote evaporation. In a like manner, the sprinkling of a hot room with water cools it.

Why is the warmth of a country promoted by draining?

The draining of a country increases its warmth, since by withdrawing the water, evaporation is diminished, and less heat is subtracted from the earth.

Why do wet feet or clothes tend to impair the health of the body?

The danger arising from wet feet and clothes is owing to the absorption of heat from the body by the evaporation from the surfaces of the wet materials; the temperature of the body is in this way reduced below its natural standard, and the proper circulation of the blood interrupted.

Why is not a tin vessel containing water exposed to the fire destroyed?

566. The absorption of heat in the process by which liquids are converted into vapor, will explain why a vessel containing a liquid that is constantly exposed to the action of fire, can never receive such a degree of heat as would destroy it. A tin kettle containing water may be exposed to the action of the most fierce furnace, and remain uninjured; but if it be exposed, without containing water, to the most moderate fire, it will soon be destroyed. The heat which the fire imparts to the kettle containing water is immediately absorbed by the steam into which the water is converted. So long as water is contained in the vessel, this absorption of heat will continue; but if any part of the vessel not containing water be exposed to the fire, the metal will be fused, and the vessel destroyed.

Under what circumstances does latent heat become sensible?

567. When vapors are condensed into liquids, and liquids are changed into solids, the latent heat contained in them is set free, or made sensible.

If water be taken into an apartment whose temperature is several degrees below the freezing point, and allowed to congeal, it will render the room sensibly warmer. It is, therefore, in accordance with this principle that tubs of water are allowed to freeze in cellars in order to prevent excessive cold.

It is from this cause that oceans, seas, and other large collections of water are most powerful agents in equalizing the temperature of the inhabited parts of the globe. In the colder regions, every ton of water converted into ice gives out and diffuses in the surrounding region as much heat as would raise a ton of water from 32° to 174°; and, on the other hand, when a rise of temperature takes place, the thawing of the ice absorbs a like quantity of heat: thus, in the one case, supplying heat to the atmosphere when the temperature falls; and, in the other, absorbing heat from it when the temperature rises.

In the winter, the weather generally moderates on the fall of snow; snow is frozen water, and in its formation heat is imparted to the atmosphere, and its temperature increased.

Why is steam especially adapted for warming and cooking?

Steam, on account of the latent heat it contains, is well adapted for the warming of buildings, or for cooking. In passing through a line of pipes, or through meat and vegetables, it is condensed, and imparts to the adjoining surfaces nearly 1000° of the latent heat which it contained before condensation.

Steam burns much more severely than boiling water, for the reason that the heat it imparts to any surface upon which it is condensed is much greater than that of boiling water.

Is the quantity of heat in all bodies the same?

568. All bodies contain incorporated with them more or less of heat; but equal weights of dissimilar substances, having the same sensible temperature, contain unequal quantities of heat.

How may this be demonstrated?

Thus if we place a pound of water and a pound of mercury over a fire, it will be found that the mercury will attain to any given temperature much quicker than the water. Or if we perform the converse of this experiment, and take two equal quantities of mercury and water, and having heated them to the same degree of temperature, allow them to cool freely in the air, it will be found that the water will require much more time to cool down to a common temperature than the mercury. The water obviously contains more heat at the elevated temperature than the mercury, and therefore requires a longer time to cool.

What is the meaning of the term specific heat?

569. Dissimilar substances require, respectively, different quantities of heat to raise their temperatures one degree; and the quantity of heat necessary to produce this effect upon a body is termed its specific heat. In like manner, the weight which a body includes under a given volume, is termed its specific gravity.

What is understood by capacity for heat?

570. A substance is said to have a greater or less capacity for heat, according as a greater or less quantity of heat is required to produce a definite change of temperature, or an elevation of temperature of one degree.

How does the capacity for heat in different substances vary?

In general, the capacity of bodies for heat decreases with their density. Thus mercury has a less capacity for heat than water, because its density is greater. Air that is rarefied, or thin, has a greater capacity for heat than dense air. This circumstance will explain, in part, the reason of the very low temperatures which exist at great elevations in the atmosphere. Persons ascending high mountains, or in balloons, find that the cold increases with the elevation. The reason of this is, that as the air expands and becomes rarefied, its capacity for heat is greatly increased, and it therefore absorbs its own sensible heat.

What is the limit of perpetual snow?

In all quarters of the globe, the temperature of the air at a certain height is reduced so low by its rarefaction, that water can not exist in a liquid state. This limit, the height of which varies, being the most elevated at the equator, and the most depressed at the poles, is called the line of PERPETUAL SNOW.*

Air forcibly expelled from the mouth feels cool; in this instance the cold is due to a sudden expansion of the air, by which its capacity for heat is increased.

The capacity for heat also increases with the temperature. Thus it requires a greater amount of heat to elevate the temperature of platinum from 212° to 213°, than from 32° to 33°.

Of all known bodies, water has the greatest capacity for heat.

How may the capacity for heat in different substances be ascertained?

There are several different ways by means of which the capacity of bodies for heat may be determined. One method consists in inclosing equal weights of different bodies heated to the same temperature, in closed cavities in a block of ice, and measuring the respective quantities of water which they produce by melting the ice.

The same result may also be obtained by what is called the method of mixtures. Thus, if we mix 1 pound of mercury at 66° with 1 pound of water at 32°, the common temperature will be 33°. Here the mercury loses 33° and the water gains 1°; that is to say, the 33° of the mercury only elevates the water 1°, therefore the capacity of water for heat is 33 times that of mercury; or, if we call the capacity or specific heat of water 1, then the capacity or specific heat of mercury will be 1-33d or .0303.

In this way the capacities for heat of a great number of bodies has been determined, and tables constructed in which they are recorded. In these tables water is taken as the unit of comparison.

What is the elasticity of vapors?

All vapors are elastic, like air.

The tendency of vapors to expand is unlimited; that is to say, the smallest quantity of vapor will diffuse itself through every part of a vacant space, be its size what it may, exercising a greater or less degree of force against any obstacle which may have a tendency to restrain it.

* The line of perpetual snow at the equator occurs at a height of about 15,000 feet; at the Straits of Magellan, it occurs at an elevation of only 4,000 feet.

The force with which a vapor expands is called its elastic force, or tension.

The elasticity or pressure of vapors is best illustrated in the case of steam, which may be considered as the type of all vapors.

In what manner is the elastic force of steam exerted?

When a quantity of pure steam is confined in a close vessel, its elastic force will exert on every part of the interior of the vessel a certain pressure directed outward, having a tendency to burst the vessel.

What is the elastic force of steam formed in an open vessel?

When steam is generated in an open vessel its elastic force must be equal to the elastic force or pressure of the atmosphère; otherwise the pressure of the air would prevent it from forming and rising. Steam, therefore, produced from boiling water at 212° F., is capable of exerting a pressure of 15 pounds upon every square inch of surface, or one ton on every square foot, a force equivalent to the pressure of the atmosphere.

How may the elastic force of steam be increased or diminished?

If water be boiled under a diminished pressure, and therefore at a lower temperature, the steam which is produced from it will have a pressure which is diminished in an equal degree. If, on the contrary, the pressure under which water boils be increased, the boiling temperature of the water and the pressure of the steam formed will be increased in a like proportion. We have, therefore, the following rule:—

To what is the elastic force of steam always equal?

571. Steam raised from water, boiling under any given pressure, has an elasticity always equal to the pressure under which the water boils.

How is steam of high elastic force generated?

Steam of a high elastic force can only be made in close vessels, or boilers. The water in a steam-boiler, in the first instance, boils at 212°, but the steam thus generated being prevented from escaping, presses on the surface of the water equally as on the surface of the boiler, and therefore the boiling point of the water becomes higher and higher; or in other words, the water has to grow constantly hotter, in order that the steam may form. The steam thus formed has the same temperature as the water which produces it.

To what extent can water be heated under pressure?

The temperature of the water in working steam-boilers is always much greater than 212°. It should also be borne in mind that water, if subjected to sufficient pressure, can be heated to any extent without boiling. There is no limit to the degree to which water may be heated, provided the vessel is strong enough to confine the vapor; but the expansive force of steam is so enormous under these circumstances, as to overcome the greatest resistance which has ever been exerted upon it.

If a boiler, containing water thus overheated many degrees beyond the boiling point, be suddenly opened, and the steam allowed to expand, the whole water is immediately blown out of the vessel as a mist by the steam formed at the same instant throughout every part of the mass. To use a common expression, "the water flashes into steam."

To what extent can steam be heated under pressure? Steam, like water, may be heated to any extent when confined and prevented from expanding with the increase of temperature; in some of the methods lately introduced for purifying oils, etc., the temperature of the steam, before its application, is required to be sufficiently elevated to enable it to melt lead.

What is Superheated Steam? 572. Steam which has been heated in a separate state to a high degree of temperature under pressure, is known as "Superheated Steam." In this condition its mechanical and chemical powers are wonderfully increased.

In the manufacture of lard on an extensive scale the carcass of the whole hog is exposed to the action of steam at very high pressure, this acting upon the mass of flesh and bones, breaks up and reduces the whole to a fat fluid mass. Ordinary steam, under the same circumstances, would dissolve nothing.

Steam has also been recently applied to the carbonization of wood. For this purpose ordinary steam is conducted through red hot pipes, whereby it attains a very high degree of temperature. It is then allowed to pass into a vessel containing wood intended to be converted into charcoal. The heated steam, penetrating into the pores of the wood, drives off the volatile portions, the water, the tar, etc., and leaves the pure carbon alone behind.

What is High-pressure Steam? 573. Steam generated by water boiling at a very high temperature, is known as High-pressure Steam. By this term we mean steam condensed not by withdrawal of heat, but by pressure, just as high-pressure air is merely condensed air. To obtain a double, triple, or greater pressure of steam, we must have twice, thrice, or more steam under the same volume.

What relation exists between sensible and latent heat? 574. The sum of the sensible heat of any vapor, and the latent heat contained in it, is always the same.

It is an established fact that the heat absorbed by vaporization is always less the higher the temperature at which this vaporization takes place, and just in proportion also as vapor or steam indicates a lower temperature by the thermometer, it contains more latent heat. Thus, if water boils at 312°, the heat absorbed in vaporization will be less by 100° than if it boiled at 212°. And again, if water be boiled under a diminished pressure at 112°, the heat absorbed in vaporization will be 100° more than the heat absorbed by water boiled at 212°.

SECTION IV.

What is a Steam-Engine?

575. The Steam-Engine is a mechanical contrivance by which coal, wood, or other fuel, is rendered capable of executing any kind of labor.*

How much mechanical force can be excited by the combustion of two ounces of coal?

The substance which furnishes the means of calling the powers of coal into activity is water; two ounces of coal, with a proper arrangement will evaporate about one pint of water; this will produce 216 gallons of steam, which can exert a mechanical force equivalent to raising a weight of 37 tons to the height of one foot.

How does the force of a man compare with the force generated by the combustion of coal?

It has been found by experiment that the greatest amount of force which a man can exert when applying his strength to the best advantage through the help of machinery, is equal to elevating one and a half millions of pounds to the height of one foot, by working on a treadmill continuously for eight hours. A well-constructed steam-engine will perform the same labor with an expenditure of a pound and a half of coal.

How much coal is equivalent to the whole active power of a man?

The average power of an able-bodied man during his active life, supposing him to work for twenty years at the rate of eight hours per day, is represented by an equivalent of about four tons of coal, since the consumption of that amount will evolve in a steam-engine, fully as much mechanical force.

The great pyramid of Egypt is five hundred feet high, and weighs twelve thousand seven hundred and sixty millions of pounds. Herodotus states that in constructing it one hundred thousand men were constantly employed for twenty years. At the present time, with the consumption of 480 tons of coal, all the materials could be raised to their present position from the ground in comparatively little time.

What is the greatest amount of work ever accomplished by a steam-engine?

The greatest work ever known to have been performed by a steam-engine, was to raise sixty thousand tons of water a foot high with the expenditure of one bushel of coal. This work was accomplished by one of the engines employed in the mines of Cornwall, England.

* "Coals are by it made to spin, weave, dye, print, and dress silks, cottons, woolens and other cloths; to make paper, and print books upon it when made; to convert corn into flour; to express oil from the olive, and wine from the grape; to draw up metals from the bowels of the earth; to pound and smelt it; to melt and mold it; to roll it and fashion it into every desirable form; to transport these manifold products of its own labor to the doors of those for whose convenience they are produced; to carry persons and goods over the waters of rivers, lakes, seas, and oceans, in opposition alike to the natural difficulties of wind and water; to carry the wind-bound ship out of port, to place her on the open deep, ready to commence her voyage; to transport over the surface of the sea and the land, persons and information from town to town, and from country to country, with a speed as much exceeding the ordinary wind, as the ordinary wind exceeds that of a pedestrian."—*Lardner*.

How is steam made available for mechanical purposes?

576. Steam is rendered useful for mechanical purposes simply by its pressure, or elastic force.

Steam can not, like wind and water, be made to act advantageously by its impulse in the open air, because the momentum of so light a fluid, unless generated in vast quantities, would be inconsiderable. The first attempts, however, to employ steam as a moving power, consisted in directing a current of steam from the mouth of a tube against the floats or vanes of a revolving wheel.

Fig. 210.

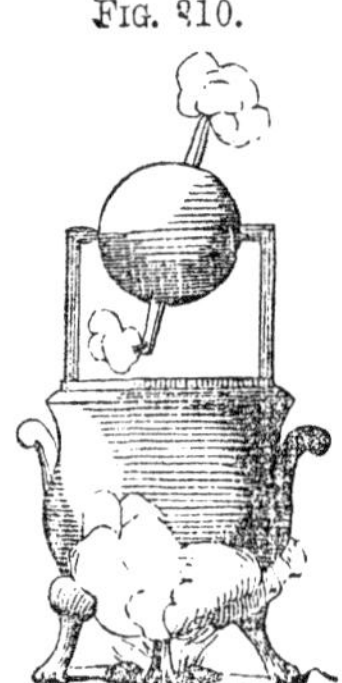

A machine of this kind, invented more than 2,000 years ago by Hero of Alexandria, is represented in Fig. 210. It consists of a small hollow sphere, furnished with arms at right angles to its axis, and whose ends are bent in opposite directions. The sphere is suspended between two columns, bent and pointed at their extremities, as represented in the figure: one of these is hollow, and conveys steam from the boiler below, into the sphere; and the escape of the vapor from the small tubes, by the reaction, produces a rotary motion.

To render the pressure of steam available in machinery, what conditions are necessary?

In order to render the pressure of steam practically available in machinery, it is necessary that it should be confined within a cavity which is air-tight, and so constructed that its dimensions or capacity can be enlarged or diminished without impairing its tightness. When the steam enters such a vessel, its elastic force pressing against some movable part, causes it to recede before it, and from this movable part motion is communicated to machinery.

How are these conditions attained?

Fig. 211.

The practical arrangement by which such a result is accomplished is by having a hollow cylinder, A B, Fig. 211, with a movable piston, D, accurately fitted to its cavity. When steam under pressure in a boiler is admitted into the cylinder below the piston, it expands, and acting upon the under surface of the piston, causes it to rise, lifting the piston-rod along with it

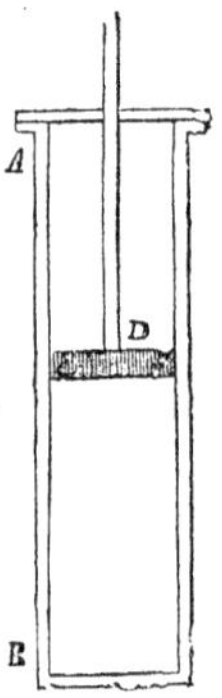

Suppose, as in Fig. 212, the cylinder to be connected at the bottom or side with a pipe, R, opening into a steam boiler, and on the other side with a pipe, B, terminating in a vessel of cold water. Suppose the valve in R to be open, and that in B to be shut; steam then passing into the cylinder from the boiler will force the piston up to the top of the cylinder. Let the valve in R then be shut, and the valve in B be opened; the steam contained in the cylinder will pass out of the pipe B, and coming in contact with cold water, in the vessel connected with it, will be condensed, and a vacuum form

FIG. 212.

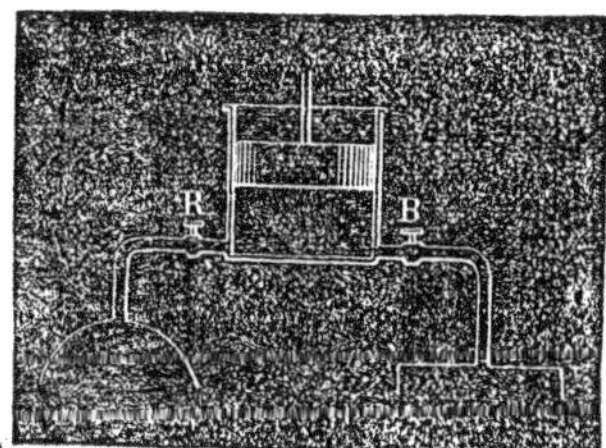

beneath the piston. The pressure of the atmosphere then acting upon the other side of the piston, will drive it down. The position of the valves in R and B being reversed, the piston may be raised anew by the admission of more steam, to be condensed in its turn, and in this manner the alternate motion may be continued indefinitely. The alternating, or reciprocating motion of the piston, is converted, by means of a lever and crank attached to the top of the piston-rod, into a rotary motion, suitable for driving-wheels, shafts, and other machinery.

Such an arrangement as described constituted the first practical steam-engine. It received the name of the atmospheric engine, from the fact that the pressure of the atmosphere was employed to press down the piston after it had been elevated by the steam.

What is the construction and operation of a condensing steam-engine?

577. In modern engines, the pressure of the atmosphere is not employed to drive the piston down. The steam is admitted into the cylinder above the piston, at the same time that it is condensed or withdrawn from below, and thus exerts its expansive force in the returning as well as in the ascending stroke.

This results in a great increase of power. By the condensation or withdrawal of the steam, a vacuum is created below the piston, and the steam admitted into the cylinder above the piston, forces it through the vacuum with an ease and rapidity far greater than would be possible if atmospheric or other resistance were to be overcome.*

The withdrawal or condensation of the steam, in order to produce a vacuum either above or below the piston, is accomplished by opening at the proper time a communication between the cylinder and a strong vessel situated at a distance from it, called the condenser. Into this vessel a jet of cold water is thrown, which instantly condenses the steam, escaping from the bottom of the cylinder, into water.

* "A proof of the extraordinary power obtained in this way, through the combustion of fuel, is presented in the following calculations:—One cubic inch of water is convertible into steam, of one atmospheric pressure, by 15⅓ grains of coal, and this expansion of the water into steam is capable of raising a weight of one ton the height of a foot. The one cubic inch of water becomes very nearly one cubic foot of steam, or 1,728 cubic inches. When a vacuum is produced by the condensation of this steam, a piston of one square inch surface, that may have been lifted 1,728 inches, or 144 feet, will fall with a velocity of a heavy body rushing by gravity down a perpendicular height of 13,500 feet. This would give the falling body a velocity, at the termination of its descent, equal to 1,300 feet per second, greater than that of the transmission of sound. From this we can form some estimate of the strength of the tempest which alternately blows the piston in its cylinder, when elastic steam of high-pressure is employed."—*Prof. H. D. Rogers.*

A steam-engine of this character is called a condensing steam-engine, because the steam which has been employed in raising or depressing the piston is condensed, after it has accomplished its object, leaving a vacuum above or below the piston. It is also called a low-pressure engine, because, on account of the vacuum which is produced alternately above and below the piston, the steam, in acting, does not expend any force in overcoming the pressure of the atmosphere. Steam, therefore, may be used under such conditions of low expansive force, or, as it is technically called, of "low-pressure."

The practical construction of the piston and cylinder, and the arrangement of connecting pipes by which the steam is admitted alternately above and below the piston, is fully shown in Fig. 213. The valves, which are of various forms, are connected by levers with the machinery, in such a way as to open and close with great accuracy at exactly the proper moment.

FIG. 213.

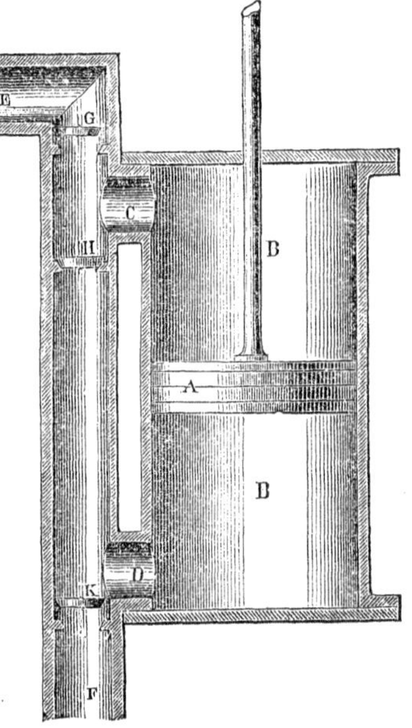

What is a high-pressure engine?

578. In some engines, the apparatus for condensing the steam alternately above or below the piston, is dispensed with, and the steam, after it has moved the piston from one end of the cylinder to the other, is allowed to escape, by the opening of a valve, directly into the air. To accomplish this, it is evident that the steam must have an elastic force greater than the pressure of the atmosphere, or it could not expand and drive out the waste steam on the other side of the piston, in opposition to the pressure of the air. An engine of this character is accordingly termed a "high-pressure" engine.

High-pressure engines are generally worked with a pressure of from fifty to sixty pounds per square inch of the piston; of this pressure, at least fifteen pounds must be expended in overcoming the pressure of the atmosphere, and the surplus only can be applied to drive machinery.

One of the most familiar examples of a high pressure engine is the locomotive used on railroads. The steam which has been employed in forcing the piston in one direction is, by the return movement of the piston, forced out of the cylinder into the smoke-pipe, and escapes into the open air with irregular puffs.

What are the advantages and disadvantages of high-pressure engines?

High-pressure engines are generally used in all situations where simplicity and lightness are required, as in the case of the locomotive; also in situations where a free supply of water for condensation can not be readily obtained. As they use steam at a much higher pressure than the condensing engines, they are more liable to accidents arising from explosions. High-pressure engines are less expensive than low-pressure, since all the apparatus for condensing the steam is dispensed with, the only parts necessary being the boiler, cylinder, piston, and valves.

When is steam said to be used expansively?

579. It is not necessary in the steam-engine that the steam should flow continuously from the boiler into the cylinder during the whole movement of the piston, but it may be cut off before it has fully completed its ascent or descent in the cylinder. The steam already in the cylinder immediately expands, and completes the movement already begun, thus saving a considerable quantity of steam at each movement. Steam employed in this way is said to be used expansively.

To carry out this plan to the best advantage, the expansive force of the steam must be greatly increased by working it under a high pressure.

Fig. 214.

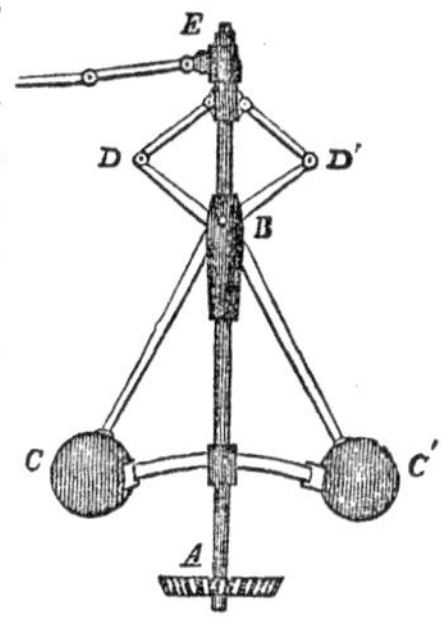

How is the motion of steam-engines regulated?

580. In many engines the supply of steam to the cylinder is regulated by an apparatus called the Governor. This consists, as is represented in Fig. 214, of two heavy balls, C and C′, connected by jointed rods, D D′, with a revolving axis. A. When the axis is made to revolve rapidly, the centrifugal force tends to make the balls diverge, or separate from one another in the same manner as the two legs of a tongs will fly apart when whirled round by the top. This divergence draws down the jointed rods, but a slower motion of the axis causes the balls, on the contrary, to approach each other, and thus push them up. These movements of the jointed rods in turn raise or lower the end of a bar, E, which acts as a lever, and moves a valve which increases or diminishes the quantity of steam admitted from the boilers into the cylinder—thus preserving the motion of the engine uniform.

In stationary engines, also, a large and heavy fly-wheel is often used, which by its momentum causes the machinery to move uninterruptedly, even if the pressure of steam be less at one point than at another.*

* Fig. 215 illustrates the principal parts of a condensing steam engine and its mode of action.

Upon the left of the figure is the cylinder, which receives the steam from the boiler. A part of the side of the cylinder is cut away in order to show the piston, which moves alternately up and down according as the steam is admitted above or below it. By the rod A the piston transmits its alternating movements to the walking-beam, L, which is an enormous lever accurately balanced on its center, and supported by four columns. The walking-beam, L, communicates its motion by means of a connecting-rod, I, to the crank,

FIG. 215.

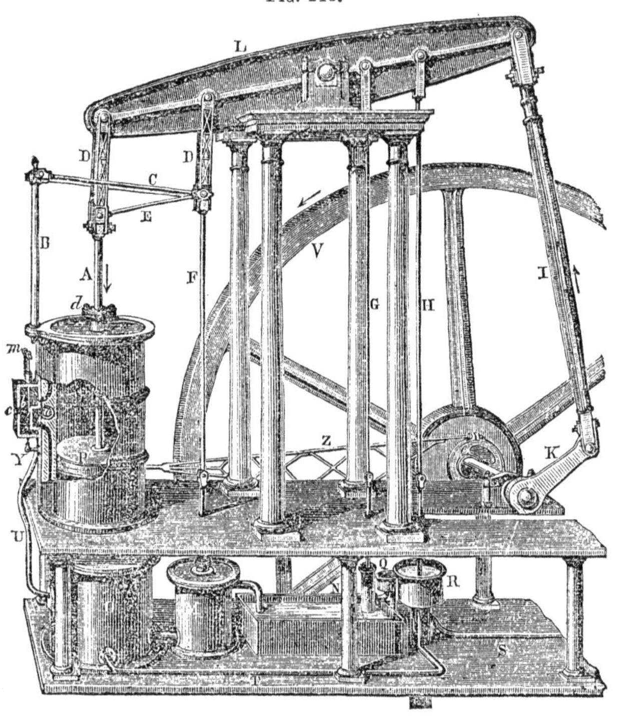

581. Steam-boilers, which, although necessary to the generation of the power, are quite independent of the engine, are constructed of thick sheets of iron or copper, strongly riveted together.

K, by which a rotary movement is communicated to the wheel, V; from this the power may be applied by other wheels, or by bands and pulleys, to effect different operations.

At the left of the cylinder is an arrangement of valves and pipes, by which the steam is allowed to act alternately above and below the piston. After the steam has completed its action by forcing the piston to the extremity of the cylinder, it is necessary that it should be withdrawn, and a vacuum formed in its place. In order to accomplish this, the steam, after having acted, is caused to pass into the cylinder, O, which contains cold water, and is termed the condenser. Here it is condensed, and a vacuum formed in the cylinder above or below the piston, as the case may be.

As the cold water of the condenser becomes quickly heated by the condensed steam withdrawn from the cylinder, it becomes necessary to constantly withdraw the hot water and replace it by cold water, in order that the condensation of the steam may take place as rapidly as possible. This is effected by means of two pumps; the one, F M, which is called the "air-pump," which withdraws the hot water from the condenser, and with it any air that may be present either in the cylinder or the condenser; the other, H R,

What are the essential requisites of a steam-boiler?

The essential requisites of a steam-boiler are, that it should possess sufficient strength to resist the greatest pressure which is ever liable to occur from the expansion of the steam, and that it should offer a sufficient extent of surface to the fire to insure the requisite amount of vaporization. In common low-pressure boilers, it requires about eight square feet of surface of the boiler to be exposed to the action of the fire and flame to boil off a cubic foot of water in an hour; and a cubic foot of water in its convertion into steam equals one-horse power.

The strongest form for a boiler, and one of the earliest which was used, is that of a sphere; but this form is the one which offers least surface to the fire. The figure of a cylinder is on many accounts the best, and is now extensively used, especially for engines of high-pressure. It has the advantage of being easily constructed from sheets of metal, and the form is of equal strength except at the ends. In such a boiler the ends should be made thicker than the other parts.

called the "cold-water pump," draws from a well or river the cold water to supply the place of the heated water withdrawn from the condenser by the air-pump. There is also a third pump, G Q, which is called the "supply" or "feed-pump," because it pumps into the boiler the hot water which the air-pump withdraws from the condenser, thus economizing the consumption of fuel.

The various parts of the engine (as shown in Fig. 215) are illustrated in detail by the following descriptive explanation:—

A—Piston-rod connected with the walking-beam, and transmitting to it the alternating movement of the piston.

B, C, D, E—Arrangements of levers and joints, intended to guide and preserve the piston-rod A in a perfectly rectilinear track during its up-and-down movements.

F—Arm or rod of the air-pump, which removes the hot water and air from the condenser.

G—Rod of the "supply" or "feed-pump," which supplies to the boiler the hot water withdrawn from the condenser.

H—Rod of the cold-water pump, which supplies the cold water necessary for condensation.

I—Connecting-rod, which transmits the motion of the walking-beam, L, to the crank, K.

M—Cylinder of the air-pump in communication with the condenser, O.

O—Condenser filled with cold water, in which the steam after acting upon the piston is condensed.

P—Piston, movable in the cylinder; it receives directly the pressure of the steam upon the upper and lower surface alternately, and transmits its movements by means of the rod A to the rest of the machinery.

S—Pipe conducting the hot water withdrawn from the condenser to the boiler.

T—Pipe discharging the cold water from the cold-water pump into the condenser, O.

U—Pipe conducting the steam from the cylinder, after it has acted upon the piston, into the condenser.

V—Fly-wheel.

Z—Connecting rod, which transmits the movements of the eccentric, *e*, through the lever, Y, to the valves, *b*. The eccentric is a wheel fixed upon the crank-shaft, as seen at *e*. It is called an eccentric from the circumstance of the wheel not being concentric, or having a common center with the crank-shaft upon which it is fixed, It becomes, therefore, a substitute for a short crank, and transmits a reciprocating movement to the rod Z, which is connected with the valves at *b* by the lever Y. These valves being alternately opened and closed by the movement of the rod Z, admit the steam alternately above or below the piston.

What is the construction of a flue-boiler?

A very great improvement was effected in the construction of steam-boilers by placing a cylindrical furnace within a cylindrical boiler, thus surrounding the heated surfaces with water upon all sides. By this method, all the heat, except what escapes up the chimney, is communicated to the water. Such boilers are known as "flue-boilers." Their general form and plan of construction are represented in Fig. 216.

Fig. 216.

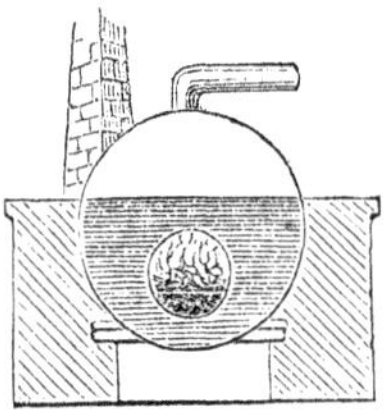

What are the peculiarities of a locomotive-boiler?

The requirements of a boiler suitable for a locomotive are, that the greatest possible quantity of water should be evaporated with the greatest rapidity in the least possible space. The quantity of fuel consumed is a secondary consideration, as this can be carried in a separate vehicle. The principle by which this has been accomplished, and the invention of which may be said to have made the present railway system, consists in carrying the hot product of the fire through the water in numerous small parallel flues or tubes, thus dividing the heated matter, and as it were filtering it through the water to be heated. In this manner the surfaces, by which the water and the heating gases communicate, are immensely increased, the whole having a resemblance to the mechanism of the lungs of animals, in which the air and the blood are divided and presented to each other at as many points, and with as little intervening matter between them, as is consistent with their separation. Fig. 217 represents the interior of the fire-box of a locomotive, showing the opening of the tubes, which extend through the whole length of the boiler, and are surrounded with water. The smoke and other products of combustion pass through these tubes, and finally escape up the smoke-pipe. It will be further observed by the examination of the figure that the fire-box is double-walled, or rather walled and roofed with a layer of water, leaving only the bottom vacant, which receives the grate-bars.

Fig. 217.

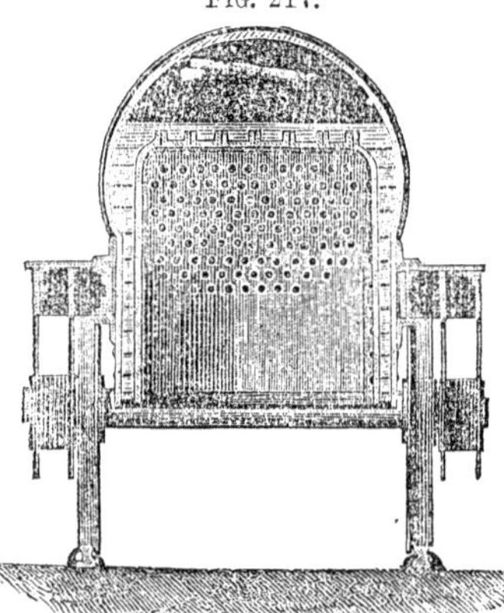

Describe the safety-valve.

582. The safety-valve is generally a conical lid fitted into the boiler, and opening outward; it is kept down by a weight, acting on the end of a lever, equal to the pressure which the boiler is capable of sustaining without danger from the steam generated within. If the amount of steam at any time exceeds the pressure,

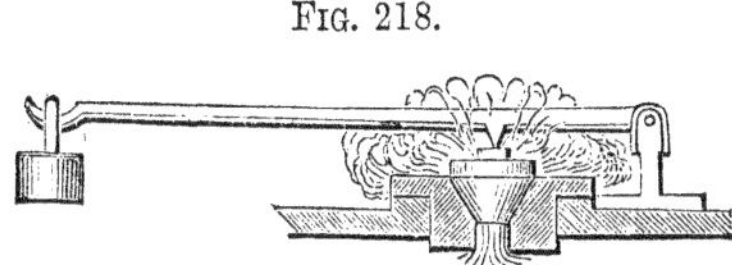

FIG. 218.

it overcomes the resistance of the weight, lifts the valve, and allows the steam to escape. When sufficient steam has escaped to diminish the pressure, the valve falls back into its place, and the boiler is as tight as if it had no such opening.

Fig. 218 represents the ordinary construction of the safety-valve.

How does a diminution of water in boilers often occasion explosions?

583. The explosion of steam-boilers, when the safety-valve is in good condition and working order, is sometimes inexplicable; but explosions often result from the engineer allowing the water to become too low in the boilers. When this occurs, the parts of the boiler which are not covered with water, and are exposed to the fire, become highly overheated. If, in this condition, a fresh supply of water is thrown into the boiler, it comes suddenly into contact with an intensely-heated metal surface, and an immense amount of steam, having great elastic force, is at once generated. In this case the boiler may burst before the inertia of the safety-valve is overcome, and the stronger the boiler the greater the explosion.

What is a steam-guage?

584. The degree of pressure which the steam exerts upon the interior of the boiler, and which is consequently available for working the engine, is indicated by means of an instrument called the "steam" or "barometer-guage." It consists simply of a bent tube, A, C, D, E, Fig. 219, fitted into the boiler at one end, and open to the air at the other. The lower part of the bend of the tube contains mercury, which, when the pressure of steam in the boiler is equal to that of the external atmosphere, will stand at the same level, H R, in both legs of the tube. When the pressure of the steam is greater than that of the atmosphere, the mercury is depressed in the leg C D, and elevated in the leg D E. A scale, G, is attached to the long arm of the tube, and by observing the difference of the levels of the mercury in the two tubes, the pressure of the steam may be calculated. Thus, when the mercury is at the same level in both legs, the pressure of the steam balances the pressure of the atmosphere, and is therefore 15 pounds per square inch. If the mercury stands 30 inches higher in the long arm of the tube, then the pressure of the steam is equal to that of two atmospheres, or is 30 pounds to the square inch, and so on.

FIG. 219.

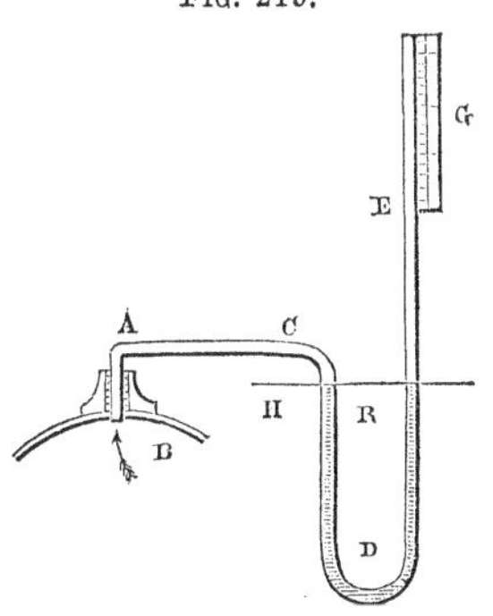

How can the pressure of steam be indicated by a thermometer?

As the pressure of steam increases with its temperature, the pressure upon the interior of the boiler may also be known by means of a thermometer inserted into the boiler. Thus it has been ascertained that steam at 212° balances the atmosphere, or exerts a pressure of 15 pounds per square inch; at 250°, 30 pounds; at 275°, 45 pounds; at 294°, 60 pounds, and so on.

Describe the steam-whistle.

585. The steam-whistle attached to locomotive and other engines is produced by causing the steam to issue from a narrow circular slit, or aperture, cut in the rim of a metal cup; directly over this is suspended a bell, formed like the bell of a clock. The steam escaping from the narrow aperture, strikes upon the edge or rim of the bell, and thus produces an exceedingly sharp and piercing sound. The size of the concentric part whence the steam escapes, and the depth of the bell part, and their distance asunder, regulate the tones of the whistle from a shrill treble to a deep bass.

SECTION V.

WARMING AND VENTILATION.

Upon what principles do the warming and ventilation of buildings depend?

586. In the warming and ventilation of buildings, the entire process, whatever expedients may be adopted, is dependent upon the expansion and contraction of air; or in other words, upon the fact that air which has been heated and expanded ascends, and air which has been deprived of heat, or contracted, descends.

What is ventilation?

587. Ventilation is the act or operation of causing air to pass through any place, for the purpose of expelling impure air and dissipating noxious vapors.

The theoretical perfection of ventilation is to render it impossible for any portion of air to be breathed twice in the same place.

Where is ventilation perfect?

In the open air, ventilation is perfect, because the breath, as it leaves the body, is warmer and lighter than the surrounding fresh air, and ascending, is immediately replaced by an ingress of fresh air ready to be received by the next respiration.

Why is air once respired unwholesome?

Common air consists of a mixture of two gases, oxygen and nitrogen, in the proportion of one fifth oxygen to four fifths nitrogen. By all the forms of respiration or breathing, and of combustion, the quantity of oxygen in atmospheric air is diminished and impaired, and to exactly the same extent is air rendered unwholesome and unsuitable to supply the wants of the animal system.

How much fresh air is required per hour by a healthy man?

It is calculated that a full-grown person of average size absorbs about a cubic foot of oxygen per hour by respiration, and consequently renders five cubic feet of air unfit for breathing, since every five cubic feet of air contain one cubic foot of oxygen. It is also calculated that two wax or sperm candles absorb as much oxygen as an adult.

To render the air of a room perfectly pure, five cubic feet of fresh air per hour, for each person, and two and a half cubic feet for each candle, should be allowed to pass in, and an equal quantity to pass out.

In what manner does a heated substance generate a current of air?

588. From every heated substance, an upward current of air is continually rising.

The existence and force of this upward current may be shown in the case of an ordinary stove, by attaching to the side of the pipe a wire on which a piece of thick paper cut in the form of a spiral is suspended, as is represented in Fig. 220. The upward current of hot air striking against the surfaces of the coil causes it to revolve rapidly around the wire.

Fig. 220.

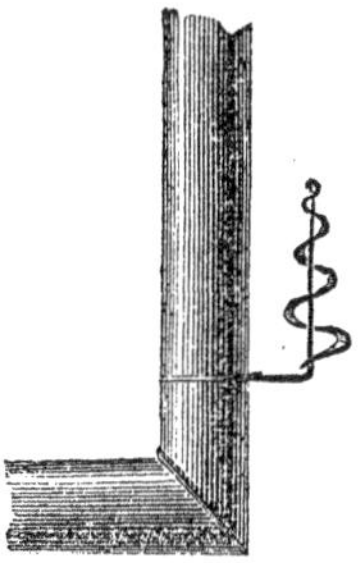

Why are stoves and grates placed near the floor?

Apart from the consideration of convenience, it is necessary that stoves and grates, intended for warming, should be located as near to the floor of the room as possible; since the heat of a fire has very little effect upon the air of an apartment below the level of the surface upon which it is placed.

Why does smoke ascend in a chimney?

589. When a fire is lighted in a stove or grate to warm a room, the smoke and other gaseous products of combustion, being lighter than the air of the room, ascend, and soon fill the chimney with a column of air lighter, bulk for bulk, than a column of atmospheric air. Such a column, therefore, will have a buoyancy proportional to its relative lightness, as compared with the external air and the air of the apartment.

Fig. 221.

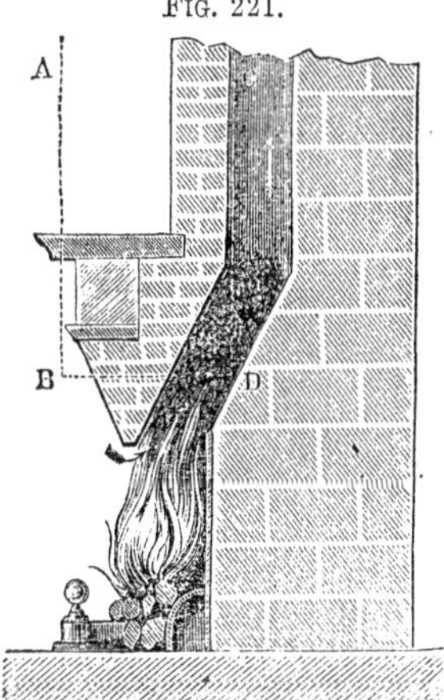

The upward tendency of a column of heated air constitutes the draft of a chimney, and this draft will be strong and effective just in the same proportion as the column of air in the chimney is kept warm.

Fig. 221 represents a section of a grate and chimney. C D represents the light and warm column of air within the chimney, and A B the cold and heavy column of air outside the chimney. The column A B being cold and heavy presses down, the column C D being light and warm rushes up, and the greater the difference between the weight of these two columns, the greater will be the draft.

How does a chimney quicken the ascent of a column of hot air?

A chimney quickens the ascent of hot air by keeping a long column of it together. A column of two feet high rises, or is pressed up, with twice as much force as a column of one foot, and so in proportion for all other lengths—just as two or more corks, strung together and immersed in water, tend upward with proportionably more force than a single cork.

In a chimney where a column of hot air one foot in height is one ounce lighter than the same bulk of external cold air, if the chimney be one hundred feet high, the air or smoke in it is propelled upward with a force of one hundred ounces.

To what is the draft of a chimney proportional?

If the fire be sufficiently hot, the draft of the chimney will be proportional to its length.

For this reason, the chimneys of large manufacturing establishments are generally very high.

How should a chimney be constructed?

A chimney should be constructed in such a way that the flue or passage will gradually contract from the bottom to the top, being widest at the bottom, and the smallest at the top.

Why should a chimney be constructed in this manner?

The reason of this will be evident from the following considerations:—At the base of the chimney, the hot column of expanded air fills the entire passage; but as the hot air ascends it gradually cools and contracts, occupying less space. If, therefore, the chimney were of the same size all the way up, the tendency would be, that the cold external air would rush down to fill up the space left by the contraction of the hot column of air. This action would still further cool the hot air of the chimney and diminish the draft.

Some persons suppose that a chimney should be made larger at the top than at the bottom, because a column of smoke ascending in the open air, expands or increases in bulk as it goes up. This, however, is owing, in great part, to the action of currents in the air, and to the fact, that a column of smoke freely exposed to the air, is more rapidly cooled than in a chimney, and losing its ascensional power, tends to float out laterally, rather than ascend perpendicularly.

The causes of "smoky chimneys" are various.

Under what circumstances will a chimney smoke?

A chimney may smoke for want of a sufficient supply of air. If the apartment is very tight, fresh air from without will not be admitted as fast as it is consumed by the fire, and in consequence a current of air rushes down the chimney to supply the deficiency, driving the smoke along with it.

A chimney will often smoke when the heat of the fire is not sufficient to rarefy all the air in the chimney; in such cases the cold air (condensed in the upper part of the flue) will sink from its own weight, and sweep the ascending smoke back into the room.

When the fire is first lighted, and the chimney is filled with cold air, there is often no draft, and consequently the flame and smoke issue into the room. This, in most cases, is remedied by the action of a "blower."

What is the use of a blower? A blower is a sheet of iron that stops up the space above the grate bars, and prevents any air from entering the chimney except that which passes through the fuel and produces combustion. This soon causes the column of air in the chimney to become heated, and a draft of considerable force is speedily produced through the fire. The increase of draft increases the intensity of the fire.

Another frequent cause of smoky chimneys is, that when the tops are commanded by higher buildings, or by a hill, the wind in blowing over them, falls like water over a dam, and beats down the smoke. The remedy in such cases is, either to increase the height of the chimney, or to fix a bonnet or cowl upon the top. The philosophy of this last contrivance consists in the fact that in whatever direction the wind blows, the mouth of the chimney is averted from it.

What is the motion of the air in a room artificially heated? In a room artificially heated, there are always two currents of air; one of hot air flowing out of the room, and another of cold air flowing into the room.

If a candle be held in the doorway of such an apartment, near the floor, it will be found that the flame will be blown inward; but if it be raised nearly to the top of the doorway, the flame will be blown outward. The warm air, in this case, flows out at the top, while the cold air flows in at the bottom.

How does a stove differ from an open fire-place in respect to ventilation? 590. An open fire-place differs greatly from a close stove in respect to ventilation, inasmuch as the former warms and ventilates an apartment, while the latter only warms, and can hardly be said to contribute at all to the ventilation. In a close stove, no air passes through the room to the flue of the chimney, except that which passes through the fuel, and the quantity of this is necessarily limited by the rate of combustion maintained in the stove. In an open fire-place, a large amount of air is continually rushing up the chimney through the opening over the grate, irrespective of what passes through the fire and maintains combustion.

In summer time, when no fire is made in the chimney, the column of air in it is generally at a higher temperature than the external air, and a current will therefore in such case be established up the chimney, so that the fire-place will still serve, even in the absence of fire, the purposes of ventilation. In very warm weather, however, when the external air is at a higher temperature than the air within the building, the effects are reversed; and the air in the chimney being cooled, and therefore heavier than the external air, a downward current is established, which produces in the room the odor of soot.

FIG. 222.

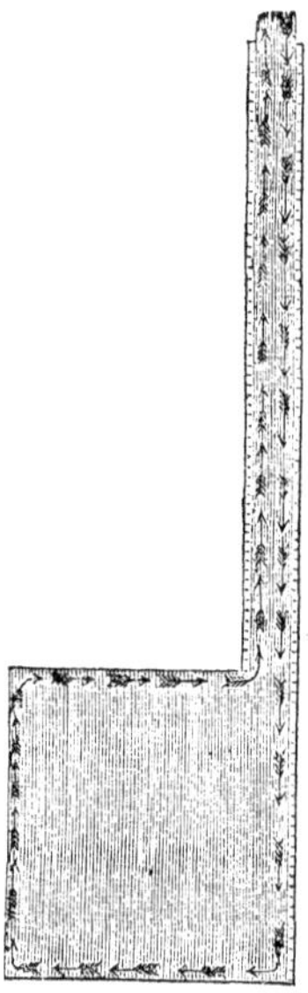

Fig. 222 represents the lines of the currents descending the chimney and circulating round an apartment.

How is a room best ventilated?

A room is well ventilated by opening the upper sash of a window; because the hot vitiated air (which always ascends toward the ceiling) can thus escape more easily. If the lower sash of the window be also partially opened, a corresponding current of cold air, flowing into the room, is created, and ventilation will be so effected more perfectly.

Why are open fire-places ill adapted for heating?

Open fire-places are ill adapted for the economical heating of apartments, because the air which flows from the room to the fire becomes heated, and passes off directly into the chimney, without having an opportunity of parting with its heat for any useful purpose. In addition to this, a quantity of the air of the room, which has been warmed by radiation, is uselessly carried away by the draft.

What are the advantages and disadvantages of stoves?

The advantages of a stove over an open fire-place are as follows:

1. Being detached from the walls of the room, the greater part of the heat produced by combustion is saved. The radiated heat being thrown into the walls of the stove, they become hot, and in turn radiate heat on all sides of the room. The conducted heat is also received by successive portions of the air of the room, which pass in contact with the stove.

2. The air being made to pass through the fuel, a small supply is sufficient to keep up the combustion, so that little need be taken out of the room; and

3. The smoke, in passing off by a pipe, parts with the greater part of its heat before it leaves the room.

Houses warmed by stoves, as a general rule, are ill-ventilated. The air coming in contact with the hot metal surfaces is rendered impure, which impurity is increased by the burning of the dust and other substances which settle upon the stove. The air is, in most cases also, kept so dry as to render it oppressive.

What is the method of warming by hot-air furnaces?

591. The method of warming houses by the common hot-air furnace is as follows:—A stove, having large radiating surfaces, is inclosed in a chamber (generally of masonry). This chamber is frequently built with double walls, that it may be a better non-conductor of heat. A current of air from without is brought by a pipe or box, and delivered under the stove. A part of this air is admitted to supply the combustion; the rest passes upward in the cavity between the hot stove and the walls of the brick chamber, and, after

becoming thoroughly heated, is conducted through passages in which its lightness causes it to ascend, and be delivered in any apartment of the house.

What two points are of special importance in the construction of furnaces? In the construction and arrangement of a furnace for heating, the two points of special importance are, to secure a perfect combustion of the fuel, and the best possible transmission of all the heat formed, into the air that is to pass into the rooms of the house.

The first of these requisites is obtained by having a good draft and a firebox which is broad and shallow, so that the coal shall form a thin stratum and burn most perfectly.

The second requisite is obtained by providing a great quantity of surface in the form of pipes, drums, or cylinders, through which the smoke and hot gases must pass on their way to the chimney, and to which their heat will be imparted, to be in turn delivered to the cold and pure air of the rooms of the house.

What is the advantage of heating by steam? 592. The great advantages of heating by steam are, that the heat can be communicated for a great distance in any direction—upward, downward, or horizontally. As the temperature of the heating surfaces, when low-pressure steam is used, never exceeds 212° F., the air in contact with them is never contaminated by the burning of dust, or the abstraction of oxygen.

Under favorable circumstances, one cubic foot of boiler will heat about two thousand cubic feet of suitably inclosed space to a temperature of 70° to 80° F.

What is fuel? 593. We apply the term fuel to any substance which serves as aliment or food for fire. In ordinary language we mean by fuel the peculiar substance of plants, or the products resulting from their decomposition, designated under the various names of wood, coal, &c.

What proportion of the weight of wood is water? In recently cut wood, from one fifth to one half of its weight is water; after wood has been dried in the air for ten or twelve months, it will even then contain from 15 to 25 per cent. of water.

The amount of moisture in wood is greatest in the spring and summer, when the sap flows freely and the influence of vegetation is the greatest. Wood, therefore, is generally cut in the winter, because at that season there is but little sap in the tissues, and the wood is drier than at any other period.

Why are woods designated as hard and soft? Woods are designated as hard and soft. This distinction is grounded upon the facility with which they are worked, and upon their power of producing heat. Hard woods, as the oak, beech, walnut, elm, and alder, contain in the same bulk more solid fiber, and their vessels are narrower and more closely packed than those of the softer kinds, such as pine, larch, chestnut, etc.

What is the comparative weight of wood?

594. The weight of wood varies greatly; from forty-four hundred pounds in a cord of dry hickory, to twenty-six hundred in a cord of dry, soft maple.

What is the comparative value of wood for fuel?

595. For fuel, the most valuable of the common kinds of wood are the varieties of hickory; after that, in order, the oak, the apple-tree, the white-ash, the dog-wood, and the beech. The woods that give out the least heat in burning are the white-pine, the white-birch, and the poplar.

Is it profitable to burn green wood?

596. The remark is sometimes made that "it is economy to burn green wood, because it is more durable, and therefore in the end more cheap." This idea is erroneous. The consumption of green wood is less rapid than dry, but to produce a given amount of heat, a far greater amount of fuel must be consumed.

The evaporation of liquids, or their conversion into steam, consumes or renders latent a great amount of caloric. When green wood or wet coal is added to the fire, it abstracts from it by degrees a sufficient amount of heat to convert its own sap or moisture into steam before it is capable of being burned. As long as any considerable part of this fluid remains unevaporated, the combustion goes on slowly, the fire is dull, and the heat feeble.

Why are coal and hard woods difficult to ignite with a match?

597. Coal and hard wood are not readily ignited by the blaze of a match, because on account of their density they are rendered comparatively good conductors, and thus carry off the heat of the kindling substance, so as to extinguish it, before they themselves become raised to the temperature necessary for combustion.

Light fuel, on the contrary, being a slow conductor of heat, kindles easily, and, from the admixture of atmospheric air in its pores and crevices, burns out rapidly, producing a comparatively temporary, though often strong heat.

CHAPTER XIII.

METEOROLOGY.

What is Meteorology?

598. Meteorology is that department of physical science which treats of the atmosphere and its phenomena, particularly in its relation to heat and moisture.

599. By climate, we mean the condition of a place in

What do we mean by the term climate?

relation to the various phenomena of the atmosphere, as temperature, moisture, etc. Thus, we speak of a warm or cold climate, a moist or dry climate, etc.

How is the average temperature of a day found?

600. The mean or average temperature of the day is found by observing the thermometer at fixed intervals of time during the twenty-four hours, and then dividing the sum of the temperatures by the number of observations.

At what time is the temperature of the day the highest and lowest?

From such a series of observations it has been found that the lowest temperature of the day occurs shortly before sunrise, and the highest a few hours after 12 at noon, somewhat later in summer and somewhat earlier in winter.

The mean annual temperature of any particular location is found by taking the average of all the mean daily temperatures throughout the year.

The mean daily temperature of any place seems to vary in a regular and constant manner, while the mean annual temperature of the same location is very nearly a constant quantity. Thus, by long observations made in Philadelphia, it has been found that the mean daily temperature of that locality is one degree less than the temperature at 9 o'clock, A. M., at the same place; while the mean annual temperature of Paris varied only 4° in thirteen years.

All the results of observation seem to show that the same quantity of heat is always annually distributed over the earth's surface, although unequally—that is to say, the average annual temperature of each place upon the earth's surface is very nearly the same. In our latitude, July is on the average the hottest month, and January the coldest; and in reference to particular days, we may on an average consider the 26th of July as the hottest, and the 14th of January as the coldest day of the year for the temperate zone of the northern hemisphere.

How does temperature vary with the latitude?

The average annual temperature of the atmosphere diminishes from the equator toward either pole.

At the equator, in Brazil, the average annual temperature is 84° Fahrenheit's thermometer; at Calcutta, lat. 22° 35′ N., the annual temperature is 78° F.; at Savannah, lat. 32° 5′ N. the annual temperature is 65° F.; at London, lat. 51° 31′ N., the annual temperature is 50° F.; at Melville Island, lat. 74° 47′ N., the average annual temperature is 1° below zero.

Why is not the temperature of all places having the same latitude alike?

601. If the whole surface of the earth were covered by water, or if it were all formed of solid plane land, possessing everywhere the same character, and having an equal capacity at all places for absorbing and again radiating heat, the

temperature of a place would depend only on its geographical latitude, and consequently all places having the same latitude would have a like climate. Owing, however, to various disturbing causes, such as the elevation and form of the land, the proximity of the sea, the direction of the winds, etc., places of the same latitude, and comparatively near each other, have very different temperatures.

In warm climates the proximity of the sea tends to diminish the heat; in cold climates, to mitigate the cold. Islands and peninsulas are warmer than continents; bays and inland seas also tend to raise the mean temperature. Chains of mountains which ward off cold winds, augment the temperature; but mountains which ward off south and west winds, lower it. A sandy soil, which is dry, is warmer than a marshy soil, which is wet and subject to great evaporation.

What is the capacity of air for moisture?

602. Air absorbs moisture at all temperatures, and retains it in an invisible state. This power of the air is termed its capacity for absorption.

The capacity of air for moisture increases with the temperature.

A volume of air at 32° can absorb an amount of moisture equal to the hundred and sixtieth part of its own weight, and for every 27 additional degrees of heat, the quantity of moisture it can absorb at 32° is doubled. Thus a body of air at 32° F. absorbs the 160th part of its own weight; at 59° F., the 80th; at 86° F., the 40th; at 113° F., the 20th part of its own weight in moisture. It follows from this that while the temperature of the air advances in an arithmetical series, its capacity for moisture is accelerated in a geometrical series.

When is air said to be saturated?

Air is said to be saturated with moisture when it contains as much of the vapor of water as it is capable of holding with a given temperature.

We say that air is dry when water evaporates quickly, or any wetted surface dries rapidly; and that it is damp when moistened surfaces dry slowly, or not at all, and the slightest diminution of temperature occasions a deposit of moisture in the form of mist and rain. These expressions do not, however, convey altogether a correct idea of the condition of the atmosphere, since air which we term "dry," may contain much more moisture than that which we distinguish as "damp." For indicating the true condition of the atmosphere in reference to moisture, we therefore use the terms "absolute" and "relative" humidity.

What is meant by absolute and relative humidity?

When we speak of the absolute humidity of the air, we have reference to the quantity of moisture contained in a given volume. By relative humidity, we refer to its proximity to saturation. Relative humidity is a state dependent upon the mutual influence of absolute humidity and temperature; for a given volume of air may be made to pass from a state of dampness to one of extreme dryness, by merely elevating its temperature, and this, too, without altering the amount of moisture it contains in the least degree.

What are Hygrometers?

Instruments designed for measuring the quantity of moisture contained in the atmosphere, are called HYGROMETERS.*

Upon what principle are hygrometers constructed?

Many organic bodies have the property of absorbing vapor, and thus increasing their dimensions. Among such may be mentioned hair, wood, whalebone, ivory, etc. Any of these connected with a mechanical arrangement by which the change in volume might be registered, would furnish a hygrometer.

A large sponge, if dipped in a solution of salt, potash, soda, or any other substance which has a strong attraction for water, and then squeezed almost dry, will, upon being balanced in a pair of scales suspended from a steady support, be found to preponderate or ascend according to the relative dampness or dryness of the weather.

The beard of the wild oat may also serve as a hygrometer, as it twists around, during atmospheric changes from dampness to dryness.

If we fix against a wall a long piece of catgut, and hang a weight to the end of it, it will be observed, as the air becomes moist or dry, to alter in length; and by marking a scale, the two extremities of which are determined by observation when the air is very dry, and when it is saturated with moisture, it will be found easy to measure the variations.

Describe the "Hair Hygrometer."

An instrument called the "Hair Hygrometer," is constructed upon this principle. It consists of a human hair, fastened at one extremity to a screw (see Fig. 223), and at the other passing over a pulley, being strained tight by a silk thread and weight, also attached to the pulley. To the axis of the pulley an index is attached, which passes over a graduated scale, so that as the pulley turns, through the shortening or lengthening of the hair, the index moves. When the instrument is in a damp atmosphere, the hair absorbs a considerable amount of vapor, and is thus made longer, while in dry air it becomes shorter; so that the index is of course turned alternately from one side to the other.

FIG. 223.

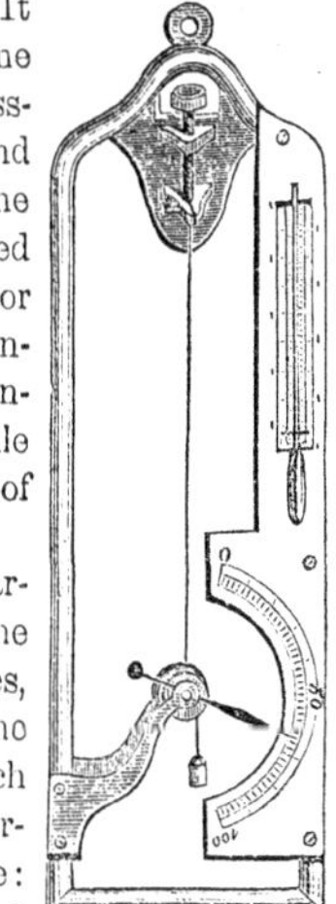

The instrument is graduated by first placing it in air artificially made as dry as possible, and the point on the scale at which the index stops under these circumstances, is the point of greatest dryness, and is marked 0. The hygrometer is then placed in a confined space of air, which is completely saturated with vapor, and under these circumstances the index moves to the other end of the scale: this point, which is that of greatest moisture, is marked

* *Hygrometer*, from the Greek words υγρος (moist) and μετρον (measure).

100. The intervening space is then divided into 100 equal parts, which indicate different degrees of moisture.

Such hygrometers are not, however, considered as altogether reliable.

SECTION I.

PHENOMENA AND PRODUCTION OF DEW.

What is Dew?

603. Dew is the moisture of the air condensed by coming in contact with bodies colder than itself.

What is the Dew-Point?

604. The temperature at which the condensation of moisture in the atmosphere commences, or the degree indicated by the thermometer at which dew begins to be deposited, is called the "Dew-Point."

Is the dew-point a constant one?

This point is by no means constant or invariable, since dew is only deposited when the air is saturated with vapor, and the amount of moisture required to saturate air of high temperature is much greater than air of low temperature.

If the saturation be complete, the least diminution of temperature is attended with the formation of dew; but if the air is dry, a body must be several degrees colder before moisture is deposited on its surface; and indeed the drier the atmosphere, the greater will be the difference between the temperature and its dew-point.

How may the production of dew be occasioned at any time?

Dew may be produced at any time by bringing a vessel of cold water into a warm room. The sides of the vessel cool the surrounding air to such an extent that it can no longer retain all its vapor, or, in other words, the temperature of the air is reduced below the dew-point; dew therefore forms upon the vessel. A pitcher of water under such circumstances is vulgarly said to "sweat."

In the same manner, moisture is deposited upon the windows of a heated apartment when the temperature of the external air is low enough to sufficiently cool the glass.

Why is dew formed in summer after sunset?

As soon as the sun has set in summer, and the earth is no longer receiving new supplies of heat, its surface begins to throw off the heat which it has accumulated during the day by radiation; the air, however, does not radiate its heat, and, in consequence, the different objects upon the earth's surface are soon cooled down from 7 to 25 degrees below the temperature of the air. The warm vapor of the air, coming in contact with these cool bodies, is condensed and precipitated as dew.

In a clear summer's night, when dew is depositing, a thermometer laid upon the grass, will sink nearly 20 degrees below one suspended in the air at a little distance above.

Upon what substances is dew deposited most freely? All bodies have not an equal capacity for radiating heat, but some cool much more rapidly and perfectly than others. Hence it follows, that with the same exposure, some bodies will be densely covered with dew, while others will remain perfectly dry.

Grass, the leaves of trees, wood, etc., radiate heat very freely: but polished metals, smooth stones, and woolen cloth, part with their heat slowly: the former of these substances will therefore be completely drenched with dew, while the latter, in the same situations, will be almost dry.

The surfaces of rocks and barren lands are so compact and hard, that they can neither absorb nor radiate much heat; and (as their temperature varies but slightly) very little dew deposits upon them. Cultivated soils, on the contrary (being loose and porous) very freely radiate by night the heat which they absorb by day; in consequence of which they are much cooled down, and plentifully condense the vapor of the air into dew. Such a condition of things is a remarkable evidence of design on the part of the Creator, since every plant and inch of land which needs the moisture of dew is adapted to collect it; but not a single drop is wasted where its refreshing moisture is not required.

What circumstances influence the production of dew? 605. Dew is deposited most freely upon a calm, clear night, since under such circumstances heat radiates from the earth most freely, and is lost in space. On a cloudy night, on the contrary, the deposition of dew is almost entirely interrupted, since the lower surfaces of the clouds turn back the rays of heat as they radiate, or pass off from the earth, and prevent their dispersion into space; the surface of the earth is not, therefore, cooled down sufficiently to chill the vapor of the air into dew.

When the wind blows briskly, also, little or no dew is formed, since warm air is constantly brought into contact with solid bodies, and prevents their reduction in temperature.

Can dew be properly said to fall? Dew is always formed upon the surface of the material upon which it is found, and does not fall from the atmosphere.

Other things being equal, dew is most abundant in situations most exposed, because the radiation of heat is not arrested by houses, trees, etc. Little dew is ever observed in the streets of cities, because the objects are necessarily exposed to each other's radiation, and an interchange of heat takes place, which maintains them at a temperature uniform with the air.

Does dew form upon the surface of water? Dew rarely falls upon the surface of water, or upon ships in mid-ocean. The reason of this is, that whenever the aqueous particles at the surface are cooled, they become heavier than those below them, and sink, while warmer and lighter particles rise to the top. These, in their turn, become heavier, and descend; and this process, continuing throughout the night, maintains the surface of the water and the air at nearly the same temperature.

Although dew does not appear upon ships in mid-ocean, it is freely deposited on the same vessels arriving in the vicinity of land. Thus, navigators who proceed from the Straits of Sunda to the Coromandel coast, know that they are near the end of the voyage when they perceive the ropes, sails, and other objects placed on the deck become moistened with dew during the night.

The exposed parts of the human body are never covered with dew, because the vital temperature, varying from 96° to 98° F., effectually prevents a loss of heat sufficient for its deposition.

Dew is produced most copiously in tropical countries, because there is in such latitudes the greatest difference between the temperature of the day and that of the night. The development of vegetation is also greatest in tropical countries, and a great part of the nocturnal cooling is due to the leaves which present to the sky an immense number of thin bodies, having large surface, well adapted to radiate heat.

Dew rarely falls upon the small islands of the Pacific; the reason is, that the air over the vast ocean in which these islands are situated, preserves a nearly uniform temperature day and night. The islands are comparatively of small extent, and the stratum of air cooled by the contact of the soil is warmed by mixing with the air that is constantly reaching it from the sea. This prevents a depression of temperature in the air sufficient to cause a deposition of dew.

What is frost?

606. Frost is frozen dew.

When the temperature of the body upon which the dew is deposited sinks below 32° F., the moisture freezes and assumes a solid form, constituting what is called "*frost.*"

Shrubs and low plants are more liable to be injured by frost than trees of a greater elevation, since the air contiguous to the surface of the ground is the most reduced in temperature.

Why does a thin covering protect objects from dew or frost?

An exceedingly thin covering of muslin, matting, etc., will prevent the deposition of dew or frost upon an object, since it prevents the radiation of heat, and a consequent cooling sufficient to occasion the production of either dew or frost.

Fig. 224, in which the arrows indicate the movements of heat, and the numerals the temperatures of the earth and air under different circumstances, will render the explanations of the phenomena of dew and frost more intelligible.

The figures in the middle of the diagram represent the temperature of the air at a distance from the surface of the earth; the figures in the margin, the temperature of the air adjoining the surface of the earth; the figures below the margin, the temperature of the earth itself. The directions of the arrows represent the radiation and reflection of the heat.

FIG. 224.

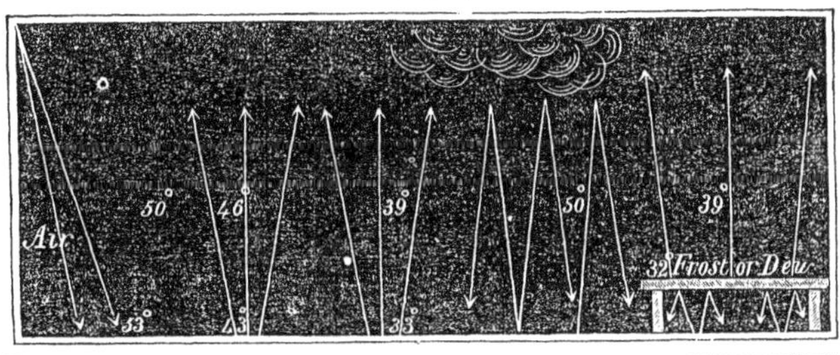

Surface of the earth, 50°.	41°. Dew.	32°. Frost.	53°. No dew or frost.	41°. No dew or frost.
In the day-time.	In clear and serene nights.		Cloudy or windy nights.	Clear night; soil protected.

SECTION II.

CLOUDS, RAIN, SNOW, AND HAIL.

What are clouds?

607. Clouds consist of vapor evaporated from the earth, and partially condensed in the higher regions of the atmosphere.

How is mist or fog occasioned?

When air, saturated with vapor, in immediate contact with the surface of the earth is cooled down rapidly, its vapor is condensed ; if the condensation, however, is not sufficient to allow of its precipitation in drops, it floats above the surface of the earth as mist or fog.

How do clouds, fog, and mist differ?

Clouds, fog, and mist differ only in one respect. Clouds float at an elevation in the air, while fogs and mists come in contact with the surface of the earth.

Mist and fog are also formed when the water of lakes and rivers, or the damp ground, is warmer than the surrounding air which is saturated with moisture. The vapors which rise in consequence of the higher temperature of the water, are immediately recondensed, as soon as they diffuse themselves through the colder air.

Mist and fog are observed most frequently over rivers and marshes, because in such situations the air is nearly saturated with vapor, and therefore the least depression of temperature will compel it to relinquish some of its moisture.

Why is the moisture of our breath visible in winter and not in summer?

The moisture contained in the air we expel from the lungs in the process of respiration, is visible in winter, but not in summer. The reason of this is, that in cold weather the vapor is condensed by the external air, but in summer the temperature of the air is not sufficiently reduced to effect condensation.

In what manner are clouds formed?

During the daily process of evaporation from the surface of the earth, warm, humid currents are continually ascending; the higher they ascend, the colder is the atmosphere into which they enter; and as they continue to rise, a point will at length be attained where, in union with the colder air, their original humidity can no longer be retained: a cloud will then appear, which increases in bulk with the upward progress of the current into colder regions.

To a person in the valley, the top of a mountain may seem enveloped in clouds; while, if he were at the summit, he would be surrounded by a mist, or fog.

Why do clouds float in the atmosphere?

The reason why clouds, which are condensed vapor, float in the atmosphere is, that they consist of very minute globules (called vesicles), which, although heavier than the surrounding air, have a great extent of surface in comparison with their weight. On account of the resistance of the air, they sink very slowly, as a soap-bubble, which greatly resembles these vesicles, sinks but slowly in a calm atmosphere. As these vesicles do, however, gradually sink, the question arises, why do not the clouds fall to the ground? The explanation of this is, that the vesicles which sink in calm weather can not reach the ground, because in their descent they soon meet with warmer strata of air which are not saturated with moisture, where they again dissolve into vapor and are lost to view: at the same time that the vesicles of vapor dissolve at the lower limits of the clouds, new ones are formed above, and thus the cloud appears to float immovably in the air.

When the atmosphere is agitated, the vesicles of vapor constituting clouds are driven in the direction of the currents of air. A wind moving in a horizontal direction will carry the clouds in the same direction; and an ascending current of air will lift them up, as soon as its velocity becomes greater than the velocity with which the vesicles would fall to the ground in a calm condition of the air. In like manner, soap-bubbles are elevated by the wind and carried to considerable distances.

How do winds affect the clouds?

Clouds frequently appear and disappear with a change in the direction and character of the wind. Thus, if a cold wind blows suddenly over any region, it condenses the invisible vapor of the air into cloud or rain; but if a warm wind blows over any region, it disperses the clouds by absorbing their moisture.

What is the average height of clouds?

The *average* height at which clouds float above the surface of the earth in a calm day, is between one and two miles. Light, fleecy clouds, however, sometimes attain an elevation of five or six miles.

What occasions the irregular and broken appearance of clouds?

When clouds are not continuous over the whole surface of the sky, various circumstances contribute to give them a rough and uneven appearance. The rays of the sun falling upon different surfaces at different angles, melt away one set of elevations and create another set of depressions; the heat also, which is liberated below in the process of condensation, the currents of warm air escaping from the earth, and of cold air descending from above, all tend to keep the clouds in a state of agitation, upheaval, and depression. Under these influences, the masses of condensed vapor composing the clouds are caused to assume all manner of grotesque and fanciful shapes.

The shape and position of clouds is also undoubtedly influenced in a considerable degree by their electrical condition.

Why do clouds frequently collect around mountain peaks?

Clouds are frequently seen to collect around mountain peaks, when the atmosphere elsewhere is clear and free from clouds. This is caused by the wind impelling up the sides of the mountains the warm, humid air of the valleys, the moisture of which, in its ascent, gradually becomes condensed by cold, and appears as a cloud.

How many kinds of clouds are recognized?

608. Clouds are generally divided into four great classes, viz.: the CIRRUS, the CUMULUS, the STRATUS, and the NIMBUS.

What is the Cirrus cloud?

The Cirrus* cloud consists of very delicate thin streaks, or feathery filaments, and is usually seen floating at great elevations in the sky during the continuance of fine weather.

It is highly probable that the cirrus cloud, at great elevations, does not consist of vesicles of mist, but of flakes of snow.

Fig. 225, *a*, represents the appearance of this variety of cloud.

What is the Cumulus cloud?

The Cumulus† cloud consists of large rounded masses of vapor, apparently resting upon a horizontal basis. When lighted up by the sun, cumulus clouds present the appearance of mountains of snow.

The cumulus is especially the cloud of day, and its figure is most perfect during the fine, warm days of summer.

Fig. 225, *b*, illustrates the appearance of the cumulus cloud.

These clouds appear in greatest number at noon, on a fine day, but disappear as evening approaches. The explanation of this is, that at noon the cur-

* From the Latin word *cirrus*—a lock of hair, or curl.

† From the Latin word *cumulus*—a mass, or pile.

rents of warm air ascending from the earth are more buoyant, larger, and rise higher, and when condensed, form large masses of clouds, each of which may be considered as the capital of a column of air, whose base rests upon the earth. As the heat of the sun diminishes in the afternoon, the strength of the currents abate, the clouds, which are buoyed up by their force, sink down into warmer regions of the atmosphere, and are either partially or wholly dissolved.

FIG. 225.

Cirrus, *a*; Cumulus, *b*; Stratus and Nimbus, *c*.

The rounded figure of the cumulus has been attributed to its method of formation; for when one fluid flows through another at rest, the outline of the figure assumed by the first will be composed of curved lines. This fact may be shown, and the appearance of the cumulus imitated, by allowing a drop of milk or ink to fall into a glass of water. The same thing is also seen in the shape of a cloud of steam as it issues from the boiler of a locomotive.

What is the Stratus cloud?

The Stratus,* or stratified cloud, consists of horizontal streaks, or layers of vapor, which float like a veil at no very great elevation from the surface of the earth. They frequently appear with extraordinary brilliancy of color at sunset.

The appearance of the stratus is represented at *c*, Fig. 225.

What is the Nimbus?

The Nimbus, or the cloud of rain, has no characteristic form. It generally covers the whole horizon, imparting to it a bluish black appearance.

The various forms of clouds gradually pass into each other, so that it is often difficult to decide whether the appearance of a cloud approaches more to one type than another. The intermediate forms are sometimes designated as cirro-stratus, cirro-cumulus, and cumulo-stratus.

What is Rain?

609. Rain is the vapor of the clouds or air condensed and precipitated to the earth in drops.

How is rain occasioned?

Rain is generally occasioned by the union of two or more volumes of humid air, differing considerably in temperature. Under such circumstances, the several portions in union are incapable of absorbing the same amount of moisture that each could retain if they had not united. The excess, if very great, falls as rain ; if of slight amount, it appears as cloud.

Upon what law does the formation of rain depend?

610. The law upon which the condensation of vapor and the formation of rain depends is, that the capacity of the air for moisture decreases in a greater ratio than the temperature.

Why does rain fall in drops?

611. Rain falls in drops, because the vesicles of vapor, in their descent, attract each other and merge together, thus forming drops of water. The size of the drop is increased in proportion to the rapidity with which the vapors are condensed.

In rainy weather the clouds fall toward the earth, for the reason that they are heavy with partially-condensed vapors, and the air, on account of its diminished density, is less able to buoy them up.

612. The quantity of rain falling at any one time or place, is measured by means of an instrument called a "Rain-Guage."

Describe the Rain-Guage.

This usually consists of a tin cylindrical vessel, M, Fig. 226, the upper part of which is closed by a cover, B, in the shape of a funnel, with an aperture in its center. The water

* Stratus, from the Latin *stratus*—that which lies low in the form of a bed or layer.

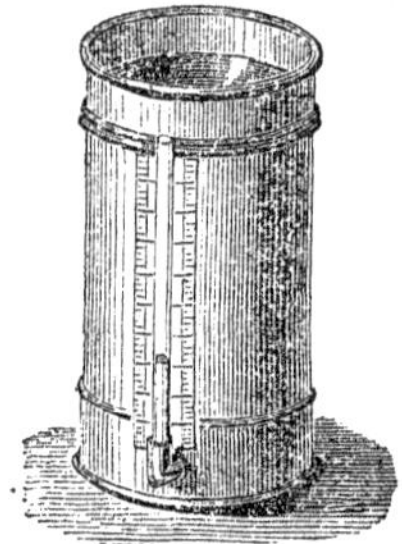

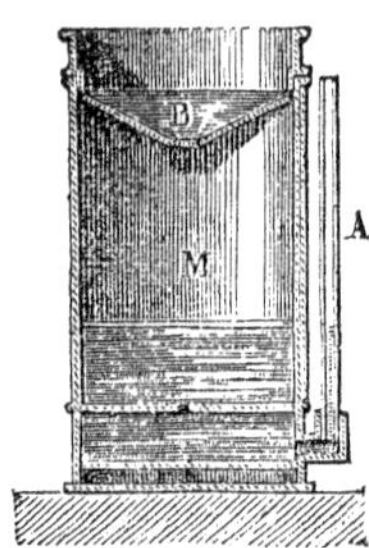

falling upon the top of the cylinder flows into the interior through the opening, and i thus protected from evaporation. From the base of the apparatus a graduated glass tube, A, ascends, in which the water rises to the same height as in the interior of the cylinder. Supposing the apparatus to be placed in an exposed situation, and at the end of a month, for example, the height of the water in the tube is five inches: this would indicate that the water in the cylinder had attained to an equal elevation, and consequently that the rain which had fallen during this interval, would, if not diminished by evaporation or infiltration, cover the earth to the depth of five inches.

In what situations is rain most abundant?

613. Rain falls most abundantly in countries near the equator, and decreases in quantity as we approach the poles. There are more rainy days, however, in the temperate zones than in the tropics, although the yearly quantity of rain falling in the latter districts is much greater than in the former.

In the northern portions of the United States, there are on an average about 134 rainy days in a year; in the Southern States the number is somewhat less, being about 103.

The reason why it rains more frequently in the temperate zones than in the tropics is because, the former are regions of variable winds, and the temperature of the atmosphere changes often; while in the tropics the wind changes but rarely, and the temperature is very constant throughout a great part of the year. In the tropics the year is divided into only two seasons, the wet, or rainy, and the dry season.

What is the average fall of rain in different countries?

The average yearly fall of rain in the tropics is ninety-five inches; in the temperate zone only thirty-five.

The greatest rain-fall, however, is precipitated in the shortest time. Ninety-five inches fall in eighty days on the equator, while at St. Petersburg the yearly rain-fall is but seventeen inches, spread over one hundred and sixty-nine days. Again, a tropical wet day is not continuously wet. The morning is clear; clouds form about ten o'clock; the rain begins at twelve, and pours till about half past four; by sunset the clouds are gone, and the nights are invariably fine.

The depth of rain which falls yearly in London is about twenty-five inches; but at Vera Cruz, in the Gulf of Mexico, rain to the amount of two hundred and seventy-eight inches is precipitated. The explanation of this is to be found in the peculiar location of the city, at the foot of lofty mountains, whose summits are covered with perpetual snow; against these the hot, humid air from the sea is driven by the winds, condensed, and its excess of moisture precipitated as rain.

614. Some countries are entirely destitute of rain; in a part of Egypt it never rains, and in Peru it rains once, perhaps, in a man's lifetime. Upon the table-land of Mexico, in parts of Guatemala and California, rain is very rare. But the most extensive rainless districts are those occupied by the great desert of Africa, and its continuation eastward over portions of Arabia and Persia to the interior of Central Asia, over the great desert of Gobi, the table-land of Thibet, and part of Mongolia. These regions embrace an area of five or six millions of square miles that never experience a shower.

The cause of this scarcity is to be sought for in the peculiar conformation of the country.

In Peru, for example, parallel to the coast, and at a short distance from the sea, is the lofty range of the Andes, the peaks of which are covered with perpetual snow and ice. The prevailing wind is an east wind, sweeping from the Atlantic to the Pacific across the continent of South America. As it approaches the west coast, it encounters this range of mountains, and becomes so cooled by them that it is forced to precipitate its moisture, and passes on to the coast almost devoid of moisture. In Egypt and other desert countries, the dry sandy plains heat the atmosphere to such an extent that it absorbs moisture, and precipitates none.

On the other hand, there are some countries in which it may be said to always rain. In some portions of Guiana, in South America, it rains for a great portion of the year. The fierce heat of the tropical sun fills the atmosphere with vapor, which returns to the earth again in constant showers as the cool winds of the ocean flow in and condense it.

What is the whole estimated yearly quantity of rain?

615. The whole quantity of water annually precipitated as rain over the earth's surface is calculated to exceed seven hundred and sixty millions of tons. This entire amount is raised into the atmosphere solely by evaporation. It has been also calculated, that the daily amount of water raised by evaporation from the sea alone, amounts to no less than one hundred and sixty-four cubic miles, or about sixty thousand cubic miles annually.

During the months of October and November, the daily amount of evaporation from the surface of the ocean, between the Cape of Good Hope and Calcutta, is known to average three quarters of an inch from the whole surface.

What curious influences are occasioned by the moisture of the atmosphere?

The amount of moisture constantly present in the atmosphere of any country, exercises an important influence upon the physical system of the inhabitants, and upon their arts and professions. The atmosphere of the northern United States is uncommonly dry, much more so than in England or Germany. To this in a great measure is owing the difference in the physical appearance of the inhabitants of these respective countries. Painters find that their work dries quicker, also, in New England than in central Europe. Cabinet-makers in the United States are obliged to use thicker glue, and watchmakers animal instead of vegetable oil. Pianos are rarely imported from Europe into the United States, because the difference in the climate of these two countries is so great, as respects moisture, that the foreign instruments shrink, and quickly become damaged.

What is Snow?

616. Snow is the condensed vapor of the air, frozen and precipitated to the earth.

How is snow probably formed?

Our knowledge in respect to the formation of snow in the atmosphere is very limited. It is probable that the clouds in which the flakes of snow are first formed, consist, not of vesicles of vapor, but of minute crystals of ice, which by the continuous condensation of vapor become larger and form flakes of snow, which continue to increase in size as they descend through the air.

When the lower regions of the air are sufficiently warm, the flakes of snow melt before they reach the ground; so that it may rain below, while it snows above.

The largest flakes of snow are formed when the air abounds with vapor, and the temperature is about 32° F.; but as the moisture diminishes, and the cold increases, the snow becomes finer.

In extreme cold weather, when a volume of cold air is suddenly admitted into a room, the air of which is saturated with moisture, it sometimes happens that the vapor of the room will be condensed and frozen at the same instant, thus producing a miniature fall of snow.

What is the physical composition of a snow-flake?

617. On examining a snow-flake beneath a microscope, it is found to consist of regular and symmetrical crystals, having a great diversity of form.

These crystals also exist in ice, but are so blended together that their symmetry is lost in the compact mass.

The crystals of snow may, under favorable circumstances, be seen with the naked eye, by placing the flake upon a dark body cooled below 32° F. Fig. 227 represents the varied and beautiful forms of snow crystals.

The bulk of recently-fallen snow is ten or twelve times greater than that of the water obtained by melting it.

What is Hail?

618. Hail is the moisture of the air frozen into drops of ice.

Can the phenomenon of hail be explained satisfactorily?

The phenomenon of hail has never been satisfactorily explained. It is difficult to conceive how the great cold is produced which causes the water to freeze under the circumstances, and also how it is possible that the hail-stones, after having once become sufficiently large to fall by their own weight, can yet remain long enough in the air to increase to so considerable a size as is sometimes seen. A hail-storm generally lasts but a few minutes, very seldom as long as a quarter of an hour; but the quantity of ice which escapes from the clouds in so short a time is very great, and masses have been observed to fall of a weight of 10 or 12 ounces.

Fig. 227.

619. Hail-stones are generally pear-shaped, and if they are divided through the center, they will be found to be composed of alternate layers of ice and snow, arranged around a nucleus, like the coats of an onion.

Hail-storms occur most frequently in temperate climates, and rarely within the tropics. They occur most frequently in northern latitudes, in the vicinity of high mountains, whose peaks are always covered with ice and snow. The south of France, which lies between the Alps and Pyrenees, is annually ravaged by hail; and the damage which it causes yearly to vineyards and standing crops has been estimated at upward of nine millions of dollars.

SECTION III.

WINDS.

What is Wind?

620. Wind is air put in motion. The air is never entirely free from motion, but the velocity with which it moves is perpetually varying.

What is the principal cause of wind?

621. The principal cause of movements in the atmosphere is the variation of temperature produced by the alternation of day and night and the succession of the seasons.

How can variations of temperature produce wind?

When, through the agency of the sun, a particular portion of the earth's surface is heated to a greater degree than the remainder, the air resting upon it becomes rarefied and ascends, while a current of cold air rushes in to supply the vacancy. Two currents, the one of warm air flowing out, and the other of cold air flowing in, are thus continually produced; and to these movements of the atmosphere we apply the designation of wind.

How do the physical features of the earth affect the winds?

If the whole surface of the earth were covered with water, the winds would always follow the sun, and blow uniformly from east to west. The direction of the wind is, however, continually subject to interruption from mountains, deserts, plains, oceans, etc.

Thus mountains which are covered with snow, condense and cool the air brought in contact with them, and when the temperature of the current of air constituting the wind is changed, its direction is liable to be changed also. The ocean is never heated to the same degree as the land, and in consequence of this, the general direction of the wind is from tracts of ocean toward tracts of land.

In those parts of the world which present an extended surface of water, the wind blows with a great degree of regularity.

What is the velocity and force of winds?

622. Every variation exists in the speed of winds, from the mildest zephyr to the most violent hurricane.

A wind which is hardly perceptible moves with a velocity of about one mile per hour, and with a perpendicular force on one square foot of ·005 pounds avoirdupois.

In a storm, the velocity of the wind is from 50 to 60 miles per hour, and the pressure from 7 to 12 pounds per square foot. In some hurricanes, the velocity has been estimated at from 80 to 100 miles per hour, with a varying force of from 30 to 50 pounds.

How is the force of wind calculated?

The force of the wind is ascertained by observing the amount of pressure that it exerts upon a given plane surface perpendicular to its own direction.

If the pressure-plate acts freely upon spiral springs, the power of the wind is denoted by the extent of their compression, which thus becomes a measure of their force, the same as in weighing by the ordinary spring-balance.

What is an Anemometer?

An instrument for measuring the force of the wind is called an Anemometer.

How may winds be divided?

623. Winds may be divided into three classes:—Constant, Periodical, and Variable winds.

What are the trade-winds?

624. In many parts of the Atlantic and Pacific oceans, the wind blows with a uniform force and constancy, so that a vessel may sail for weeks without altering the position of a sail or spar. Such winds have received the designation of trade-winds, inasmuch as they are most convenient for navigation, and always blow in one direction.

What is the cause of the trade-winds?

The trade-winds are caused by the movements of vast currents of air which are continually flowing between the poles and the equator. Thus the air which has been greatly heated by the sun in regions near to the equator, rises and runs over toward either pole in two grand upper currents, under which there flows from north and south two other currents of colder air to occupy the space vacated, and to restore the equilibrium.

What occasions the direction of the trade-winds?

625. In the northern hemisphere the trade-winds blow from the north-east, and in the southern hemisphere from the south-east.

The reason they do not blow from the direct north and south is owing to the revolution of the earth. The circumference of the earth being larger at the equator than at the poles, every spot of the equatorial surface must move much faster than the corresponding one at the poles: when, therefore, a current of air from the poles flows toward the equator, it comes to a part of the earth's surface which is moving faster than itself; in consequence of which it is left behind, and thus produces the effect of a current moving in the opposite direction.

The region over which the trade-winds prevail extends for about 25 degrees of latitude, on each side of the equator, in the Atlantic and Pacific oceans.

The reason the trade-winds do not blow uninterruptedly from the equator to each pole is owing to the change which takes place in their temperature as they move north and south. Thus in the northern hemisphere the hot air that ascends from the equator and passes north, gradually cools, and becomes denser and heavier, running as it does over the cold current below. The cold air from the pole, too, gradually becomes warmer and lighter as it passes south, so that in the temperate climates there is a constant struggle as to which shall have the upper and which the lower position. In these regions, consequently, there are no uniform winds.*

What are monsoons?

626. Monsoons are periodical currents of air which in the Arabian, Indian, and China seas blow for nearly six months of the year in one direction, and for the other six in a contrary direction.

They are called monsoons from an Arabic word signifying season; they are also called periodical winds, to distinguish them from the trade-winds which are constant.

What is the theory of the monsoons?

The theory of the monsoons is as follows:—During six months of the year, from April to October, the air of Arabia, Persia, India, and China, is so rarefied by the enormous heat of their summer sun, that the cold air from the south rushes toward these

* The existence of a great current of air in the upper regions of the atmosphere, flowing in an nearly contrary direction to the trade-winds, has been confirmed by the observations of travelers who have ascended the Peak of Teneriffe, or some of the high mountains in the islands of the Southern Pacific Ocean. At a height of about 12,000 feet a wind is encountered, blowing constantly in an opposite direction to that which prevails at the level of the sea below.

countries, across the equator, and produces a south-west wind. When the sun, on the other hand, has left the northern side of the equator for the southern, the southern hemisphere is rendered hotter than the northern, and the direction of the wind is reversed, or the monsoon blows north-east from October to April.

The monsoons are more powerful than the trade-winds, and very often amount to violent gales. They are also more useful than the trade-winds, since the mariner is able to avail himself of their periodic changes to go in one direction during one half of the year, and return in the opposite direction during the other half.

What is the explanation of land and sea breezes?

627. In some parts of the world, as on coasts and islands, the heating action of the sun produces daily periodical winds, which are termed land and sea-breezes.

During the day, the land becomes much more highly heated by the sun than the adjacent water, and consequently the air resting upon the land is much more heated and rarefied than that upon the water. The cooler and denser air, therefore, flows from the water toward the land, constituting a sea-breeze, and, displacing the warmer and lighter air over the land, forces it into a higher region, along which it flows in an upper current seaward.

At night a contrary effect is produced. After sunset the land cools much more rapidly than the water, and the air over the shore becoming cooler, and consequently heavier than that over the sea, flows toward the water and forms the land-breeze.*

The phenomena of land and sea-breezes may be well illustrated by a simple experiment. Fill a large dish with cold water, and place in the middle of it a saucer full of warm water; let the dish represent the ocean, and the saucer an island heated by the sun, and rarefying the air above it; blow out a candle, and if the air of the room be still, on applying it successively to every side of the saucer, the smoke will be seen moving toward it and rising over it, thus indicating the course of the air from sea to land. On reversing the experiment, by filling the saucer with cold water, and the dish with warm, the land-breeze will be shown by holding the smoking wick over the edge of the saucer; the smoke will then be wafted to the warmer air over the dish.

In what regions do variable winds prevail?

628. In the temperate zones, the winds have little of regularity, and these latitudes are known as the regions of "variable winds."

In the tropics, the great aerial currents known as the trade-winds exist in all their power, and control most of the local influences; but in the temperate zones, where the force of the trade-winds is diminished, a perpetual contest

* Advantage is taken of these breezes by coasters, which, drawing less water than larger vessels, can approach the coast within those limits where the sea and land-breezes first begin to operate. Thus a ship of war may not be able to take advantage of these winds, while sloops and schooners may be moving along close to the shore under a press of canvas, and be out of sight before the larger vessel is released from the calm bordering those breezes, and fringing for some time the beach only.

occurs between the permanent and temporary currents, giving rise to constant fluctuations in the strength and direction of the winds.

What is the character of the winds of the United States?

629. The driest winds of the United States are west and north-west winds, since they blow over great tracts of land, and have little opportunity of absorbing moisture.

The south winds are generally warm and productive of rain, since coming from tropical countries, they are highly heated, and readily absorb moisture as they pass over the ocean. As soon, however, as they reach a cold climate they are condensed, and can no longer hold all their vapor in suspension; in consequence of which some of it is deposited as rain.

630. Other disturbances of the air occasion a variety of phenomena known as "Simoons," "Hurricanes," "Tornadoes," "Water-Spouts," etc.

What is a Simoon?

631. The Simoon is an intensely hot wind that prevails upon the vast deserts of Africa and the arid plains of Asia, causing great suffering, and often destruction of whole caravans of men and animals when encountered. Its origin is to be sought in the peculiarities of the soil and the geographical position of the countries where it occurs.

"The surface of the deserts of Africa and Asia is composed of dry sand, which the vertical rays of the sun render burning to the touch. The heat of these regions is insupportable, and their atmosphere like the breath of a furnace. When, under such circumstances, the wind rises and sweeps over these plains, it is intensely hot and destitute of moisture, and at the same time bears aloft with it great clouds of fine sand and dust—a dreadful visitant to the traveler of the desert."

What is a Hurricane?

The Hurricane is a remarkable storm wind, peculiar to certain portions of the world. It rarely takes its rise beyond the tropics, and it is the only storm to dread within the region of the trade-winds.

Hurricanes are especially distinguished from all other kinds of tempests by their extent, irresistible power, and the sudden changes that occur in the direction of the wind.

At what times and locations do hurricanes most frequently occur?

In the northern hemisphere, the hurricane most frequently occurs in the regions of the West Indies; in the southern hemisphere, it occurs in the neighborhood of the Island of Mauritius, in the Indian Ocean. They also seem to be confined to particular seasons; thus the West Indian occur from August to October; the Mauritian from February to April.

What is the nature of the hurricane?

Recent investigations have proved the hurricanes to consist of extensive storms of wind, which revolve round an axis either upright or inclined to the horizon; while at the same time the body of the storm has a progressive motion over the surface of the ocean.

Thus it is the nature of a hurricane to travel round and round as well as forward, much as a corkscrew travels through a cork, only the circles are all flat, and described by a rotary wind upon the surface of the water. A ship revolving in the circles of a hurricane, would find, in successive positions, the wind blowing from every point of the compass.*

The effect produced by a hurricane upon the atmosphere is very singular. As it consists of a body of air rotating in a vast circle, its center is the point of least motion. Mariners who have been caught in such a center, describe the unnatural calm that prevails as awful—an apparent lull of the tempest, which seems to have rested only to gather strength for greater efforts. The mass of air, however, which constitutes the body of the storm will be driven outward from the center toward the margin, just as water in a pail which is made to revolve rapidly flies from the center and swells up the sides. But the pressure of the atmosphere beyond the whirl, checking and resisting the centrifugal force, at length arrests the outward progress of the mass of air, and limits the storm.

What is known respecting the velocity and spaces traversed by hurricanes?

The progressive velocity of hurricanes is from seventeen to forty miles per hour; but distinct from the progressive velocity is the rotary velocity, which increases from the exterior boundary to the center of the storm, near which point the force of the tempest is greatest, the wind sometimes blowing at the rate of one hundred miles per hour.

The distance traversed by these terrible tempests is also immense. The great gale of August, 1830, which occurred at St. Thomas, in the West Indies, on the 12th, reached the Banks of Newfoundland on the 19th, having traveled more than three thousand nautical miles in seven days; the track of the Cuba hurricane of 1844 was but little inferior in length.

The surface simultaneously swept by these tremendous whirlwinds is a vast circle varying from one hundred to five hundred miles in diameter.

Mr. Redfield has estimated the great Cuba hurricane of 1844 to have been not less than eight hundred miles in breadth, and the area over which it prevailed during its whole length was computed to be two million four hundred thousand square miles—an extent of surface equal to two thirds of that of all Europe.

* In 1845, a ship encountered a hurricane near Mauritius. The wind, as the ship sailed in the circuit of the storm, changed five times completely round in one hundred and seventeen hours. The whole distance sailed by the vessel was thirteen hundred and seventy-three miles, and at the termination of the storm she was only three hundred and fifty-four miles from the place where the storm commenced.

What are Tornadoes?

632. Tornadoes may be regarded as hurricanes, differing chiefly in respect to their continuance and extent.

Tornadoes usually last from fifteen to seventy seconds; their breadth varies from a few rods to several hundred yards, and the length of their course rarely exceeds twenty miles.

The tornado is generally preceded by a calm and sultry state of the atmosphere, when suddenly the whirlwind appears, prostrating every thing before it. Tornadoes are usually accompanied with thunder and lightning, and sometimes showers of hail.

How are tornadoes produced?

Tornadoes are supposed to be generally produced by the lateral action of an opposing wind, or the influence of a brisk gale upon a portion of the atmosphere in repose.

Similar phenomena are seen in the eddies, or little whirlpools formed in water, when two streams flowing in different directions meet. They occur most frequently at the junction of two brooks or rivers.

Whirlwinds on a small scale are often produced at the corners of streets in cities, and are occasioned by a gust of wind sweeping round a building, and striking the calm air beyond.

The whirl of a tornado, or whirlwind, appears to originate in the higher regions of the atmosphere; it increases in velocity as it descends, its base gradually approaching the earth, until it rests upon the surface.

Great conflagrations sometimes produce whirlwinds, in consequence of a strong upward current, which is produced by the expansion of the heated air. A remarkable example of this is recorded to have happened at the burning of Moscow, in 1812, where the air became so rarefied by heat, that the wind rose to a frightful hurricane.

It has been noticed as one of the curious effects of a tornado, that fowls and birds overtaken by it and caught in its center, are often entirely stripped of their feathers. In a theory propounded some years since to the American Association for the Promotion of Science, by an eminent scientific authority, it was supposed that in the vortex, or center of the tornado, there was a vacuum, and the fowls being suddenly caught in it, the air contained in the barrel of their quills expanded with such force as to strip them from the body.

What is a Water-spout?

633. A water-spout is a whirlwind over the surface of water, and differs from a whirlwind on land in the fact that water is subjected to the action of the wind, instead of objects on the surface of the earth. In diameter the spout at the base ranges from a few feet

to several hundreds, and its altitude is supposed to be often upward of a mile.

When an observer is near to the spout, a loud hissing noise is heard, and the interior of the column seems to be traversed by a rushing stream.

FIG. 228.

The successive appearances of a water-spout are as follows:—At first it appears to be a dark cone, extending from the clouds to the water; then it becomes a column uniting with the water. After continuing for a little time, the column becomes disunited, the cone reappears, and is gradually drawn up into the clouds. These various changes are represented in Fig. 228. It is a common belief that water is sucked up by the action of the spout into the clouds; but there is reason to suppose that water rather descends from the clouds, as water which has fallen from a spout upon the deck of a vessel has been found to be fresh. There is no evidence, furthermore, that a continuous column of water exists within the whirling pillar.

SECTION IV.

METEORIC PHENOMENA.

What are Meteorites?

634. Meteorites are luminous bodies, which from time to time appear in the atmosphere, moving with immense velocity, and remaining visible but for a few moments. They are generally accompanied by a luminous train, and during their progress explosions are often heard.

What are Aerolites?

635. The term aerolite is given to those stony masses of matter which are sometimes seen to fall from the atmosphere.*

What is known respecting the weight and velocity of aerolites?

The weight of those aerolites which have been known to fall from the atmosphere varies from a few ounces to several hundred pounds, or even tons.

The height above the earth's surface at which they are supposed to make their appearance has been estimated to vary from 18 to 80 miles.

* Aerolite is derived from the Greek words αερ (atmosphere) and λιθος (a stone). A meteor is distinguished from an aerolite by the fact that it bursts in the atmosphere, but leaves no residuum, while the aerolite, which is supposed to be a fragment of a meteor, comes to the ground.

The estimated velocity of these bodies is somewhat more than three hundred miles per minute, though one meteor of immense size, which is supposed to have passed within twenty-five miles of the earth, moved at the rate of twelve hundred miles per minute. Owing, however, to the short time the meteor is visible, and its great velocity, accurate observations can not be made upon it; and all estimates respecting their distance, size, etc., must be considered as only approximations to the truth.

What is known respecting the constitution of aerolites?

Very many of the meteorites which have fallen at different times and in different parts of the globe, resemble each other so closely, that they would seem to have been broken from the same piece or mass of matter.

Most of them are covered with a black shining crust, as if the body had been coated with pitch. When broken, their color is ash-gray, inclining to black. They consist for the most part of malleable iron and nickel, but they often contain small quantities of other substances. They do not resemble in composition any other bodies found upon the surface of the earth, but have a character of their own so peculiar that it enables us to decide upon the meteoric origin of masses of iron which are occasionally found scattered up and down the surface of the earth, as in the south of Africa, in Mexico, Siberia, and on the route overland to California. Some of these masses are of immense weight, and undoubtedly fell from the atmosphere.

What is the supposed origin of meteoric bodies?

636. Four hypotheses have been advanced to account for the origin of these extraordinary bodies: 1. That they are thrown up from terrestrial volcanoes; 2. That they are produced in the atmosphere from vapors and gases exhaled from the earth; 3. That they are thrown from lunar volcanoes; 4. That they are of the same nature as the planets, either derived from them, or existing independently.

The fourth of these suppositions most fully explains the facts connected with the appearance of meteorites, and the third likewise has some strong evidence in its favor.

How do shooting-stars differ from meteors?

637. Shooting-stars differ in many respects from meteors. Their altitude and velocity are greater; they are far more numerous and frequent, and are unaccompanied by any sound or explosion. Their brilliancy is also much inferior to that of the meteor, and no portion of their substance is ever known to have reached the earth.

At what height do shooting-stars appear?

The altitude of shooting-stars is supposed to vary from six to four hundred and sixty miles, the greatest number appearing at a height of about seventy miles. Owing to their number and frequency of occurrence, many careful observations have been made upon them, with a view of determining these facts.

Their velocity is supposed to range from sixty to fifteen hundred miles per minute.

Some of these meteoric appearances may be seen every clear night, but they appear to fall in great numbers at certain periodical epochs. The periods when they may be noticed most abundantly are on the 9th and 10th of August, and the 12th and 13th of November.*

The majority of shooting stars appear to radiate from a particular part of the heavens, viz., a point in the constellation Perseus, undoubtedly far beyond the limits of our atmosphere.

What theories have been proposed to account for the origin of shooting stars?

In order to account for the origin of shooting stars, it has been supposed by Prof. Olmstead, that they are derived from a body composed of matter exceedingly rare, like the tail of a comet, revolving around the sun within the orbit of the earth, in a space little less than a year; and that at times the body approaches so near the earth that the extreme portions become detached and drawn to the earth by virtue of its great attraction. It has been further supposed that the matter of which these bodies is composed is combustible, and becomes ignited on entering the earth's atmosphere.

The nearest approach of the central body to the earth is supposed to be about 2,000 miles. Bodies falling from this distance would enter the earth's atmosphere at a height of at least 50 miles above the surface, with a velocity generated by the force of gravity above 4 miles per second—a velocity ten times greater than the utmost speed of a cannon-ball.

When common air is compressed in a tight cylinder to the extent of one fifth of its volume, sufficient heat is generated to ignite tinder. If we suppose that the fragments descend with such velocity as to compress the rarefied atmosphere at the height of 30 miles to such an extent only as to make it as dense as ordinary air, the temperature would be raised as high as 46,000° F. —a heat far more intense than can be generated in any furnace. Unless, therefore, the mass of matter comprising the body was very large, it must be dissipated by heat long before it reaches the surface of the earth.

Another theory has been proposed by the eminent astronomer Chaldini, who supposes that, in addition to the planets and their satellites which revolve about the sun, there are innumerable smaller bodies; and that these occasionally enter within the atmosphere of the earth, take fire, or descend to its surface.

* They have also been noticed in unusual abundance on the 18th of October, the 6th and 7th of December, the 2d of January, the 23d and 24th of April, and from the 18th to the 20th of June.

Four most remarkable meteoric showers have been noticed, viz., in 1797, 1831, 1832, and 1833, all in the month of November. In the shower of 1833, the meteors, in many parts of the United States, appeared to fall as thick as snow flakes.

SECTION V.

POPULAR OPINIONS CONCERNING THE WEATHER.

Do changes in the weather occur in accordance with fixed laws?

638. There is no reason to doubt that every change in the weather is in strict accordance with some definite physical agencies, which are fixed and certain in their operations. We can not, however, foretell with any degree of certainty the character of the weather for any particular time, because the laws which govern meteorological changes are as yet imperfectly understood.

Are the popular ideas respecting changes in the weather founded on fact?

There are, however, in all countries, certain ideas and popular proverbs respecting changes in the weather, the influence of the moon, the aurora borealis, etc., which are wholly erroneous and unworthy of belief; since, when tested by long-continued observations, they are invariably found to be unsupported by evidence.

Thus an examination of meteorological records, kept in different countries, through many years, proves conclusively that the popular notions concerning the influence of the moon on the weather has no foundation in any well-established theory, and no correspondence with observed facts.

There is, however, some reason for supposing that rain falls more frequently about four days before full moon, and less frequently about four or five days before new moon, than at other parts of the month; but this can not be considered as an established fact. In other respects, the changes of the moon can not be shown to have influenced in any way the production of rain.

There is also a current belief among many persons that timber should be cut during the decline of the moon. To test the matter, an experiment, on an extensive scale, was made some years since in France, when it was found that there was no difference in the quality of any timber felled in different parts of the lunar month.

It is also supposed that bright moonlight hastens, in some way, the putrefaction of animal and vegetable substances. The facts in respect to this supposition are, that on bright, clear nights, when the moon shines brilliantly, dew is more freely deposited on these substances than at other times, and in this way putrefaction may be accelerated. With this result the moon has no connection.

It is a traditional idea with many that a long and violent storm usually accompanies the period of the equinoxes, especially the autumnal; but the examination of weather records for sixty-four years has shown that no particular day can be pointed out in the month of September (when the "equinoctial storm" is said to occur) upon which there ever was, or ever will be, a so-called equinoctial storm. The fact, however, should not be concealed, that, taking the average for the five days embracing the equinox for the period above stated, the amount of rain is greater than for any other five days, by three per cent., throughout the month.

Observations recorded for a long period have proved that the phenomenon of the aurora borealis, which is said to precede a storm, is as often followed by fair, as by foul weather.

Meteorological records, kept for eighty years at the observatory of Greenwich, England, seem to show that groups of warm years alternate with cold ones in such a way as to render it probable that the mean annual temperatures rise and fall in a series of curves, corresponding to periods of about fourteen years.

There is little doubt that some animals and insects are able to foretell changes in the weather, when man fails to perceive any indications of the same. Thus some varieties of the land-snail only make their appearance before a rain. Some other varieties of land crustaceous animals change their color and appearance twenty-four hours before a rain.

For a light, short rain, some trees have been observed to incline their leaves, so as to retain water; but for a long rain, they are so arranged as to conduct the water away.

The admonition given several thousand years ago, is equally sound in its philosophy at the present day: "He that observeth the winds shall not sow; and he that regardeth the clouds shall not reap."—*Eccles.*, xi. 4.

CHAPTER XIV.

LIGHT.

What is Light?

639. LIGHT is the physical agent which occasions, by its action upon the eye, the sensation of vision.

What is the Science of Optics?

640. Optics is the name given to that department of physical science which treats of vision, and of the laws and properties of light.*

Between the eye and any visible object a space of greater or less extent intervenes. In some instances, as when we look at a star, the extent of the space existing between the eye and the object seen is so great, that the mind is unable to form any adequate conception of it. Yet we recognize the existence of objects at such distances, by the physical effect which they produce on our organs of vision.

* From the Greek word "Οπτομαι," to see.

What theories of light have been proposed?

641. In order to explain how such a result is possible, or in other words, to account for the origin of light, two theories have been proposed, which are called the CORPUSCULAR and the UNDULATORY Theories.

What is the Corpuscular Theory of Light?

The CORPUSCULAR THEORY supposes that a distant object becomes visible to us by emitting particles of matter from its surface, which particles of matter, passing through the intervening space between the visible object and the eye, enter the eye, and striking upon the nervous membrane, so affect it as to produce the sensation of light, or vision.

According to this theory, there is a striking analogy or resemblance between the eye and the organs of smelling. Thus, we recognize the odor of an object in consequence of the material particles which pass from the object to the organs of smelling, and there produce a sensation. In the same manner, a visible object at any distance may be supposed to send forth particles of light, which move to the eye and produce vision, by acting mechanically on its nervous structure, as the odoriferous particles of a rose produce a sensible effect upon the organs of smelling.

What is the Undulatory Theory?

The UNDULATORY THEORY supposes that there exists throughout all space an ethereal, elastic fluid, which, like the air, is capable of receiving and transmitting undulations, or vibrations. These, reaching the eye, affect the optic nerve, and produce the sensation which we call light.

According to this theory, there is a striking analogy between the eye and the ear; the vibrations, or undulations of the ethereal medium being supposed to pass along the space intervening between the visible object and the eye in the same manner that the undulations of the air, produced by a sounding body, pass through the air between it and the ear.

Which of the two theories of light is generally received?

The Corpuscular Theory was sustained by Newton, and was for a long time generally believed. At the present day it is almost entirely discarded, and the Undulatory Theory is now received by scientific men as substantially correct; since it explains in a satisfactory manner nearly all the phenŏmena of light, which the Corpuscular Theory does not.

If the Corpuscular Theory be correct, a common candle is able to fill for hours, with particles of luminous matter, a circle four miles in diameter, since it would be visible, under favorable circumstances, in every portion of this space. Light, moreover, has no weight; the largest possible quantity collected in one point and thrown upon the most sensitive balance, does not affect it in the slightest degree.

What are the chief sources of light?

The chief sources of light are the sun, the stars, fire or chemical action, electricity, and phosphorescence.

Under the head of chemical action are included all the forms of artificial light which are obtained by the burning of bodies. Examples of light produced by phosphorescence, as it is called, are seen in the glow of old and decayed wood, and in the light emitted by fire-flies and some marine animals.

642. All bodies are either luminous or non-luminous.

What is a luminous body?

Luminous bodies are those which shine by their own light; such, for example, as the sun, the flame of a candle, metal rendered red hot, etc.

All solid bodies, when exposed to a sufficient degree of heat, become luminous. It has been recently proved* that all solids begin to emit light at the same degree of heat, viz., 977° of Fahrenheit's thermometer. As the temperature rises, the brilliancy of the light rapidly increases, so that at a temperature of 2600° it is almost forty times as intense as at 1900°. Gases must be heated to a much greater extent before they begin to emit light.

What is a non-luminous body?

Non-luminous bodies are those which produce no light themselves, but which may be rendered temporarily luminous by being placed in the presence of luminous bodies.

Thus, the sun, or a candle, renders objects in an apartment luminous, and therefore visible; but the moment the sun or candle is withdrawn, they become invisible.

What are transparent bodies?

Transparent bodies are those which do not interrupt the passage of light, or which allow other bodies to be seen through them. Glass, air, and water are examples of very transparent bodies.

What are opaque bodies?

Opaque bodies are those which do not permit light to pass through them. The metals, stone, earth, wood, etc., are examples of opaque bodies.

Transparency and opacity exist in different bodies in very different degrees. We can not clearly explain what there is in the constitution of one mass of matter, as compared with another, which fits the one to transmit light, and the other to obstruct it; but the arrangement of the particles has undoubtedly much influence.

Strictly speaking, there is no body which is perfectly transparent, or perfectly opaque. Some light is evidently lost in passing even through space, and still more in traversing our atmosphere. It has been calculated that the atmosphere, when the rays of the sun pass perpendicularly through it, inter-

* By Prof. J. W. Draper.

cept from one fifth to one fourth of their light: but when the sun is near the horizon, and the mass of air through which the solar rays pass is consequently vastly increased in thickness, only 1-212th part of their light can reach the surface of the earth. If our atmosphere, in its state of greatest density, could be extended rather more than 700 miles from the earth's surface, instead of 40 or 50, as it is at present, the sun's rays could not penetrate through it, and our globe would roll on in darkness. Bodies, on the contrary, which are considered as perfectly opaque, will, if made sufficiently thin, allow light to pass through them. Thus, gold-leaf transmits a soft, green light.

In what manner is light propagated?

643. Light, from whatever source it may be derived, moves, or is propagated in straight lines, so long as the medium it traverses is uniform in density.

If we admit a sunbeam through a small opening into a darkened chamber, the path which the light takes, as defined by means of the dust floating in the air, is a straight line.

What practical applications are made of the movement of light in straight lines?

It is for this reason that we are unable to see through a bent tube, as we can through a straight one.

In taking aim, also, with a gun or arrow, we proceed upon the supposition that light moves in straight lines, and try to make the projectile go to the desired object as nearly as possible by the path along which the light comes from the object to the eye.

Fig. 229.

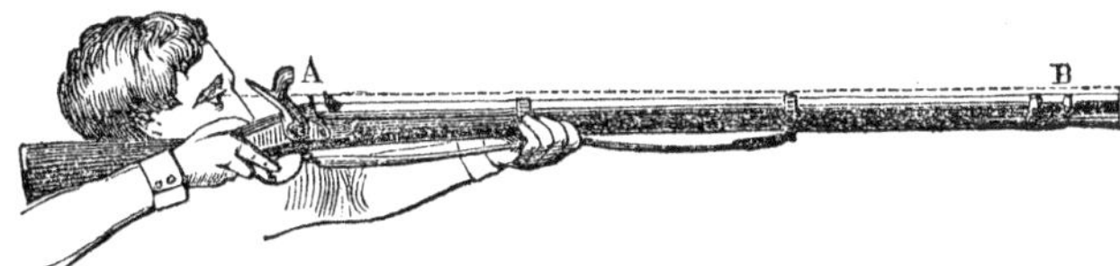

Thus, in Fig. 229, the line A B, which represents the line of sight, is also the direction of a line of light passing in a perfectly straight direction from the object aimed at to the eye of the marksman.

A carpenter depends upon this same principle for the purpose of determining the accuracy of his work. If the edge of the plank be straight and uniform, the light from all points of its surface will come to the eye regularly and uniformly; if irregularities, however, exist, they will cause the light to be irregular, and the eye at once notices the confusion and the point which occasions it.

What is a ray of light?

644. A ray of light is a line of particles of light, or the straight line along which light passes from any luminous body.

A luminous body is said to radiate its light, because the light issues from it in every direction in straight lines.

Explain the divergence of rays of light.

When rays of light radiate from any luminous body, they diverge from one another, or they spread over more space as they recede from their source.

Fig. 230 represents the manner of the divergence.

FIG. 230.

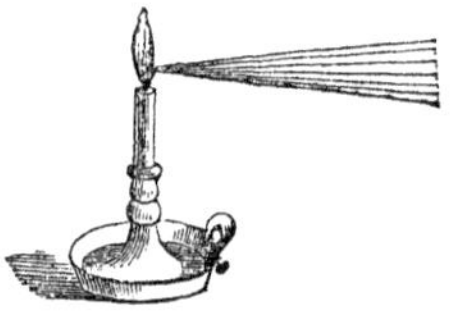

What is the law of divergence?

The surfaces covered, or illuminated by rays of light diverging from a luminous center, increase as the squares of the distances.

Thus, a candle placed behind a window will illuminate a certain space on the wall of a house opposite. If the wall is twice as far from the candle as from the window, the space illuminated by it will be four times as large as the window. If the wall be removed to three times the distance, the surface covered by the rays of light will be nine times as large, and so on.

A collection of radiating rays of light, as shown in Fig. 230, constitutes what is called a "pencil of light."

Why are a great number of persons able to see the same object at the same time?

A thousand, or any number of persons, are able to see the same object at the same time, because it throws off from its surface an infinite number of rays in all directions; and one person sees one portion of these rays, and another person another.

Any number of rays of light are able to cross each other, in the same space, without jostling or interfering. If a small hole be made from one room to another through a thin screen, any number of candles in one room will shine through this opening, and illuminate as many spots in the other room as there are candles in this, all their rays crossing in the same opening, without hinderance or diminution of intensity; just as sounds of different character proceed through the air and communicate to the ear, each its own particular tone, without materially interfering with each other.

When are rays said to be diverging, converging, and parallel?

Rays of light which continually separate as they proceed from a luminous source, are called Diverging Rays. Rays which continually approach each other and tend to unite at a common point, are called Converging Rays. Rays which move in parallel lines, are called Parallel Rays.

What is a shadow?

645. When rays of light, radiated from a luminous point, through the surrounding space, encounter an opaque body, they will (on account of their transmission in straight lines) be excluded from

the space behind such a body. The comparative darkness thus produced is called a shadow.

When the light-giving surface is greater than the body casting the shadow, a cross section of the shadow thrown upon a plane surface will be less than the body; and less, moreover, the further this surface is from the body, for the shadowed space terminates in a point.

When the luminous center is smaller than the opaque body casting the shadow, the shadow will gradually increase in size with the distance, without limit; thus the shadow of a hand held near a candle, and between a candle and the wall, is gigantic.

Under what circumstances will the size of a shadow be increased or diminished?

If the shadow of any object be thrown on a wall, the closer the opaque body is held to the light-producing center, as a candle, for example, the larger will be its shadow. The reason of this is, that the rays of light diverge from the center in straight lines, like lines drawn from the center of a circle; and therefore the nearer the object is held to the center, the greater the number of rays it intercepts. Thus, in Fig. 231, the arrow A, held close to the candle, intercepts a large number of rays, and produces the shadow B F; while the same arrow held at C, intercepts a smaller number of rays, and produces only the little shadow D E.

FIG. 231.

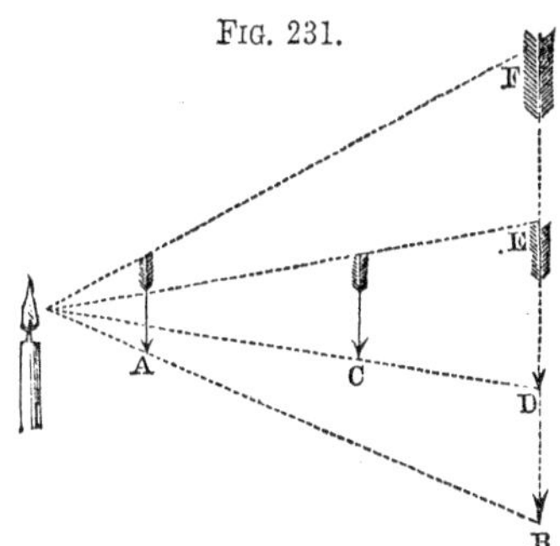

When two or more luminous objects, not in the same straight line, shine upon the same object, each one will produce a shadow.

How does the intensity of light vary?

646. The intensity of light which issues from a luminous point diminishes in the same proportion as the square of the distance from the luminary increases.

Thus, at a distance of two feet, the intensity of light will be one fourth of what it is at one foot; at three feet the intensity will be one ninth of what it is at one foot. In other words, the amount of illumination at the distance of one foot from a single candle would be the same as that from four, or nine candles at a distance of two or three feet, the numbers four and nine being the squares of the distances two, and three, from the center of illumination.

Upon what principle may the relative intensities of different luminous bodies be ascertained?

647. This law, therefore, may be made available for measuring the relative intensities of light proceeding from different sources. Thus, in order to ascertain the relative quantities of light furnished by two different candles, as, for example, a wax and a tallow candle, place two discs or sheets of white

paper, a few feet apart on a wall, and throw the light of one candle on one disc, and the light of the other candle upon the other disc. If they are of unequal illuminating power, the candle which affords the most light must be moved back until the two discs are equally illuminated. Then, by measuring the distance between each candle and the disc it illuminates, the luminous intensities of the two candles may be calculated, their relative intensities being as the squares of their distances from the illuminated discs. If, when the discs are equally illuminated, the distance from one candle to its disc is double the distance of the other candle from its disc, then the first candle is four times more luminous than the second; if the distance be triple, it is nine times more luminous, and so on.

Instruments called "Photometers," operating in a similar manner, have also been constructed for measuring the relative intensity of two luminous bodies. Their arrangement and plan of operation is substantially the same as in the method described.

What is the most intense light known?

648. The light of the sun greatly exceeds in intensity that derived from any other luminous body.

The most brilliant artificial lights yet produced, are very far inferior to the splendor of the solar light, and when placed between the disc of the sun and the eye of the observer, appear as black spots.

Dr. Wollaston has calculated that it would require twenty thousand millions of the brightest stars like Sirius to equal the light of the sun, or that that orb must be one hundred and forty thousand times further from us than he is at present, to be reduced to the illuminating power of Sirius.

The light of the full moon has also been estimated as three hundred thousand times less intense than that of the sun.

During the day the intensity of the sun's light is so great as to entirely eclipse that of the stars, and render them invisible; and for the same reason, we only notice the light emitted by fire-flies and phosphorescent bodies in the dark.

Are the movements of light instantaneous?

649. Light does not pass instantaneously through space, but requires for its passage from one point to another a certain interval of time.

With what velocity does light travel?

The velocity of light is at the rate of about one hundred and ninety-two thousand miles in a second of time.

What are illustrations of the velocity of light?

Light occupies about eight minutes in traveling from the sun to the earth. To pass, however, from the planet Uranus to the earth, it would require an interval of three hours.

The time required for light to traverse the space intervening between the nearest fixed star and the earth, has been estimated at 3¼ years; and from the farthest nebulæ, a period of several hundred years would be requisite, so immense is their distance from our earth. If, therefore, one of the remote fixed stars were to-day blotted from the heavens, several generations on the earth would have passed away before the obliteration could be known to man.

The following comparison between the velocity of light and the speed of a locomotive engine has been instituted:—Light passes from the sun to the earth in about eight minutes; a locomotive engine, traveling at the rate of a mile in a minute, would require upward of one hundred and eighty years to accomplish the same journey.

Who first ascertained the velocity of light?

650. The velocity of light was first determined by Von Roemer, an eminent Danish astronomer, from observations on the satellites of Jupiter.

Explain the method by which the velocity of light was determined from the eclipse of Jupiter's satellites.

The method by which Von Roemer arrived at this result may be explained as follows:—The planet Jupiter is surrounded by several satellites, or moons, which revolve about it in certain definite times. As they pass behind the planet, they disappear from the sight of an observer on the earth, or in other words, they undergo an eclipse.

The earth also revolves in an orbit about the sun, and in the course of its revolution is brought at one time 192 millions of miles nearer to Jupiter than it is at another time, when it is in the most remote part of its orbit. Suppose, now, a table to be calculated by an astronomer, at the time of year when the earth is nearest to Jupiter, showing, for twelve successive months, the exact moment when a particular satellite would be observed to be eclipsed at that point. Six months afterward, when the earth, in the course of its revolution, has attained a point 192 millions of miles more remote from Jupiter than it formerly occupied, it would be found that the eclipse of the satellite would occur sixteen minutes, or 960 seconds, later than the calculated time. This delay is occasioned by the fact that the light has had to pass over a greater distance before reaching the earth than it did when the earth was in the opposite part of its orbit, and if it requires sixteen minutes to pass over 192 millions of miles, it will require one second to move over 200,000 miles. When, on the contrary, the earth at the end of the succeeding six months has assumed its former position, and is 192 millions of miles nearer Jupiter, the eclipse will occur sixteen minutes earlier, or at the exact calculated time given in the tables. The velocity of light, therefore, in round numbers, may be considered as 200,000 miles per second.* A more exact calculation, founded on perfectly accurate data, gives as the true velocity of light 192,500 miles per second.

* The explanation above given will be made clear by reference to the following diagram, Fig. 232. S represents the sun, *a b* the orbit of the earth, and T T′ the position of the earth at different and opposite points of its orbit, J represents Jupiter, and E its satellite, about to be eclipsed by passing within the shadow of the planet. Now the time of the commencement or termination of an eclipse of the satellite, is the instant at which the satellite would appear, to an observer on the earth, to enter, or emerge from the

Several other plans have been devised for determining the velocity of light, the results of which agree very nearly with those obtained by the observations on the satellites of Jupiter.*

When is light reflected?

651. When a ray of light strikes against a surface, and is caused to turn back or rebound in a direction different from whence it proceeded, it is said to be reflected.

What is absorption of light?

652. When rays of light are retained upon the surface upon which they fall, they are said to be absorbed; in consequence of which their presence is not made sensible by reflection.

The question as to what becomes of the light which is absorbed by a body, can not be satisfactorily answered. In all probability it is permanently retained within the substance of the absorbing body, since a body which absorbs light by continued exposure, does not radiate or distribute it again in any way, as it might do if it had absorbed heat.

shadow of the planet. If the transmission of light were instantaneous, it is obvious that an observer at T′, the most remote part of the earth's orbit, would see the eclipse begin and end at the same moment as an observer at T, the part of the earth's orbit nearest to Jupiter. This, however, is not the case, but the observer at T′ sees the eclipse 960 seconds later than the observer at T; and as the distance between these two stations is 192 millions of miles, we have, as the velocity of light in one second, 192,000,000÷960= 200,000.

FIG. 232.

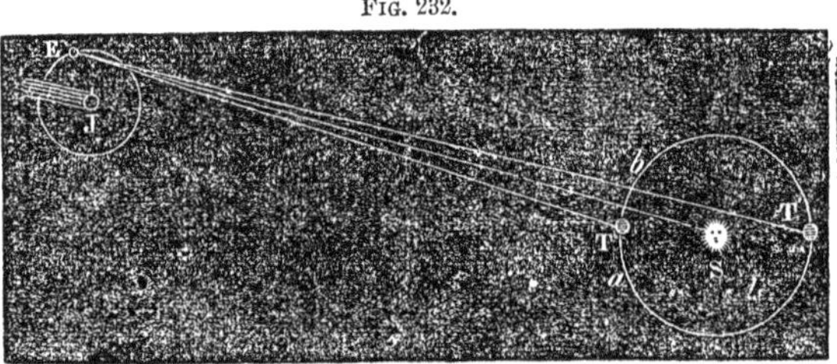

* A very ingenious plan was devised a few years since by M. Fizeau of Paris, by which the velocity of artificial light was determined and found to agree with that of solar light. A disc, or wheel, carrying a certain number of teeth upon its circumference, was made to revolve at a known rate: placing a tube behind these, and looking at the open spaces between the teeth, they become less evident to sight, the greater the velocity of the moving wheel, until, at a certain speed, the whole edge appears transparent. The rate at which the wheel moves being known, it is easy to determine the time occupied while one tooth passes to take the place of the one next to it. A ray of light is made to traverse many miles through space, and then passes through the teeth of the revolving disc. It moves the whole distance in just the time occupied in the movement of a single tooth to the place of another at a certain speed.

SECTION I.

REFLECTION OF LIGHT.

What occurs when light falls upon any surface?

653. When rays of light fall upon any surface, they may be reflected, absorbed, or transmitted. Only a portion of the light, however, which meets any surface is reflected, the remainder being absorbed, or transmitted.

When does a body appear white and when dark?

654. When the portion of light reflected from any surface, or point of a surface, to the eye is considerable, such surface, or point, appears white; when very little is reflected, it appears dark-colored; but when all, or nearly all the rays are absorbed, and none are reflected back to the eye, the surface appears black.

Thus, charcoal is black, because it absorbs all the light which falls upon it, and reflects none. Such a body can not be seen unless it is situated near other bodies which reflect light to it.

According to a variation in the manner of reflecting light, the same surface which appears white to an eye in one position, may appear to be black from another point of view, as frequently happens in the case of a mirror, or of any other bright, or reflecting surface.

What are good reflectors of light?

Dense bodies, particularly smooth metals, reflect light most perfectly. The reflecting power of other bodies decreases in proportion to their porosity.

How are non-luminous bodies rendered visible?

655. All bodies not in themselves luminous, become visible by reflecting the rays of light.

It is by the irregular reflection of light that most objects in nature are rendered visible; since it is by rays which are dispersed from reflecting surfaces, irregularly and in every direction, that bodies not exposed to direct light are illuminated. If light were only reflected regularly from the surface of non-luminous bodies, we should see merely the image of the luminous object, and not the reflecting surface.* In the day-time, the image of the sun would be reflected from the surface of all objects around us, as if they were composed of looking-glass, but the objects themselves would be invisible. A room in which artificial lights were placed would reflect these lights from the walls and other objects as if they were mirrors, and all that would be visible would be the multiplied reflection of the artificial lights.

* In a very good mirror we scarcely perceive the reflecting surface intervening between us and the images it shows us.

What effect has the atmosphere upon the diffusion of light?

The atmosphere reflects light irregularly, and every particle of air is a luminous center, which radiates light in every direction. Were it not for this, the sun's light would only illuminate those spaces which are directly accessible to its rays, and darkness would instantly succeed the disappearance of the sun below the horizon.

What is a Mirror?

656. Any surface which possesses the power of reflecting light in the highest degree is called a MIRROR.

Into how many classes are mirrors divided?

Mirrors are divided into three general classes, without regard to the material of which they consist, viz., Plane, Concave, and Convex Mirrors.

These three varieties of mirrors are represented in Fig. 233; A, being plane, like an ordinary looking-glass; B, concave, like the inside of a watch-glass; and C, convex, like the outside of a watch-glass.

FIG. 233.

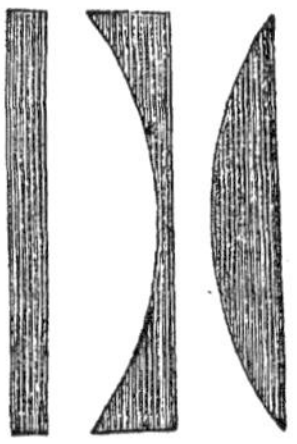

A B C

What is the great law of the reflection of light?

657. When light falls upon a plane and polished surface, the angle of reflection is equal to the angle of incidence.

This is the great general law which governs the reflection of light, and is the same as that which governs the motion of elastic bodies.

Thus, in Fig. 234, let A B be the direction of an incident ray of light, falling on a mirror, F C. It will be reflected in the direction B E. If we draw a line, D B, perpendicular to the surface of the mirror, at the point of reflection, B, it will be found that the angle of incidence, A B D, is precisely equal to the angle of reflection, E B D.

FIG. 234.

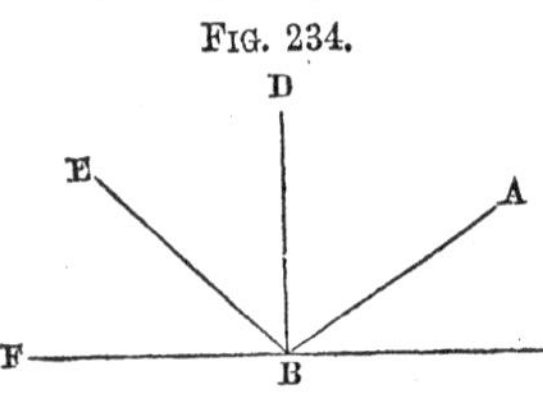

The same law holds good in regard to every form of surface, curved as well as plane, since a curve may be supposed to be formed of an infinite number of little planes.

FIG. 235.

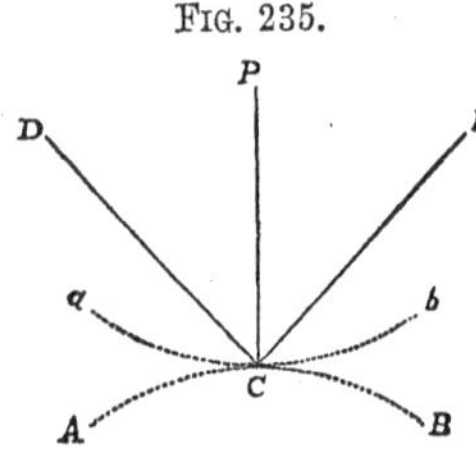

Thus, in Fig. 235, the incident ray, E C, falling upon the concave surface, *a* C *b*, will still be reflected, in obedience to the same law, in the direction C D, the angle being reckoned from the perpendicular to that point of the curve where the incident ray falls. The same will also be true of the convex surface, A C B.

What is meant by an image? 658. An image, in optics, is the figure of any object made by rays proceeding from the several points of it.

What is a common looking-glass? 659. A common looking-glass consists of a glass plate, having smooth and parallel surfaces, and coated on the back with an amalgam* of tin and quicksilver.

How are the images formed in a looking-glass? The images formed in a common looking-glass are mainly produced by the reflection of the rays of light from the metallic surface attached to the back of the glass, and not from the glass itself.

The effect may be explained as follows:—A portion of the light incident upon the anterior surface is regularly reflected, and another portion irregularly. The first produces a very faint image of an object placed before the glass, while the other renders the surface of the glass itself visible. Another, and much greater portion, however, of the light falling upon the anterior surface passes into the glass and strikes upon the brilliant metallic coating upon the back, from which it is regularly reflected, and returning to the eye, produces a strong image of the object. There are, therefore, strictly speaking, two images formed in every looking-glass—the first a faint one by the light reflected regularly from the anterior surface, and the second a strong one by the light reflected from the metallic surface; and one of these images will be before the other at a distance equal to the thickness of the glass. In good mirrors, the superior brilliancy of the image produced by the metallic surface will render the faint image produced by the anterior surface invisible, but in glasses badly silvered, the two images may be easily seen.

If the surfaces of the mirror could be so highly polished as to reflect regularly all the light incident upon it, the mirror itself would be invisible, and the observer, receiving the reflected light, would perceive nothing but the images of the objects before it. This amount of polish it is impossible to effect artificially, but in many of the large plate-glass mirrors manufactured at the present time, a high degree of perfection is attained. Such a mirror placed vertically against the wall of a room, appears to the eye merely as an opening leading into another room, precisely similar and similarly furnished and illuminated; and an inattentive observer is only prevented from attempting to walk through such an apparent opening by encountering his own image as he approaches it.

In what manner does a plane mirror reflect rays of light? 660. A plane mirror only changes the direction of the rays of light which fall upon it, without altering their relative position. If they fall upon it perpendicularly, they will be

* An amalgam is a mixture or compound of quicksilver and some other metal.

reflected perpendicularly; if they fall upon it obliquely, they will be reflected obliquely; the angle of reflection being always equal to the angle of incidence.

When will the image in a looking-glass appear distorted?

If the two surfaces of mirrors are not parallel, or uneven, then the rays of light falling upon it will not be reflected regularly, and the image will appear distorted.

How is an apparent change of place caused by reflection?

661. We always seem to see an object in the direction from which its rays enter the eye. A mirror, therefore, which, by reflection, changes the direction of the rays proceeding from an object, will change the apparent place of the object.

Thus, if the rays of a candle fall obliquely upon a mirror, and are reflected to the eye, we shall seem to see the candle in the mirror in the direction in which they proceed after reflection.

If we lay a looking-glass upon the floor, with its face uppermost, and place a candle beside it, the image of the candle will be seen in the mirror, by a person standing opposite, as inverted, and as much below the surface of the glass as the candle itself stands above the glass. The reason of this is, that the incident rays from the candle which fall upon the mirror are reflected to the eye in the same direction that they would have taken, had they really come from a candle situated as much below the surface of the glass, as the first candle was above the surface. This fact will be clearly shown by referring to Fig. 236.

FIG. 236.

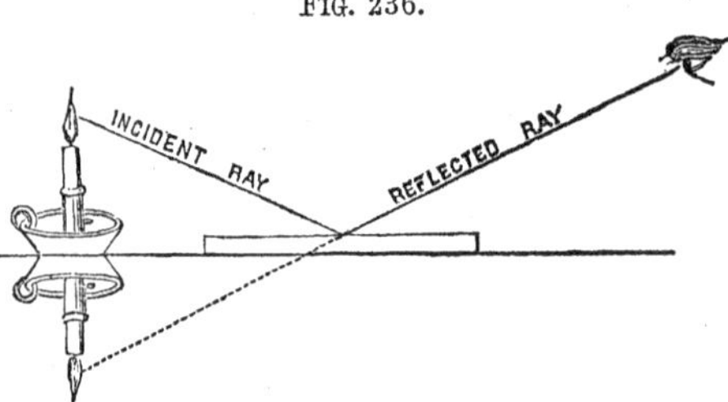

When we look into a plane mirror (the common looking-glass) the rays of light which proceed from each point of our body before the mirror will, after reflection, proceed as if they came from a point holding a corresponding position behind the mirror; and therefore produce the same effect upon the eye of the observer as if they had actually come from that point. The image in the glass, consequently, appears to be at the same distance behind the surface of the glass, as the object is before it.

Let A, Fig. 237, be any point of a visible object placed before a looking-glass, M N. Let A B and A C be two rays diverging from it, and reflected from B and C to an eye at O. After reflection they will proceed as if they had issued from a point, *a*, as far behind the surface of the looking-glass as A is before it—that is to say, the distance A N will be equal to the distance N *a*.

For this reason our reflection in a mirror seems to approach us when we walk toward it, and to retire from us as we retire.

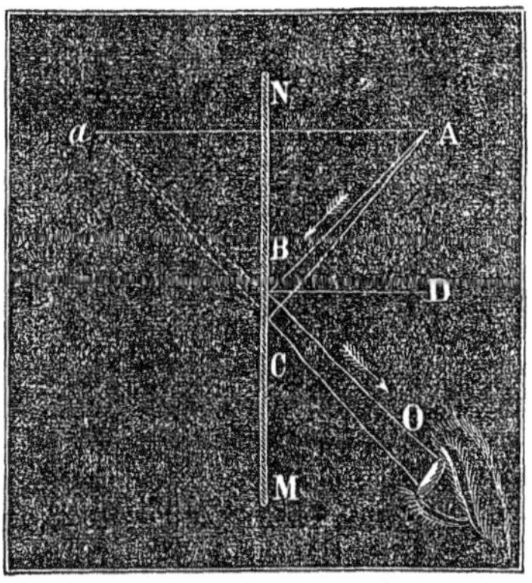

FIG. 237.

Upon the same principle, when trees, buildings, or other objects are reflected from the horizontal surface of a pond, or other smooth sheet of water, they appear inverted, since the light of the object, reflected to our eyes from the surface of the water, comes to us with the same direction as it would have done, had it proceeded directly from an inverted object in the water.

In Fig. 238, the light proceeding from the arrow-head, A, strikes the water at F, and is reflected to D, and that from the barb, B, strikes the water at E, and is reflected to C. A spectator standing at G will see the reflected rays, E G and F G, as if they proceeded directly from C and D, and the image of the arrow will appear to be located at C D.

FIG. 238.

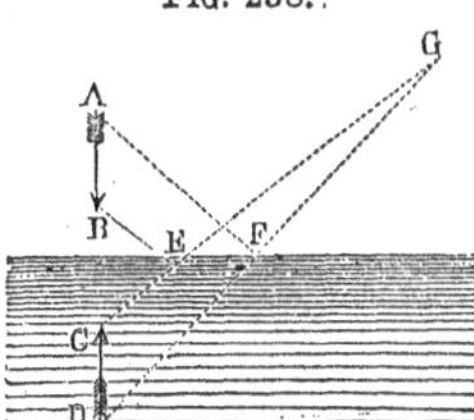

It is in accordance with the law that the angles of incidence are equal to the angles of reflection, that a person is enabled to see his whole figure reflected from the surface of a comparatively small mirror. Thus, in Fig. 239, let a person, C D, be placed at a suitable distance from a mirror, A B. The rays of light, C A, proceeding from the head of the person, fall perpendicularly upon the mirror, and are therefore reflected back perpendicularly, or in the same line; the rays B D proceeding from the feet, however, fall obliquely upon the mirror, and are therefore reflected obliquely, and reach the eye in the same direction they would have taken had they proceeded from the point F behind the mirror.

FIG. 239.

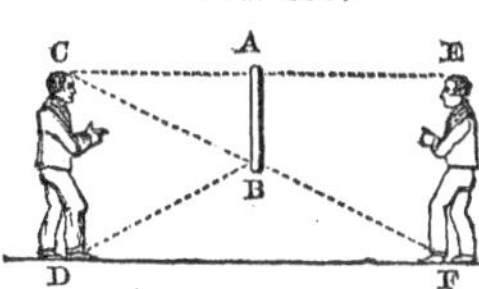

Is the same quantity of light reflected at all angles?

662. The quantity of light reflected from a given surface, is not the same at all angles, or inclinations. When the angle or inclination with which a ray of light strikes upon a reflecting surface is great, the amount of light reflected to the eye will be

considerable ; when the angle, or inclination is small, the amount of light reflected will be diminished.

Thus, for example, when light falls perpendicularly upon the surface of glass, 25 rays out of 1,000 are returned; but when it falls at an angle of 85°, 550 rays out of 1,000 are returned.

Thus, a surface of unpolished glass produces no image of an object by reflection when the rays fall on it nearly perpendicularly; but if the flame of a candle be held in such a position that the rays fall upon the surface at a very small angle, a distinct image of it will be seen.

We have in this an explanation of the fact, that a spectator standing upon the bank of a river sees the images of the opposite bank and the objects upon it reflected in the water most distinctly, while the images of nearer objects are seen imperfectly, or not at all. Here the rays coming from the distant objects strike the surface of the water very obliquely, and a sufficient number are reflected to make a sensible impression upon the eye; while the rays proceeding from near objects strike the water with little obliquity, and the light reflected is not sufficient to make a sensible impression upon the eye.

This fact may be clearly seen by reference to Fig. 240.

Fig. 240.

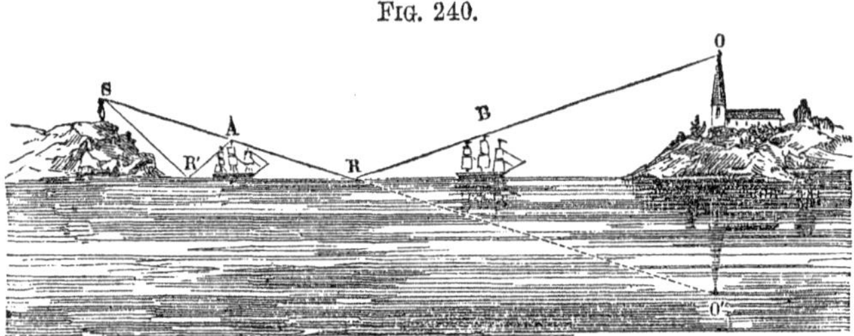

Let S be the position of the spectator; O and B the position of distant objects. The rays O R and B R which proceed from them, strike the surface of the water very obliquely, and the light which is reflected in the direction R S is sufficient to make a sensible impression upon the eye. But in regard to objects, such as A, placed near the spectator, they are not seen reflected, because the rays A R′ which proceed from them strike the water with but little obliquity; and consequently, the part of their light which is reflected in the direction R′ S, toward the spectator, is not sufficient to produce a sensible impression upon the eye.

What is the effect of two parallel plane mirrors?

663. If an object be placed between two plane mirrors, each will produce a reflected image, and will also repeat the one reflected by the other—the image of the one becoming the object for the other. A great number of images are thus pro

duced, and if the light were not gradually weakened by loss at each successive reflection, the number would be infinite.

If the mirrors are placed so as to form an angle with each other, the number of mutual reflections will be diminished, proportionably to the extent of the angle formed by the mirrors.

Describe the Kaleidoscope.

The construction of the optical instrument called the Kaleidoscope is based simply upon the multiplication of an image by two or more mirrors inclined toward each other. It consists of a tube containing two or more narrow strips of looking-glass, which run through it lengthwise, and are generally inclined at an angle of about 60°. If at one end of the tube a number of small pieces of colored glass and other similar objects are placed, they will be reflected from the mirrors in such a way as to form regular and most elegant combinations of figures. An endless variety of symmetrical combinations may be thus formed, since every time the instrument is moved or shaken the objects arrange themselves differently, and a new figure is produced.

Why does the sun appear at noon to shine at only one point upon the surface of water?

Upon the surface of smooth water the sun, when it is nearly vertical, as at noon, appears to shine upon only one spot, all the rest of the water appearing dark. The reason of this is, that the rays fall at various degrees of obliquity on the water, and are reflected at similar angles; but as only those which meet the eye of the spectator are visible, the whole surface will appear dark, except at the point where the reflection occurs.

FIG. 241.

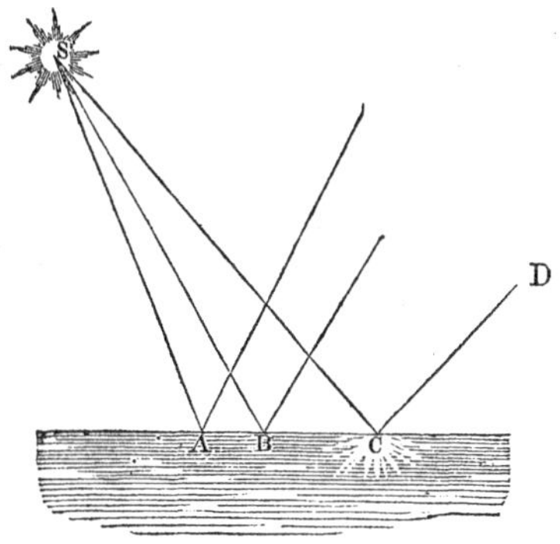

S A B C D

Thus, in Fig. 241, of the rays S A, S B, and S C, only the ray S C meets the eye of the spectator, D. The point C, therefore, will appear luminous to the spectator D, but no other part of the surface.

Another curious optical phenomenon is seen when the rays of the sun, or moon fall at an angle upon the surface of water gently agitated by the wind. A long, tremulous path of light seems to be formed toward the eye of the spectator, while all the rest of the surface appears dark. The reason of this appearance is, that every little wave, in an extent perhaps of miles, has some part of its rounded surface with the direction or obliquity which, according to the required relation of the angles of incidence and reflection, fits it to reflect the light to the eye, and hence every wave in that extent sends its momentary gleam, which is succeeded by others.

What is a Concave Mirror?

664. A concave mirror may be considered as the interior surface of a portion, or segment of a hollow sphere.

This is clearly shown in Fig. 242.

A concave mirror may be represented by a bright spoon, or the reflector of a lantern.

How are parallel rays reflected from a concave mirror?

When parallel rays of light fall upon the surface of a concave mirror, they are reflected and caused to converge to a point half way between the center of the surface and the center of the curve of the mirror. This point in front of the mirror is called the principal focus of the mirror.

Thus, in Fig. 242, let 1, 2, 3, 4, etc., be parallel rays falling upon a concave mirror; they will, after reflection, be found converging to the point *o*, the principal focus, which is situated half way between the center of the surface of the mirror and the geometrical center of the curve of the mirror, *a*.

FIG. 242.

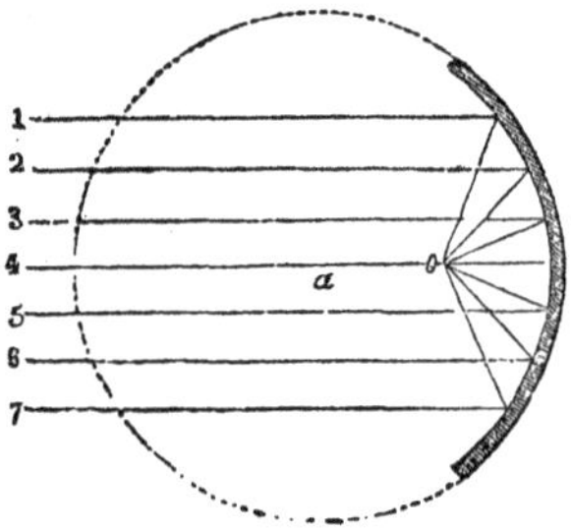

Why are concave mirrors called burning mirrors?

665. Concave mirrors are sometimes designated as "Burning Mirrors," since the rays of the sun which fall upon them parallel, are reflected and converged to a focus (fire-place), where their light and heat are increased in as great a degree as the area of the mirror exceeds the area of the focus.*

In what manner are diverging rays reflected from a concave mirror?

666. Diverging rays of light issuing from a luminous body placed at the center of the curve of a concave spherical mirror, will be reflected back to the same point from which they diverged.

* A burning mirror, 20 inches in diameter, constructed of plaster of Paris, gilt and burnished, has been found capable of igniting tinder at a distance of 50 feet. It is related that Archimedes, the philosopher of Syracuse, employed burning mirrors 200 years before the Christian era, to destroy the besieging navy of Marcellus, the Roman consul; his mirror was probably constructed of a great number of flat pieces. The most remarkable experiments, however, of this nature, were made by Buffon, the eminent French naturalist, who had a machine composed of 168 small plane mirrors, so arranged that they all reflected radiant heat to the same focus. By means of this combination of reflecting surfaces he was able to set wood on fire at the distance of 209 feet, to melt lead at 100 feet, and silver at 50 feet.

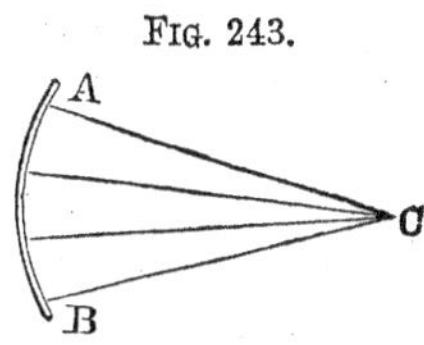

Fig. 243.

Thus, if A B, Fig. 243, were a concave spherical mirror, of which C were the center, rays issuing from C would, in obedience to the law that the angle of incidence and reflection are equal, meet again at C.

Diverging rays falling on a spherical concave mirror, if they issue from the principal focus, half way between the center of the surface and the center of the curve of the mirror, will be reflected in parallel lines.

Fig. 244.

Thus, in Fig. 244, if F represent a candle placed before a concave mirror, A B C, half way between the center of its surface, B, and the center of its curve, C, its rays, falling upon the mirror, will be reflected in the parallel lines *d e f g h*.

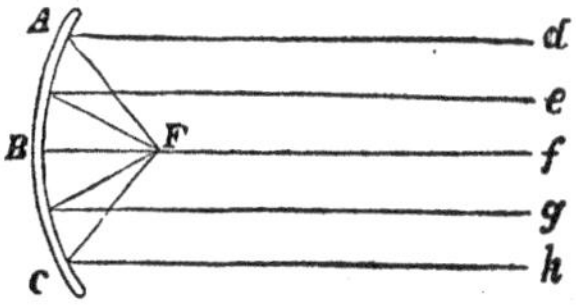

This principle is taken advantage of in the arrangement of the illuminating and reflecting apparatus of light-houses. The lamps are placed before a concave mirror, in its principal focus, and the rays of light proceeding from them are reflected parallel from the surface of the mirror.

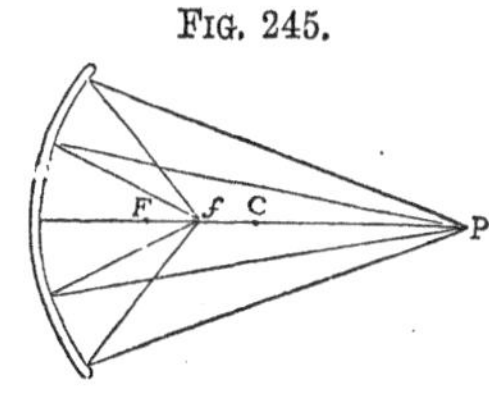

Fig. 245.

When the rays issue from a point, P, Fig. 245, beyond the center, C, of the curve of the mirror, they will, after reflection, converge to a focus, *f*, between the principal focus, F, and the center of the curve, C.

On the contrary, if the rays issue from a point between the principal focus, F, and the surface of the mirror, they will diverge after reflection.

How are images formed by concave mirrors?

667. Images are formed by concave mirrors in the same manner as by plane ones, but they are of different size from the object, their general effect being to produce an image larger than the object.

When will the image formed by a concave mirror be magnified?

When an object is placed between a concave mirror and its principal focus, the image will appear larger than the object, in an erect position and behind the mirror.

This will be apparent from Fig. 246. Let *a* be an object situated within the focus of the mirror. The rays from its extremities will fall divergent on the mirror, and be reflected less divergent to the eye at *b*,

Fig. 246.

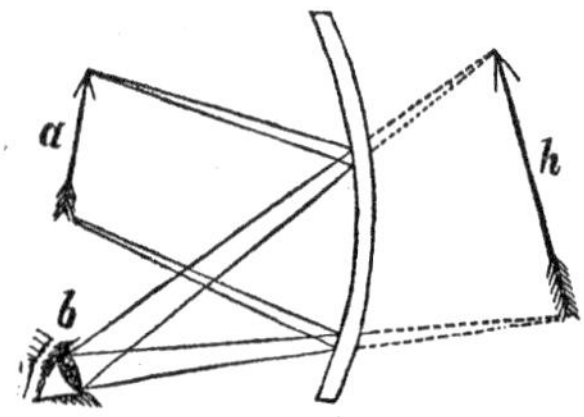

as though they proceeded from an object behind the mirror, as at *h*. To an eye at *b* also, the image will appear larger than the object *a*, since the angle of vision is larger.

If the rays proceed from a distant body, as at E D., Fig. 247, beyond the center, C, of a spherical concave mirror, A B, they will, after reflection, be converged to a focus in front of the mirror, and somewhat nearer to the center, C, than the principal focus, and there paint upon any substance placed to receive it, an image inverted, and smaller than the object; this image will be very bright, as all the light incident upon the mirror will be gathered into a small space. As the object approaches the mirror, the image recedes from it and approaches C; and when situated at C, the center of the curve of the mirror, the image will be reflected as large as the object; when it is at any point between C and *f*, supposing *f* to be the focus for parallel rays, it will be reflected, enlarged, and more distant from the mirror than the object, this distance increasing, until the object arrives at *f*, and then the image becomes infinite, the rays being reflected parallel.*

Fig. 247.

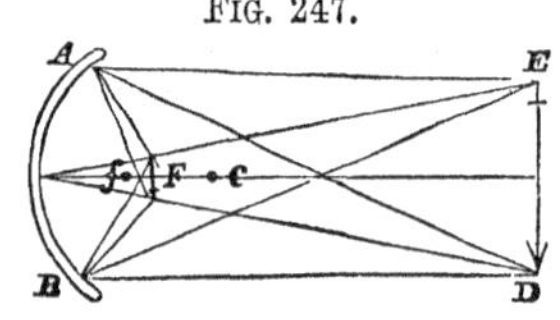

When will the images reflected from a concave mirror appear inverted, and when erect?

668. When an object is further from the surface of a concave mirror than its principal focus, the image will appear inverted; but when the object is between the mirror and its principal focus, the image will be upright, and increase in size in proportion as the object is placed nearer to the focus.

The fact that images are formed at the foci of a concave mirror, and that by varying the distance of objects before the surface of the mirror, we may vary the position and size of the images formed at such foci, was often taken advantage of in the middle ages to astonish and delude the ignorant. Thus, the mirror and the object being concealed behind a curtain, or a partition, and the object strongly illuminated, the rays from the object might be reflected from the mirror in such a manner as to pass through an opening in the screen, and come to a focus at some distance beyond, in the air. If a cloud of smoke

* In all the cases referred to, of the reflection of light from concave mirrors, the aperture or curvature of the mirror is presumed to be inconsiderable. If it be increased beyond a certain limit, the rays of light incident upon it are modified in their reflection from its surface.

from burning incense were caused to ascend at this point, an image would be formed upon it, and appear suspended in the air in an apparently supernatural manner. In this way, terrifying apparitions of skulls, daggers, etc., were produced.

What is a Convex Mirror?

669. A Convex Mirror may be considered as any given portion of the exterior surface of a sphere.

Where is the principal focus of a convex mirror?

The principal focus of a convex mirror lies as far behind the reflecting surface as in concave mirrors it lies before it. (See § 664.) The focus in this case is called the virtual focus, because it is only an imaginary point, toward which the rays of reflection appear to be directed.

FIG. 248.

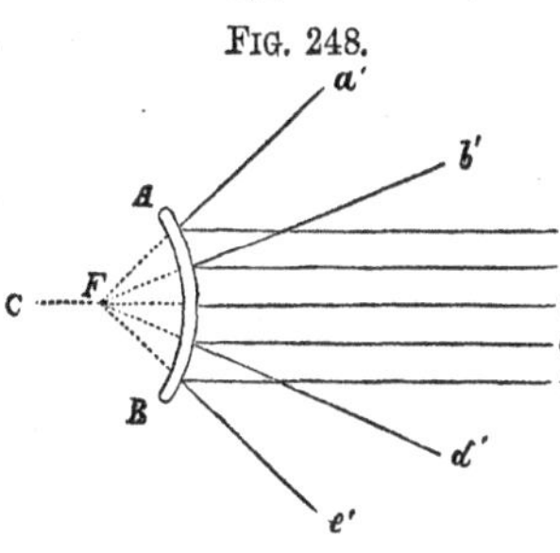

Thus, let $a\ b\ c\ d\ e$, Fig. 248, be parallel rays incident upon a convex mirror, A B, whose center of curvature is C. These rays are reflected divergent, in the directions $a'\ b'\ c'\ d'\ e'$, as though they proceeded from a point, F, behind the mirror, corresponding to the focus of a concave mirror.

If the point C be the geometrical center of the curve of the mirror, the point F will be half way between C and the surface of the mirror; as this focus is only apparent, it is called the virtual focus.

How are diverging and converging rays reflected from a convex mirror?

Rays of light falling upon a convex mirror, diverging, are rendered still more divergent by reflection from its surface; and convergent rays are reflected, either parallel or less convergent.

What is the nature of the images formed by convex mirrors?

670. The general effect of convex mirrors is to produce an image smaller than the object itself.

FIG. 249.

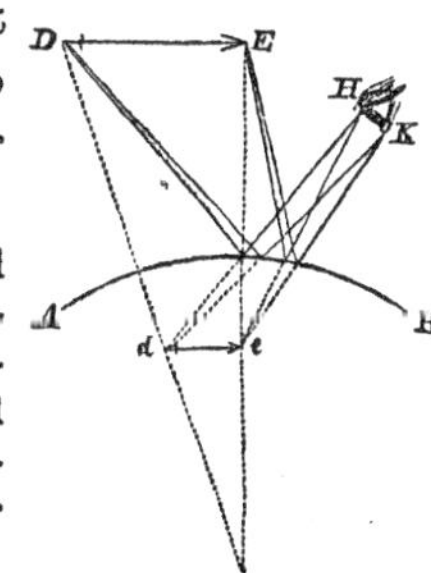

Thus, in Fig. 249, let D E be an object placed before a convex mirror, A B; the rays proceeding from it will be reflected from the convex surface to the eye at H K, as though they proceeded from an object, $d\ e$, behind the mirror, thus presenting an image smaller, erect, and much nearer the mirror than the object.

Thus the globular bottles filled with colored liquid, in the window of a drug-store, exhibit all the variety of moving scenery without, such as carriages, carts, and people moving in different directions: the upper half of each bottle exhibiting all the images inverted, while the lower half exhibits another set of them in the erect position.

Convex mirrors are sometimes called dispersing mirrors, as all the rays of light which fall upon them are reflected in a diverging direction.

What is Catoptrics? 671. That department of the science of optics which treats of reflected light, is often designated as CATOPTRICS.

SECTION II.

REFRACTION OF LIGHT.

What is meant by the refraction of light? Light traverses a given transparent substance, such as air, water, or glass, in a straight line, provided no reflection occurs and there is no change of density in the composition of the medium; but when light passes obliquely from one medium to another, or from one part of the same medium into another part of a different density, it is bent from a straight line, or refracted.

What is a medium in optics? 672. A medium, in optics, is any substance, solid, liquid, or gaseous, through which light can pass.

A medium, in optics, is said to be dense or rare, according to its power of refracting light, and not according to its specific gravity. Thus alcohol, olive oil, oil of turpentine, and the like substances, although of less specific gravity than water, have a greater refractive power; they are, therefore, called denser media than water.

673. The fundamental laws which govern the refraction of light may be stated as follows:

What laws govern the refraction of light? When light passes from one medium into another, in a direction perpendicular to the surface, it continues on in a straight line, without altering its course. When light passes obliquely from a rarer into a denser medium, it is refracted toward a perpendicular to the surface, and this refraction is increased or diminished in proportion as the rays fall more or less obliquely upon the refracting surface.

When light passes obliquely out of a denser into a rarer medium, it passes through the rarer medium in a more oblique direction, and further from a perpendicular to the surface of the denser medium.

FIG. 250.

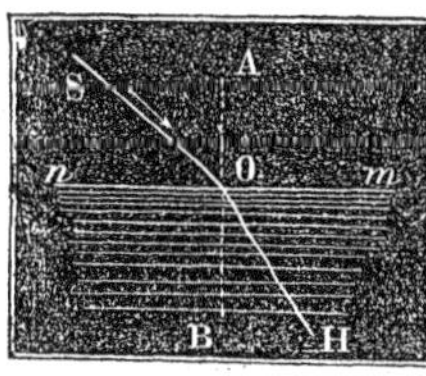

Thus, in Fig. 250, suppose *n m* to represent the surface of water, and S O a ray of light striking upon its surface. When the ray S O enters the water, it will no longer pursue a straight course, but will be refracted, or bent toward the perpendicular line, A B, in the direction S O. The denser the water or other fluid may be, the more the ray S O H will be refracted, or turned toward A B. If, on the contrary, a ray of light, H O, passes from the water into the air, its direction after leaving the water will be further from the perpendicular A O, in the direction O S.

FIG. 251.

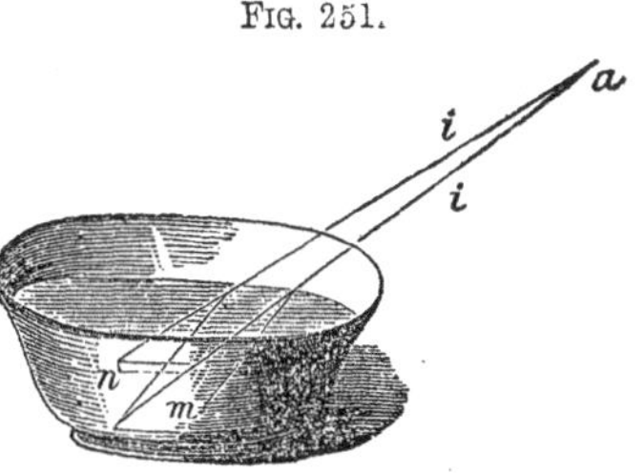

The effects of the refraction of light may be illustrated by the following simple experiment:—Let a coin or any other object be placed at the bottom of a bowl, as at *m*, Fig. 251, in such a manner that the eye at *a* can not perceive it, on account of the edge of the bowl which intervenes and obstructs the rays of light. If now an attendant carefully pours water into the vessel, the coin rises into view, just as if the bottom of the basin had been elevated above its real level. This is owing to a refraction by the water of the rays of light proceeding from the coin, which are thereby caused to pass to the eye in the direction *i i*. The image of the coin, therefore, appears at *n*, in the direction of these rays, instead of at *m*, its true position.

A straight stick, partly immersed in water, appears to be broken or bent at the point of immersion. This is owing to the fact that the rays of light proceeding from the part of the stick contained in the water are refracted, or caused to deviate from a straight line as they pass from the water into the air; consequently that portion of the stick immersed in the water will appear to be lifted up, or to be bent in such a manner as to form an angle with the part out of the water.

FIG. 252.

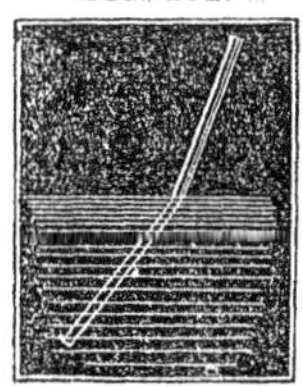

The bent appearance of the stick in water is represented in Fig. 252. For the same reason, a spoon in a glass of water, or an oar partially immersed in water, always appears bent.

On account of this bending of light from objects under water, a person who endeavors to strike a fish with a spear, must, unless directly above the fish, aim at a point apparently below it, otherwise the weapon will miss, by passing too high.

A river, or any clear water viewed obliquely from the bank, appears more shallow than it really is, since the light proceeding from the objects at the bottom, is refracted as it emerges from the surface of the water. The depth of water, under such circumstances, is about one third more than it appears, and owing to this optical deception, persons in bathing are liable to get beyond their depth.

What is atmospheric refraction?

Light, on entering the atmosphere, is refracted in a greater or less degree, in proportion to the density of the air; consequently, as that portion of the atmosphere nearest the surface of the earth possesses the greatest density, it must also possess the greatest refractive power.

What effect has refraction upon the position of the sun?

From this cause the sun and other celestial bodies are never seen in their true situations, unless they happen to be vertical; and the nearer they are to the horizon, the greater will be the influence of refraction in altering the apparent place of any of these luminaries.

This forms one of the sources of error to be allowed for in all astronomical observations, and tables are calculated for finding the amount of refraction, depending on the apparent altitude of the object, and the state of the barometer and thermometer. When the object is vertical, or nearly so, this error is hardly sensible, but increases rapidly as it approaches the horizon; so that, in the morning, the sun is rendered visible before he has actually risen, and in the evening, after he has set.

What is the cause of twilight?

For the same reason, morning does not occur at the instant of the sun's appearance above the horizon, or night set in as soon as he has disappeared below it. But both at morning and evening, the rays proceeding from the sun below the horizon are, in consequence of atmospheric refraction, bent down to the surface of the earth, and thus, in connection with a reflecting action of the particles of the air, produce a lengthening of the day, termed twilight.

In what manner is light reflected by the atmosphere?

As the density of the air diminishes gradually upward from the earth, atmospheric refraction is not a sudden change of direction, as in the case of the passage of light from air into water, but the ray of light actually describes a curve, being refracted more and more at each step of its progress. This applies to the light received from a distant object on the surface of the earth, which is lower or higher than the eye, as well as to that received from a celestial object, since it must pass through air constantly increasing or diminishing in density. Hence, in the engineering operation of leveling, this refraction must be taken into consideration.

Explain the phenomena of Mirage.

674. The application of the laws of refraction of light account for many curious deceptive appearances in the atmosphere, which are included under the general name of Mirage. In these phenomena, the images of objects far remote are seen at an elevation in the atmosphere, either erect or inverted. Thus travelers upon a desert, where the surface of the earth is highly heated by the sun, are often deceived by the appearance of water in the distance, surrounded by trees and villages. In the same manner at sea, the images of vessels at a great distance and below the horizon, will at times appear floating in the atmosphere. Such appearances are frequently seen with great distinctness upon the great American lakes. These phenomena appear to be due to a change in the density of the strata of air which are immediately in contact with the surface of the earth. Thus it often happens that strata resting upon the land may be rendered much hotter, and those resting upon the water much cooler, by contact with the surface, than other strata occupying more elevated positions. Rays, therefore, on proceeding from a distant object and traversing these strata, will be unequally reflected, and caused to proceed in a curvilinear direction; and in this way an object situated behind a hill, or below the horizon, may be brought into view and appear suspended in the air. This may be readily understood by reference to Fig. 253. Suppose the rays of light from the ship, S, below the horizon to reach the eye, after assuming a curvilinear direction by passing through strata of air of varying density; then, as an object always appears in the direction in which the last rays proceeding from it enter the eye, two images will be seen in the direction of the dotted lines, one of them being inverted.

FIG. 253.

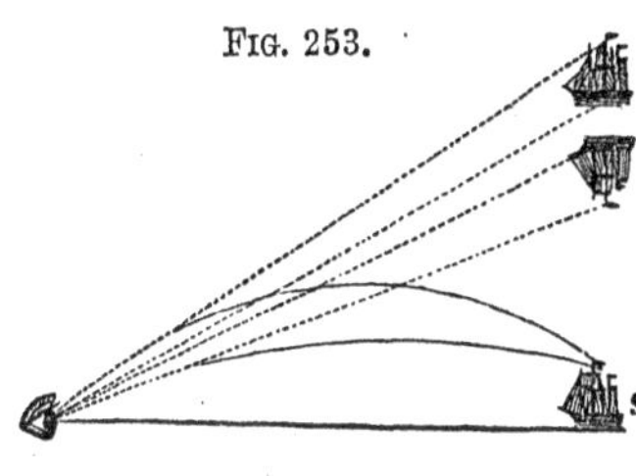

These phenomena may be sometimes imitated. Thus, if we look along a red hot bar of iron, or a mass of heated charcoal at some image, a short distance from it, an inverted reflection of it will be seen. In the same manner, if we place in a glass vessel liquids of different densities, so that they float one above another, and look through them at some object, it will be seen distorted and removed from its true place, by reason of the unequal refractive and reflective powers of the liquids employed.

Is the angle of refraction equal to the angle of incidence?

675. The angle of refraction of light is not, like the angle of reflection, equal to the angle of incidence; but it is nevertheless subject to a definite law, which is called the law of sines.

What is a sine?

A sine is a right line drawn from any point in one of the lines inclosing an angle, perpendicular to the other line.

FIG. 254.

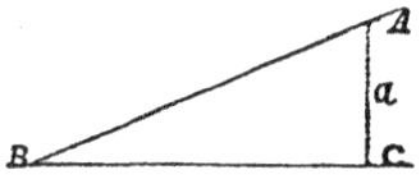

Thus, in Fig. 244, let A B C be an angle; then *a* will be the sine of that angle, being drawn from a point in the line A B, perpendicular to the line B C. Two angles may be compared by means of their sines, but whenever this is done, the lengths of the sides of the angles must be made equal, because the sine varies in length according to the length of the lines forming the angle.

The general law of refraction is as follows:—

What is the general law of refraction?

When a ray of light passes from one medium to another, the sine of the angle of incidence is in a constant ratio to the sine of the angle of refraction.

The proportion or relation between these sines differs when different media are used; but for the same medium it is always the same.

FIG. 255.

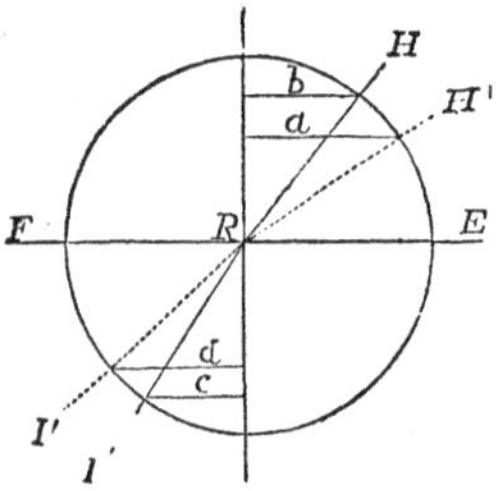

Thus, in Fig. 255, let F E be the surface of some refracting medium, as water, and H R, H′ R, rays incident upon it, at different angles; the former will be refracted in the direction R I′; *a* and *b* will be the sines of the angle of incidence, and *c d* the sines of the angle of refraction; and the quotient arising from dividing *b* by *c*, is the same as that from dividing *a* by *d*. In the case of air and water, the sine of the angle of incidence in the air will be to the sine of the angle of refraction in water as 4 is to 3; in any two other media, a different ratio would be observed with equal constancy.

What is the index of refraction?

The quotient found by dividing the sine of the angle of incidence by the sine of the angle of refraction, is called the index of refraction.

As different bodies have different refractive powers, they will present different indices, but in the same substance it is always constant. Thus, the refractive index of water is 1.335, of flint glass, 1.55, of the diamond, 2.487.

Is light ever wholly transmitted?

No surface ever transmits all the light which falls upon it, but a portion is always reflected. If, in a dark room, we allow a sunbeam to fall on the surface of water, the division of the light into a reflected and refracted ray will be clearly perceptible.

Under what circumstances will total reflection of light occur?

When the obliquity of an incident ray passing through a denser medium toward a rarer (as through water into air), is such that the sine of its refracting angle is equal to 90°, it ceases to pass out, and is reflected from the surface of the denser medium back into it again. This constitutes the only known instance of the total reflection of light. The phenomenon may be seen by looking through the sides of a tumbler containing water, up to the surface in an oblique direction, when the surface will be seen to be opaque, and more reflective than any mirror, appearing like a sheet of burnished silver.

What circumstances influence the refractive power of bodies?

No law has yet been discovered which will enable us to judge of the refractive power of bodies from their other qualities. As a general rule, dense bodies have a greater refractive power than those which are rare; and the refractive power of any particular substance is increased or diminished in the same ratio as its density is increased or diminished. Refractive power seems to be the only property, except weight, which is unaltered by chemical combination; so that by knowing the refractive power of the ingredients, we can calculate that of the compound.

All highly inflammable bodies, such as oils, hydrogen, the diamond, phosphorus, sulphur, amber, camphor, etc., have a refractive power from ten to seven times greater than that of incombustible substances of equal density.

Of all transparent bodies the diamond possesses the greatest refractive or light-bending power, although it is exceeded by a few deeply-colored, almost opaque minerals. It is in great part from this property that the diamond owes its brilliancy as a jewel.

Many years before the combustibility of the diamond was proved by experiment, Sir Isaac Newton predicted, from the circumstance of its high refractive power, that it would ultimately be found to be inflammable.

If the surface of any naturally transparent body is made rough and irregular, the rays of light which fall upon it are refracted and reflected so irregularly, that they fail to penetrate and pass through the substance of the body, and its transparency is thus destroyed.

Glass made rough on its surface loses its transparency; but if we rub a ground glass surface with wax, or any other substance of nearly the same optical density, we fill up the irregularities and restore its transparency. Horn is translucent, but a horn shaving is nearly opaque. The reason of this is that the surface of the shaving has been torn and rendered rough, and the rays of light falling upon it are too much reflected and refracted to be transmitted, and thereby render it translucent. On the same principle, by filling up the pores and irregularities of the surface of white paper, which is opaque, with oil, we render it nearly transparent.

How is refraction accounted for?

According to the undulatory theory of light, refraction is supposed to be due to an alteration in the velocity with which the ray of light travels. According to the corpuscular theory, it is accounted for on the supposition that different substances exert different attractive influences on the particles of light coming in contact with them.

What is Dioptrics? That department of the science of optics which treats of the refraction of light is termed Dioptrics.

What ensues when light passes through media with parallel surfaces? 676. When a ray of light passes through a transparent medium whose sides where the ray enters and emerges are parallel, it will suffer no permanent change of direction by refraction, since the second surface exactly compensates for the refractive effect of the first.

FIG. 256.

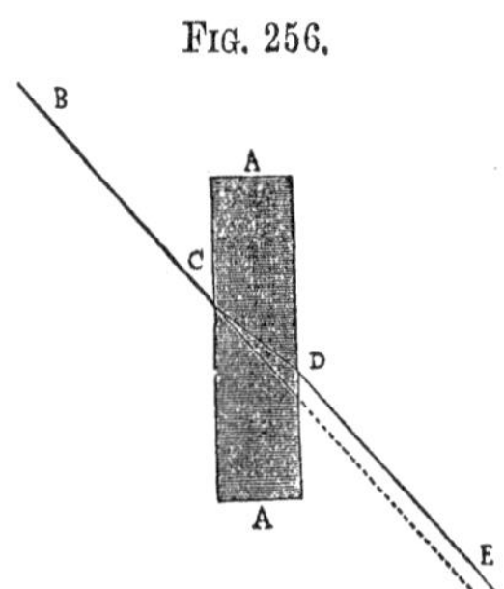

Thus let A A, Fig. 256, be a plate of glass, whose sides are parallel, and B C a ray of light incident upon it; it will be refracted in the direction C D, and on leaving the glass will be refracted again, emerging in the line D E, parallel to the course it would have pursued if it had not been refracted at all, and which is shown by the dotted line. A small lateral displacement is, however, occasioned in the path of the ray, depending on the thickness of the glass plate.

This explains the reason why a plate of glass in a window whose surfaces are perfectly parallel, occasions no distortion, or alteration of the position of objects seen through it, by reason of its refractive power. The rays suffer two refractions in contrary directions, which produce the same effect as if no refraction had taken place.

What happens when light passes through media whose surfaces are not parallel? If the surfaces of the medium through which light passes are not parallel, the direction of every ray passing through it is permanently altered, the change being greater as the inclination of the two surfaces is greater.

Thus window-glass of unequal thickness displaces and distorts all objects seen through it. Hence the singular distortion of objects viewed through that swelling, or lump of glass known as the "bull's eye," which is sometimes seen in the center of very coarse panes of glass, and which remains where the glass-blower's instrument was attached.

What is a Prism? 677. Any glass having two plane surfaces not parallel, is called a PRISM.

As ordinarily constructed, a prism is an oblong, triangular, or wedge-shaped piece of glass, with sides inclined at any angle, as is represented in Fig. 257.

FIG. 257.

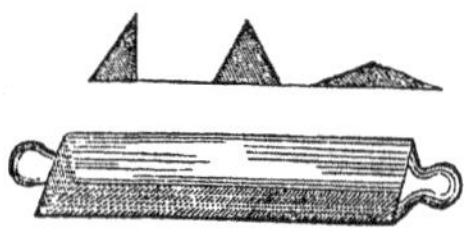

Explain the action of the prism.

On looking through a prism, all objects are seen removed from their true place. Thus, let C A B, Fig. 258, be a prism, and D E a ray of light incident upon it; it will be refracted in the direction E F, and on emerging, will again be refracted in the direction F H; and as objects always appear in the direction in which the last ray enters the eye, the object D will appear at G, in the direction of the dotted line, elevated above its real position. If the refracting angle, A C B, had been placed downward, the object would have appeared as much depressed.

FIG. 258.

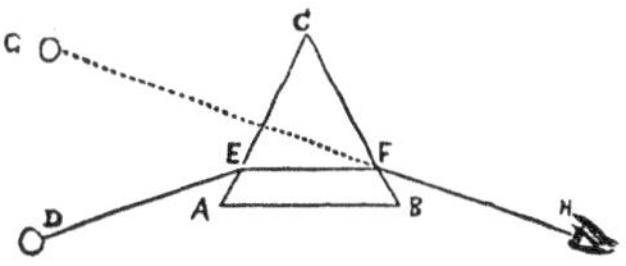

The prism, although of simple construction, is one of the most important of optical instruments, and to its agency we are indebted for most of the information we possess respecting the nature and constitution of light. The beautiful and complicated results of its practical application belong to that department of optics which treats of the phenomena of color.

What is a Lens?

678. A LENS is a piece of glass or other transparent substance, bounded on both sides by polished spherical surfaces, or on the one side by a spherical, and on the other by a plane surface. Rays of light passing through it are made to change their direction, and to magnify or diminish the appearance of objects at a certain distance.

How many kinds of simple lenses are there?

There are six different kinds of simple lenses, all of which may be considered as portions of the external or internal surface of a sphere. Four of these lenses are bounded by two spherical surfaces, and two by a plain and spherical surface.

Fig. 259 represents sectional views of the six varieties of simple lenses.

Explain the different kinds of lenses.

A double convex lens is bounded by two convex spherical surfaces, as at A, Fig. 259.

To this figure the appellation of lens was first applied from its resemblance to a lentil seed (in Latin, *lens*).

A plano-convex, or single convex lens has one side

bounded by a plane surface, and the other by a convex surface. It is represented at B, Fig. 259.

FIG. 259.

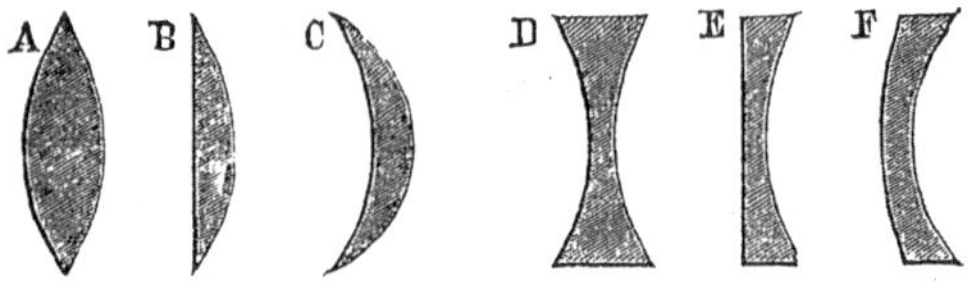

A meniscus, or concavo-convex lens is convex on one side and concave on the other, as at C, Fig. 259.

To this kind of lens the term "periscopic" has recently been applied, from the Greek, signifying to view on all sides.

A double concave lens is concave upon both sides, as at D, Fig. 259.

A plano-concave, or single concave lens, is bounded on one side by a plane, and on the other by a concave surface, as at E, Fig. 259.

A concavo-convex lens is bounded on one side by a concave, and on the other by a convex surface, as at F, Fig. 259.

Into how many classes may lenses be divided?

The six varieties of simple lenses are divided into two classes, which are denominated converging and diverging lenses, since the one class renders parallel rays of light falling upon them convergent, and the other class renders them divergent.

In Fig. 259 A B C are converging, or collecting lenses, and D E F diverging, or dispersing lenses. The former are thickest at the center; the latter are thinner at the center than at the edges.

In the first class it is sufficient to consider only the double-convex lens, and in the second class only the double-concave lens, since the properties of each of these lenses apply to all the others of the same class.

For optical purposes lenses are generally made of glass, but in some instances other substances are employed, such as rock-crystal, the diamond, etc.

What is the optical center of a lens?

In all the various kinds of lenses there must be a point through which rays of light passing experience no deviation; or in other words, the incident and emergent rays are parallel. Such a point is called the optical center of a lens.

What is the axis of a lens?

The axis of a lens is a straight line passing through the center perpendicular to the surface of the lens.

When is a lens considered exactly centered?

On this line will be situated the geometrical centers of the two surfaces of the lens, or rather of the spheres of which they form portions.

A lens is said to be truly or exactly centered when its optical center is situated at a point on the axis equally distant from corresponding parts of the surface in every direction; as then objects seen through the lens will not appear altered in position when it is turned round perpendicularly to its axis.

In what manner are parallel rays affected by a convex lens?

679. Parallel rays of light falling upon a double-convex lens are converged to a focus at a distance varying with the curvature of its sides.

FIG. 260.

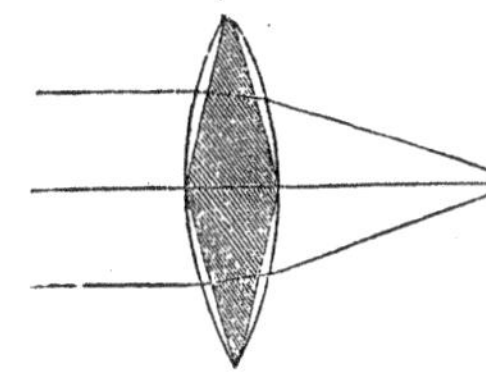

The double-convex lens may be regarded as two prisms, with curved surfaces, united at their bases, as is represented in Fig. 260; and as in a prism the ray of light refracted by it is always turned toward its back, or thicker part (whether that be turned upward, downward, or to either side), it follows that when parallel rays fall upon a double-convex lens, or two prisms united at their bases, they will converge to a point.

What is the principal focus of a convex lens?

The point where parallel rays of light falling upon one side of a convex lens unite by refraction upon the opposite side, is called the principal focus of a lens.

What is the focal distance of a lens?

The distance from the middle of a lens to its principal focus, is called the focal distance of a lens.

This in a single convex lens is equal to the diameter of the sphere of which the lens is a portion; in a double-convex lens it is equal to the radius, or semi-diameter of the sphere of which the lens is a portion.

The focal distance of parallel rays falling upon a convex lens is represented at A, Fig. 261. If the rays are converging, as at B, they will come to a focus sooner, and if diverging, as at C, the focus will be further from the lens than for parallel rays.

The focus of a convex lens may be easily found by allowing the rays of the sun to fall perpendicularly upon one side of it, while a sheet of paper is held on the other. A bright ring of light will be observed on the paper, diminishing or increasing in size according to the distance of the paper from the glass. If the former is held in such a manner that the ring of light is reduced to a dazzling luminous point, as is represented in Fig. 262, it is then situated in the focus of the glass.

FIG. 261.

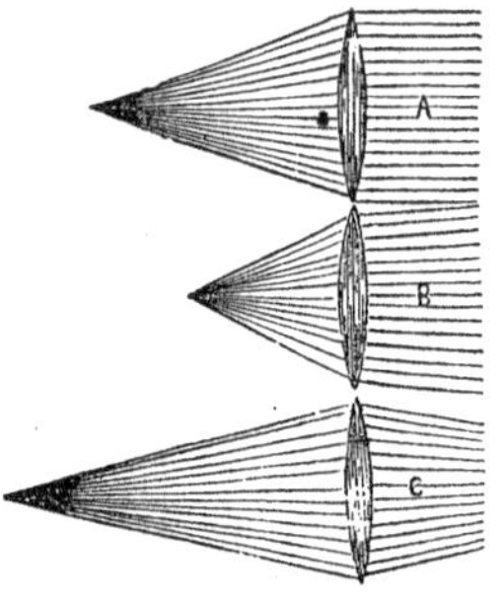

On what principle may convex lenses be used as burning-glasses?

680. From their property of converging parallel rays to a focus, convex lenses, like concave mirrors, may be used for the production of high temperatures, by concentrating the rays of the sun.

FIG. 262.

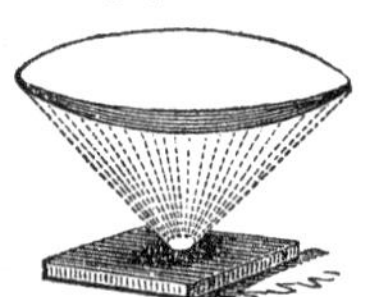

The ordinary burning, or sun-glass, as is represented in Fig. 262, is simply a double-convex lens. By the employment of very large lenses, a degree of heat may be produced far exceeding that of the best constructed furnace.*

How does the heat at the focus of a burning-glass compare with the heat of the sun?

In the employment of convex lenses as burning-glasses, the heat concentrated at the focus is to the common heat of the sun, as the area of the surface of the lens is to the area of the focus.

Thus, if a lens four inches in diameter collects the sun's rays into a focus at the distance of twelve inches, the focus will not be more than one tenth of an inch in diameter; its surface, therefore, is 1,600 times less than the surface of the lens, and consequently the heat will be 1,600 times greater at the focus than at the lens.

681. The properties of a concave lens are greatly different from those of a convex lens.

What is the course of rays falling upon a double concave lens?

Rays falling upon a concave lens are so refracted in passing through it, that they diverge on emerging from the lens, as though they issued from a focus behind it. The focus,

* A lens of this character was constructed many years since in England, three feet in diameter, with a focal distance of six feet eight inches. Exposed to the heat concentrated in the focus of this powerful instrument, the metals were instantly melted, and even volatilized, while quartz, flint, and the most refractory earthy substances, were readily liquefied and caused to boil.

therefore, of a concave lens is not real, but virtual, as is the case with a convex mirror.

Fig. 263.

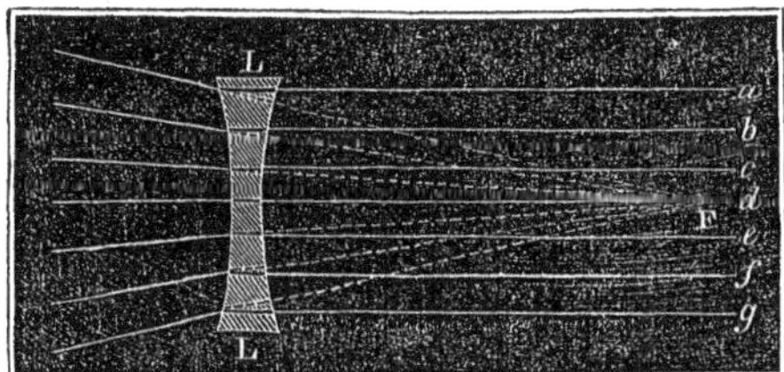

Thus, in Fig. 263, the parallel rays, *a b c d e*, etc., falling upon the double concave lens, L L′, are so refracted in passing through it, that they are made to diverge, as though proceeding from the point F, behind the lens.

In a similar manner convergent rays are rendered less convergent, or even parallel.

Do convex lenses give rise to the formation of images?

682. Images are formed in the foci of convex lenses in the same way as in the foci of concave mirrors.

Thus, if we take a convex lens and place behind it, at a proper distance, a sheet of paper, there will be depicted upon the paper beautifully clear and distinct images of all the objects in front of the lens, in an inverted position. The manner in which they are formed is illustrated in Fig. 264.

Describe the formation of images by the convex lens.

Fig. 264.

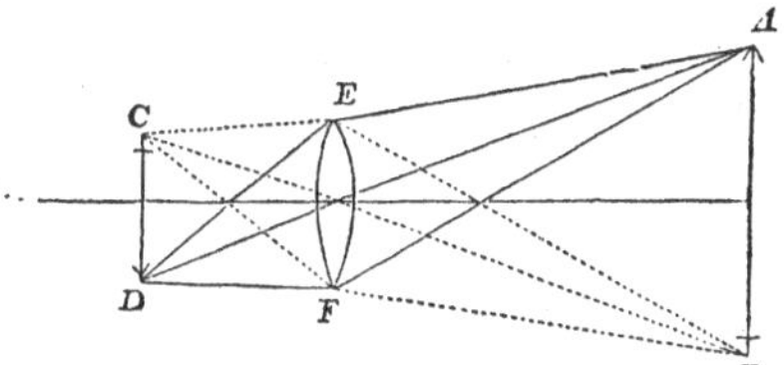

Thus, let A B represent an object placed before a double convex lens, E F. The rays proceeding from A, the top of the object, will be converged by the lens and brought to a focus at D, where they will form an image; the rays proceeding from B, the base of the object, will also be converged and brought to a focus at C; and so each point of the object, A B, will have its corresponding image between C D. In this way a complete image will be formed.

Why are the images formed by convex lenses inverted?

The image formed by a convex lens will appear inverted, because the rays of light from the several points of the object cross each other in proceeding to the corresponding points of the image.

Thus, in Fig. 264, the ray, A E, proceeding from the top of the object and falling obliquely upon the lens, is refracted into the course E D, and in like manner the ray B F is refracted in the direction F C; and as these rays cross each other, the image of the arrow appears inverted. The central ray of light proceeding from the object in the direction of the axis G, and falling perpendicularly upon the surface of the lens, undergoes no refraction, but continues on in a direct course.

How may images formed by convex lenses be made visible?

The images thus formed by convex lenses may be rendered visible by being received upon white screens, or any suitable objects, or directly by the eye, when placed in a proper position to receive the rays.

When, by the employment of the convex lens as a burning-glass, we concentrate on any suitable surface, the sun's rays to a focus, the little luminous spot, or circle formed, is really an image, or picture of the sun itself.

Why are convex lenses called Magnifying Glasses?

683. Convex lenses, as ordinarily used, are called magnifying-glasses, because they increase the apparent size of the objects seen through them.

Why does a convex lens magnify?

The reason of this is, that the lens so alters, by refraction, the direction of the rays of light proceeding from an object, that they enter the eye as if they came from points more distant from each other than is actually the case, and hence the object appears larger, or magnified.

Why does a concave lens diminish the apparent size of an object?

On the contrary, the concave lens, which produces an exactly opposite effect upon the rays of light, causes the image of an object seen through it to appear smaller.

On the same principles also, concave mirrors magnify, and convex mirrors diminish the images of objects reflected from their surfaces.

What is said of the magnifying or diminishing power of lenses?

Hence the magnifying or diminishing power of lenses is not, as is often popularly supposed, due merely to the peculiar nature of the glass of which they are made, but to the figure of their surfaces.

The double convex lens, inclosed in a convenient setting of metal or horn, is extensively employed by watch-makers, engravers, etc., with whom it passes under the general name of lens.

How may convex lenses render distant objects visible?

684. In addition to the effect which convex lenses produce by magnifying the images of objects, they are also capable of rendering distant objects visible which would be invisible to the

naked eye, by causing a greater number of rays of light proceeding from them to enter the eye.

Explain more fully the action of the convex lens in this respect?

The light which produces vision, as will be more fully explained hereafter, enters the eye through a circular opening called the pupil, which is the black circular spot surrounded by a colored ring, appearing in the center of the front of the eye. Now, as the rays of light proceeding from an object diverge or spread out in every direction, the number which will enter the eye will be limited by the size of the pupil. At a great distance from an object, as will be seen in Fig. 265, few rays will enter the eye; but if, as in Fig. 266, we place before the eye a convex lens of moderate size, a large number of the diverging rays will be collected and concentrated into a single point or focus behind it, and thus afford to the eye occupying a proper position sufficient light to enable it to see the distant object distinctly.

FIG. 265.

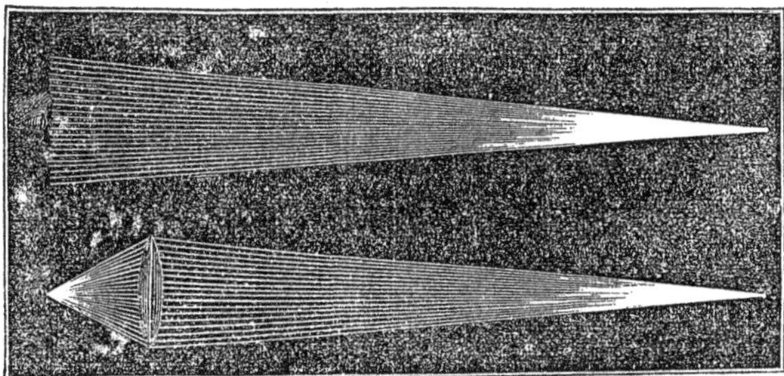

FIG. 266.

In like manner a concave mirror, by causing divergent rays which fall upon the surface to become convergent, may be used to produce the same effect, as is shown in Fig. 267.

FIG. 267.

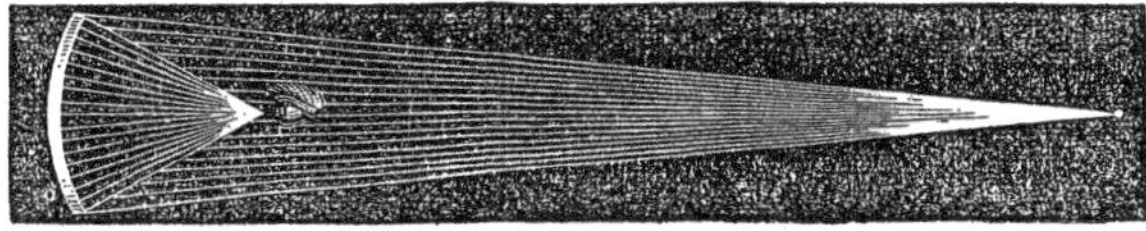

SECTION III.

THE ANALYSIS OF LIGHT.

685. It has, up to this point, been assumed, that light is a simple substance, and that all its rays, or parts, are refracted in precisely the same manner, and therefore suffer the same changes when acted upon by transparent media. This, however, is not its constitution.

What is the composition of white light? White light, as emitted from the sun, or from any luminous body, is composed of seven different kinds of light, viz., red, orange, yellow, green, blue, indigo, and violet.

What is the origin of Color? The seven different kinds of light produce seven different colors, viz., red, orange, yellow, green, blue, indigo, and violet. These seven colors are called primary colors, since by the union or mixture of some, two or more of them, all other colors, or varieties of color are produced.

How is light analyzed? The separation of white light into its several parts is effected by means of a prism.

When a ray of white light is made to pass through a prism, each of the seven rays of which it is composed are refracted, or bent out of their course differently, and form on an opposite screen or wall an image composed of bands of the seven different colors.

What is the Spectrum? 686. The image formed by a ray of white light passing through a prism, is called the Solar, or Prismatic Spectrum.

FIG. 268.

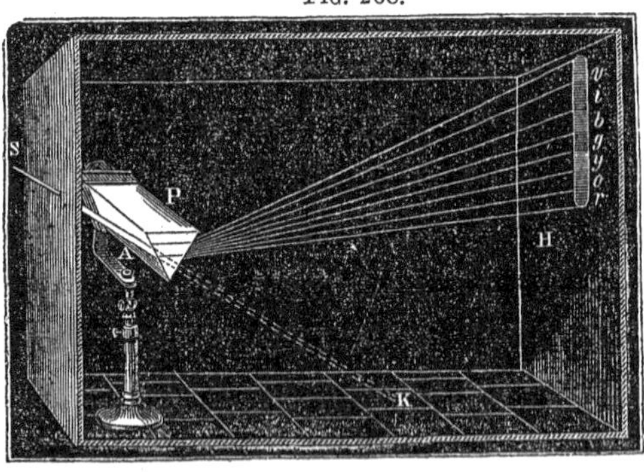

The separation of a ray of solar light into different colored rays, by refraction, is represented in Fig. 268. A ray of light, S A, is admitted through an aperture in a shutter into a darkened chamber, and caused to fall on a prism, P. The ray thus entering would, if allowed to pass unobstructedly, have moved in a straight line to the point K, on the floor of the room, and there formed a circular disc of white light; but by the interposition of the prism the ray spreads out in a fan-shape, and forms an oblong colored image on the opposite wall. This image, called the solar spectrum, is divided horizontally into seven colored spaces, or bands, of unequal extent, which succeed each other in an invariable order, viz., red, orange, yellow, green, blue, indigo, violet.

Upon what does the separation of white light depend?

The separation of the seven different rays composing white light from one another, depends entirely upon a difference in their refrangibility in passing through the prism; those which are refracted the least falling upon the lowest part of the screen, and those which are refracted the most upon the upper part.

Thus the red rays, which are the least refracted, or the least turned from their course by the prism, always occur at the bottom of the spectrum, while the violet, which is the most refracted, occurs at the top; the remaining colors being arranged in the intermediate space in the order of their refrangibility.

What additional proof have we of the composition of white light?

The seven different rays of light, when once separated and refracted by a prism, are not capable of being further analyzed by refraction; but if by means of a convex lens they are collected together and converged to a focus, they will form white light.

If the spectrum formed by a prism of glass be divided into three hundred and sixty parts, it is found that the red ray, or color, occupies forty-five of those parts, the orange twenty-seven, the yellow forty-eight, the green sixty, the blue sixty, the indigo forty, and the violet eighty.

If we take a circle of paper and paint upon it in divisions of proportionate size the seven colors of the spectrum, and then cause it to rotate rapidly about a center, the colors by combination will impart to it a white appearance.* From this and other experiments, therefore, it is inferred that light which we call colorless, or white (as that coming immediately from the sun), really contains light of all possible colors so mixed as to neutralize each other.

687. The separation of the different rays of light which takes place in their passage through a prism, is designated by the term Dispersion.

Explain what is meant by the dispersive power of different substances.

The order of refrangibility of the seven different rays of light, or the arrangement of the seven colors in the spectrum, is always the same and invariable, whatever way the prism may be turned; the lower end of the spectrum being

* It is very common to find it stated in books of science that by mixing powders of the seven different colors together a white, or grayish-white compound may be produced. The experiment, is not, however, satisfactory

red, which passes upward into orange, then into yellow, then green, blue, indigo, and violet, which is at the upper end.

Dissimilar substances, however, produce spectra of different lengths, on account of a difference in their refractive properties. Thus a ray of light traversing a prism of flint-glass, will have its red and violet colors separated on a screen twice as widely as those of a ray passing through a similar prism of crown-glass. This difference is expressed by saying that the dispersive power of the two substances is different, or that flint-glass has twice the dispersive power of crown-glass.

Why will not an ordinary lens produce a perfect image?

As a lens may be considered as a modification of the prism, it follows that when light is refracted through a lens, it is separated into the different colors, precisely as by a prism; and as every ray contained in white light is refracted differently, every lens, of whatever substance made, will have a different focus for every different color. The images, therefore, of such lenses will be more or less indistinct, and bordered with colored edges. This imperfection is termed chromatic aberration.

For this reason the focus of a burning-glass, which is an optical image of the sun, is never perfectly distinct, but always confused by a red, or blue border, since the various-colored rays of which sunlight is composed, can not all be brought to the same focus at once. In a like manner, if we point a common telescope at a blue and red hand-bill at a short distance, we shall have to draw out the tube of the instrument to a greater length in order to read the red than the blue letters.

Explain the construction of an achromatic lens.

These fringes of color are a most serious obstacle to the perfection of optical instruments, especially in astronomical telescopes, where great nicety of observation is required; and to prepare a lens in such a way that it would refract light without at the same time dispersing it into colors, was long considered an impossibility.

FIG. 268.

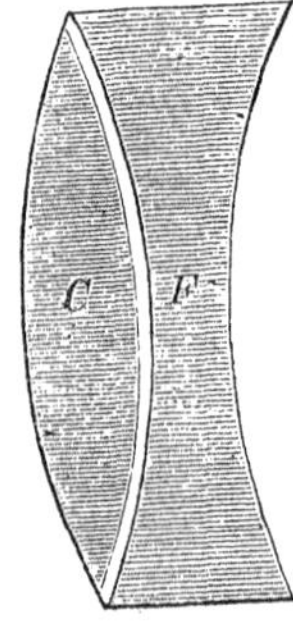

The discovery was, however, made by Mr. Dollond, an Englishman, that by combining two lenses, formed of materials which refract light differently, the one might be made to counteract the effects of the other; on the same principle as by combining two metals together which expand unequally, we may construct a pendulum whose length never varies.

Such a combination is represented in Fig. 268, where a convex lens of crown glass is united with a concave lens of flint glass, so as to destroy each the dispersive power of the other, while at the same time the refracting, or converging power of the convex lens is preserved. A lens of this character is called Achromatic,* since it produces images in their natural colors.

* Achromatic, from α, not, and $\chi\rho\omega\mu\alpha$, color.

What is spherical aberration? Lenses are also subject to another imperfection, which is called spherical aberration. This arises from the fact that the curved surface of a lens is at unequal distances from the object and from the screen which receives the image formed at its focus; and hence, if one point of the image is perfect, another point is less so, owing to a difference in the convergence of the rays coming from the center and the edges of the lens.

Thus, if the image is received on a screen of ground glass, it will be found that when the picture is well defined at the center, it will be indistinct at the edges; but by bringing the lens nearer the screen, the edges of the image will be more sharply defined, but the middle is indistinct. To make the image perfect, therefore, the marginal portions of the lens should be covered with a circlet of paper, so as to permit those rays only to pass which lie near the axis of the lens. This plan, however, impairs the brightness of the image.

When the image formed by the lens is small, the effect of spherical aberration is scarcely noticed, and by combination of lenses of different refractive powers, it may be almost entirely overcome.

Are all the rays of light equally brilliant? 688. The various rays composing solar light are not all equally luminous, that is to say, they do not appear to the eye equally brilliant. The color most visible to the human eye is yellow.

The luminous intensity of the different colored rays of light may be expressed numerically as follows:—Red, 94; orange, 640; yellow, 1,000; green, 480; blue, 170; indigo, 31; violet, 6.*

689. According to some authorities, white solar light consists of only three colors—red, yellow and blue, which, by combining, produce the other four colors, orange, green, indigo and violet.

What are sometimes called the simple colors? Red, yellow, and blue, are, therefore, sometimes called the simple colors.

Thus, by the union of red and yellow, we may produce orange; by yellow and blue, green; by blue and red, violet; indigo being considered as merely a shade of blue. Red, yellow, and blue, on the contrary, can not be produced by the mingling of any two other colors.

When blue and yellow powders are mixed together, blue and yellow rays are reflected to the eye from the minute particles, but the two colors are so

* It would appear, from numerous observations, that soldiers are shot during battle according to the color of their dress in the following proportion:—red, 12; dark green, 7; brown, 6; bluish gray, 5. Red is therefore the most fatal color, and a light gray the least so.

mingled that the eye only notices the combined effect, which is green. If we now examine the same mixture with a microscope, the blue and yellow particles will be seen separately, and the green color will disappear.

Why do natural objects exhibit colors?

690. The natural color which an object exhibits when exposed to the light, depends upon the nature and arrangement of the particles of matter of which it is composed, and is not the result of any quality inherent in the object itself.

Bodies which naturally exhibit color have, by reason of a certain peculiar arrangement of their surfaces, or molecular structure, a greater preference for some qualities of light than for others. If the body is not transparent, it will reflect certain rays of light from its surface, and appear of the color of the light it reflects ; if the body is transparent, it will allow only certain rays to pass through its structure, and will consequently appear of the color of the light it transmits.

Thus a red body appears red because it reflects or transmits the red ray of solar light to the eye; and a yellow body appears yellow because yellow light is reflected or transmitted by its surface or structure more powerfully than light of any other color; and so on through all the colors.

It is not, however, to be understood that colored bodies reflect or transmit only pure rays of one color, and perfectly absorb all others; on the contrary, it has been found that a colored body reflects, in great abundance, those rays of light which determine its particular color, and also the other rays which make up white light in a greater or less degree, in proportion as they more or less resemble its color in the order of their refrangibility.

When is a body colorless, when white, and when black?

Some substances have no preference for any one quality of light more than another, but reflect or absorb them all equally; such are called neutral, or colorless bodies. Those substances which reflect all the rays of light which fall upon them appear white ; those which absorb all the rays appear black.

In the dark there is no color, because there is no light to be absorbed or reflected, and therefore none to be decomposed.

A glass is called red because it allows the red rays of light to penetrate through a greater thickness of its substance than the other rays; but at a certain thickness, even the red rays would be absorbed like the rest, and we should call the glass black.

No body, unless self-luminous, can appear of a color not existing in the light which it receives. This may be proved by holding a colored body in a ray of light which has been refracted by a prism, when the body will appear of the color of the ray in which it is placed; for since it receives but one colored ray, it can reflect no other.

May the color of bodies be changed by changing their molecular structure?

691. By changing the structure or molecular arrangement of a body, the color which it exhibits may be often changed also.

Illustrations of this principle are frequently seen in chemical compounds. The iodide of mercury is a beautiful scarlet compound, which, when gently heated, becomes a bright yellow, and so remains when undisturbed. If, however, it is touched, or scratched with a hard substance, as with the point of a pin, its particles turn over, or readjust themselves, and resume their original red color. Chameleon mineral is a solid substance produced by fusing manganese with potash; when dissolved in water, it changes, according to the amount of dilution, from green to blue and purple. Indigo also, spread on paper and exposed to heat, becomes red.

692. Some bodies have the power of reflecting from their surfaces one color while they transmit another.

This is the case with the precious opal. A solution of quinine in water containing a little sulphuric acid, is colorless and transparent to the eye looking through it, but by looking at it, it appears intensely blue. An oil obtained in the distillation of resin transmits yellow light, but reflects violet light. Smoke reflects blue light, but transmits red light. These phenomena result from a peculiar action of the surface or outer layer of the substance of the body on some of the rays of light entering it, and have received the name of *epipolic*, or surface dispersion.

Deepness of color proceeds from a deficiency, rather than from an abundance of reflected rays: thus, if a body reflects only a few of the red rays, it will appear of a dark red color. When a great number of rays are reflected, the color will appear bright and intense.

If the objects of the material world had been illuminated only with white light, all the particles of which possessed the same degree of refrangibility, and were equally acted upon by all substances, the general appearance of nature would have been dull, and all the combinations of external objects, and all the features of the human countenance would have exhibited no other variety than that which they posses in a pencil sketch or India-ink drawing.

What are complementary colors?

693. Any two colors which are able, by combining, to produce white light, are termed complementary colors.

Each color of the solar ray has its complementary color, for if it be not white, it is deficient in certain rays that would aid in producing white. And these absent rays compose its complementary color.

The relative position of complementary colors in the prismatic spectrum may be determined as follows: Thus, if we take half the length of a spectrum by a pair of compasses, and fix one leg on any color, the other leg will fall upon its complementary color, or upon the one which added to the first will produce white light. The complementary color of red is bluish green; of orange is blue; of yellow is indigo; of green is reddish violet; of blue is orange red; of indigo is orange yellow; of violet is yellow green; of black is white; of white is black.

Complementary colors may be seen by fixing the eye steadily upon any colored object, such as a wafer upon a sheet of white paper. A ring of colored light will play round the wafer, and this ring will be complementary to the color of the wafer. A red wafer will give a green ring, a blue wafer an orange-colored ring, and so on. Or if, after having regarded the colored wafer steadily for a few moments, the eye be closed, or turned away, it will retain the impression of the wafer, not in its own, but in its complementary color; thus a red wafer will give a green ray, and so on.

In like manner, if we look at a red hot fire for a few minutes, every object as we turn away appears tinged with bluish green.

The art of harmonizing and contrasting colors is intimately connected with the principles of complementary colors.

How do colors affect each other in appearance?

Every color placed beside another color is changed, and appears differently from what it does when seen alone; it equally modifies, moreover, the color with which it is in proximity.

As a general rule, two colors will appear to the best advantage when one is complementary to the other.

Thus, if a dress is composed of cloths of two colors, the one complementary to the other, as red and green, orange and blue, yellow and violet, they will mutually heighten the effect of each, and make each portion appear to the best advantage. For this reason, a dress composed of cloths of different colors, looks well for a much longer time, although worn, than one of a single color, the character of the fabric being the same in both instances.

A suit of clothes of one color can be worn to advantage only when it is new, because as soon as one portion of the suit loses its freshness from having been worn longer than another, the difference will increase by contrast. Thus a pair of new black pantaloons worn with a vest of the same color, which is old and rusty, will make the tinge of the latter appear more conspicuous, and at the same time the black of the pants will appear more brilliant. White and other light-colored pantaloons would produce a contrary effect.

In printing letters on colored paper, the best effect will be produced when the color of the paper is complementary to the ink; blue should be put upon orange, and red upon green.

Stains will be less visible on a dress of different colors than on one composed of only a single color, since there exists in general a greater contrast among the various parts of the first-named dress, than between the stain and the adjacent part, and this difference renders the stain less apparent to the eye.

In the grouping of flowers in gardens, and in the preparation of bouquets, the most pleasing effects will be produced by placing the blue flowers next to the orange, and the violet next to the yellow. White, red, and pink flowers are never seen to greater advantage than when surrounded with green leaves, or white flowers; on the other hand, we should always separate pink flowers from those that are either scarlet or crimson; orange, from orange-yellow flowers; yellow flowers from greenish-yellow flowers; blue from violet-blue, red from orange, pink from violet.

By grouping colors together which are not complementary, or which do not rightly contrast with each other, we produce a discordant effect upon the eye, analogous to the discord which is produced upon the ear by instruments out of tune. It is always necessary that, if one part of the dress be highly ornamented, or consists of various colors, a portion should be plain, to give repose to the eye.

Black being the complementary color of white, the effect of black drapery upon the color of the skin or face is to make it appear pale, or whiter than it usually is.

The optical effect of dark and black dresses is to make the figure appear smaller; hence it is a suitable color for stout persons. On the contrary, white and light-colored dresses make persons appear larger. Large patterns or designs upon dress, make the figure appear shorter: longitudinal stripes, if not too wide, add to the height of the figure; horizontal stripes have a contrary tendency, and are very ungraceful.*

What is a Rainbow? 694. The Rainbow is a semicircular band or arch, composed of the seven different colors, generally exhibited upon the clouds during the occurrence of rain in sunshine.

How is a rainbow produced? The rainbow is produced by the refraction and reflection of the solar rays in the drops of falling rain.

* The following curious facts are known to persons employed in trade:—"When a purchaser has for a considerable time looked at a yellow fabric, and is then shown orange or scarlet stuffs, he considers them to be amaranth-red, or crimson, for there is a tendency in the eye, excited by yellow, to see violet, whence all the yellow of the scarlet or orange cloth disappears, and the eye sees red, or red tinged with scarlet. Again, if there are presented to a buyer, one after another, fourteen pieces of red cloth, he will consider the last six or seven less beautiful than those first seen, although the pieces be identically the same. Now what is the cause of this error in judgment? It is that the eyes having seen seven or eight red pieces in succession, are in the same condition as if they had regarded fixedly during the same period of time a single piece of red cloth; they have then a tendency to see the complementary color of red, that is to say, green. This tendency goes, of necessity, to enfeeble the brilliancy of the red of the pieces seen later. In order that the merchant may not be the sufferer by this failing of the eyes of his customer, he must take care after having shown the latter seven pieces of red, to present to him some pieces of green cloth, to restore the eyes to their natural state. If the sight of the green be sufficiently prolonged to exceed the normal state, the eyes will acquire a tendency to see red; then the last seven pieces will appear more beautiful than the others."—*Chevreul on Color.*

695. Rainbows are also formed when the sun shines upon drops of water falling in quantity from fountains, waterfalls, paddle-wheels, etc.

What experiments prove the decomposition of light by drops of water?

That the rainbow results from the decomposition of the solar rays by drops of water, may be proved by the following simple experiment:—If we take a glass globe filled with water, and suspend it at a certain height in the solar rays above the eye, a spectator standing with his back to the sun will see the refraction and reflection of red light; if, then, the globe be lowered slowly, the observer retaining his position, the red light will be replaced by orange, and this in its turn by yellow, and so on, the globe at different heights presenting to the eye the seven primitive colors in succession. If now, in the place of the globe occupying different positions, we substitute drops of water, we have a ready explanation of the phenomena of the rainbow.

Drops of rain, suspended to grass or bushes, may be frequently found to appear to the eye of a bright red; and by slightly changing the position of the eye, the colors of the drop may be made to appear successively yellow, green, blue, violet, and also colorless. This also proves that rays of light, falling in certain directions upon drops of water, are refracted thereby and decomposed into colored rays that become visible to the eye when it is situated in the proper direction.

FIG. 269.

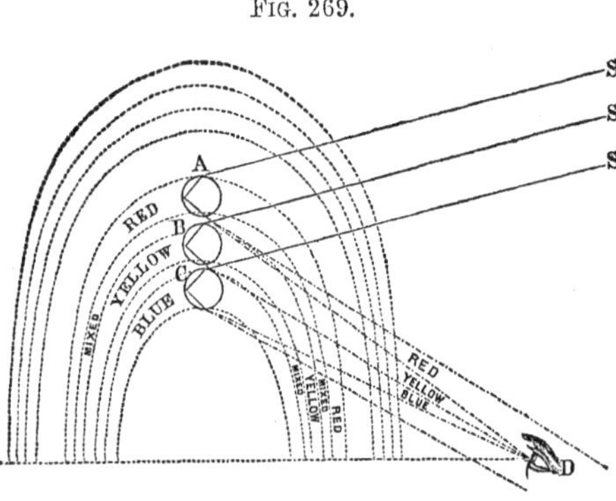

The principles of the formation of the rainbow may be further illustrated by Fig. 269. Let A B and C be three drops of rain; S A, S B, and S C, three rays of the sun. The ray S A, by refraction, is divided into three colors; the blue and yellow are bent above the eye, D, and the red enters it.

The ray, S B, is divided into three colors; the blue is bent above the eye, and the red falls below the eye D, but the yellow enters it.

The ray, S C, is also divided into three colors. The blue (which is bent most) enters the eye, and the other two fall below it. Thus the eye sees the blue of C, and of all drops in the position of C; the yellow of B, and of all drops in the position of B; and the red of A, and of all drops in the position of A. The same may be also inferred respecting the other four colors of the spectrum; and thus the eye sees a rainbow.

What are the conditions necessary in order to see a rainbow?

The rainbow can be seen only when it rains, and in that point of the heavens which is opposite to the sun.

Hence a rainbow is always observed to be situated in the west in the morning, and in the east in the afternoon.

It is also necessary for the production of a rainbow that the height of the sun above the horizon should not exceed forty-two degrees.

Hence we generally observe this phenomenon in the morning, or toward evening; and it is only in the winter, when the sun stands very low, that the rainbow is sometimes seen at hours approaching noon.

Is the same rainbow seen alike by all persons?

As the rays of light differ greatly in refrangibility, only a single and different-colored ray from each drop will reach the eye of a spectator; but as in a shower there is a succession of drops in all positions relative to the eye, the eye is enabled to receive the different-colored rays refracted at different inclinations. This is clearly illustrated in Fig. 270, in which S represents rays of the sun falling upon successive drops, R, O, Y, G, B, I, V; but a single colored ray, and a different one for each drop, will reach the eye. As no two spectators can occupy exactly the same position, no two can see the same color reflected from the same drop; and consequently no two persons see the same rainbow.

FIG. 270.

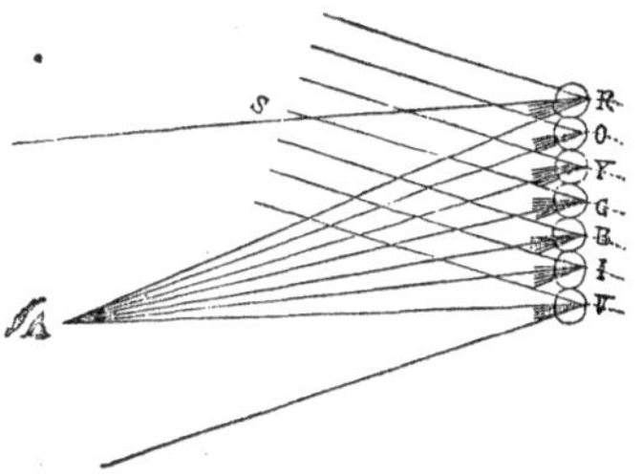

Why is a rainbow circular.

In the formation of a rainbow each colored ray reflected from the falling drops of rain, enters the eye at a different inclination or angle. But the several positions of those drops, which alone are capable of reflecting the same color at the same angle, to the eye constitute a circle,—and hence the bands of color which make up a rainbow, appear circular.

What are primary and secondary rainbows?

Two rainbows are not unfrequently observed at the same time, the one being exterior to, and less strongly developed than the other. The inner arch, which is the brightest, is called the primary bow, and the outer, or fainter arch, the secondary bow. The order of colors in the inner bow is also the reverse of that in the outer bow.

How is the primary rainbow formed?

The inner, or primary rainbow, which is the one ordinarily seen, is formed by two refractions of the solar ray, and one reflection, the ray of light entering the drops at the top, and being reflected to the eye from the bottom.

Fig. 271.

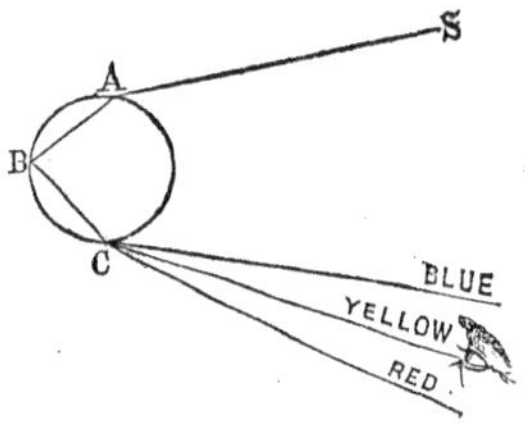

Thus, in Fig. 271, the ray S A of the primary rainbow strikes the drop at A, is refracted, or bent to B, the back part of the inner surface of the drop; it is then reflected to C, the lower part of the drop, when it is refracted again, and so bent as to come directly to the eye of the spectator.

How is the secondary rainbow formed?

The secondary, or outer rainbow, is produced by two refractions of the solar ray, and two reflections, the ray of light entering the drops at the bottom, and being reflected to the eye from the top.

Fig. 272.

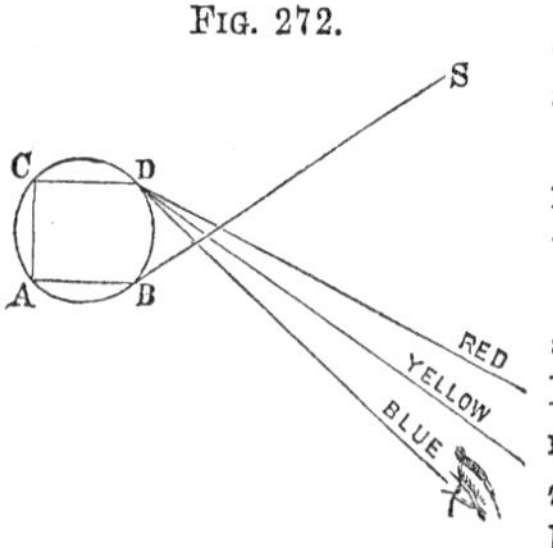

Thus, in Fig. 272, the ray S B of the secondary bow strikes the bottom of the drop at B, is refracted to A, is then reflected to C, is again reflected to D, when it is again refracted or bent, till it reaches the eye of the spectator.

The position and formation of the primary and secondary rainbows are represented in Fig. 273. Thus, in the formation of the primary bow, the ray of light S strikes the drop *n* at *a*, is refracted to *b*, reflected to *g*, and leaving the drop at this point, is refracted to the eye of the spectator at O. In the formation of the secondary bow, the ray S′ strikes the drop *p* at the bottom at the point *i*, is refracted to *d*, reflected to *f*, and thence to *e*, and refracted from the top of the drop, proceeds to the eye of the spectator at O.

The reason the outer bow is paler than the inner is because it is formed by rays which have undergone a second internal reflection, and after every reflection light becomes weaker.

What are Halos?

Halos are colored rays which are sometimes seen surrounding luminous bodies, especially the sun and moon. They are occasioned by the refraction and decomposition of light by particles of moisture, or crystals of ice floating in the higher regions of the atmosphere, and are never seen when the sky is perfectly clear.

FIG. 273.

The production of halos may be illustrated experimentally, by crystallizing various salts upon plates of glass, and looking through the plates at the sun, or a candle. A few drops of a saturated solution of alum, spread over a glass so as to crystallize quickly, will cover it with an imperfect crust of crystals, scarcely visible to the eye. Upon looking at a luminous body through the glass plate, with the smooth side next the eye, three fine halos will be perceived encircling the source of light.

The fact that halos, or rings round the moon, are more frequently observed than solar halos, is dependent upon the circumstance that the sun's light is too intense and dazzling to allow the halo to be recognized. Halos may be observed most frequently in the winter season, and in high northern latitudes.

What is the occasion of the red appearance of the clouds at sunrise and sunset?

696. The beautiful crimson appearance of the clouds after sunset in the western horizon, is due in a great measure to the fact that the red rays of the solar light are less refrangible than any of the other colored rays, and in consequence of this, they are not bent out of their course so much as the blue and yellow rays, and are the last to disappear. For the same rea-

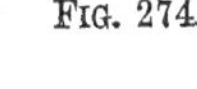
FIG. 274.

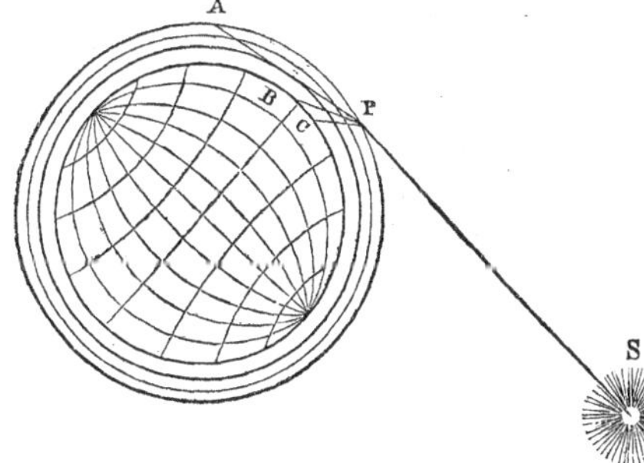

son they are the first to appear in the morning when the sun rises, and impart to the morning clouds red or crimson colors.

Let us suppose, as in Fig. 274, a ray of light proceeding from the sun, S, to enter the earth's atmosphere at the point P. The red rays, which compose in part the solar beam, being the least refrangible, or the least deviated from their course, will reach the eye of a spectator at the point A; while the yellow and blue rays, being refracted to a greater degree, will reach the surface of the earth at the intermediate points B and C. They will, consequently, be quite invisible from the point A.

The red and golden appearance of the clouds at morning and evening is also due in part to the fact, that aqueous vapor on the point of being condensed, only allows the red and yellow rays of light to pass through it. For this reason, if the sun be viewed through a column of steam escaping from a boiler, it appears of a deep red, or crimson color. The same thing may be noticed during a drought in summer, when the air is filled with dry exhalations.

What is Twinkling?

697. The irregular brilliancy of the stars, known as twinkling, is supposed to be due to unequal reflections of light occasioned by inequalities and undulations in the atmosphere.

How is color explained by the undulatory theory of light?

698. Light, according to the undulatory theory, is occasioned by the vibrations or undulations of a certain elastic medium diffused throughout all space, called Ether. Color, according to this theory, depends on the number of vibrations which are made in a certain time; those vibrations which are the most rapid, producing upon the eye the sensation of violet, and those which are the slowest, the sensation of red.

What analogy is there between color and the notes of music?

The analogy between sound and light, according to the undulatory theory, is perfect, even in its minutest circumstances. When a certain number of vibrations of a musical chord are caused in a given time, we produce a required sound; as the vibrations of the chord vary from a quick to a slow rate, we produce sounds sharp or grave. So with light; if the rate at which the ray undulates is altered, a different sensation is made upon the organs of vision.

The number of aerial vibrations per second required to produce any particular note in music has been accurately calculated, and it is also known that the ear is able to detect vibrations producing sound, through a range commencing with 15, and reaching as far as 48,000 in a second. So also in the case of light, the frequency of vibrations of the ether required for the production of any particular color has been determined, and the length of the waves corresponding to these vibrations.

What relation exists between the wave-lengths and vibrations of the different colors?

The waves requisite to produce red are the largest; orange comes next; then yellow, green, blue, indigo, and violet, succeed each other, the waves of each being less than the preceding. The rapidity of vibration is in the same order, the waves producing red light vibrating with the least rapidity, and the waves producing violet with the greatest rapidity.

To produce red light it is necessary that 40,000 waves or undulations should be comprised within the space of a single inch, and that 480 billions of vibrations should be executed in one second of time; while, for the production of violet, 60,000 waves within an inch, and 720 billions of vibrations per second are required.*

Can waves of light be made to interfere?

699. As two sets of sound-waves or vibrations may so combine as to modify or destroy each other, and thus produce partial or total silence, so two waves or vibrations of light may be made to interfere and produce various colors, or entire darkness.

* It may perhaps be asked, with something of incredulity, how such a result could possibly have been arrived at, with any degree of scientific accuracy. The problem, however, is not a difficult one.

In the first place, Newton, by a series of perfectly satisfactory and beautiful experiments, ascertained the number of waves or undulations of the different colored rays comprised within the space of an inch.

Let us now suppose an object of any particular color, a red star, for example, to be viewed from a distance. From the star to the eye there proceeds a continuous line of waves; these waves enter the pupil, and impinge upon the retina; for each wave which thus strikes the retina, there will be a separate pulsation of that membrane. Its rate of pulsation, or the number of pulsations which it makes per second, will therefore be known, if we can ascertain how many luminous waves enter the eye per second.

It has been already shown that light moves at the rate of about 200,000 miles per second; it follows, that a length of ray amounting to 200,000 miles must enter the pupil each second; the number of times, therefore, per second, which the retina will vibrate, will be the same as the number of the luminous waves contained in a ray 200,000 miles long.

Let us take the case of red light. In 200,000 miles there are, in round numbers, 1,000,000,000 feet, and therefore 12,000,000,000 inches. In each of these 12,000,000,000 of inches there are 40,000 waves of red light. In the whole length of the ray, therefore, there are 480,000,000,000,000 waves. Since this ray, however, enters the eye in one second, and the retina must pulsate once for each of these waves, we arrive at the astounding conclusion, that when we behold a red object, the membrane of the eye trembles at the rate of 480,000,000,000,000 of times between every two ticks of a common clock!

In the same manner, the rate of pulsation of the retina corresponding to other tints of colors is determined; and it is found that when violet is perceived, it trembles at the rate of 720,000,000,000,000 of times per second.—*Lardner.*

How may the interference of light produce darkness?

If we stand at the junction of two streams of water, it will be noticed that when the waves from each meet in the same state of vibration, the resulting wave will be equal to the two combined; if, however, one wave is half an undulation behind the other, the crest of one will meet the hollow of the other, and comparatively smooth water will be the result. So if two pencil rays of light, radiating from two points, reach a point of interference at the same degree of elevation, a spot of double the luminous intensity of either will be produced; but if one is half a vibration behind the other, the result will be, that a dark instead of a light spot will be apparent.

How is color produced by the interference of light?

The brilliant tints of soap bubbles, and thin plates of different transparent bodies, are examples of the interference of light; for the undulations reflected from the first surface interfere with those reflected from the second, and thus produce the various colors.

The varying play of colors exhibited by films of oil on the surface of water, and the iridescent appearance of mother-of-pearl, the scales of fishes, and the wings of some insects, are all phenomena resulting from the interference of light.

What is double refraction?

700. Double refraction is a property which certain transparent substances possess, of causing a ray of light in passing through them to undergo two refractions; that is, the single ray of light is divided into two separate rays.

FIG. 275.

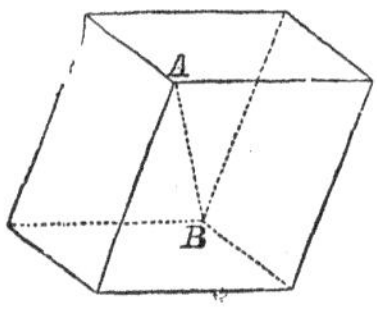

A very common mineral called "Iceland spar," which is a crystallized form of carbonate of lime, is a remarkable example of a body possessing double refracting properties. It is usually transparent and colorless, and its crystals, as shown in Fig. 275, have the geometrical form of a rhomb, or rhomboid;—this term being applied to a solid bounded by parallel faces, inclined to each other at an angle of 105°.

Illustrate the phenomenon of double refraction.

The manner in which a crystal of Iceland spar divides a ray of light into two separate portions is clearly shown in Fig. 276; in which S T represents a ray of light, falling upon a surface of a crystal of Iceland spar, A D E C, in a perpendicular direction. Instead of passing through without any refraction, as it would in case it had fallen perpendicularly upon the surface of glass, the ray is divided into two separate rays, the one, T O, being in the direction of the original ray, and the other, T E, being bent or refracted. The first of these rays, or the one which follows the ordinary

FIG. 276.

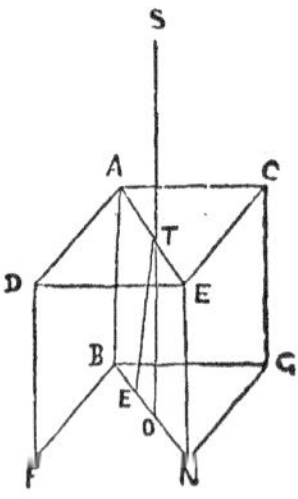

law of refraction, is called the "ordinary" ray; the second, which follows a different law, is called the "extraordinary" ray.

FIG. 277.

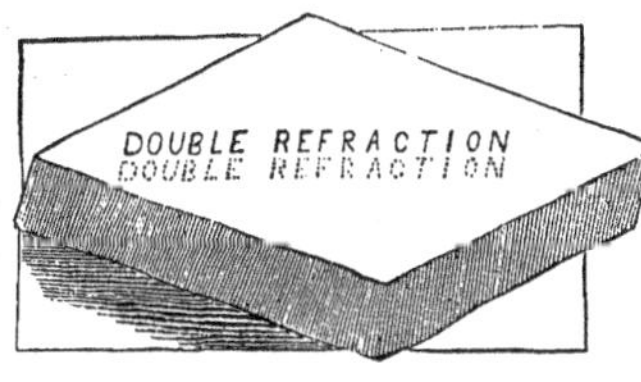

If we look at a small object, as a dot, a letter, or a line, through a plate of glass, it appears single; but if a plate of Iceland spar be substituted, a double image will be perceived, as two dots, two letters, two lines, etc. This result of double refraction is represented in Fig. 277. Crystals of many other substances, such as mica, the topaz, gypsum, etc., possess the property of double refraction, but not in so remarkable a degree as Iceland spar.

What are the axes of double refraction? In all these crystals, there are one or more directions along which objects when viewed through them appear single; these directions are termed the lines, or axes of double refraction. In the case of Iceland spar, there is one axis of double refraction, *i. e.*, one direction along which objects when viewed appear single; this is in the direction of the line A B, Fig. 275, which joins the two obtuse three-sided angles. If the summits A and B be ground down and polished, no double refraction will occur in looking through the crystal in this direction.

To what is the phenomenon of double refraction due? That the phenomenon of double refraction is due entirely to the molecular structure of the medium through which light passes, is proved by taking a cube of regularly annealed glass, which produces but one refracted ray, and heating it unequally, by subjecting it to pressure: a change is thereby affected in the arrangement of its parts, and double refraction takes place.

What is polarized light? 701. When a ray of light has been reflected from the surface of a body under certain special conditions, or transmitted through certain transparent crystals, it undergoes a remarkable change in its properties, so that it is no longer reflected and refracted as before. The effect thus produced upon it has been called polarization, and the ray or rays of light thus affected are said to be polarized.

What are the poles of a body? The name poles is given in physics in general to the sides or ends of any body which enjoy, or have acquired any contrary properties.

Thus, the opposite ends or sides of a magnet have contrary properties, inasmuch as each attracts what the other repels. The opposite ends of an electric or galvanic arrangement are, for like reasons, denominated poles. So also in the case of light, the rays which have been reflected or transmitted under peculiar conditions are said to possess poles, because in some positions they can be reflected and in others they can not, and these positions are at right angles to one another.

Explain the discovery and phenomena of polarized light.

702. The phenomenon of polarized light was discovered in 1808, by Malus, a young engineer officer of Paris. On one occasion, as he was viewing through a double refracting prism of Iceland spar the light of the sun reflected from a glass window in one of the French palaces, he observed some very peculiar effects. The window accidentally stood open like a door on its hinges at an angle of 54°, and Malus noticed that the light reflected from this angle was entirely altered in its character.

This alteration in the character of the light reflected from the glass window, which was thus first observed by Malus, may be made clear by the following experiment:—Suppose we have a cylinder with a mirror at one end of it. If we point this to the sun, and receive the image on a distant screen, we may turn the cylinder round on its axis, and the reflected ray will be found to revolve constantly with it. But if now, instead of receiving the ray direct from the sun, we allow a beam reflected from a glass plate, at an angle of about 54°, to fall upon the mirror, and then be reflected on the screen, it will be found that the point of light will not have the same properties as that previously examined; it will be altered in its degree of intensity as the cylinder turns round; will have points where it is very bright, and others where it will entirely disappear. It is thus proved that light reflected from glass at an angle of about 54°, has undergone some peculiar modification, or, as it has been termed, has become polarized.

Certain minerals, especially those called "tourmalines," have the property of polarizing a ray of light transmitted through them.

FIG. 278.

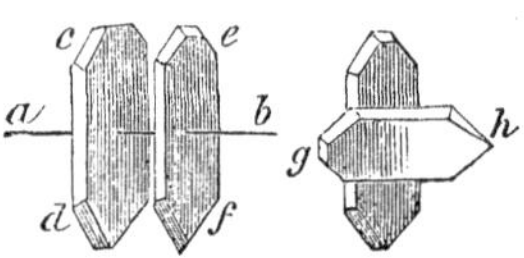

If a ray of light be caused to pass through a thin plate of tourmaline, as *c d*, Fig. 278, in the direction of the line *a b*, and be received upon a second plate, *e f*, placed symmetrically with the first, it passes through both without difficulty; but if the second plate be turned a quarter round, as in the direction *g h*, the light is totally cut off.

How is the polarization of light explained?

According to the undulatory theory, the difference between common and polarized light may be explained by supposing that in common light the vibrations of the ether which produce it take place in every possible direction, transverse to the path of the ray; but in polarized light they take place in only one direction, or are all in one plane.

Thus, in the passage of a ray of light through the plate of tourmaline, only one set of vibrations is transmitted, while the others are absorbed.

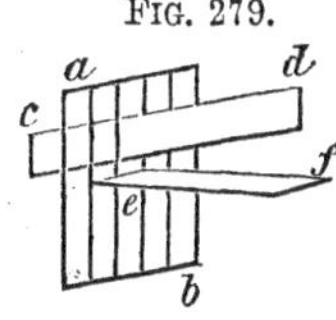

FIG. 279.

The transmitted ray, having all its vibrations in one direction, readily passes through a second plate of tourmaline, the structural arrangement of which is symmetrical with that of the first; but if this arrangement be altered by turning the plate partially round, the vibrations are intercepted. In the same manner a sheet of paper, *c d*, Fig. 279, may be slipped through a grating, *a b*, its plane coinciding with the length of the bars; but can no longer go through when it is turned, as at *e f*, a quarter round.

Is light polarized by reflection from other substances than glass?

Light is polarized by reflection from many different substances, such as glass, water, air, ebony, mother-of-pearl, surfaces of crystals, etc., etc., provided that the light falls at a certain angle peculiar to each surface. This angle is called the polarizing angle.*

What are some of the practical applications of polarized light?

Since the discovery of polarized light, its principles have been applied to the determination of many practical results. Thus, it has been found that all reflected light, come from whence it may, acquires certain properties which enable us to distinguish it from direct light; and the astronomer, in this way, is enabled to determine with infallible precision whether the light he is gazing on (and which may have required hundreds of years to pass from its source to the eye), is inherent in the luminous body itself, or is derived from some other source by reflection. It has been also ascertained by Arago that light proceeding from incandescent bodies, as red-hot iron, glass, and liquids, under a certain angle, is polarized light; but that light proceeding, under the same circumstances, from an inflamed gaseous substance, such as is used in street illumination, is always in a natural state, or unpolarized. Applying these principles to the sun, he discovered that the light-giving substance of this luminary was of the nature of a gas, and not a red-hot solid or liquid body.

In a similar manner the chemist is able to determine, by the manner in which light is reflected or polarized by a crystallized body, whether it has been adulterated by the addition of foreign substances.

What three principles are included in solar light?

703. Solar light, in addition to the luminous principle which produces the phenomena of color and is the cause of vision, contains two other principles, viz., heat and actinism, or the chemical principle. These principles are invisible to the eye, and have only been discovered by their effects on other bodies.

* The phenomena of polarized light are so abstruse, and depend to so great an extent on experimental illustration for their proper comprehension, that an extended description of them in an elementary work is impossible.

The constitution of the solar ray may be compared to a bundle of three sticks, one of which represents heat, another light, and a third the actinic principle.

How do we know that solar light contains three principles?

We know that these three principles exist in every ray of solar light, because we are able to separate them in a great degree from each other. Thus the luminous principle passes readily through a transparent plate of alum, but nearly all the heat is absorbed. Certain dark-colored bodies, on the contrary, allow nearly all the heat to pass, but obstruct the light. A blue glass obstructs nearly all the light and heat of the solar ray, but allows the chemical principle to pass freely; while a yellow glass allows light and heat to pass, but obstructs the passage of the chemical influence.

How are the three principles of solar light affected by a prism?

When we decompose a ray of solar light by means of a prism, and throw the spectrum upon a screen, the luminous, the calorific, and the actinic radiations will each assume a different position. All will be refracted by passing through the prism, but in different degrees.

The calorific, or heat radiations will be refracted least, and their maximum point will be found but slightly thrown out of the right line which the solar ray would have traversed had it not been intercepted by the prism. The heat diminishes with much regularity on each side of this line.

The luminous radiations are subject to a greater degree of refraction; their point of maximum intensity being in the yellow ray, lying considerably above the point of greatest heat. The light diminishes on each side of it, producing orange, red, and crimson colors below the maximum point, and green, blue, and violet above it.

The radiations which produce chemical action are more refrangible than either the calorific or luminous radiations, and the maximum of chemical power is found at that point of the spectrum where light is feeble, and where scarcely any heat can be detected.

The positions in the spectrum of the heat and actinic radiations, which are invisible to the eye, may be found by experiment. Thus, if we place a delicate thermometer in the different rays of the spectrum (§ 686, Fig. 268), it will be found that the indigo and violet rays scarcely affect it all, while the yellow ray, which is the most luminous, is inferior in heating action to the red ray, which, yielding but little light, possesses the greatest amount of heat. If now, the thermometer be carried a little below and just out of the red ray, into the darkened space, it will exhibit the greatest increase in temperature, thus proving the presence of a heating ray in solar light, independent of the luminous ray. In a like manner, by substituting a chemically prepared surface, as a piece of photographic paper, for the thermometer, the presence of a chemical ray can be proved in the darkened space at the other end of the spectrum, and near to the blue and violet rays.

704. Those rays of solar light which are less refrangible than any of the

visible colored rays of the spectrum, have all the properties of radiant heat coming from bodies of a lower temperature than 800° F. Such heat is much less refrangible than red light; but if the temperature of the radiating body be increased, it emits, in addition to the rays previously emitted, others of a higher refrangibility, until at last some few of its rays become as refrangible as the least refrangible rays of light. The body then appears of the same color as the least refrangible rays of light, and is said to be *red* hot. If it be heated more, it emits, in addition to the red, still more refrangible rays, viz., orange; then (at a higher temperature) yellow rays are added, and so on, until when the body is *white* hot, it emits all the colors visible to us; and in some instances (of very intense heat), even the invisible chemical rays, more refrangible than the violet, are emitted, though in less quantity than in the solar rays. Thus light appears to be nothing more than visible heat, and heat invisible light—the constitution of the eye being such that it can perceive one and not the other, in the same way as the ear can appreciate vibrations of sound more rapid than sixteen per second, but not those which are less rapid.

What curious fact has the study of the chemical principle of light evolved?

705. The study of the chemical principle contained in the rays of solar light has rendered probable the curious fact, that no substance can be exposed to the sun's rays without undergoing a chemical change; and from numerous examples it would seem that the changes in the molecular condition of bodies which sunlight effects during the daytime, is made up during the hours of night, when the action is no longer influencing them. Thus darkness appears to be essential to the healthy condition of all organized and unorganized forms of matter.

Upon what does the production of photographic pictures depend?

The process of forming Daguerreotype and other photographic pictures, depends solely upon the actinic, or chemical influence of the solar ray.

The term "photography," signifying light drawing, which is the general name given to this art, is unfortunate and ill-chosen, for not only does light not exercise any influence in producing the pictures, but it tends to destroy them.

What are the essential steps of the Daguerreotype process?

The essential steps of the process of forming a Daguerreotype picture consist in coating a suitable plate of metal with some chemical compound easily affected by the action of the solar ray. Such a coating is usually a compound of the elementary body *Iodine*. The plate is then exposed to the image formed by the lens of a camera obscura. Relatively, the quantity of light and actinism reflected from any object are the same; therefore as the light and shadows of the luminous image vary, so will the power of producing change upon the plate vary, and the result will be the production of an image which will be a faithful copy of nature, with reversed lights and shadows; the lights darkening the plate, while the shadows preserve it white, or unaltered.

If the plate were then left without further care, the image formed would soon fade away, and leave no trace on its surface. In practice, the plate is not exposed to the influence of light sufficiently long to form upon its surface an image visible to the eye, but the picture is developed, or brought out and rendered permanent by exposure to the vapor of mercury. This metal, in a state of very fine division, is condensed upon and adheres to those portions of the surface of the plate which have been subjected to the influence of the chemical action. Where the shadows are deep, there is scarcely a trace of mercury; but where the lights are strong, the metallic dust is deposited of considerable thickness. This deposition of mercury essentially completes and fixes the picture.

The reason why the vapor of mercury attaches itself only to those portions of the plate which have been affected by the chemical influence of light is not definitely known: in all probability, we have involved the action of several forces. It is not, however, necessary that a surface should be chemically prepared to exhibit these results. A polished plate of metal, a piece of marble, of glass, or even wood, when partially exposed to the action of light, will, when breathed upon, or presented to the action of mercurial vapor, show that a disturbance has been produced upon the portions which were illuminated; whereas no change can be detected upon the parts kept in the dark.

What experiment shows that light is not necessary for the production of a photographic result?

That the luminous principle is not necessary for the success of the photographic process, may be proved by the experiment of taking a daguerreotype in absolute darkness. This can be accomplished in the following manner:—A large prismatic spectrum is thrown upon a lens fitted into one side of a dark chamber; and as the actinic power resides in great activity at a point beyond the violet ray, where there is no light, the only rays allowed to pass the lens into the chamber are those beyond the limit of coloration, and non-luminous; these are directed upon any object, and from that object radiated upon a highly sensitive photographic surface. In this way a picture may be formed by radiations which produce no effect upon the eye.

What influence do the three principles of the solar ray exert on vegetation?

706. There are many reasons for supposing that each of the three principles, light, heat, and actinism, included in the solar ray, exercise a distinct and peculiar influence upon vegetation. Thus the luminous principle controls the growth and coloration of plants, the calorific principle their ripening and fructification, and the chemical principle the germination of seeds. Seeds which ordinarily require ten or twelve days for germination, will germinate under a blue glass in two or three. The reason of this is, that the blue glass permits the chemical principle of light to pass freely, but excludes, in a great measure, the heat and the light. On the contrary, it is nearly impossible to make seeds germinate under a yellow glass, because it excludes nearly all the chemical influence of the solar ray.

SECTION IV.

THE EYE, AND THE PHENOMENA OF VISION.

If an opening be made in the side of a dark chamber how will images of external objects be represented?

707. If we make a small aperture through the shutter of a darkened room, the images of external objects will be pictured indistinctly, and in an inverted position, upon the opposite wall. The reason of this will appear evident from an inspection of Fig. 280. It will be seen that the rays of light diverging from the top and bottom of the object cross each other in passing through the aperture, and consequently form an inverted image. This image is rendered more distinct with a small aperture than with a large one, since, in the first case, the rays which proceed from any particular part of the object fall only upon the corresponding part of the image, and are not scattered indiscriminately over the whole picture, as they would be if the aperture was larger.

FIG. 280.

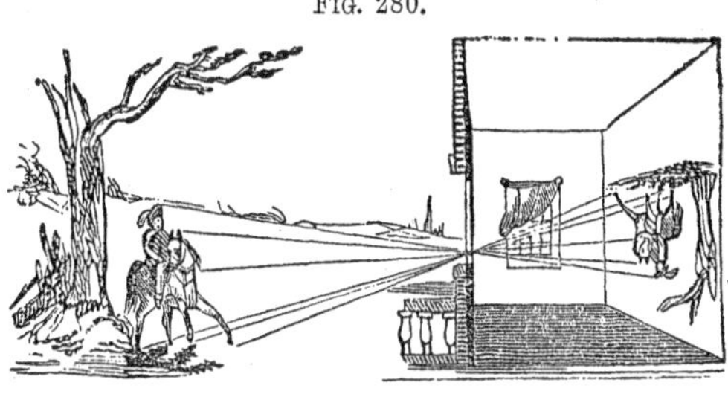

Describe the construction of the Camera Obscura.

If, in the place of the room with an aperture in the shutter, we substitute a dark box, with a double-convex lens fitted into one side, a picture will be formed on the opposite side of the box, or upon a screen placed at the focal distance of the lens. This picture will represent, with great beauty and distinctness, whatever is in front of the lens, all the objects having their proper relations of light and shadow, and their proper colors. Such an apparatus is called a CAMERA OBSCURA.

Fig. 281 represents the ordinary construction of the camera obscura. It consists of a wooden rectangular box, into which the rays of the light penetrate through a convex lens placed at the termination of the tube B. These rays, if unobstructed, will form an image upon the opposite side of the box O, but if they are received upon a mirror, M, inclined at an angle of 45°, their direction is changed, and the image will be formed upon a screen, or plate of ground glass, N, placed at the top of the box, By placing upon this screen a sheet of tracing paper, the outlines of the image may be readily copied.

Such a modification of the camera is very convenient for artists and travelers in sketching landscapes, etc.

FIG. 281.

How does the eye resemble the camera obscura?

708. The mechanical arrangement of the eye in man and the higher animals is the same as that of the camera obscura, being simply a double-convex lens, fitted into one side of a spherical chamber, through which the rays of light pass to form an inverted picture upon the back of the chamber.*

What is the general structure of the eye in man?

In man, the organs of vision consist of two hollow spheres, each about an inch in diameter, filled with certain transparent liquids, and deposited in cavities of suitable magnitude and form, in the upper part of the front of the head on each side of the nose.

How are we enabled to move the eye in different directions?

The sphere of the eye, or the eye-ball, is moved in its socket by muscles attached to different points of its surface, so that it is capable of being moved within certain limits in every direction.

* This may be proved by taking the eye of a recently-killed bullock and cutting a small hole in the upper part of the ball, looking into the interior.

FIG. 282.

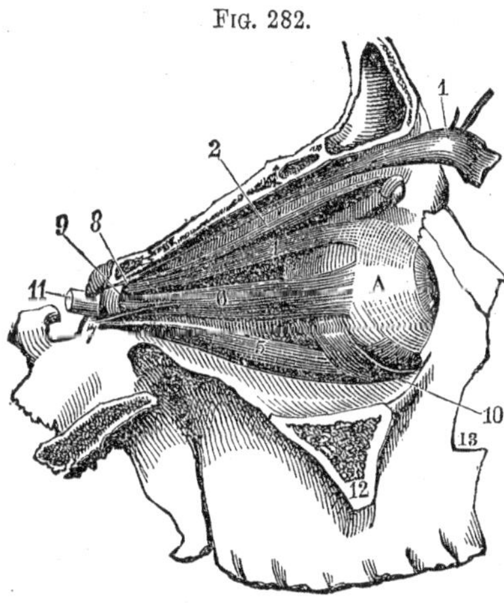

The arrangement of these muscles is shown in Fig. 282, where the external bones of the temple are supposed to be removed in order to render them visible. The muscle, 1, raises the eyelid, and is constantly in action while we are awake. During sleep, the muscle being in repose and relaxed, the eye-lid falls and protects the eye from the action of light. The muscle, 4, turns the eye upward; 5, downward; 6, outward; and a corresponding one on the inside, not seen in the figure, turns it inward. Nos. 2 and 10 turn the eye round its axis.

Of what parts does the eye consist?

The eye consists essentially of four coats, or membranes, called the SCLEROTIC coat, the CHOROID coat, the CORNEA, and the RETINA; and these coats inclose three transparent liquids, called humors—the AQUEOUS humor, the VITREOUS humor, and the CRYSTALLINE humor, the last of which has the form of a lens.

Describe the Sclerotic coat.

The Sclerotic coat is the external coat of the eye, and the one upon which the maintenance of the form of the eye chiefly depends.

It is a strong, tough membrane, and to it the muscles which move the eye are attached. It covers about four fifths of the external surface of the eye-ball, leaving, however, two circular openings, one before and the other behind the eye. Its position is shown at *i*, Fig. 283.

FIG. 283.

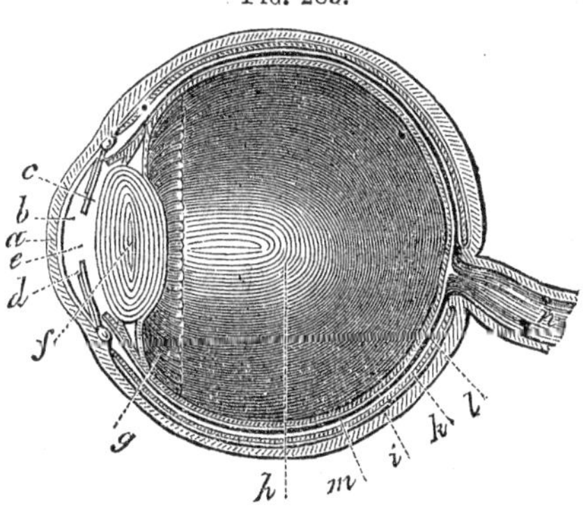

What is the Cornea?

The Cornea is the clear, transparent coat which

forms the front of the eye-ball. It is firmly united to, or fixed in the sclerotic coat, like the glass in the case of a watch.

The Cornea is represented at *a*, Fig. 283.

What is the Choroid Coat? The Choroid coat is a delicate membrane, lining the inner surface of the sclerotic coat, and covered on the interior with a black pigment.

It is represented at *k*, Fig. 283.

What is the Retina? The Retina is a delicate, transparent membrane which spreads over the chief part of the internal surface of the eye-ball, and is situated immediately within and close to the choroid coat.

The position of the Retina is shown at *m*, Fig. 283.

How is the retina formed? The retina is formed by the expansion of a nerve called the optic nerve, which proceeds from the back of the eye through the bones of the skull into the brain, and conveys to the brain the impressions made by external objects on the organs of vision. If this nerve were divided, notwithstanding the eye might be in other respects perfect, the sense of sight would be destroyed.

No. 11, Fig. 282, and *n*, Fig. 283, exhibit the relative position of the optic nerve.

What is the Iris? In looking into the eye from without, we perceive a flat, circular membrane, which, in different eyes, is of a black, blue, or gray color. This membrane is called the IRIS, and divides the eye into two very unequal portions.

The Iris is represented at *c d*, Fig. 283.

What is the Pupil of the eye? The Pupil of the eye is the circular black opening in the center of the iris, and is the space through which light is admitted into the interior of the eye.

The open space between *c* and *d*, Fig. 283, represents the pupil. It is, properly speaking, the window of the eye, and appears black, only because the chamber within and behind it is dark. When a small quantity of light enters the eye the pupil widens or expands; but when a large quantity enters, it closes or contracts.

The two parts into which the iris divides the eye are called the *anterior* and *posterior* chambers.

What are the aqueous and vitreous humors?

The anterior chamber, or the space before the iris, is filled with a fluid resembling pure water, and therefore called the aqueous humor; and the posterior chamber, or the space behind the iris, is filled with a thick liquid, somewhat resembling the white of an egg, called the vitreous humor.

In Fig. 283, *b e* represents the aqueous humor, and *h* the vitreous humor, this last occupying all the interior of the chamber of the eye.

The crystalline lens is composed of a more solid substance than either the aqueous or vitreous humor. It is inclosed within a transparent bag, or capsule, having the form of a double-convex lens, and is suspended immediately behind the iris, and between the aqueous and vitreous humors.

Its form and position are represented at *f*, Fig. 283.

How do we by the organs of the eye perceive objects?

709. Rays of light proceeding from an object and entering the eye, are refracted by the cornea and crystalline lens, and made to converge to a focus at the back of the eye, and form an image upon the retina. This image, by producing a sensation upon the optic nerve, conveys in some unknown way to the mind a perception and knowledge of the external object.

FIG. 284.

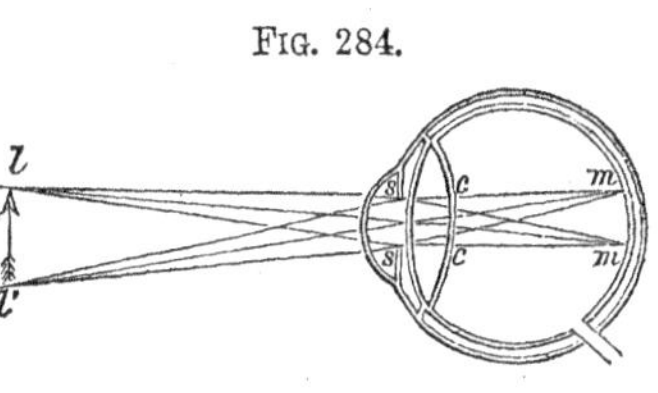

Fig. 284 represents the manner in which the image is formed upon the retina in the perfect eye. The curvature of the cornea, *s s*, and of the crystalline lens, *c c*, is just sufficient to cause the rays of light proceeding from the image, *l l'*, to converge to the right focus, *m m*, upon the retina.

When does distinct vision take place?

Distinct vision can only take place in the eye when the cornea and crystalline lens have such convexities as to bring the rays of light proceeding from an object to an exact focus upon the retina.

How is the eye enabled to see objects distinctly at different distances?

As the rays of light proceeding from distant objects enter the eye at different angles, they will naturally tend to meet at different foci after refraction by the crystalline lens, and thus form indistinct images. This is remedied by a power which the eye possesses of adapting itself to the direction of the light proceeding from various distances, so that in the healthy eye, rays coming from near and distant objects are all equally converged to a focus on the same point of the retina. How the eye effects this is not certainly known, but it is supposed to be by increasing or diminishing the sphericity of the crystalline lens and cornea.

What is the cause of near-sightedness?

A person is said to be near-sighted when the curvature of the cornea and crystalline lens is so great, that the rays of light which form the image are brought to a focus before they reach the retina, or the back part of the eye. The object, therefore, is not distinctly seen.

Fig. 285 represents the manner in which the image is formed in the eye of a near-sighted person. The curvature of the cornea, *s s*, and of the crystalline lens, *c c*, is so great that the image is formed at *m m*, in advance of the retina.

FIG. 285.

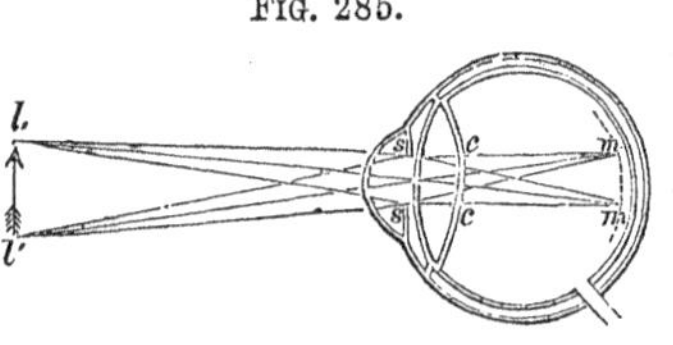

How is short-sightedness remedied?

Short-sightedness is remedied either by holding the object nearer to the eye, or by the employment of spectacles the glasses of which are concave lenses. In both cases the rays proceeding from the object enter the eye with a greater degree of divergence, and therefore do not converge so soon to a focus.

What is the cause of far-sightedness?

A person is said to be far-sighted when, on account of a flattening of the cornea and the crystalline lens, the rays of light do not converge sufficiently to form a distinct image upon the retina.

FIG. 286.

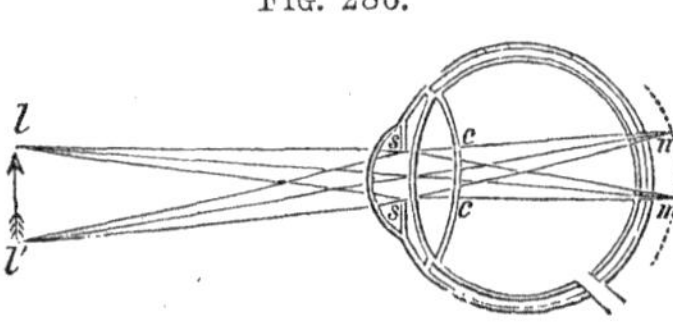

Fig. 286, represents the manner in which the image is formed in the eye, when the cornea or crystalline lens is flattened. The perfect image would be produced at *m m*, behind the retina, and, of course, beyond the point necessary to secure distinct vision.

How may long-sightedness be remedied?

Long-sightedness may be remedied by the employment of spectacles, the glasses of which are convex lenses. These, by increasing the convergence of rays of light passing through them, bring them sooner to a focus in the eye, and thus produce the image upon the right point of the retina.*

Most persons of advanced age are troubled with long-sightedness, and are obliged to use spectacles. The reason of this is, that as the physical organization of the body becomes enfeebled, the humors of the eye dry up, or are absorbed, and in consequence of this, the cornea and crystalline lens shrink and become flattened.

Beside these defects of the eye, a person may have the sense of vision impaired or destroyed by an injury or disease of the optic nerve, or by a diminution of the transparency of the crystalline lens; the first of these cases is called *amaurosis*, and is incurable—the second, which is called *cataract*, may be cured.

As the images on the retina are inverted, why do we not see them upside down?

The images formed by the rays of light upon the retina are inverted. It may, therefore, be asked why all visible objects do not appear upside down? The explanation of this curious point, which has formed the subject of much dispute, appears to be this: an object appears to be inverted only as it is compared with some other objects which are erect. If all objects hold the same relative position, none can be properly said to be inverted. Now, since all the images produced upon the retina hold, with relation to each other, the same position, none are inverted with respect to others; and as such images alone can be the object of vision, no one object of vision can be inverted with respect to any other object of vision; and, consequently, all being seen in the same position, that position is called the erect position.

What is the optic axis of the eye?

710. The optic axis of the eye is a line drawn perpendicularly through the center of the cornea, and center of the eye-ball.

Why with two eyes do we not see the same point of an object double?

The reason why with two eyes we do not see double is, because the axis of both eyes is turned to one point, and therefore the same impression is made on the retina of each eye.

The law of vision for visible objects is entirely different from that for points. A visible object can not, in all its parts, be seen single at the same instant of time, but the two eyes converge their axes to the near and the remote parts of it in succession, and thus give an idea of the different distances of its parts. Any defect which will prevent the two eyes from moving together conjointly, and from converging their optic axes upon every point of an object in succession, will be fatal to distinct vision.

* Birds of prey are enabled to adjust their eyes so as to see objects at a great distance, and again those which are very near. The first is accomplished by means of a muscle in the eye, which permits them to flatten the cornea by drawing back the crystalline lens; and to enable them to perceive distinctly very near objects, their eyes are furnished with a flexible bony rim, by which the cornea is thrown forward at will, and the eye thus rendered near-sighted.

How may double vision be produced?

Double vision may be produced by pressing slightly from the side upon the ball of either eye while viewing an object; the pressure of the finger prevents the ball of one eye from following the motion of the other, and the axis of vision in each eye being rendered different, we see two images.

Strabismus, or squinting, is caused by the inability of one eye to follow the motions of the other, and persons so affected always see double; practice, however, gives them power of attending to the sensation of only one eye at a time.

It is from this inability of the eye to fix its optical axis that drunkards see double.

How do we judge of the distance and size of an object?

711. We judge of the distance and size of an object by the relative direction of lines drawn from the object to the eye, and by the angle which the intersection of these lines makes with the eye. This angle is called the angle of vision.

FIG. 287.

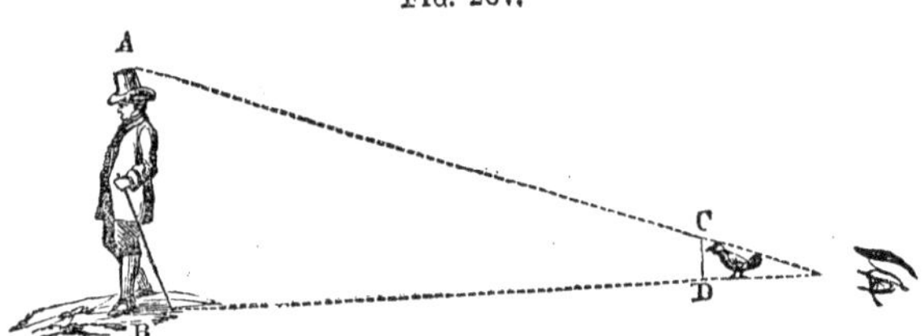

Explain the angle of vision.

The student will bear in mind that an angle is simply the inclination of two lines without any regard to their length. Thus, in Fig. 287, the lines drawn from A and B, C and D, which may be supposed to represent rays of light, meet at the eye, and form an angle at the point of intersection. This angle is the angle of vision.

If A B, Fig. 287, represent a man on a distant mountain, or on a church steeple, and C D a crow close by, the angle formed by the inclination of the lines proceeding from the two objects will be equal, or the line A B, which is the height of the man, will subtend the same angle as the line C D, which is the height of the crow; and therefore the man appears at such a distance no larger than a crow.

How is the angle of vision affected by distance?

The nearer an object is to the eye, the greater must be the inclination of the lines drawn from its extremities to intersect and form an angle at the eye, and consequently the greater will be its angle of vision. On the contrary, the more remote an object is from the eye, the less will be the inclination of the lines, and the less the angle of vision. The nearer an object is to the eye, therefore, the larger it will appear.

FIG. 288.

Thus the trees and houses far down a street or avenue appear smaller than those near by, and the size of a vessel seen at sea diminishes with the increase of distance, as is shown in Fig. 288. The moon, on account of its proximity, appears much larger than any of the stars or planets, although it is, in fact, very much smaller.

FIG. 289.

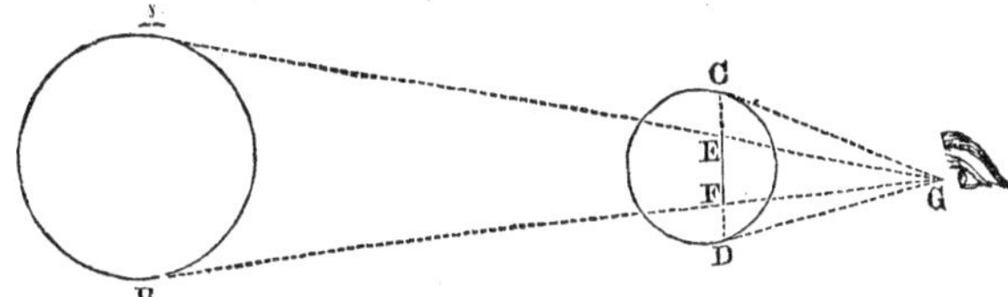

Let A B, Fig. 289, represent a planet, and C D the moon. The angle of vision which the planet A B makes with the eye at G, is evidently less than the angle which the moon subtends at the same point. To a spectator at G, therefore, A B, though much the larger body, will appear no larger than E F; whereas the moon, C D, will appear as large as the line C D.

When will an object appear as a mere point?

712. When an object is so remote, or so small, that lines drawn from its extremities form no appreciable angle at the eye, the object appears as a mere speck or point.

How small an object is visible to the eye?

The eye, with an ordinary amount of light, can see an object which occupies in the field of view a space of only the sixtieth of a degree (or one minute).

This space is about the 100th of an inch in a circle of twelve inches diameter, the eye being supposed to be in the center of the circle. Now a body smaller than this at six inches from the eye, or any thing, however large, placed so far from the eye as to occupy in the field of view less space than this, is invisible to ordinary sight. At four miles off, a man becomes thus invisible, and a pin-head near by will hide a house on a distant hill.*

What do we mean when we say we see an object?

713. When we say we see an object, we mean that the mind is taking cognizance of a picture or image of the object formed on the retina. The manner in which the sensation is conveyed by the optic nerve to the brain, and a knowledge of the external object imparted to the mind, is entirely unknown.

Does the sense of sight give immediate perception of form, size, position, etc.?

As the picture, or image on the retina, is formed on a comparatively flat surface, the sense of sight can not of itself afford any immediate perception of the distance, size, or position of external objects. This knowledge we gain by experience derived from continued observation, and from the other senses.

A young child has no conception of distance, and grasps at the moon as if it were an object immediately within its reach. Persons born blind and restored to sight by surgical operations, although able to see distinctly, can not properly comprehend any object or prospect before them. "I see men as trees walking," said the man born blind when restored to sight. Individuals thus situated acquire the correct sense of vision only by degrees, like infants, and it is by experience that they learn to walk about among the objects around them, without the continual apprehension of striking themselves against every thing they behold.

What is Perspective?

Perspective is the name given to that science which teaches how to draw on a plane surface true pictures of objects as they appear to the eye from any distance and in any position.

The skill of the artist consists in rightly applying the laws and principles of perspective; and a picture is perfect to the extent in which it agrees with our experience of the objects it represents.

714. Many optical and mental delusions are occasioned in estimating the size, figure, and position of objects, by

* "The smallest particle of a white substance distinguishable by the naked eye upon a black ground, or of a black substance upon a white ground, is about the 1-400th of an inch square. It is possible, by the closest attention, and by the most favorable direction of light, to recognize particles that are only 1-540th of an inch square, but without any sharpness or certainty. But particles which strongly reflect light may be seen when not half the size of the least of the foregoing: thus, gold dust of the fineness of 1-1125th of an inch may be discerned by the naked eye in common daylight. When particles that can not be distinguished by themselves with the naked eye are placed in a row, they become visible; and hence the delicacy of vision is greater for lines than for single particles. Thus, opaque threads of no more than 1-4900th of an inch across, or about half the diameter of the silkworm's fiber, may be discerned with the naked eye when they are held toward the light."—*Dr. Carpenter.*

an erroneous application of the experience which in ordinary cases supplies true and accurate conclusions.

Why do we misjudge the distance of a fire in the night?

Thus, to most persons a conflagration at night, however distant, appears as if very near. The explanation of this mistake is as follows:—Light radiating from a center rapidly weakens as the distance from the center increases, being, for instance, only one fourth part as intense at double the distance. The eye learns to make these allowances, and by the clearness and intensity of the light proceeding from the object, judges with considerable accuracy of the comparative distance. But a fire at night appears uncommonly brilliant, and therefore seems near.

The evening-star rising over a hill-top, appears as if situated directly over the top of the eminence. The reason of this also is, that in judging we make brightness and clearness to depend on contiguity, as it ordinarily does; and as the star is bright, we unconsciously think it near us.

Why do the sun and moon appear larger when rising and setting than at other times?

In consequence of terrestrial objects being placed in close comparison, the sun and moon appear larger at their rising and setting than at any other time. This illusion is wholly a mental one, since the organs of vision do not present to us a larger image of the sun or moon in the horizon than when in the zenith, or overhead.

Why does the moon, a sphere, appear like a flat surface?

The moon, although a sphere, appears to be a flat surface, since it is so remote that we are unable to distinguish any difference between the length of the rays reflected from the circumference, and those reflected from the center.

Thus the rays A D and C D, Fig. 290, appear to be no longer than the ray B D; but if all the rays seem of the same length, the part B will not seem to be nearer to us than A and C; and therefore the curve A B C will look like a flat, or horizontal surface. The rays A D and C D are 240,000 miles long. The ray B D is 238,910 miles long.

Fig. 290.

What two things are essential for distinct vision?

715. In order that the eye may see distinctly, the picture formed upon the retina must be illuminated to the right degree, and it must also remain sufficiently long upon the retina to produce a sensation upon the optic nerve.

The image of an object on the retina may be illuminated too much or too little to produce a sensible perception of its form. Thus, we can gain no idea of the form of the sun by viewing it in the clear sky, because the degree of illumination is so great, that the sense of vision is overpowered, just as sounds are sometimes so intense as to be deafening. That it is the intense splendor alone which prevents a distinct perception of the sun's figure, is rendered evident by the fact that when a portion of the light is cut off by a colored glass, or a thin cloud, the image of the sun is seen distinctly. On the contrary, we fail to perceive many stars at night, because the images they produce on the retina are too faintly illuminated to produce sensation. That some light from such stars actually enters the eye, is proved by the fact that if we place a lens before the eye, and collect a greater quantity of their light upon the retina, they at once become visible.

Can the eye adapt itself to degrees of illumination?

The eye possesses a limited power of accommodating itself to various degrees of illumination. In the dark, the pupil of the eye enlarges its opening, and allows a greater number of rays to fall upon the retina; in the light, the pupil contracts in proportion to the intensity of the illumination, and diminishes the number of rays falling upon the retina.

Why, in going from the light into the dark, do we find it difficult at first to see any thing?

This change does not take place instantaneously. When we leave a brilliantly illuminated apartment at night and go into the dark street, we are unable for a few moments to see any thing distinctly. The reason of this is, that the pupil of the eye, which has become contracted in the light, is unable to collect sufficient rays from the objects in the dark to see them distinctly. In a few moments, however, the pupil dilates, allows more rays to pass through its aperture, and we see more distinctly. The reverse of this takes place when we go from the dark into the light. Cats, owls, and some other animals are able to see distinctly in the dark, because they have the power of enlarging the pupils of their eyes so as to collect the scattered rays of light.

Every impression made by light remains for a certain length of time on the retina of the eye, according to the intensity of its effects, and a measurable period is necessary to produce a sensation.

What facts prove the continuance of the image upon the retina after the object has disappeared?

We are unable, when riding rapidly on a railroad, to count the posts of an adjoining fence, because the light from each post falls upon the eye in such rapid succession, that the different images become confused and blended, and we do not obtain a distinct vision of the particular parts.

If we rotate a stick, lighted at one end, somewhat rapidly, it seems to produce a complete circle of fire; the reason of this is, that the eye retains the image of any bright object for some little time after the object is withdrawn; and as the light of the stick returns to each particular point of its path before the image previously formed has faded from the retina, it seems to form a complete circle of fire.

Why is it not dark when we wink?

This continuance of the impression of external objects on the retina after the light proceeding from them has ceased to act, is the reason also why we are not sensible of darkness when we wink.

The apparent motion of certain colored figures in worsted work, known by the name of the "dancing mice," is due to the fact that when the surface is moved in a particular direction, as from side to side, the impression of the color on the retina remains for an appreciable interval after the figures have moved, and this gives to them an apparent motion. This effect will not, however, take place unless the colors of the figures and the ground-work are very brilliant and complementary of each other, as red upon a green ground.

When is motion imperceptible to the eye?

716. No motion is perceptible to the eye which has a less apparent velocity than one degree per minute.

It is for this reason that the motions of the heavenly bodies are invisible, notwithstanding their immense velocity. The apparent motion of the sun, moon, and stars, owing to the revolution of the earth, is one quarter of a degree a minute; but if the earth revolved on its axis in six hours instead of twenty-four, then the celestial bodies would have a motion of one degree per minute, and their movements would be distinctly perceptible.

For the same reason, the motions of the hands of a clock are not perceptible to the eye.

On the contrary, when a body moves with such rapidity from one position to another, that its image does not remain long enough upon one point of the retina to sufficiently impress it, it becomes invisible. Hence it is that a ball discharged from a cannon, and passing transversely across the eye, is not seen.

How is apparent motion affected by distance?

Apparent motion is affected by distance, and the motion of a body which is visible at one distance may be invisible at another, inasmuch as the angular velocity will be increased as the distance is diminished.

Thus, if an object at a distance of 57½ feet from the eye move at the rate of a foot per second, it will appear to move at the rate of one degree per second, inasmuch as a line one foot long at 57½ feet distance subtends an angle of one degree. Now if the eye be removed from such an object to a distance of 115 feet, the apparent motion will be half a degree, or thirty minutes per second; and if it be removed to thirty times that distance, the apparent motion will be thirty times slower. Or if, on the other hand, the eye be brought nearer to the object, the apparent motion will be accelerated in exactly the same proportion as the distance of the eye is diminished.

A cannon-ball moving at 1,000 miles an hour transversely to the line of vision, and at a distance of fifty yards from the eye, will be invisible, since it will not remain a sufficient time in any one position to produce perception. The moon, however, moving with more than double the velocity of the cannon-ball, being at a distance of 240,000 miles, has an apparent motion so slow as to be imperceptible to the unassisted eye.

SECTION V.

OPTICAL INSTRUMENTS.

Describe the portable camera obscura.

717. The portable camera obscura, such as is ordinarily used for photographic purposes, consists of a pair of achromatic double convex lenses, set in a brass mounting (see Fig. 291), into a box consisting of two parts, one of which slides within the other. The total length of the box is adjusted to suit the focal distance of the lens. In the back of the box, which can be opened, there is a square piece of ground glass which receives the images of the objects to which the lens is directed, and by sliding the movable part of the box in or out, the ground glass can be brought to the precise focus. The interior of the box is blackened all over to extinguish any stray light.

Fig. 291.

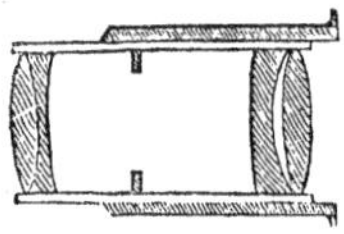

The appearance of the camera as described is represented by Fig. 292.

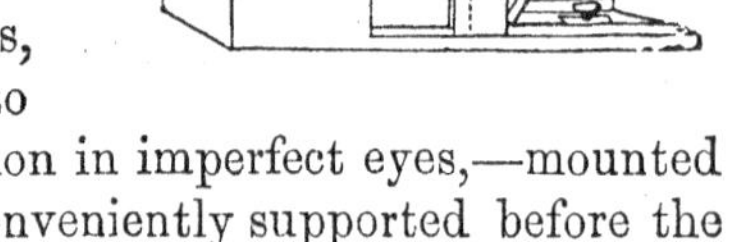

Fig. 292.

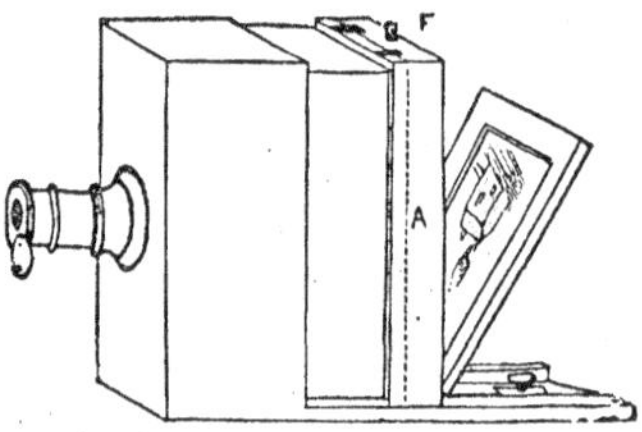

What are Spectacles?

718. Spectacles consist of two glass or crystal lenses, of such a character as to remedy the defects of vision in imperfect eyes,—mounted in a frame so as to be conveniently supported before the eyes.

What are the two varieties of spectacles?

Spectacles are of two kinds, namely those with convex glasses, which magnify objects, or bring their images nearer to the eyes; and those with concave glasses, which diminish the apparent size of objects, or extend the limits of distinct vision.

Some persons, in order to protect the eye from excessive light, use blue glasses as spectacles; they are, however, more mischievous than useful, since they absorb different parts of the spectrum unequally, and transmit the violet and blue rays.

What is a Microscope?

719. A Microscope is any instrument which magnifies the images of minute objects, and enables us to see them with greater distinctness. This result is produced by enlarging the angle of vision under

which the object is seen—since the apparent magnitude of every body increases or diminishes with the size of this angle.

Microscopes are of two kinds—simple and compound.

What are the two varieties of microscopes?

In the simple microscope, the object under examination is viewed directly, either by a simple or compound converging lens.

In the compound microscope, an optical image of the object, produced upon an enlarged scale, is thus viewed.

The simple microscope is generally a simple convex lens, in the focus of which the object to be examined is placed. Little spheres of glass, formed by melting glass threads in the flame of a candle, form very powerful microscopes.

FIG. 293.

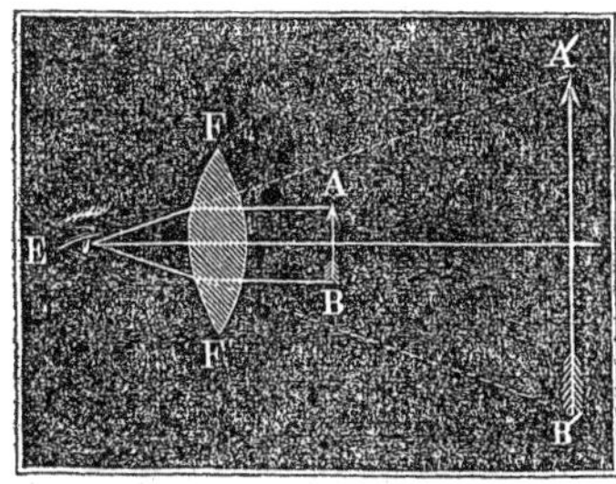

Fig. 293 represents the magnifying principle of the microscope. An eye at E would see the arrow A B, under the visual angle A E B; but when the lens, F F′ is interposed, it is seen under the visual angle at A′ E B′, and hence it appears much enlarged, as shown in the image A′ B′.

Fig. 294 represents the most improved form of mounting a simple microscope. A horizontal support, capable of being elevated or depressed by means of a screw and ratch-work, D, sustains a double-convex lens, A. The object to be viewed is placed upon a piece of glass, C, upon a standard, B, immediately below the lens. As it is desirable that the object to be magnified should be strongly illuminated, a concave mirror of glass, M, is placed at the base of the instrument, inclined at such an angle as to reflect the rays of light which fall upon it directly upon the object.

FIG. 294.

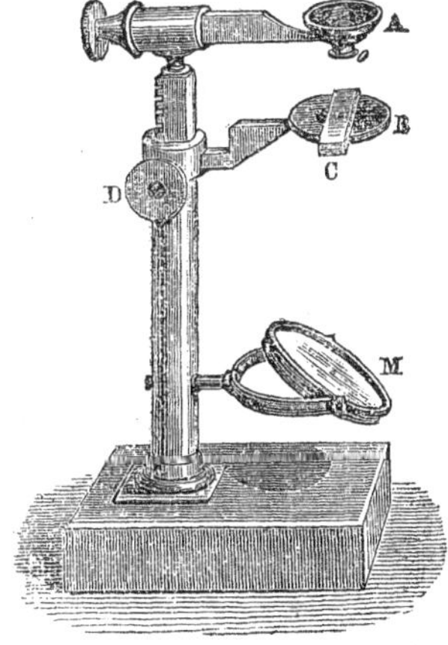

What is the construction of the Compound Microscope?

720. The Compound Microscope, in its most

simple form, consists of two lenses, so arranged that the second lens magnifies the image formed by the first lens, or simple microscope. In this way the image of the object is examined by the eye, and not the object itself.

How are the lenses of a compound microscope designated?

The first of these lenses is called the object-glass, or objective, since it is always directed immediately to the object, which is placed very near it; and the latter the eye-glass, or eye-piece, inasmuch as the eye of the observer is applied to it to view the magnified image of the object.

FIG. 295.

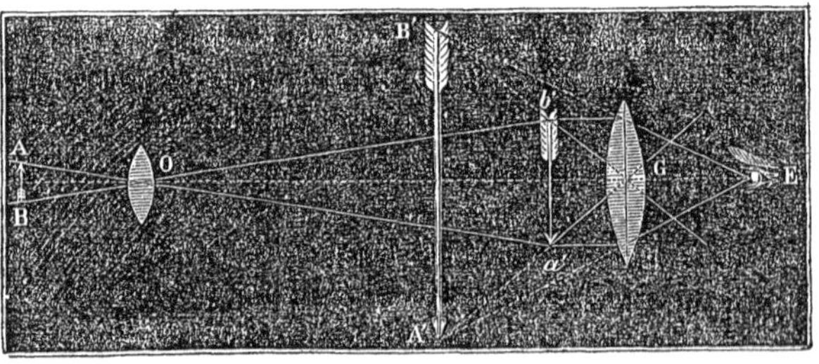

Fig. 295 illustrates the magnifying principle of the compound microscope. O represents the object-glass placed near the object to be viewed, A B, and G, the eye-glass placed near the eye of the observer, E. The object-glass, O, presents a magnified and inverted image, *a b*, of the object at the focus of the eye-glass, G. The image thus formed, by means of the second lens or eye-glass, G, is magnified and brought to the eye at E, so as to appear under the enlarged visual angle, A′ E B′. If we suppose the object-glass, O, to have a magnifying power of 25—that is, if the image *a b* equals 25 A B, and the eye-glass, G, to have a magnifying power of 4—then the total magnifying power of the microscope will be 4 times 25, or 100; that is to say, the image will appear 100 times the size of the object.

Fig. 296 represents the most approved form of mounting the lenses which compose a compound microscope. The tube, A, which contains in its upper part the eye-glass, slides into another tube, B, in the bottom of which the object-glass is fixed; this last tube also moves up and down in the stand, C, and in this way the lenses in the tubes may be adjusted to the proper distance from each other and the object. M is a mirror for reflecting light upon the object, and S a support on which the object to be examined is placed

Fig. 296.

What is a Telescope?

721. A Telescope is any instrument which magnifies and renders visible to the eye the images of distant objects. This result is effected in the same manner as in the microscope, viz., by enlarging the visual angle under which the objects are seen.

How many kinds of telescopes are there?

Telescopes are of two kinds, refracting telescopes and reflecting telescopes; the principle of construction in both being the same as that of the compound microscope.

What is a Refracting Telescope?

722. The Refracting Telescope consists essentially of two convex lenses, the object-glass and the eye-glass. An inverted image of an object, as a star, is produced by the object-glass, and magnified by the eye-glass.

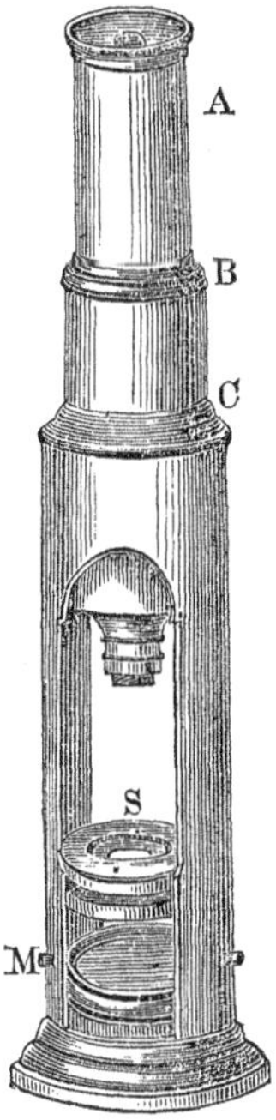

Fig. 297 represents the principle of construction of the astronomical refracting telescope. O is an object-glass placed at the end of a tube, which collects the rays proceeding from a distant object and forms an inverted image of the same at *o o′*, in the focus of the eye-glass, G. By this the image is magnified and viewed by the eye at E.

Fig. 297.

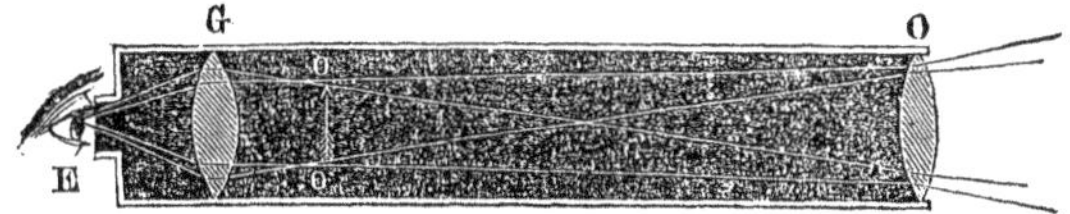

What is an Equatorial Telescope?

723. When a telescope is mounted on an axis inclined to the latitude of a place, so that it can follow a star, or planet, in its diurnal revolution, by a single motion, it is called an Equatorial Telescope.

Such an instrument is generally moved by clock-work, and is accurately counterbalanced by an arrangement of weights. A small telescope called the finder, is attached near the eye end of the large one; this is so adjusted that when the object is seen through it, it appears in the field of the large telescope, thus saving much trouble in directing the instrument toward any particular object.

The mounting and attachments of an equatorial telescope are represented in Fig. 298.

FIG. 298.

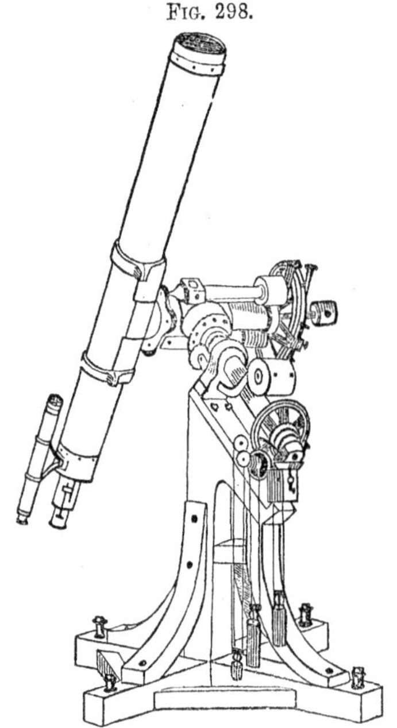

What is a Spy-glass?

724. A spy-glass, or terrestrial telescope, differs from an astronomical telescope only in an adjustment of lenses, which enables the observer to see the images of objects erect instead of inverted. This is effected by the addition of two lenses placed between the eye and the image.

The arrangement of the lenses, and the course of the rays of light, in a common spy-glass, are represented in Fig. 299. O is the object-glass, and C L M the eye-glasses, placed at distances from each other equal to double their focal length. The progress of the rays through the object-glass, O, and the first eye-glass, C, is the same as in the astronomical telescope, and an inverted image is formed; but the second lens, L, reverses the image, which is viewed therefore, in an erect position by the last eye-glass, M.

FIG. 299.

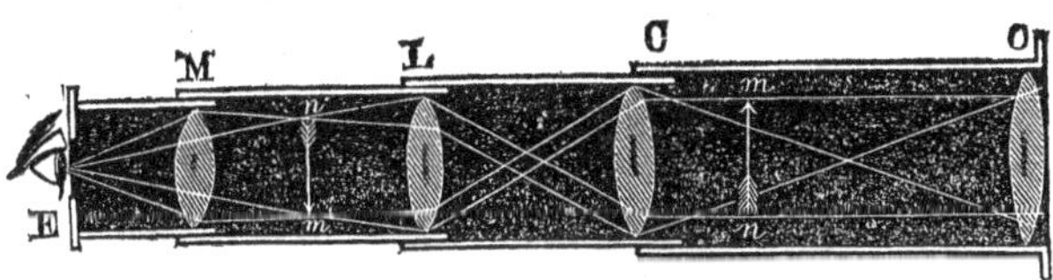

What is the construction of the Opera-glass?

725. The common opera-glass, also called the Galilean telescope from Galileo, its inventor, consists of a single convex object-glass and a concave eye-glass.

FIG. 300.

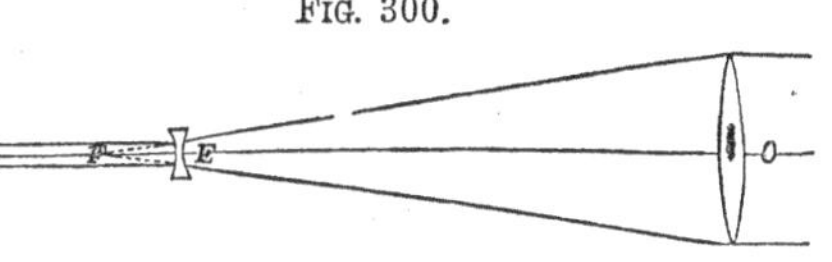

Fig. 300 represents the construction of this form of telescope. O is a single convex object-glass, in the focus of which an inverted image of the object would be naturally formed, were it not for the interposition of the double-concave lens, E. This receiving the converging rays of light, causes them to diverge and enter the eye parallel, and form an erect image.

What is a Reflecting Telescope?

726. A Reflecting Telescope consists essentially of a concave mirror, the image in which is magnified by means of a lens. The mirror employed in reflecting telescopes is made of polished metal, and is termed a *speculum*.

The manner in which the rays of light falling upon the concave speculum of a reflecting telescope are caused to converge to a focus is clearly shown in Fig. 301. The image formed at this focus is viewed through a double-convex lens.

FIG. 301.

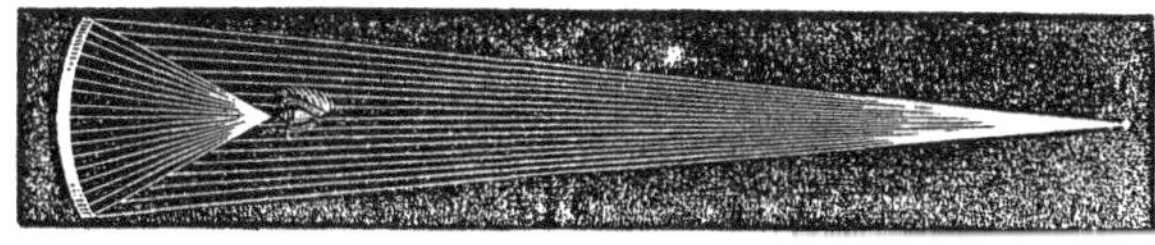

Fig. 302 represents one of the earliest forms of the reflecting telescope, called from its inventor, Mr. Gregory, the "Gregorian Telescope." It consists of a concave metallic speculum, A B, with a hole in its center, and a convex eye-glass, E, the whole being fitted into a tube. An inverted image, $n'\ m'$, of a distant object is formed by the speculum, A B; this image is again

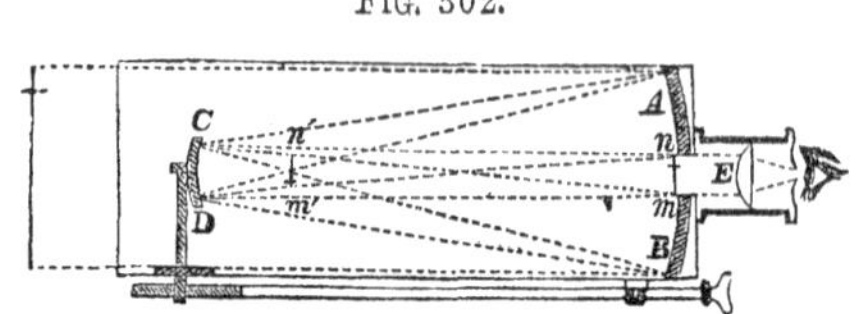

FIG. 302.

reflected by a small mirror, C D, and forms an erect image at *n m*, which is magnified by the lens, E, when observed by the eye.

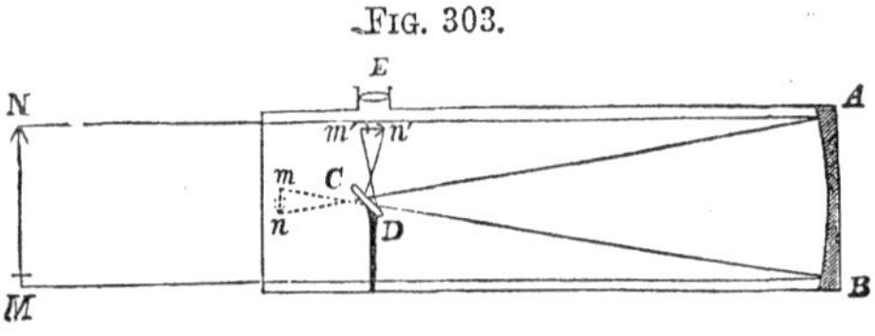

FIG. 303.

Another form of the reflecting telescope, called the Newtonian, is represented in Fig. 303. It consists of a large concave speculum, A B, set in one end of a tube, and a small plane mirror, C D, placed obliquely to the axis of the tube. The image of a distant object formed by the speculum, A B is reflected by the mirror, C D, to a point, *m′ n′*, on the side of the tube, and is there viewed through an eye-glass, E.

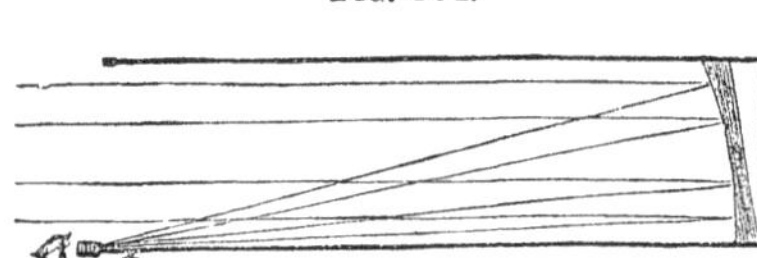
FIG. 304.

Large reflecting telescopes, at the present day, are so constructed as to dispense with the small mirror. This is accomplished by slightly inclining the large speculum, so as to throw the image on one side where it is viewed by an eye-glass, as is represented in Fig. 304.

FIG. 305.

The largest telescope ever constructed is that made by Lord Rosse. This instrument, which is a reflecting telescope, is located at Parsonstown, in Ireland. Its external appearance and method of mounting is represented in Fig. 305. The diameter of the speculum is 6 feet, and its weight about 4 tons. The tube in which it is placed is of wood hooped with iron, 52 feet in length, and 7 feet in diameter. It is counterpoised in every direction, and moves between two walls, 24 feet distant, 72 feet long, and 48 feet high. The observer stands on a platform which rises or falls, or at great elevation upon sliding galleries which draw out from the wall.

This telescope commands an immense field of vision, and it is said that objects as small as 100 yards' cube, can be distinctly observed by it in the moon at a distance of 240,000 miles.*

What is a Magic Lantern?

727. The Magic Lantern is an optical instrument adapted for exhibiting pictures painted on glass in transparent colors, on a large scale, by means of magnifying lenses.

FIG. 306.

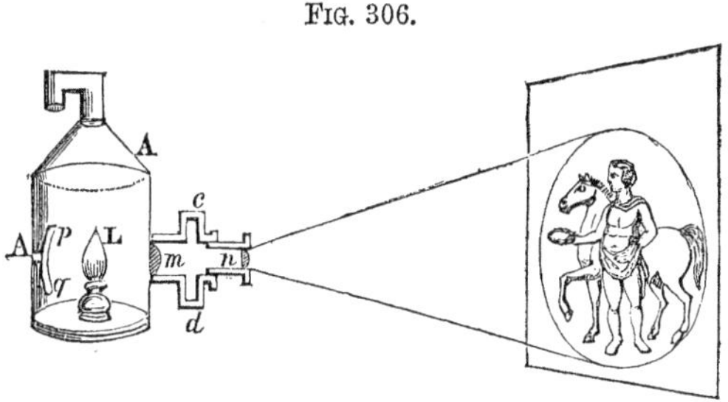

It consists of a metallic box, or lantern, A A′, Fig. 306, containing a lamp, L, behind which is placed a metallic concave mirror, *p q*. In front of the lamp are two lenses, fixed in a tube projecting from the side of the lantern, one of which, *m*, is called the illuminator, and the other the magnifier. The objects to be exhibited are painted on thin plates of glass, which are introduced by a narrow opening in the tube, *c d*, between the two lenses. The mirror and the first lens, *m*, serve to illuminate the painting in a high degree, for the lamp being placed in their foci, they throw a brilliant light upon it, and the magnifying lens, *n*, which can slide in its tube a little backward and forward, is placed in such a position as to throw a highly magnified image of the drawing upon a screen, several feet off, the precise focal distance being adjusted by sliding the lens. The further the lantern is withdrawn from the

* By the aid of this mighty instrument, "one of the most wonderful contributions of art and science the world has yet seen," what astronomers have before called nebula, on account of their cloud-like appearance, have been discovered to be stars, or suns, analogous, in all probability, in constitution, to our own sun. In the constellations Andromeda and the sword-hilt of Orion, both of which are visible to the naked eye, these cloud-like patches have been seen as clusters of stars.

screen, the larger the image will appear; but when the distance is considerable the image becomes indistinct.

What are Dissolving Views?

728. The beautiful optical combinations known as Dissolving Views are produced by means of two magic lanterns of equal power, so placed as to throw pictures of precisely equal magnitude on the same part of the same screen. By gradually closing the aperture of one lantern and opening that of the other, a picture formed by the first may seem to be dissolved away and changed into another.

Thus, if the picture produced by one lantern represents a day landscape, and the picture produced by the other the same landscape by night, the one may be changed into the other so gradually as to imitate with great exactness the appearance of approaching night.

What is a Solar Microscope?

729. The Solar Microscope is an optical instrument constructed on the principle of the magic lantern, but the light which illuminates the object is supplied by the sun instead of a lamp.

This result is effected by admitting the rays of the sun into a darkened room, through a lens placed in an aperture in a window shutter, the rays being received by a plane mirror fixed obliquely, outside the shutter, and thrown horizontally on the lens. The object is placed between this lens and another smaller lens, as in the magic lantern; and the magnified image formed is received upon a screen. In Fig. 307, which represents the construction of the solar microscope, C is a plane mirror, A the illuminating lens, and B the magnifying lens. The objects to be magnified are placed between the lenses A and B. In consequence of the superior illumination of the object by the rays of the sun, it will bear to be magnified much more highly than with the lantern. Hence this form of microscope is often employed to represent, on a very enlarged scale, various minute natural objects, such as animalculæ existing in various liquids, crystallization of various salts, and the structure of vegetable substances.

FIG. 307.

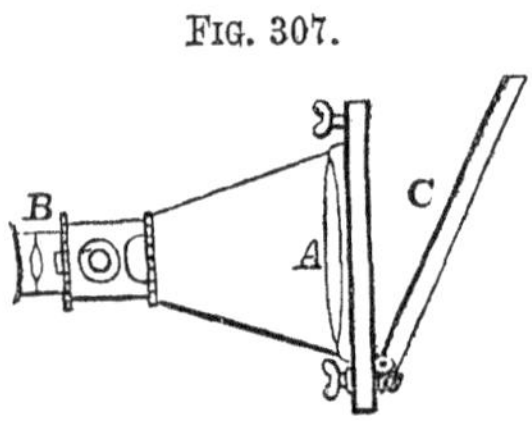

CHAPTER XV.

ELECTRICITY.

What is Electricity?

730. ELECTRICITY is one of those subtle agents without weight, or form, that appear to be diffused through all nature, existing in all substances without affecting their volume or their temperature, or giving any indication of its presence when in a latent, or ordinary state. When, however, it is liberated from this repose, it is capable of producing the most sudden and destructive effects, or of exerting powerful influences by a quiet and long-continued action.

How may electricity be excited?

731. Electricity may be excited, or called into activity by mechanical action, by chemical action, by heat, and by magnetic influence.

We do not know any reason why the means above enumerated should develop electricity from its latent condition, neither do we know whether electricity is a material substance, a property of matter, or the vibration of an ether. The general opinion at the present day is that electricity, like light and heat, is the result of vibrations of an ether pervading all space.

How is electricity most easily excited?

732. The most ordinary and the easiest way of exciting electricity is by mechanical action —by friction.

How does electricity excited by friction manifest itself?

If we rub a glass rod, or a piece of sealing-wax, or resin, or amber, with a dry woolen, or silk substance, these substances will immediately acquire the property of attracting light bodies, such as bits of paper, silk, gold-leaf, balls of pith, etc.

This attractive force is so great, that even at the distance of more than a foot, light substances are drawn toward the attracting body. The cause of this attraction is called electricity.

Thales, one of the seven wise men of Greece, noticed and recorded the fact more than two thousand years ago, that amber when rubbed would attract light bodies; and the name *electricity*, used to designate such phenomena has been derived from the Greek word ηλεκτρον, electron, signifying *amber.*

What other effects beside attraction are noticed in exciting electricity by friction?

If the friction of the glass, wax, amber, etc., is vigorous, small streams of light will be seen, a crackling noise heard, and sometimes a remarkable odor will be perceived.

When is a body said to be electrified?

733. When, by friction or other means, electricity is developed in a body, it is said to be electrified, or electrically excited.

What is electric attraction?

The tendency which an electrified body has to move toward other bodies, or of other bodies toward it, is ascribed to a force called electric attraction.

What is electric repulsion?

Every electrified body, in addition to its attractive force, manifests also a repulsive force. This is proved by the fact that light substances, after touching an electrified body, recede from it just as actively as they approached it before contact. Such action is ascribed to a force called electric repulsion.

Fig. 308.

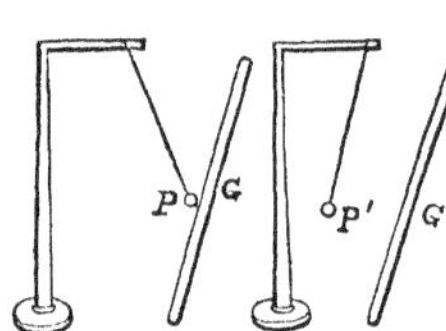

Thus, if we take a dry glass rod, rub it well with silk, and present it to a light pith ball, or feather, P, suspended from a support by a silk thread, the ball or feather will be attracted toward the glass, as seen at G, Fig. 308. After it has adhered to it a moment, it will fly off, or be repelled, as P' from G'.

The same thing will happen if sealing-wax be rubbed with dry flannel, and a like experiment made; but with this remarkable difference, that when the glass repels the ball, the sealing-wax attracts it, and when the wax repels, the glass will attract. Thus if we suspend a light pith ball, or feather, by a silk thread, as in Fig. 309, and present a stick of excited sealing-wax, S, on one side, and a tube of excited glass, G, on the other, the ball will commence vibrating like a pendulum from one to the other, being alternately attracted and repelled by each, the one attracting when the other repels; hence we conclude that the electricities excited in the glass and wax are different.

Fig. 309.

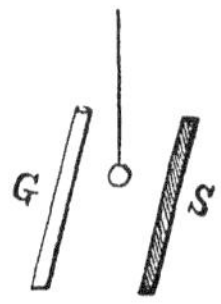

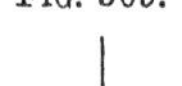

Is there more than one kind of electricity?

734. As the electricity developed by the friction of glass and other like substances is essentially different from that developed by

the friction of resin, wax, etc., it has been inferred that there are two kinds or states of electricity—the one called vitreous, because especially developed on glass, and the other resinous, because first noticed on resinous substances.

What is the general law of electrical attraction and repulsion?

The fundamental law which governs the relation of these two electricities to each other, and which constitutes the basis of this department of physical science, may be expressed as follows:—

Like electricities repel each other, unlike electricities attract each other.

Thus, if two substances are charged with vitreous electricity, they repel each other; two substances charged with resinous electricity also repel each other; but if one is charged with vitreous, and the other with resinous electricity, they attract each other.

When is a body non-electrified?

735. When a body holds its own natural quantity of electricity undisturbed, it is said to be non-electrified.

When an electrified body touches one non-electrified, what occurs?

When an electrified body touches one that is non-electrified, the electricity contained in the former is transferred in part to the latter.

Thus, on touching the end of a suspended silk thread with a piece of excited wax or glass, electricity will pass from the wax or glass into the silk, and render it electrified; and the silk will exhibit the effects of the electricity imparted to it, by moving toward any object that may be placed near it.

What two theories have been formed to account for electrical action?

736. Two theories, based upon the phenomena of attraction and repulsion, have been formed to account for the nature and origin of electricity. These two theories are known as the theory of two fluids, and the theory of the single fluid; or the theory of Du Fay, an eminent French electrician, and the theory of Dr. Franklin.

What is the theory of two fluids?

737. The theory of two fluids, or the theory of Du Fay, supposes that all bodies, in their natural state, are pervaded by an exceedingly thin subtle fluid, which is composed of two constituents,

or elements, viz., the vitreous and the resinous electricities. Each kind is supposed to repel its own particles, but attract the particles of the other kind.

When these two fluids pervade a body in equal quantities, they neutralize each other in virtue of their mutual attraction, and remain in repose; but when a body contains more of one than of the other, it exhibits vitreous or resinous electricity, as the case may be.

What is the theory of a single fluid?

738. The theory of a single fluid, or the theory propounded by Dr. Franklin, supposes the existence of a single subtile fluid, without weight, equally distributed throughout nature; every substance being so constituted as to retain a certain quantity, which is necessary to its physical condition.

When a substance pervaded by this single fluid is in its natural state or condition, it offers no evidence of the presence of electricity; but when its natural condition is disturbed it appears electrified. The difference between the electricity developed by glass and that by resin is explained by this theory, by supposing electrical excitation to arise from the difference in the relative quantities of this principle existing in the body rubbed and the rubber, or in their powers of receiving and retaining electricity. Thus one body becomes overcharged by having abstracted this principle from the other.

What are positive and negative electricities?

739. The two different conditions of electricity, which were called by Du Fay vitreous and resinous electricities, were designated by Dr. Franklin as positive and negative, or plus and minus. Thus a body which has an overplus of electricity is called positive, and one that has less than its natural quantity is called negative.

The theory of a single fluid has, until quite recently, been generally adopted by scientific men, and the terms positive and negative electricities are universally used in the place of vitreous and resinous. Within the last few years, however, some discoveries have been made which seem to indicate that the theory of two fluids is the one which approaches nearest to the truth.

What is Professor Faraday's theory of electricity?

In addition to these two theories respecting the nature of electricity, another has been proposed by Professor Faraday, of England. He considers electricity to be an attribute, or quality of matter, like what we conceive of the attraction of gravitation.*

* It is not easy to perfectly explain to a beginner the view which has been taken by Professor Faraday (who is at present the highest recognized authority on this subject) respecting the nature of electricity. The following statement, as given by a late writer (Robert Hunt), may be sufficiently comprehensive and clear: "Every atom of matter is

Is there any connection between light, heat, and electricity?

740. Light, heat, and electricity appear to have some properties in common, and each may be made, under certain circumstances, to produce or excite the other. All are so light, subtle, and diffusive, that it has been found impossible to recognize in them the ordinary characteristics of matter. Some suppose that light, heat, and electricity are all modifications of a common principle.

What are the electrical divisions of all substances?

741. Electricity exists in, or may be excited in all bodies. There are no exceptions to this rule, but electricity is developed in some bodies with great ease, and in others with great difficulty. All substances, therefore, have been divided into two classes, viz., Electrics, or those which can be easily excited, and Non-electrics, or those which are excited with difficulty. Such a division is, however, of little practical value in science, and at present is not generally recognized.

There is no certain test which will enable us to determine, previous to experiment, which of two bodies submitted to friction will produce positive, and which negative electricity. Of all known substances, a cat's fur is the most susceptible of positive, and sulphur of negative electricity. Between these extreme substances others might be so arranged, that any substance in the list being rubbed upon any other, that which holds the highest place will be positively electrified, and that which holds the lower place negatively electrified. For instance, smooth glass becomes positively electrified when rubbed with silk or flannel, but negatively electrified when excited by the back of a living cat. Sealing-wax becomes positive when rubbed with the metals, but negative by any thing else.

Can one electricity be excited without setting free the other?

In no case can electricity of one kind be excited without setting free a corresponding amount of electricity of the other kind ; hence, when electricity is excited by friction, the rubber always exhibits the one, and the electric, or body rubbed, the other.

What are conductors and non-conductors of electricity?

742. Bodies differ greatly in the freedom with which they allow electricity to pass over or through them. Those substances which

regarded as existing by virtue of certain properties or powers, these being merely peculiar affections, which may be regarded as being of a similar nature to vibrations. It is assumed that the electric state is but a mode or form of one of these affections. One particle of matter, having received this form of disturbance, communicates it to all contiguous particles—that is, those which are next to it, although not in contact—and this communication of force takes place more or less readily, the communicating particles assuming a polarized state—which may be explained as a state presenting two dissimilar extremities. When the communication is slow, the polarized state is highest, and the body is said to be an insulator: insulation being the result. If the particles communicate their condition readily, they are termed conductors: conduction is the result. The phenomena of induction, or the production of like effects in contiguous bodies, is, therefore, according to this view, but something analogous to the communication of tremors, or vibrations."

facilitate its passage are called conductors; those that retard, or almost prevent it, are called non-conductors.

No substance can *entirely* prevent the passage of electricity, nor is there any which does not oppose some resistance to its passage.

What substances are good conductors of electricity? Of all bodies, the metals are the most perfect conductors of electricity; charcoal, the earth, water, moist air, most liquids, except oils, and the human body, are also good conductors of electricity.

What is the velocity of electricity? 743. The velocity with which electricity passes through good conductors is so great, that the most rapid motion produced by art appears to be actual rest when compared to it. Some authorities have estimated that electricity will pass through copper wire at the rate of two hundred and eighty-eight thousand miles in a second of time—a velocity greater than that of light. The results obtained, however, by the United States Coast Survey, with iron wire, show a velocity of from 15,000 to 20,000 miles per second.

What substances are non-conductors of electricity? Gum shellac and gutta percha are the most perfect non-conductors of electricity; sulphur, sealing-wax, resin, and all resinous bodies, glass, silk, feathers, hair, dry wool, dry air, and baked wood are also non-conductors.

Electricity always passes by preference over the best conductors.

Thus, if a metallic chain or wire is held in the hand, one end touching the ground and the other brought into contact with an electrified body, no part of the electricity will pass into the hand, the chain being a better conductor than the flesh of the hand. But if, while one end of the chain is in contact with the conductor, the other be separated from the ground, then the electricity will pass into the hand, and will be rendered sensible by a convulsive shock.

When is a body insulated? 744. When a conductor of electricity is surrounded on all sides by non-conducting substances, it is said to be *insulated;* and the non-conducting substances which surround it are called *insulators.*

When is a body said to be charged with electricity?

When a conducting body is insulated, it retains upon its surface the electricity communicated to it, and in this condition it is said to be charged with electricity.

A conductor of electricity can only remain electric as long as it is insulated, that is, surrounded by perfect non-conductors. The air is an insulator, since, if it were not so, electricity would be instantly withdrawn by the atmosphere from electrified substances. Water and steam are good conductors, consequently, when the atmosphere is damp, the electricity will soon be lost, which, in a dry condition of the air, would have adhered to an insulated conductor for a long period of time.

Thus a globe of metal supported on a glass pillar, or suspended by a silken cord, and charged with electricity, will retain the charge. If, on the contrary, it were supported on a metallic pillar, or suspended by a metallic wire, the electricity would immediately pass away over the metallic surface and escape.

In the experiments made with the pith balls (§ 733, Fig. 308), the silk thread by which they were suspended acts as an insulator, and the electricity with which they become charged is not able to escape.

Does electricity accumulate upon the surface or the interior of bodies?

745. When electricity is communicated to a conducting body it resides merely upon the surface, and does not penetrate to any depth within.

Fig. 310.

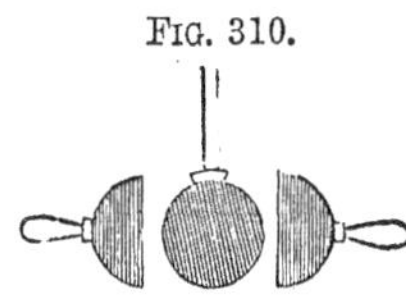

Thus, if a solid globe of metal suspended by a silken thread, or supported upon an insulated glass pillar, be highly electrified, and two thin hollow caps of tin-foil or gilt paper, furnished with insulating handles, as is represented in Fig. 310, be applied to it, and then withdrawn, it will be found that the electricity has been completely taken off the sphere by means of the caps.

An insulated hollow ball, however thin its substance, will contain a charge of electricity equal to that of a solid ball of the same size, all the electricity in both cases being distributed upon the surface alone.

How does the form of a body influence its electrical condition?

In the case of a spherical body charged with electricity, the distribution is equal all over the surface; but when the body to which the electricity is communicated is larger in one direction than the other, the electricity is chiefly found at its longer extremities, and the quantity at any point of its surface is proportional to its distance from the center.

The shape of a body also exercises great influence in retaining electricity: it is more easily retained by a sphere than by a spheroid or cylinder; but it readily escapes from a point, and a pointed object also receives it with the greatest facility.

What is the great reservoir of electricity?

746. The earth is considered as the great general reservoir of electricity.

When by means of a conducting substance a communication is established between a body containing an excess of electricity and the earth, the body will immediately lose its surplus quantity, which passes into the earth and is lost by diffusion.

What is electrical induction?

747. When a body charged with electricity of one kind is brought into proximity with other bodies, it is able to induce or excite in them, without coming in contact, an opposite electrical condition. This phenomenon is called Electrical Induction.

Explain the phenomena of induction.

This effect arises from the general law of electrical attraction and repulsion. A body in its natural condition contains equal quantities of positive and negative electricities, and when this is the case, the two neutralize each other, and remain in a state of equilibrium. But when a body charged with electricity is brought into proximity with a neutral body, disturbance immediately ensues. The electrified body, by its attractive and repulsive influence, separates the two electricities of the neutral body, repelling the one of the same kind as itself, and attracting the other, which is unlike, or opposite. Thus, if a body electrified positively be brought near a neutral body, the positive electricity of the neutral body will be repelled to the most remote part of its surface, but the negative electricity will be attracted to the side which is nearest the disturbing body. Between these two regions a neutral line will separate those points of the body over which the two opposite fluids are respectively distributed.

FIG. 311.

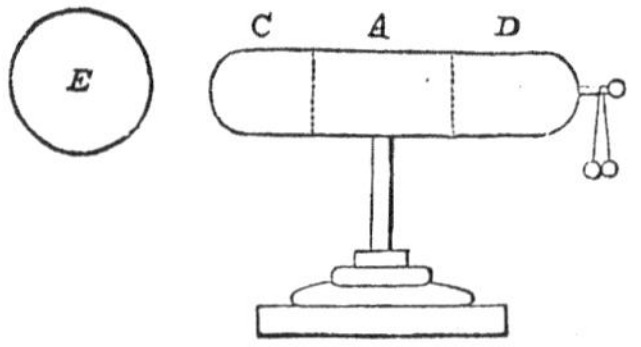

Let C A D, Fig. 311, be a metallic cylinder placed upon an insulating support, with two pith balls suspended at one end, as at D. If now an electrified body, E, be brought near to one end of the cylinder, the balls at the other extremity will immediately diverge from one another, showing the presence of free electricity. This does not arise from a transfer of any of the electric fluid from E to C, for upon withdrawing the electrified body, E, the balls will fall together, and appear unelectrified as before; but the electricity in E decomposes by its proximity the combination of the two electricities in the cylinder, C A D, attracting the kind opposite to itself toward the end nearest to it, and repelling the same kind to the further end. The middle part of the cylinder, A, which intervenes between the two extremities, will remain neutral, and exhibit neither positive nor negative electricity.

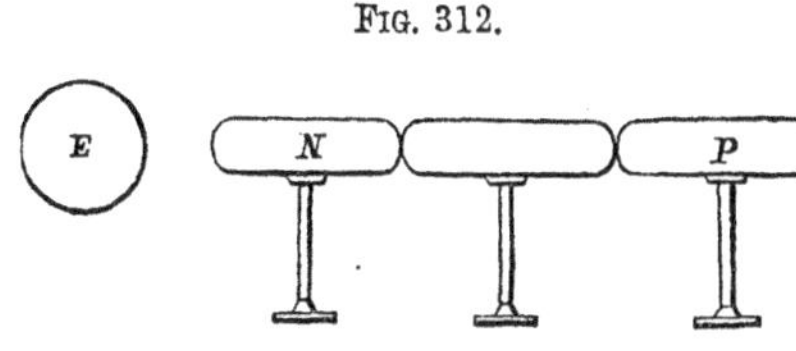

Fig. 312.

If three cylinders are placed in a row, touching one another, as in Fig. 312, and a positively electrified body, E, be brought in proximity to one extremity, the electricities of the cylinders will be decomposed, the negative being accumulated in N, and the positive repelled to P. If in this condition the cylinder P be first removed, and then the electrified body, the separate electricities will not be able to unite, as in the former experiment, but N will remain negatively, and P positively electrified.

Explain the reason why an electrified surface attracts a neutral, or unelectrified body.

These experiments explain why an electrified surface attracts a neutral, or unelectrified body, such as a pith ball. It is not that electricity causes attractions between excited and unexcited bodies, the same as between bodies oppositely excited; but that the pith ball is first rendered opposite by induction, and attracted in consequence of this opposition. A pith ball at a few inches distance from an electrified surface, is charged with electricity by induction; and the kind being contrary to that of the surface, attraction ensues; when the two touch, they become of the same kind by conduction.

A person may also receive an electric shock by induction. Thus, if a person stand close to a large conductor strongly charged with electricity, he will be sensible of a shock when this conductor is suddenly discharged. This shock is produced by the sudden recomposition of the fluids in the body of the person, decomposed by the previous inductive action of the conductor.

What is an electrical machine?

748. An electrical machine is an apparatus, by means of which electricity is developed and accumulated, in a convenient manner for the purposes of experiment.

Of what essential parts does an electrical machine consist?

All electrical machines consist of three principal parts, the rubber, the body on whose surface the electric fluid is evolved, and one or more insulated conductors, to which this electricity is transferred, and on which it is accumulated.

Describe the two varieties of electrical machines in common use.

Electrical machines are of two kinds, the plate and cylinder machines. They derive their names from the shape of the glass employed to yield the electricity.

FIG. 313.

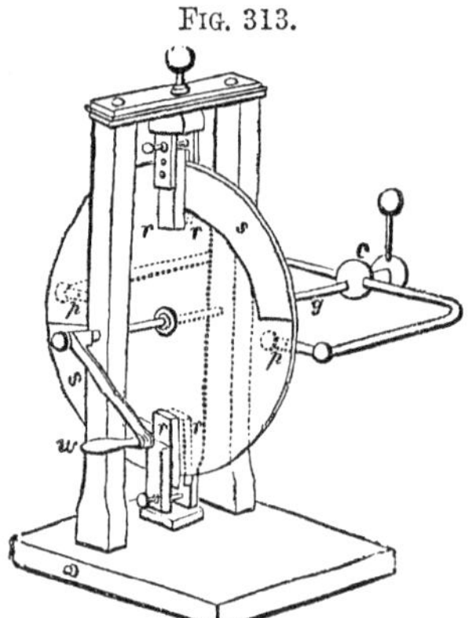

The plate electrical machine, which is represented in Fig. 313, consists of a large circular plate of glass mounted upon a metallic axis, and supported upon pillars fixed to a secure base, so that the plate can, by means of a handle, *w*, be turned with ease. Upon the supports of the glass, and fixed so as to press easily but uniformly on the plate, are four rubbers, marked *r r r r* in the figure; and flaps of silk, *s s*, oiled on one side, are attached to these, and secured to fixed supports by several silk cords. When the machine is put in motion, these flaps of silk are drawn tightly against the glass, and thus the friction is increased, and electricity excited. The points *p p* collect the electricity from the glass as it revolves, and convey it to the prime conductor, *c*, which is insulated and supported by the glass rod, *g*.

The cylinder electrical machine represented by Fig. 314, consists of a glass cylinder, so arranged that it can be turned on its axis by a crank, and supported by two uprights of wood, dried and varnished. F S indicates the position and arrangement of the rubber and silk, and Y that of the prime conductor. The principle of the construction of the cylinder machine is, in every respect, the same as that of the plate machine.

FIG. 314.

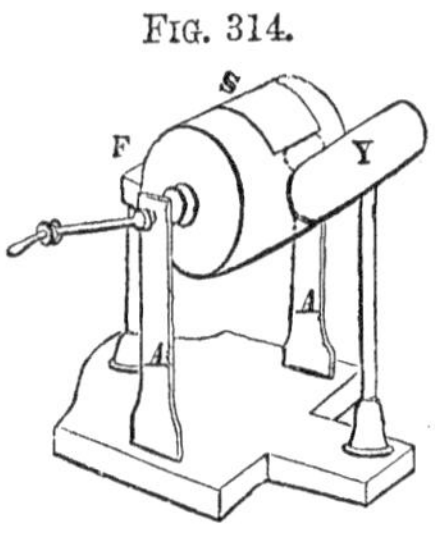

What is the construction of the rubber?

The rubber of an electrical machine consists of a cushion stuffed with hair, and covered with leather, or some substance which readily generates electricity by friction. The efficiency of the machine is greatly increased by covering the cushion with an amalgam, or mixture of mercury, tin, and zinc.*

In the ordinary working of the machine, the rubber is connected by a chain with the ground, from whence the supply of electricity is derived.

* The best composition of the amalgam is two parts, by weight, of zinc, one of tin, and six of mercury. The mercury is added to the mixture of the zinc and tin when in a fluid state, and the whole is then shaken in a wooden box until it is cold; it is then reduced to a powder, and mixed with a sufficient quantity of lard to reduce it to the consistency of paste. A thin coating of this paste is spread over the cushion; but before this is done, all darts of the machine should be carefully cleaned and warmed.

What is the conductor of an electrical machine?

The receiver of electricity from an electrical machine is called the prime conductor. It usually consists of a thin brass cylinder, or a brass rod, mounted on a glass pillar, or some other insulating material.

To put the electrical machine in good order, every part must be dry and clean, because dust or moisture would, by their conducting power, diffuse the electric fluid as fast as accumulated. As a general rule, it is highly essential that the atmosphere should be in a dry state when electrical experiments are made, as the conducting property of moist air prevents the collection of a sufficient amount of electricity for the production of striking effects. In the winter, the experiments succeed best when performed in the vicinity of a fire; and it is advisable to place the apparatus in front of the fire for some time before it is employed.

Explain the method in which an electrical machine develops electricity?

Electricity is developed by the action of an electrical machine in essentially the same manner as it is in a simple glass tube by friction. When the glass cylinder or plate is turned round by the handle, the friction between the glass and the rubber excites electricity; positive electricity being developed upon the glass, and negative upon the rubber. When the points of the prime conductor are presented to the revolving glass plate or cylinder, the positive electricity is immediately transferred to it, and it emits sparks to any conducting substance brought near. The electricity thus abundantly excited is supplied from the earth to the rubber (by means of a chain extending to the ground), and the rubber is continually having its supply drawn from it by the force called into action by friction with the glass. That the electricity is derived from this source is evident from the fact that but a small quantity of electricity can be excited when the metallic connection between the rubber and the ground is removed. For this reason the chain must always be attached to the rubber when it is desired to develop positive electricity, and to the prime conductor when negative electricity is required.

According to the theory of a single fluid, the excitement of electricity is as follows:—the friction of the glass and silk, by disturbing the electrical equilibrium deprives the rubber of its natural quantity of electricity, and it is therefore left in a negative state, unless a fresh quantity be continually drawn from the earth to supply its place. The surplus quantity is collected on the prime conductor, which thereby becomes charged with positive electricity. On the hypothesis of two electric fluids, the same frictional action causes the separation of the vitreous from the resinous electricity in the rubber, which therefore remains resinously charged, unless there be a connection with the earth to restore the proportion of vitreous electricity of which the rubber has been deprived.

Can a steam-boiler be used as an electrical machine?

Various other arrangements have been devised for the production and accumulation of electricity. High-pressure steam escaping from a steam-boiler carries with it minute particles of water, and the friction of these against the surface of the jet from which the steam issues produces electricity in great abundance. A steam-boiler, properly arranged and insulated, therefore constitutes a most powerful electrical machine; and by means of an apparatus of this character, constructed some time since in London, flashes of electricity were caused to emanate from the prime conductor more than 22 inches in length.

What is an Insulating Stool?

749. The Insulating Stool, which is a usual appendage to an electrical machine, consists of a board of hard-baked wood, supported on glass legs covered with varnish. (See Fig. 315.) It is useful for insulating any body charged with electricity; and a person standing upon such a stool, and in communication with a prime conductor, will become charged with electricity.

FIG. 315.

What are Discharging Rods?

Discharging Rods are brass rods terminating with balls, or with points, fixed to glass handles. With these rods electricity may be taken from a conductor without allowing the electrical charge to pass through the body of the operator. Their construction is represented in Fig. 316.

FIG. 316.

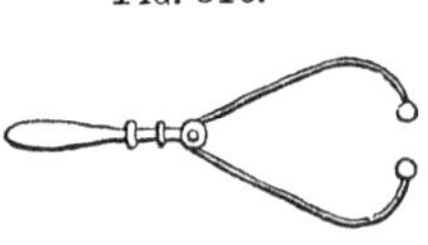

FIG. 317.

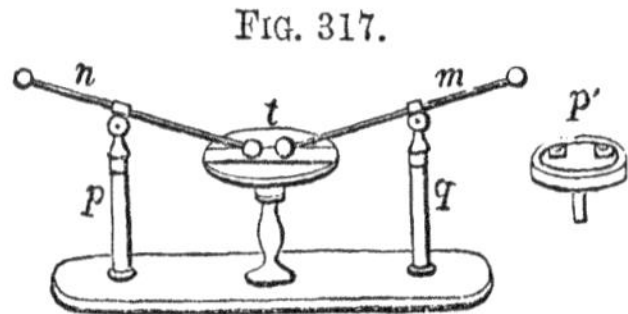

An instrument called the "Universal Discharger," used to convey strong charges of electricity through various substances, is represented by Fig. 317. It consists of two glass standards, through the top of which two metallic wires slide freely; these wires are pointed at the end, *t*, but have balls screwed upon them; the other ends are furnished with rings. The balls rest on a table of boxwood, into which a slip of ivory, or thick glass, is inlaid. Sometimes a press, *p'*, is substituted for the table, between which any substance necessary to be pressed, during the discharge, is held firm.

What is an Electrophorus?

750. An Electrophorus is a simple apparatus, in which a small charge of electricity may be generated by induction; and this, communicated successively to an insulated conductor, may produce a charge of indefinite amount.

Describe the action of the electrophorus.

It consists of a circular cake of resin (shell-lac), *r*, Fig. 318, laid upon a metallic plate; upon this cake, the surface of which has been negatively electrified by rubbing it with dry silk or fur, is placed a metallic cover, M, somewhat smaller in diameter, and furnished with a glass insulating handle, *h*. The negative electricity of the resin, by acting inductively upon the two electricities combined in the cover, separates them—the positive being attracted to the under surface, and the negative repelled to the upper, on touching the cover with the finger, all the negative

FIG. 318.

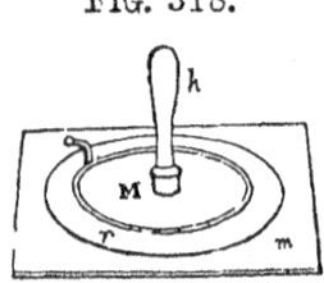

electricity will escape, and the positive electricity alone remains, which is combined with the negative electricity of the cake of resin, so long as the cover is in contact with it. If we now remove the cover by its insulating handle, the positive electricity, which was before held at the lower part of the cover by the inductive action of the resin, will become free, and may be imparted to any insulated conductor adapted to receive it. The same process may be repeated indefinitely, as the resinous cake loses none of its electricity, but simply acts by induction, and thus an insulated conductor may be charged to any extent.

What is an Electroscope? 751. An Electroscope is an instrument employed to indicate the presence of free electricity.

What is the construction of an electroscope? It usually consists of two light conducting bodies freely suspended, which in their natural state hang vertically and in contact. When electricity is imparted to them, they repel each other, and the amount of their divergence is proportioned to the quantity of electricity diffused on them.

The simplest form of the electroscope, called the "pith-ball electroscope," consists of two pith-balls suspended by silk threads. When an excited body is presented, the balls will be first attracted, but immediately acquiring the same degree of electricity as the exciting body, they repel each other. Another form of the pith-ball electroscope, represented at B, Fig. 319, consists of two pith-balls suspended by conducting threads within a glass jar, and connected with the brass cap, *m*. On touching the brass cap with an electrified body, the two balls being similarly electrified, will repel each other. C, Fig. 319, represents a more delicate electroscope; two slips of gold leaf, *g g′*, being substituted for the pith-balls. If an excited substance, *e*, be brought near the cap of brass, the leaves will instantly diverge. The best electrometers are carefully insulated, so that the electricity communicated to the balls or leaves may not be too soon dissipated.

FIG. 319.

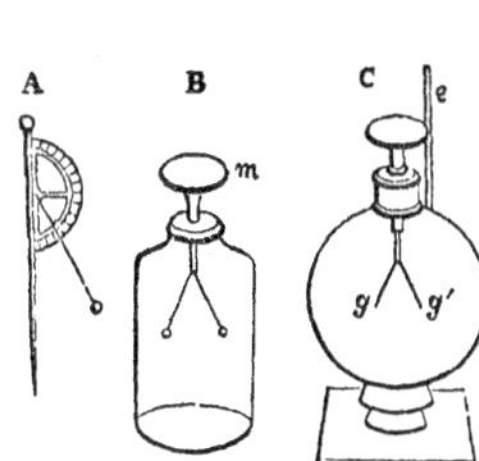

Electroscopes merely indicate the presence of an electrically excited body: they do not measure the quantity, either relatively or absolutely, of the electricity in action.

What is an Electrometer? 752. An Electrometer is an instrument for measuring the quantity of electricity.

The most simple form of the electrometer is represented at A, Fig. 319. It consists of a semicircle of varnished paper, or ivory, fixed upon a vertical rod. From the center of the semicircle a light pith-ball is suspended, and the number of degrees through which the ball is attracted or repelled by any body brought in proximity to it, indicates in a degree the active quantity of electricity present. No very accurate results, however, can be obtained with this apparatus; and for accurate investigation, instruments of more ingenious and complicated construction are used.

The electrometer usually employed for measuring with great accuracy small quantities of electricity, is that of Coulomb's, usually called the *Torsion Balance.*

Explain the construction of the Torsion Balance.

The construction of this instrument is as follows:—A needle, or stick of shell-lac, bearing upon one end a gilded pith-ball, is suspended by a fiber of silk within a glass vessel—the needle being so balanced, that it is free to turn horizontally around the point of suspension in every direction. When the pith-ball is electrified by induction, the repellent force causes the needle to turn round, and this produces a degree of torsion, or twist in the fiber which suspends it; and the tendency of the fiber to untwist, or return to its original position, measures the force which turns the needle. Within the glass vessel, which is cylindrical, a graduated circle is placed, which measures the angle through which the needle is deflected. In the cover of the vessel an aperture is made, through which the electrified body may be introduced, whose force it is desired to indicate and measure by the apparatus. Fig. 320 represents the construction and appearance of the torsion balance.

FIG. 320.

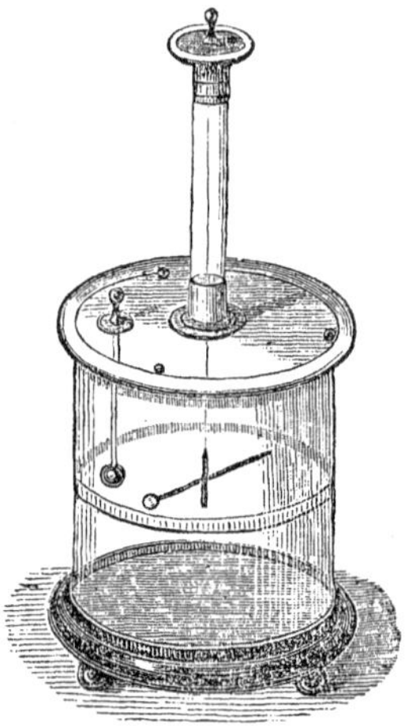

What important law of electricity has been proved by the torsion balance?

By means of the torsion balance, Coulomb proved that the law of electrical attraction and repulsion, as influenced by distance, is the same as the law of gravitation; that is, the force varies inversely as the square of the distance.

What is a Leyden Jar?

753. The Leyden Jar is a glass vessel used for the purpose of accumulating electricity derived from electrically excited surfaces.

Explain the action and construction of the Coated Pane.

The principle of the Leyden Jar may be best explained by describing what is called the "coated," or "fulminating pane." This consists of a glass plate, Fig. 321, *a*, having a square leaf of tin-foil, *b*, attached to each side. If the plate be laid upon a table, and a chain from the prime conductor of an electrical machine be brought in contact with the tin-foil upon one side, the plate will become charged—the upper side with positive, and the under with negative electricity.

FIG. 321.

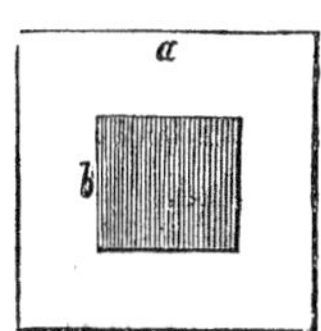

How may a coated pane produce an electric shock?

If two such conductors, as the plates of tin-foil attached to a pane of glass, be strongly charged with electricity in the manner described, and then, by means of the human body, be put in communication—which may be done by touching one plate with the fingers of one hand, and the other with the fingers of the other hand—the two electric fluids in rushing together, pass through the body, and produce the phenomenon known as the electric shock.

In what manner was the principle of the Leyden Jar first made known?

754. The Leyden Jar is constructed upon the same principle as the coated pane, and its discovery, accompanied with the first experience of the nervous commotion known as the electric shock, occurred in this way: In 1746, while some scientific gentlemen at Leyden, in Holland, were amusing themselves with electrical experiments, it occurred to one of them to charge a tumbler of water with electricity, and learn by experiment whether it would affect the taste. Accordingly, having fixed a metallic rod in the cork of a bottle filled with water, he presented it to the electrical machine for the purpose of electrifying the water, holding at the same time the bottle in his hand by its external surface, without touching the metallic rod by which the electricity was conducted to the water. The water, which is a conductor, received and retained the electricity, since the glass, a non-conductor, by which it was surrounded, prevented its escape. The presence of free electricity in the water, however, induced an opposite electricity on the outside of the glass, and when the operator attempted to remove the rod out of the bottle, he brought the two electricities into communication by means of his hand, and received, for the first time, a severe electric shock. Nothing could exceed the astonishment and consternation of the operator at this unexpected sensation, and in describing it in a letter immediately afterward to the French philosopher Reaumur, he declared that for the whole kingdom of France he would not repeat the experiment.

The experiment, however, was soon repeated in different parts of Europe, and the apparatus by which it was produced received a more convenient form, the water being replaced by some better conducting substances, as metal filings, for which tin-foil was afterward substituted.

Describe the construction of the Leyden jar.

The Leyden Jar, as usually constructed, consists of a glass jar, Fig. 322, having a wide mouth, and coated, externally and internally, to within two or three inches of the mouth, or to the line *a b*, with tin-foil. A wooden cover, well varnished, is fitted into the mouth of the jar, through which a stout brass wire, furnished with a ball, passes, having a chain or wire attached to its lower end, so as to be in contact with the inside coating.

FIG. 322.

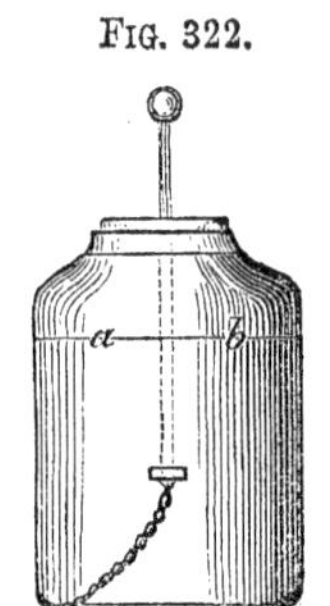

How is a Leyden jar charged?

A Leyden jar is charged by presenting the brass ball at the end of the rod of the jar to a prime conductor of an electrical machine in action, or to any other excited surface. To charge a jar strongly, it is necessary that the outside coating should be directly or indirectly connected with the ground.

How is a Leyden jar discharged?

A Leyden jar is discharged by effecting a communication between the outer and inner surfaces by means of a good conductor.

If, when we have charged the jar, we hold the exterior coating in one hand and touch the knob with the other, a spark is observed, and the peculiar sensation of the electric shock experienced.

Any number of persons can receive a shock at the same time by forming a chain by holding each other's hands—the first person in the circle touching the external coating of the jar, and the last the knob.

Where does the electricity of a Leyden jar reside?

When a Leyden jar is charged, the electricity resides wholly on the surface of the glass; the metallic coatings having no other effect than to conduct the electricity to the surface of the glass, and, when there, afford it a free passage from point to point.

The power of a Leyden jar will therefore depend upon its size, or extent of surface.

As very large jars are inconvenient and expensive, very strong charges of electricity are obtained by combining a number of jars together.

FIG. 323.

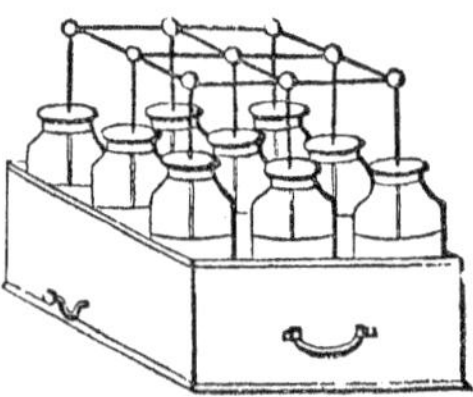

What is an Electrical Battery?

A combination of Leyden jars, so arranged that they may be all charged and discharged together, constitutes an Electricaı Battery. This may be effected by forming a connection between all the wires proceeding from the interiors of the jars, and also connecting all their exterior coatings.

Such an arrangement is represented by Fig. 323. The discharge of electricity from such a combination is accompanied by a loud report; and when the number of the jars is considerable, animals may be killed, metal wires be melted, and other effects produced analogous to those of lightning.

What experiments illustrate the attractive and repulsive forces of electricity?

755. By means of an electrical machine and the Leyden Jar, many interesting and amusing electrical experiments may be performed.

The phenomenon of the repulsion of substances similarly electrified, may be illustrated by means of a doll's head covered with long hair. When this is attached to the prime conductor of an electrical machine, the hairs stand erect, and give to the head a most exaggerated appearance of fright. See Fig. 324.

FIG. 324.

The same thing may be shown by placing a person on a stool with glass legs, so that he be perfectly insulated, and making him hold in his hand a brass rod, the other end of which touches the prime conductor; then on turning the machine, the hairs of the head will diverge in all directions.

If a small number of figures are cut out in paper, or carved out of pith, and an excited glass tube be held a few inches above them on a table, the figures will immediately commence dancing up and down, assuming a variety of droll positions. The experiment can be shown better by means of an electrical machine than with the excited tube, by suspending horizontally from the prime conductor a metal disc a few inches above a flat metal surface connected with the earth, on which the figures are placed. On working the machine, the figures will dance in a most amusing manner, being alternately attracted and repelled by each plate. See Fig. 325.

FIG. 325.

What is the experiment of the electrical bells?

The electrical bells, Fig. 326, which are rung by electric attraction and repulsion, are good illustrations of these forces. Where three bells are employed, the two outer bells A and B, are suspended by chains, but the central one and the two clappers hang from silken strings. The middle bell is connected with the earth by a chain or wire.

Upon working the machine, the outer bells become positively electrified, and the middle one, which is insulated from the prime conductor, becomes negative by induction. The little clappers between them are alternately attracted and repelled by the outer and inner bells, producing a constant ringing as long as the machine is in action.

Fig. 326.

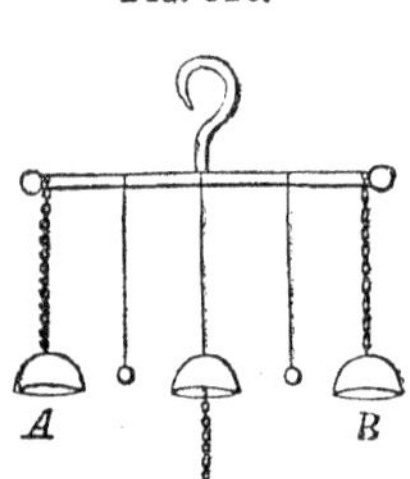

It was by attaching a set of bells of this kind to his lightning-conductor, that Dr. Franklin received notice, by their ringing, of the passage of a thunder-cloud over his apparatus.

Let a skein of linen thread be tied in a knot at each end, and let one end of it be attached to some part of the conductor of a machine. When the machine is worked the threads will become electrified, and will repel each other, so that the skein will swell out into a form resembling the meridians drawn upon a globe.

If we ignite the extremity of a stick of sealing-wax, and bring the melted wax near to the prime conductor of a machine, numerous fine filaments of wax will fly to the conductor, and will adhere to it, forming upon it a sort of network like wool. This is a simple case of electrical attraction. The experiment will succeed best if a small piece of wax is attached to the end of a metal rod.

What effect has electricity upon a conductor?

756. When a current of electricity passes through a good conductor of sufficient size to carry off the whole quantity of electricity easily, the conductor is not apparently affected by its passage; but if the conductor is too small, or too imperfect to transmit the electric fluid readily, very striking effects are produced—the conductor being not unfrequently shivered to pieces in an instant.

What experiments illustrate the mechanical effects of electricity?

The mechanical effects exerted by electricity in passing through imperfect conductors, may be illustrated by many simple experiments.

If we transmit a strong charge of electricity through water, the liquid will be scattered in every direction.

A rod of wood half an inch thick may be split by a strong charge from a Leyden jar, or battery, transmitted in the direction of its fibers.

If we place a piece of dry writing-paper upon the stand of a universal discharger, and then transmit a charge through it, the electricity, if sufficiently strong, will rupture the paper.

If we hold the flame of a candle to a metallic point projecting from the prime conductor of an electrical machine in action, the current of air caused by the issuing of a current of electricity from the point, will be sufficient to deflect the flame, and even blow it out.

How does electricity evolve heat? 757. The passage of electricity from one substance to another is generally attended with an evolution of heat, and a current of electricity passing over an imperfect conductor, raises its temperature.

The temperature of a good conductor of sufficient size to allow the electric fluid to pass freely, is not affected by the transmission of a current of electricity; but if its size is disproportionate to the quantity of fluid passing over it, it will be heated to a greater or less degree.

If a small charge of electricity be passed through small metal wire a few inches in length, its temperature will be sensibly elevated; if the charge be increased, the wire may be made red hot, and even melted and vaporized.

The worst conductors of electricity suffer much greater changes of temperature by the same charge than the best conductors. The charge of electricity which only elevates the temperature of one conductor, will sometimes render another red hot, and will volatilize a third.

The heat developed in the passage of electricity through combustible or explosive substances, which are imperfect conductors, causes their combustion or explosion.

If gunpowder be scattered over dry cotton loosely wrapped round one end of a discharging-rod, it may be ignited by the discharge of a Leyden jar.

In the same way powdered resin may be inflamed.

Ether or alcohol may be also fired by passing through it an electric discharge. Let cold water be poured into a wine-glass, and let a thin stratum of ether be carefully poured upon it. The ether being lighter will float on the water. Let a wire or chain connected with the prime conductor of a machine be immersed in the water, and, while the machine is in action, present a metallic ball to the surface of the ether. The electric charge will pass from the water through the ether to the ball, and will ignite the ether.

If a person standing on an insulated stool touches the prime conductor with one hand, and with the other transmits a spark to the orifice of a gas-pipe from which a current of gas is escaping, the gas will be ignited.

By the friction of the feet upon a dry woolen carpet, sufficient electricity may be often excited in the human body to transmit a spark to a gas-burner, and thus ignite the gas.

If we bring a candle with a long snuff, that has just been extinguished, near to a prime conductor, so that the spark passes from the conductor, through the smoke, to the candle, it may be relighted.

Is the electric fluid luminous? The electric fluid is not itself luminous; but its motion over imperfect conductors, or from one conducting substance to another, is generally attended with an exhibition of light.

Must light be regarded as a property of electricity?

The strongest electric charges that can be accumulated in a body will never afford the least appearance of light so long as a state of electric equilibrium exists, and the electric fluids are at rest. Light, therefore, must not be regarded as a property of electricity, but as the result of a disturbance occasioned by electricity.

Why does the fur of a cat sparkle?

The fur of a cat sparkles when rubbed with the hand in cold weather. The reason of this is, that the friction between the hand and the fur produces an excitation of negative electricity in the hand, and positive in the fur, and an interchange of the two is accompanied with a spark, or appearance of light.

What is the form of the electric spark?

FIG. 327

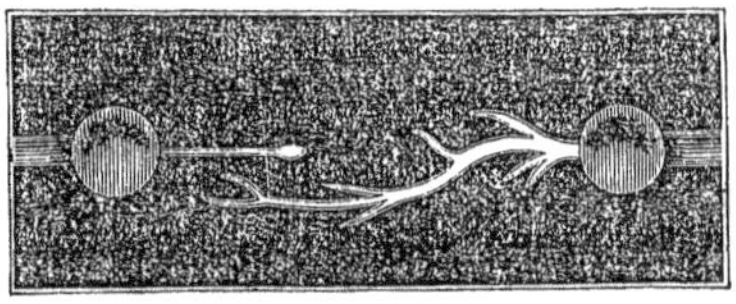

When the finger, or a brass ball at the end of a rod, is presented to the prime conductor of an electrical machine in action, a spark is produced by the passage of the fluid from the conductor to the finger or the metal. This spark has an irregular zigzag form, resembling, more or less, the appearance of lightning, as shown in Fig. 327.

Upon what does the length of the electric spark depend?

The length of the electric spark will vary with the power of the machine. A very powerful machine will so charge its prime conductor, that sparks may be taken from it at the distance of 30 inches.

How does a point influence the appearance of the spark?

FIG. 328.

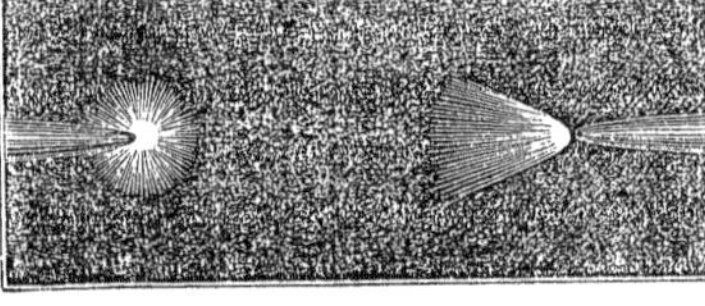

If the part of either of the electrically excited bodies which is presented to the other has the form of a point, the electric fluid will escape, not in the form of a spark, but as a brush, or pencil of light, the diverging rays of which have sometimes a length of two or three inches. Fig. 328 represents this appearance.

A substance parting with electricity generally exhibits an irregular spark, or flash of light; while a substance absorbing electricity exhibits a brush or glow of light.

What is the duration of the electric spark?

The rapidity of the electric light is marvelous; and it has been experimentally shown that the duration of the light of the spark does not exceed the one-millionth part of a second.*

When the continuity of a substance conducting electricity is interrupted, a spark will be produced at every point where the course of the conductor is broken.

A great variety of beautiful experiments may be performed to illustrate this principle. Thus, upon a piece of glass may be placed at a short distance from each other any number of bits or pieces of tin-foil, as is represented by Fig. 329; when the metal at either end is connected with the prime conductor of an electrical machine, the sparks will pass from one piece of tin-foil to the other, and form a stream of beautiful light. By varying the position of the pieces of tin-foil, letters, or any other devices may be exhibited at the pleasure of the operator.

FIG. 329.

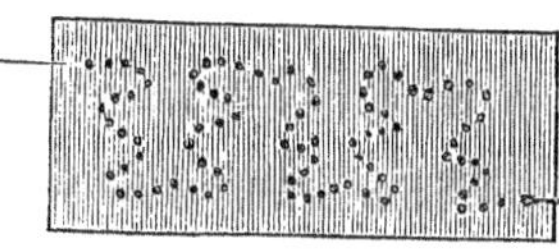

In a like manner, by fastening by means of lac-varnish a spiral line of pieces of tin-foil upon the interior of a tube, as is represented in Fig. 330, a serpentine line of fire may be made to pass from one end of the tube to the other.

FIG. 330

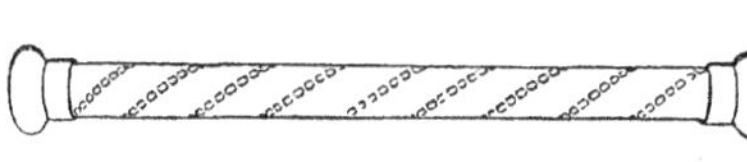

* The arrangement by which this fact was demonstrated by Mr. Wheatstone of England, may be described as follows:—Considerable lengths of copper wire (about half a mile being employed), are so arranged, that three small breaks occur in its continuity—one near the outer coating of a Leyden jar, one near the connection with the inner coating, and another exactly in the middle of the wire—so that three sparks are seen at every discharge, one at the break near the source of excitation, another in the middle of its path, and the third close to the point of returning connection; these, by bending the wire, are brought close together. Exactly opposite to this was placed a metallic speculum, fixed on an axis, and made to revolve parallel to the line of the three sparks. When a spark of light is viewed in a rapidly revolving mirror, a long line is seen instead of a point. It will be obvious that three lines of light will be seen in the revolving mirror every time a discharge takes place, and that if the first or the last differ in the smallest portion of time, these lines must begin at different points on the speculum.

When the mirror revolved slowly, the position of the lines was uniform, thus ; but when the velocity was increased, they appeared thus ; those produced by the sparks at either end of the wire being constantly coincident, but the spark evolved at the break in the middle being slightly behind the other two. From this, it appears that the disturbance commences simultaneously at either end of a circuit, and travels toward the middle. This has been adduced in proof of the two electricities. It was thus determined that electricity moves through copper wire at a rate beyond 288,000 miles in a second. It will be evident to any one considering the subject, that the length of the line seen in the speculum depends on the duration of the spark. When the mirror was made to revolve 800 times in a second, the image of the spark, at 10 feet distance, appeared to the eye of the observer to make an arc of about half a degree, and from this its duration was calculated.—*Hunt.*

Upon what does the intensity of the electric light depend?

758. The intensity of the electric light depends both upon the density of the accumulated electricity, and the density and nature of the aerial medium through which the spark passes.

Thus, the electric light, in condensed air, is very bright, and in a rarefied atmosphere it is faint and diffusive, like the light of the aurora borealis; in carbonic acid gas the light is white and intense; it is red and faint in hydrogen, yellow in steam, and green in ether or alcohol.

How may the auroral light be imitated?

If, by means of an air-pump, the air is exhausted from a long cylindrical tube closed at each end with a metallic cap, and a current of electricity passed through it, an imitation of the appearance of the aurora borealis is produced. When the exhaustion of the tube is nearly perfect, the whole length of the tube will exhibit a violet red light. If a small quantity of air be admitted, luminous flashes will be seen to issue from points attached to the caps. As more and more air is admitted, the flashes of light which glide in a serpentine form down the interior of the tube will become more thin and white, until at last the electricity will cease to be diffused through the column of air, and will appear as a glimmering light at the two points.

759. The crackling noise, or sound which is produced by the electric discharge, is attributed to the sudden displacement of the particles of air, or other medium through which the electric fluid passes.

760. The electric shock, or convulsive sensation occasioned by the passage of the electric fluid through the body of a man, or animal, is supposed to arise from a momentary derangement of the organs of the body, owing to an imperfection, or difference in the conducting power of the solids and fluids which compose them.

If this derangement does not exceed the power of the parts to recover their position and organization, a convulsive sensation is felt, the violence of which is greater or less according to the force of electricity and the consequent derangement of the organs; but if it exceeds this limit, a permanent injury, or even death, may ensue.

What are the most active agents in nature in exciting electricity?

761. In the processes hitherto described electricity has been developed by friction. In nature the agents which are undoubtedly the most active in producing and exciting electricity, are the light and heat of the sun's rays.

The change of form or state in bodies is also one of the most powerful methods of exciting electricity

Water, in passing into steam by artificial heat, or in evaporating by the action of the sun or wind, generates large quantities of electricity. The crystallization of solids from liquids, all changes of temperature, the growth and decay of vegetables, are also instrumental in producing electrical phenomena.

Does vital and muscular action excite electricity?

Recent investigations have shown that vital action and all muscular movements in man and animals, develop or produce electricity; it may also be shown by direct experiment that a person can not even contract the muscles of the arm without exciting an electrical action.

Certain animals are gifted with the extraordinary power of producing at pleasure considerable quantities of electricity in their system, and of communicating it to other animals, or substances. Among these the electrical eel and the torpedo are most remarkable, the former of which can send out a charge sufficient to knock down and stun a man, or a horse. The electricity generated by these animals appears to be the same in character as that produced by the electrical machine.

Is there any reason in ascribing unknown phenomena to electricity?

762. It has of late become the habit with many to regard electricity as the agent of all phenomena in the natural world, the cause of which may not be apparent. For this there is no good reason. Electricity is diffused through all matter, and is ever active, and many of its phenomena can not be satisfactorily explained; but it is governed, like all other forces of nature, by certain fixed laws, and it is by no means a necessary agent in all the operations of nature. It therefore argues great ignorance to refer without examination every mysterious phenomenon to the influence of electricity.

SECTION I.

ATMOSPHERIC ELECTRICITY.

Does electricity exist in the atmosphere?

763. Electricity is always found in the air, and appears to increase in strength and quantity with the altitude.

What kind of electricity is diffused through the atmosphere?

It is sometimes different in the lower regions from what it is in the upper, being positive in one and negative in the other; but in the ordinary state of the atmosphere, its electricity is invariably positive.

When the sky is overcast, and the clouds are moving in different directions, the atmosphere is subject to great and sudden variations, rapidly changing from positive to negative, and back again in the space of a few minutes.

What is supposed to occasion electricity in the atmosphere? The principal causes which are supposed to produce electricity in the atmosphere are, evaporation from the earth's surface, chemical changes which take place upon the earth's surface, and the expansion, condensation, and variation of temperature of the atmosphere and of the moisture contained in it.

When a substance is burning, positive electricity escapes from it into the atmosphere, while the substance itself becomes negatively electrified. Thus the air becomes the receptacle of a vast amount of positive electricity generated in this manner.

When is the atmosphere most highly charged with electricity? The atmosphere is most highly charged with electricity when hot weather succeeds a series of wet days, or wet weather follows a succession of dry days.

There is more electricity in the atmosphere during the cold of winter than in the summer months.

Lightning is accumulated electricity, generally discharged from the clouds to the earth, but sometimes from the earth to the clouds.

Who first established the identity of lightning and electricity? 764. The identity of lightning and electricity was first established by Dr. Franklin, at Philadelphia, in 1752.

Describe Franklin's experiment. The manner in which this fact was demonstrated was as follows:—Having made a kite of a large silk handkerchief stretched upon a frame, and placed upon it a pointed iron wire connected with the string, he raised it upon the approach of a thunder-storm. A key was attached to the lower end of the hempen string holding the kite, and to this one end of a silk ribbon was tied, the other end being fastened to a post. The kite was now insulated, and the experimenter for a considerable time awaited the result with great solicitude. Finally, indications of electricity began to appear on the string; and on Franklin presenting his knuckles to the key, he received an electric spark. The rain beginning to descend, wet the string, increased its conducting power, and vivid sparks in great abundance flashed from the key. Franklin afterward charged Leyden jars with lightning, and made other experiments, similar to those usually performed with electrical machines.

Why was this experiment one of great danger?

The experiment, as thus performed, was one of great risk and danger, since the whole amount of electricity contained in the thunder-cloud was liable to pass from it, by means of the string, to the earth, notwithstanding the use of the silk insulator.*

What is the cause of lightning?

From whatever cause electricity is present in the air, the clouds appear to collect and retain it; and when a cloud over-charged with electric fluid approaches another which is under-charged, the fluid rushes from the former into the latter. In a like manner, the fluid may pass from the cloud to the earth, and in such cases elevated objects upon the earth's surface, as trees, steeples, etc, appear to govern its direction.

Under what circumstances does lightning pass from the earth to the clouds?

When a cloud highly charged with electricity is near to the earth, the surface of the earth, for a great extent, may also become highly charged by induction; and when the tension of the electricity becomes sufficiently great, or the two electric surfaces come sufficiently near, a flash of lightning not unfrequently passes from the earth to the clouds. In this way an equilibrium of the two elements is restored.

Lightning clouds are sometimes greatly elevated above the surface of the earth, and sometimes actually touch the earth with one of their edges; they are, however, rarely discharged in a thunder-storm when they are more than 700 yards above the surface of the earth.

How many kinds of lightning are there?

765. Lightning has been divided into three kinds, viz., zigzag, or chain-lightning, sheet-lightning, and ball-lightning.

Explain the cause of the diverse appearance of lightning?

The zigzag, or forked appearance of lightning, is believed to be occasioned by the resistance of the air, which diverts the electric current from a direct course. The globular form of lightning sometimes observed, is not satisfactorily accounted for. What is called "sheet," or "heat" lightning, is sometimes the reflection in the atmosphere of lightning very remote, or not distinctly visible; but generally this phenomenon is occasioned by the play of silent flashes of electricity between the clouds, the amount of electricity developed not being sufficient to produce any other effects than the mere flash of light.

What is the cause of thunder?

766. The usual explanation of thunder is, that it is due to a sudden displacement of the particles of air by the electrical current. Others have supposed that the passage of the electricity creates

* When the experiment was subsequently repeated in France, streams of electric fire, nine and ten feet in length, and an inch in thickness, darted spontaneously with loud reports from the end of the string confining the kite. During the succeeding year, Prof. Richman of St. Petersburg, in making experiments somewhat similar, and having his apparatus entirely insulated, was immediately killed.

a vacuum, and that the air rushing in to fill it produces the sound. Every explanation that has yet been offered is somewhat unsatisfactory.

The rolling of the thunder has been ascribed to the effect of echo, but this undoubtedly is not the only cause. The rolling of thunder is heard as perfectly at sea as upon land, but there none of the causes which are generally supposed to produce echo, as mountains, hills, buildings, etc., etc., are present. Another, and perhaps the true reason is, that the sound is developed by the lightning in passing through the air, and consequently separate sounds are produced at every point through which the lightning passes.

Where do thunder storms most prevail? Thunder-storms prevail most in the torrid zone, and decrease in frequency toward either pole. In the arctic regions thunder-storms seldom or never occur. As respects time, they are most frequent in the summer months.

What is called a thunder-storm may be considered to be merely an effort of nature to effect an equilibrium of forces which have become disturbed.

When were lightning conductors first introduced? 767. A knowledge of the laws of electricity has enabled man to protect himself from its destructive influences. Lightning-rods, or conductors, were first introduced by Dr. Franklin. He was induced to recommend their adoption as a means of protection to buildings, etc., from observing that electricity could be quietly and gradually withdrawn from an excited surface by means of a good conductor, which was pointed at its extremity.

What is a lightning-rod? As ordinarily constructed, a lightning-conductor consists of a metal rod fixed in the earth, running up the whole height of a building and rising to a point above it.

How should a lightning-rod be constructed? The best metal that can be used for a lightning-rod is copper; if iron is used, the rod should not be less than three quarters of an inch in diameter. When only one rod is used, it should be continuous from the top to the bottom, and an entire metallic communication should exist throughout its whole length. This law is violated when the joints of the several parts that form the conductor are imperfect, and when the whole is loosely put together.

The rod should also be of the same dimensions throughout. The rod is best fastened to the building by wooden supports. If there are masses of metal about the building, as gutters, pipes, etc., they should be connected with the rod by strips of metal, and directly, if possible, with the ground. The lower end of the rod, where it enters the ground, should be divided into two or three branches, and turned from the building.

It ought also to extend so far below the surface of the ground as to reach water, or earth that is permanently damp. It is, moreover, a good plan to bury the end of the lightning-rod in powdered charcoal, since this preserves in a measure the iron from rust, and facilitates the passage of the electricity.

A building will be most perfectly protected when the lightning-conductor has several branches, with pointed rods projecting freely in the air from distant summits of the building, and connected with the main rod.

Professor Faraday advises that lightning-conductors should be arranged upon the inside of buildings rather than upon the outside.

What space will a lightning-rod protect?

A lightning-conductor of sufficient size is believed to protect a circle the diameter of which is four times the length of that part of the rod which rises above the building. Thus, if the rod rises two feet above the house, it will protect the building for (at least) eight feet all round.

When may a lightning-rod be productive of harm?

A lightning-conductor may be productive of harm in two ways; if the rod be broken or disconnected, the electric fluid, being obstructed in its passage, may enter the building; and if the rod be not large enough to conduct the whole current to the earth, the lightning will fuse the metal and enter the building.

A lightning-conductor protects a building even when no visible discharge takes place, by attracting the electricity of an approaching cloud, and causing it to pass off silently and quietly into the earth. This process commences as soon as the cloud has approached a position vertically over the rod.

What places are safe and what dangerous in a thunder-storm?

768. As regards safety in a thunder-storm, it is prudent, if out of doors, to avoid trees and elevated objects of every kind, which the lightning would be likely to strike in its passage to the earth. A stream of water, being a good conductor, should be avoided.

If within doors, the middle of a carpeted room is tolerably safe, provided there is no lamp hanging from the ceiling. It is prudent to avoid the neighborhood of chimneys, because lightning may enter the room by them, soot being a good conductor. For the same reason, a person should remove as far as possible from metals, mirrors, and gilt articles. The safest position that can be occupied is to lie upon a bed in the middle of a room—feathers and hair being excellent non-conductors. In all cases, the position of safety is that in which the body can not assist as a conductor to the lightning. The position of surrounding bodies must therefore be attended to, whether a person be insulated or not.

The apprehension and solicitude respecting lightning are proportionate to the magnitude of the evils it produces, rather than the frequency of its occurrence. The chances of an individual being killed by lightning are infinitely less than those which he encounters in his daily walks, in his occupation, or even during his sleep from the destruction of the house in which he lodges by fire.

How are the mechanical effects of lightning accounted for?

769. The mechanical power exerted by lightning is enormous and difficult to account for. Arago supposed that the heat of the lightning in passing through any substance, instantly converted all the moisture contained in it into steam of a highly explosive character, and that the great mechanical effects observed are due to this agent rather than to the direct effect of the electric current. A temperature that can instantly render iron red hot, is known to be sufficient to generate steam of such an elastic force that it would overcome all obstacles, and if the water contained in the pores of bodies is at once converted into steam of this character, its force would be capable of producing any of the mechanical effects witnessed in lightning discharges.

Another theory supposes that the natural electricities of non-conducting bodies are forcibly decomposed by the presence of the electric fluid which forms the lightning, and that their violent separation forces every thing asunder which tends to confine them.

What is the cause of the aurora borealis?

770. The phenomenon of the aurora borealis is supposed to be due to the passage of electric currents through the higher regions of the atmosphere—the different colors manifested being produced by the passage of the electricity through air of different densities.

Where does the aurora appear?

In the northern hemisphere the aurora always appears in the north, but in the southern hemisphere it appears in the south; it seems to originate at or near the poles of the earth, and is consequently seen in its greatest perfection within the arctic and antarctic circles.*

The aurora is not a local phenomenon, but is seen simultaneously at places widely remote from each other, as in Europe and America.

The general height of the aurora is supposed to be between one and two hundred miles above the surface of the earth; but it sometimes appears within the region of the clouds.

Auroras occur more frequently in the winter than in the summer, and are only seen at night. They affect in a peculiar manner the magnetic needle and the electric telegraph, and as the disturbances occasioned in these instruments are noticed by day as well as by night, there can be no doubt of the occurrence of the aurora at all hours. The intense light of the sun, however, renders the auroral light invisible during the day.

Fig. 331.

The accompanying figure represents one of the most beautiful of the auroral phenomena.

It has often been asserted, and on good authority, that sounds have been heard attending the phenomena of the aurora, like the rustling of silk, or the sound and crackling of a fire. On this point, however, there is great difference of opinion.

Auroras appear to be subject to some variation in their appearance, extending through a circle of years. Thus, from 1705 to 1752, the northern lights became more and more frequent, but after that for a period they were seen but rarely. Since 1820 they have been quite frequent and brilliant.

* In the arctic and antarctic circles, when the sun is absent, the aurora appears with a magnificence unknown in other regions, and affords light sufficient for many of the ordinary out-door employments.

CHAPTER XVI.

GALVANISM.

What is Galvanic Electricity?

771. Electricity excited or produced by the chemical action of two or more dissimilar substances upon each other is termed Galvanic, or Voltaic Electricity, and the department of physical science which treats of this form of electrical disturbance is called Galvanism.

What simple experiment illustrates the production of galvanic electricity?

The most simple method of illustrating the production of galvanic electricity is by placing a piece of silver (as a coin) on the tongue, and a piece of zinc underneath. So long as the two metals are kept asunder no effect will be noticed, but when their ends are brought together, a distinct thrill will pass through the tongue, a metallic taste will diffuse itself, and, if the eyes are closed, a sensation of light will be evident at the same moment.

This result is owing to a chemical action which is developed the moment the two metals touch each other. The saliva of the tongue acts chemically upon, or oxydizes a portion of the zinc, which excites electricity, for no chemical action ever takes place without producing electricity. Upon bringing the ends of the two metals together, a slight current passes from one to the other.

If a living fish, or a frog, having a small piece of tin-foil on its back, be placed upon a piece of zinc, spasms of the muscles will be excited whenever a metallic communication is made between the zinc and the tin-foil.

When and how was galvanic electricity discovered?

The production of electricity by the chemical action of two metals when brought in contact, was first noticed by Galvani, professor of anatomy at Bologna, Italy, in 1790.

His attention was directed to the subject in the following manner:—Having occasion to dissect several frogs, he hung up their hind legs on some copper hooks, until he might find it necessary to use them for illustration, In this manner he happened to suspend a number of the copper hooks on an iron balcony, when, to his great astonishment, the limbs were thrown into violent convulsions. On investigating the phenomenon, he found that the mere contact of dissimilar metals with the moist surfaces of the muscles and nerves, was all that was necessary to produce the convulsions.

FIG. 332.

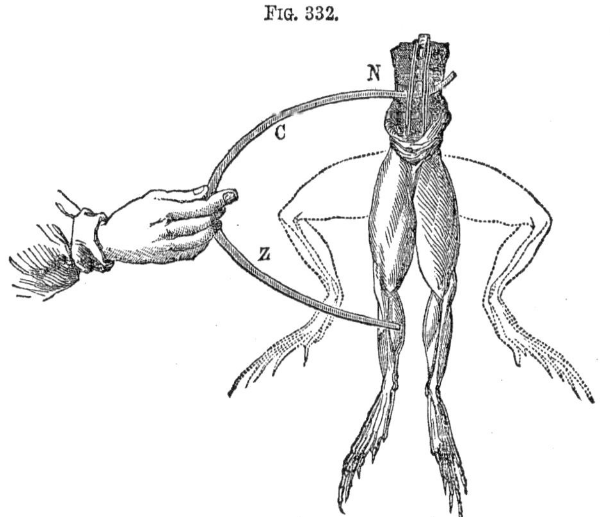

This singular action of electricity, first noticed by Galvani, may be experimentally exhibited without difficulty. Fig. 332 represents the extremities of a frog, with the upper part dissected in such a way as to exhibit the nerves of the legs, and a portion of the spinal marrow. If we now take two thin pieces of copper and zinc, C Z, and place one under the nerves, and the other in contact with the muscles of the leg, we shall find that so long as the two pieces of metal are separated, so long will the limbs remain motionless; but by making a connection, instantly the whole lower extremities will be thrown into violent convulsions, quivering and stretching themselves in a manner too singular to describe. If the wire is kept closely in contact, these phenomena are of momentary duration, but are renewed every time the contact is made and broken.

To what did Galvani attribute these phenomena?

Galvani attributed these movements of the muscles to a kind of nervous fluid pervading the animal system, similar to the electric fluid, which passed from the nerves to the muscles, as soon as the two were brought in communication with each other, by means of the metallic connection, in the same way as a discharge takes place between the external and internal coatings of a Leyden jar. He therefore called the supposed fluid animal electricity.

What was determined by Volta? The experiments of Galvani were repeated by Volta, an eminent Italian philosopher, who found that no electrical or nervous excitement took place unless a communication between the muscles and the nerves was made by two different metals, as copper and iron, or copper and zinc. He considered that electricity was produced by simple contact of the dissimilar metals, positive electricity being evolved from the one and negative electricity from the other.

What is the true cause of electricity developed by contact of different metals? The true cause of electrical excitement occasioned by the contact of dissimilar metals is now fully ascertained to be chemical action; and recent researches have also proved that no chemical action ever takes place without the development of free electricity.

The electricity produced by chemical action has been termed Galvanic, or Voltaic Electricity, in honor of Galvani and Volta, who first developed its phenomena.

How does galvanic differ from ordinary electricity? 772. Galvanic electricity, or the electricity developed by chemical action, differs from frictional, or ordinary electricity, chiefly in its continuance of action. The electricity developed by friction from a glass plate, or the cylinder of an electrical machine, exhibits itself in sudden and intermittent shocks, accompanied with a sort of explosion; whereas that which is generated by chemical action is a steady, flowing current.

The fundamental principle which forms the basis of the science of galvanic electricity is as follows:

What principle forms the basis of the science of galvanic electricity? Any two metals, or more generally, any two different bodies which are conductors of electricity, when placed in contact, develop electricity by chemical action—positive electricity flowing from the metal which is acted upon most powerfully, and negative electricity from the other.

What are electro-positive and electro-negative elements? In general, that metal which is acted upon most easily is termed the electro-positive metal, or element; and the other the electro-negative metal, or element.

The electrical force or power generated in this way is called the electro-motive force.

773. Different bodies placed in contact manifest different electro-motive forces, or develop different quantities of electricity.

How may bodies capable of exciting electro-motive forces be classified?

Bodies capable of developing electricity by contact may be arranged in a series in such a manner that any one placed in contact with another holding a lower place in the series, will receive the positive fluid, and the lower one the negative fluid; and the more remote they stand from each other in the order of the series, the more decidedly will the electricity be developed by their contact.

The most common substances used for exciting galvanic electricity may be arranged in such a series as follows:—zinc, lead, tin, antimony, iron, brass, copper, silver, gold, platinum, black lead or graphite, and charcoal.

Thus, zinc and lead, when brought in contact, will produce electricity, but it will be much less active than that produced by the union of zinc and iron, or the same metal and copper, and the last less active than zinc and platinum or zinc and charcoal.

What is the practical method of exciting galvanic electricity?

774. In the production of galvanic electricity for practical purposes, it is necessary to have a combination of three different conductors, or elements, one of which must be solid and one fluid, while the third may be either solid or fluid.

The process usually adopted is to place between two plates of different kinds of metal a liquid capable of exciting some chemical action on one of the plates, while it has no action, or a different action upon the other. A communication is then formed between the two plates.

What is a Galvanic Circuit?

When two metals capable of exciting electricity are so arranged and connected that the positive and negative electricities can meet and flow in opposite directions, they are said to form a galvanic circuit, or circle.

Describe a simple Galvanic Battery.

A very simple, and at the same time an active galvanic circuit may be formed by an arrangement as represented in Fig. 333. C and Z are thin plates of copper and zinc immersed in a glass vessel containing a very weak solution of sulphuric acid and water. Metallic contact can be made between the plates by wires, X and W, which are soldered to them. If now the wires are connected, as at Y, a galvanic circuit will be formed; positive electricity passing from the zinc through the liq-

FIG. 333.

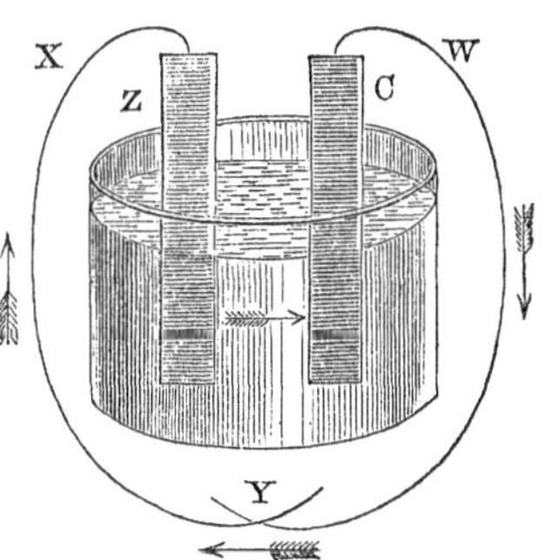

uid, to the copper, and from the copper along the conducting-wires to the zinc, as indicated by the arrows in the figure. A current of negative electricity at the same time traverses the circuit also, from the copper to the zinc, in a direction precisely reversed.

Such an arrangement is called a simple galvanic battery.

What are the poles of a galvanic battery?

The two metals forming the elements of the battery are generally connected by copper wires; the ends of these wires, or the terminal points of any other connecting medium used, are called the poles of the battery.

Thus, when zinc and copper plates are used, the end of the wire conveying positive electricity from the copper would be the positive pole, and the end of the wire conveying negative electricity from the zinc plate would be the negative pole. Faraday describes the poles of the battery as the doors by which electricity enters into or passes out of the substance suffering decomposition, and in accordance with this view he has given to the positive pole the name of *anode*, or ascending way, and to the negative pole the name of *cathode*, or descending way.

At what point of the circuit is electricity especially manifested?

The manifestations of electricity will be most apparent at that point of the circuit where the two currents of positive and negative electricity meet.

When is a circuit said to be closed?

When the two wires connecting the metal plates of a battery are brought in contact, the galvanic circuit is said to be closed. No sign of electrical excitement is then visible; the action, nevertheless, continues. The opposite electricities collected at the poles, in particular, neutralize each other perfectly on meeting; every trace of electricity must therefore vanish, as when a Leyden jar is discharged, if a fresh quantity were not continually produced by the pairs of plates. If the wires which conduct the two electricities be slightly disconnected, a spark will be observed at the point of interruption.

Explain the theory of the production of galvanic electricity.

In the formation of a galvanic circuit, by the employment of two metals and a liquid, the chemical action which gives rise to the electricity takes place through a decomposition of the liquid. It is, therefore, essential to the formation of an active galvanic circuit, that the liquid employed should be capable of being decomposed. Water is most conveniently applicable for this purpose. When a plate of zinc and copper are immersed in water, the elements of the water, oxygen and hydrogen, are separated from each other, in consequence of the greater attraction which the oxygen has for the zinc. The oxygen, therefore, unites with the zinc, and by so doing produces an alteration in the electrical condition of the metal. The zinc communicating its natural share of electricity to the liquid, becomes negatively electrified. The copper attracting the same electricity from the liquid, becomes positively electrified, ; at the same time the hydrogen, which is the other element of the water, is also attracted to the copper, and appears in minute bubbles upon its surface. If the two metal plates be now connected with metallic wires, positive electricity will flow from the copper and negative electricity from the zinc, and by the union of these two an electric current will be formed.*

With water alone and two metals, the quantity of electricity excited is very small, but by the addition of a small quantity of some acid, the excitement is greatly increased.

What is the necessity of two metals in a galvanic circuit?

Although two metal plates are employed in the arrangement described, only one of them is active in the excitement of electricity, the other plate serving merely as a conductor to collect the force generated. A metal plate is generally used for this purpose, because metals conduct electricity much better than other substances exposing an equal surface to the fluids in which they are immersed; but other conductors may be used, and when a proportionately larger surface is exposed to compensate for inferior conducting power, they answer as well, and in some instances better, than metal plates. Thus charcoal is very often employed in the place of copper, and a very hard material obtained from the interior of gas retorts, called graphite, is considered one of the best conductors.

Two metals are not absolutely essential to the formation of a simple galvanic circuit. A current may be obtained from one metal and two liquids, provided the liquids are such that a stronger chemical action takes place on one side of the metal plate than on the other.

In some electric batteries also, two metals and two dissimilar liquids are employed.

How may galvanic action be increased?

775. The electricity developed by a simple galvanic circuit, whether it be composed of two metals and a liquid, or any other combination, is exceedingly feeble. Its power can, however, be increased to any extent by a repetition of the simple combinations.

* The terms "electric fluid" and "electric current," which are frequently employed in describing electrical phenomena, are calculated to mislead the student into the supposition that electricity is known to be a fluid, and that it flows in a rapid stream along the wires. Such terms, it should be understood, are founded merely on an assumed analogy of the electric force to fluid bodies. The nature of that force is unknown, and whether its transmission be in the form of a current, or by vibrations, or by any other means, is undetermined.

In a discussion which took place some years since at a meeting of the British Association for the Advancement of Science, respecting the nature of electricity, Professor Faraday expressed his opinion as follows:—"There was a time when I thought I knew something about the matter; but the longer I live, and the more carefully I study the subject, the more convinced I am of my total ignorance of the nature of electricity."

"After such an avowal as this," says Mr. Bakewell, "from the most eminent electrician of the age, it is almost useless to say that any terms which seem to designate the form of electricity are merely to be considered as convenient conventional expressions."

Describe the pile of Volta.

The first attempt to increase the power of a galvanic circuit by increasing the number of the combinations, was made by Volta. He constructed a pile of zinc and copper plates with a moistened cloth interposed between each. He commenced with a zinc plate, upon which he placed a copper plate of the same size, and on that a circular piece of cloth previously soaked in water slightly acidulated. On the cloth was laid another plate of zinc, then copper, and again cloth, and so on in succession, until a pile of fifty series of alternate metal plates and moistened cloths was formed, the terminal plate of the series at one end being copper and at the other end zinc. A metallic wire attached to the highest copper plate will constitute the positive pole, and another to the lowest zinc plate the negative pole of such a series.

FIG. 334.

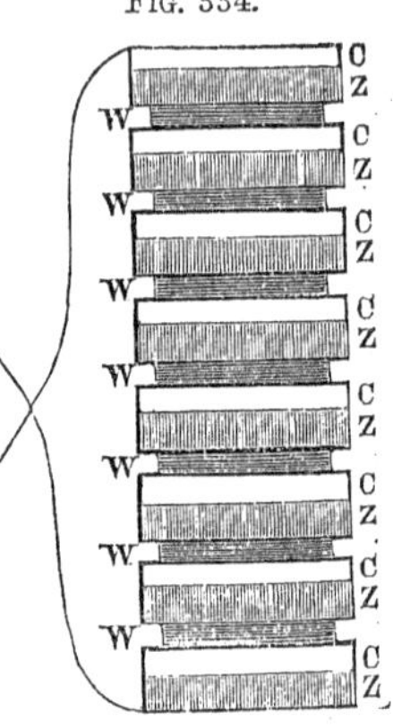

Fig. 334 represents Volta's arrangement of metal plates and wet cloths, with the metallic wires, which constitute the poles.

Such combinations are denominated Voltaic Piles, or Voltaic Batteries, and very often Galvanic Batteries.

As two different metals and an interposing liquid are generally employed for this purpose, it has been usual to call these combinations *pairs* or *elements;* so that the battery is said to consist of so many pairs or elements, each pair or element consisting of two metals and a liquid.

Of what substances have voltaic piles been constructed?

776. Voltaic piles or batteries have been composed and constructed in a great variety of forms, by combining together in a series various substances which excite electricity when acted upon chemically.

FIG. 335.

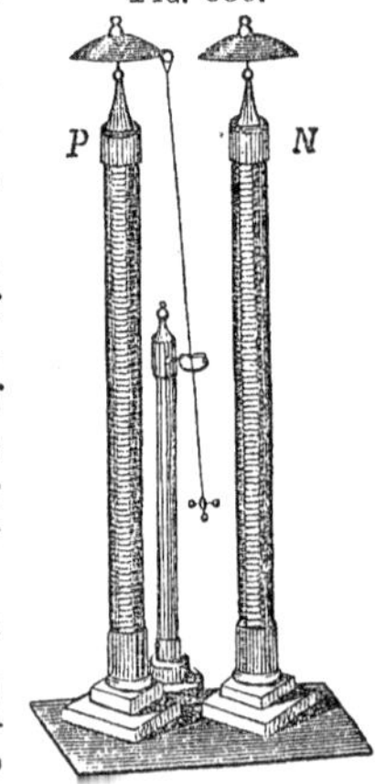

Thus, they have been constructed entirely of vegetable substances, without resorting to the use of any metal, by placing discs of beet-root and walnut-wood in contact. With such a pile, and a leaf of grass as a conductor, convulsions in the muscles of a dead frog are said to have been produced. Other experimentalists have formed voltaic piles wholly of animal substances.

Describe Zamboni's Pile.

A perfectly dry voltaic pile, known from its inventor as Zamboni's Pile, may be formed of sheets of gilded paper and sheet zinc. If several thousands of these be packed together in a glass tube, so that their similar metallic faces shall all look the same way, and be pressed tightly together at each end by metallic plates, it will be found that one extremity of the pile is positive and the other negative. Such a series will last more than twenty years, but it requires as many as 10,000 pairs to afford sparks visible in daylight, and to charge the Leyden jar.

Fig. 335 represents a pair of these piles, so arranged as to produce what has been called a perpetual motion. Two piles, P N, are placed in such a position that their poles are reversed, and between them a light pendulum, vibrating on an axis and insulated on a glass pillar. This pendulum is alternately attracted to one and then to the other, and thus rings two little bells connected with the positive and negative poles.

The galvanic batteries in practical use at the present time differ considerably in form and efficiency, but the principle of construction in all is the same as that of the original voltaic pile.

Describe the trough battery.

FIG. 336.

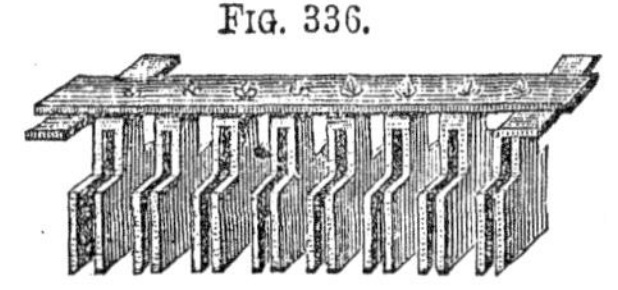

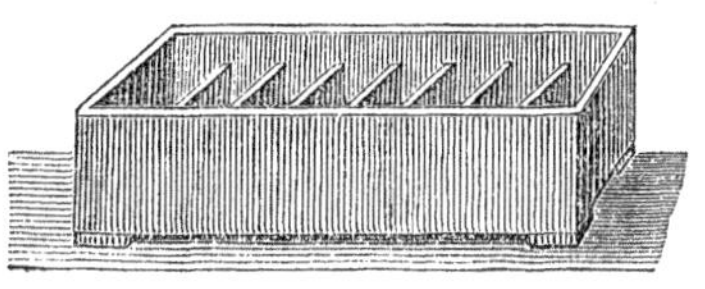

A very effective arrangement known as the trough battery, is represented in Fig. 336. This consists of a trough of wood divided into water-tight cells, or partitions, each cell being arranged to receive a pair of zinc and copper plates. The plates are attached to a bar of wood, and connected with one another by metallic wires, in such a way that every copper plate is connected with the zinc plate of the next cell. The battery is excited by means of dilute sulphuric acid poured into the cells, and the current of electricity is directed by wires soldered to the extreme plates. When the battery is not in use the plates may be raised from the trough by means of the wooden bar.

The battery by which Sir Humphrey Davy effected his splendid chemical discoveries was of this form, and consisted of two thousand double plates of copper and zinc, each plate having a surface of thirty-two square inches. Now, however, by improved arrangements, we can produce with ten or twenty pairs of plates, effects every way superior.

FIG. 337.

In other and more efficient compound galvanic circuits, the exciting liquid is placed in a series of separate cups, or glasses, arranged in a circle, or in parallel lines. Each cup contains one zinc and one copper plate, not immediately in connection with each other, but every zinc plate of one cup is connected with the copper plate of the preceding, by a copper band, or wire. This arrangement is represented in Fig. 337, the copper plate, and the direction of the positive current being indicated by the sign +, and the zinc plate and the direction of the negative current by the sign —.

Describe Smee's battery.

The simplest form of galvanic battery at present used is that invented by Mr. Smee, and known as Smee's battery (See Fig. 338.) It consists of a plate of silver coated with platinum, suspended between two plates of zinc, *z z*, the surfaces of which last have been coated with mercury, or amalgamated, as it is called.* The three are attached to a wooden bar, which serves to support the whole in a tumbler, G, partially filled with a weak solution of sulphuric acid and water. The wires, or poles for directing the current of electricity are connected with the zinc and platinum plates by small screw-cups, S and A.

FIG. 338.

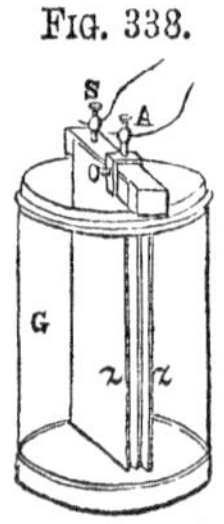

What is the sulphate of copper battery?

Another form of battery, called the sulphate of copper battery, from the fact that a solution of sulphate of copper (blue vitriol) is used as the exciting liquid, is represented by Fig. 339. It consists of two concentric cylinders of copper tightly soldered to a copper bottom, and a zinc cylinder, Z, fitting in between them. The zinc cylinder, when let down into the solution, is prevented from touching the copper by means of three pieces of wood or ivory, shown in the figure. Two screw-cups for holding the connecting wires are attached, one to the outer copper cylinder, and the other to the zinc.

FIG. 339.

What is the principal imperfection of the galvanic battery?

The principal imperfection of the galvanic battery is the want of uniformity in its action. In all the various forms the strength of the electric current excited continually diminishes from the moment the battery action commences. In the sulphate of copper battery, especially, the power is reduced to almost nothing in a comparatively brief space of time. This is chiefly owing to the circumstance that the metallic plates soon become coated with the products of the chemical decomposition, the result of the chemical action, whereby the electricity is developed.

This difficulty is obviated, in a great degree, by the use of a diaphragm, or porous partition, between the two metallic plates, which allows a free contact

* It is found that by coating the zinc with mercury, the waste of the zinc is greatly diminished. It is not well understood in what way the mercury contributes to this effect. We have a parallel to it in the rubber of the electrical machine, which, when coated with an amalgam of zinc and tin, acts with greater efficiency than under any other circumstances.

of the liquid on each side, within its pores, but prevents the solid products of decomposition from passing from one plate to the other.

Describe Daniel's constant battery.

Daniel's constant battery, constructed according to this principle, and represented in Fig. 340, maintains an effective galvanic action longer than any other; *a* is a hollow cylinder of copper; *z*, a solid rod of amalgamated zinc; and *e*, a porous tube of earthenware separating the two. Diluted sulphuric is placed in the porous tube, and a saturated solution of sulphate of copper in the copper cylinder.

Fig. 340.

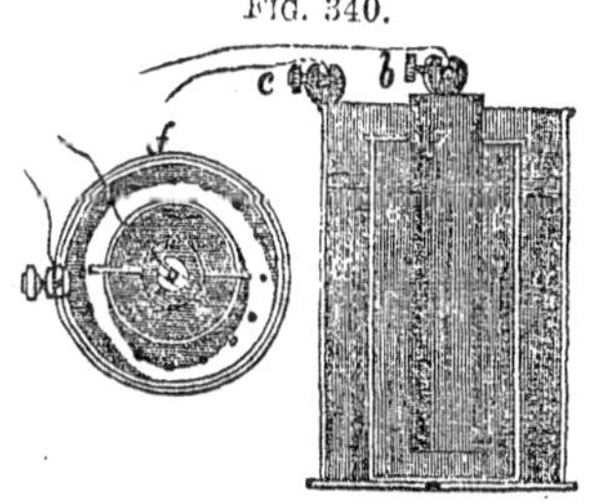

What is the construction of Grove's battery?

One of the most efficient batteries is that known as Grove's battery, from its inventor, and is the form generally used for telegraphing and for other purposes in which powerful galvanic action is required. It consists of a plain glass tumbler, in which is placed a cylinder of amalgamated zinc, with an opening on one side to allow a free circulation of the liquid Within this cylinder is placed a porous cup, or cell, of earthenware, in which is suspended a strip of platinum fastened to the end of a zinc arm projecting from the adjoining zinc cylinder. The porous cup containing the platinum is filled with strong nitric acid, and the outer vessel containing the zinc with weak sulphuric acid. Fig. 341 represents a series of these cups, arranged to form a compound circuit, with their terminal poles, P and Z. This form of battery is objectionable on account of the corrosive character of the acids employed, and the deleterious vapors that arise from it when in action.

Fig. 341.

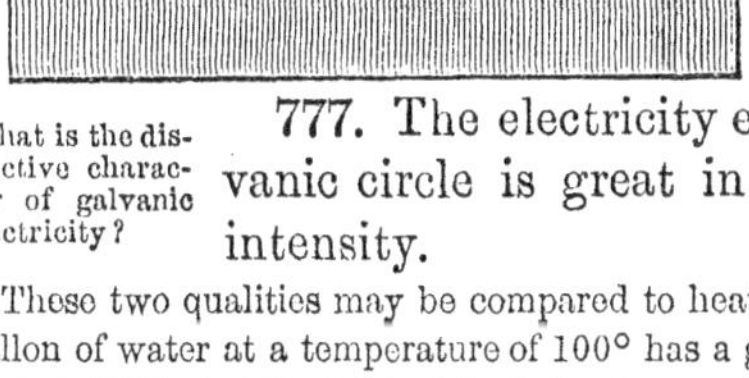

What is the distinctive character of galvanic electricity?

777. The electricity evolved by a single galvanic circle is great in quantity, but weak in intensity.

These two qualities may be compared to heat of different temperatures. A gallon of water at a temperature of 100° has a greater quantity of heat than a pint at 200°; but the heat of the latter is more intense than that of the former.

What is the distinctive character of frictional electricity?

The electricity, on the contrary, produced by friction, or that of the electrical machine, is small in quantity, but of high tension, or intensity.

Illustrate the differences between the two electricities.

Frictional electricity is capable of passing for a considerable distance through or over a non-conducting or insulating substance, which galvanic electricity can not do. Thus, the spark from a prime conductor will leap toward a conducting substance for some distance through the air, which is a non-conductor; but if a current of galvanic electricity is resisted by the slightest insulation, or the interposition of some non-conducting substance, the action at once stops. Galvanic electricity will traverse a circuit of 2,000 miles of wire, rather than make a short circuit by overleaping a space of resisting air not exceeding one hundredth part of an inch. Frictional electricity, on the other hand, will force a passage across a considerable interval, in preference to taking a long circuit through a conducting wire, or at least the greater portion of it will pass through the air, though some part of the charge will always traverse the wire.

Frictional electricity produces very slight chemical or heating effects; galvanic electricity produces very powerful effects.

A proper and simple arrangement of a zinc plate and a little acidulated water, will produce as much electricity in three seconds of time as a Leyden jar battery charged with thirty turns of a large and powerful plate electrical machine in perfect action. The shock received by transmitting this quantity of galvanic electricity through the animal system would be hardly perceptible, but received from a Leyden jar, would be highly dangerous, and perhaps fatal. A grain of water may be decomposed and separated into its two elements, oxygen and hydrogen, by a very simple galvanic battery, in a very short time; but 800,000 such charges of a Leyden jar battery, as above referred to, would be required to supply electricity sufficient to accomplish the same result. Such a quantity of electricity sent forth from a Leyden jar would be equal to a very powerful flash of lightning.

Upon what does quantity in galvanic electricity depend?

The quantity of electricity excited in a galvanic circuit is directly proportional to the amount of chemical action that takes place—as between the zinc and the acid. By increasing the amount of surface exposed to chemical action, we therefore increase the quantity of electricity evolved.

Hence, gigantic plates have been constructed for the purpose of obtaining an immense quantity.

Upon what does intensity depend?

The intensity of the electricity evolved depends upon the number of plates, and is greatest when the voltaic pile is made up of a great number of small plates.

Supposing an equal amount of surface of copper and zinc employed, the shock, and other indications of a strong charge, would be greater if it were cut up into many small circles, than if it formed a few large ones. But the actual quantity of excitement would be greatest with the large plates.

How may voltaic action be interrupted and renewed?

778. When the wire from one end of a voltaic battery is connected with the wire from the opposite end, voltaic action instantly commences; and it as instantaneously ceases when the connection is interrupted. The rapidity with which the electric circuit may be completed and broken has no ascertained limit; nor does it appear to be controlled by resistance caused by traversing miles of wire.

What are the most ordinary effects of galvanic electricity?

779. The most ordinary effects produced by the developed electricity of a large galvanic battery, are the production of sparks and brilliant flashes of light, the heating and fusing of metals, the ignition of gunpowder and other inflammable substances, and the decomposition of water, saline compounds, and metallic oxyds.

When does galvanic electricity evolve heat?

Heat is evolved whenever a galvanic current passes over a conducting body, the amount of which will depend on the quantity and intensity of the electricity transmitted, and upon the resistance which the body offers to the passage of the current.

The metals differ greatly in their conducting power. Thus, if we link together pieces of copper, iron, silver, and platinum wire, and pass a galvanic current along them, they will be found to be unequally heated, the platinum being the most, and the copper the least.

How may the heating effects of galvanic electricity be illustrated?

The easiest method of showing by experiment the heating power of the galvanic current is to connnect the poles of a battery by means of a fine platinum wire. If the wire is very long it may become hot; shorten it to a certain extent, and it will become red-hot; shorten it still more, and it will become white-hot, and finally melt. If such a wire is carried through a small quantity of salt water on a watch-glass, the liquid will boil; if through alcohol, ether, or phosphorus, they will be inflamed; if through gunpowder, it will be exploded.

What practical application has been made of this power?

This power has been applied to the purpose of firing blasts, or mines of gunpowder, an operation which may be effected with equal facility under water. The process is as follows:— The wires from a sufficiently powerful battery are connected by a piece of fine platinum wire, which is placed in a mass of gunpowder contained in a cavity of a rock, or inclosed in a vessel beneath the surface of water. The wire may be of any length, but the moment connection is made with the battery the current passes, renders the platinum red-hot, and explodes the the powder.*

How may the greatest artificial heat be produced?

The greatest artificial heat man has yet succeeded in producing has been through the agency of the galvanic battery. All the metals, including platinum, which can not be fused by any furnace heat, are readily melted. Gold burns with a blueish light, silver with a bright green flame, and the combustion of the other metals is always accompanied with brilliant results. All the earthy minerals may be liquefied by being placed between the poles of a sufficiently large battery. Sapphire, quartz, slate, and lime, are readily melted; and the diamond itself fuses, boils, and becomes converted into coal.

How are the luminous effects of the galvanic battery manifested?

780. The luminous effects of the galvanic battery are no less remarkable than its heating effects. A very small voltaic arrangement is sufficient to produce a spark of light every time the circuit is closed or opened. If the two ends of wires proceeding from the opposite poles of a battery are brought nearly together, a bright spark will pass from one to the other, and this takes place even under water, or in a vacuum.

How may the most intense artificial light be produced?

The most splendid artificial light known is produced by fixing pieces of pointed charcoal to the wires connected with opposite poles of a powerful galvanic battery, and bringing them within a short distance of each other. The space between the points is occupied by an arch of flame that nearly equals in dazzling brightness the rays of the sun.

How does the electric light differ from all other artificial lights?

This light, which is termed the electric light, differs from all other forms of artificial light, inasmuch as it is independent of ordinary combustion. The charcoal points appear to suffer no change, and the light is equally strong and brilliant in a vacuum, and in such gases as do not contain oxygen, where

* In the course of the construction of a railway recently in England, it became necessary to detach a large mass of rock from a cliff on the sea-coast in order to avoid the expense of a long tunnel. To have done this by the direct application of human labor and the ordinary operations of blasting, would have been attended with an immense expenditure of time and money. It was accordingly resolved to blow it up with gunpowder, ignited by the galvanic battery. Nine tons of powder were accordingly deposited in chambers at from 50 to 70 feet from the face of the cliff, and fired by a conducting wire connected with a powerful battery, placed at 1,000 feet from the mine. The explosion detached 600,000 tons' weight of chalk from the cliff. It was proved that this might have been equally effected at the distance of 3,000 feet. This bold experiment saved eight months' labor and $50,000 expense.

all other artificial lights would be extinguished. It may even be produced under water. To excite the electricity, however, which occasions this light, zinc or some other metal must be oxydized, or what is the same thing burnt, the same as oil in our lamps, or coal in the gas retorts for the production of other species of artificial light.

What are the physiological effects of galvanic electricity?

The effects of the galvanic battery upon the nerves and muscles of the animal system are of the same character as those produced by ordinary electricity.

On grasping the two ends of the connecting wires of a battery of some force with wet hands, a peculiar tremor will be felt in the joints of the arm and hand, accompanied by a slight contortion of the muscles, and increasing to a violent shock. This shock is repeated every time a contact between the hand and the wire is broken and renewed. The concussion of the nerves of the body is, therefore, produced by the entrance and exit of the currents of electricity; for they evidently must pass through the body the moment it forms the connecting link between the two poles.

By a particular arrangement, the circuit may be closed or interrupted at pleasure, and in such a manner that the current may be made to pass alternately through the wires and the body; the latter being thus exposed to a series of shocks which are considered particularly adapted for the cure of diseases arising from the injury or derangement of the nervous system. It is, moreover, a highly valuable remedy in cases of suffocation, drowning, paralysis, etc.; and numerous arrangements have been at various times proposed for the construction of medico-galvanic machines.

The effects of galvanic electricity on bodies recently deprived of life is very remarkable, and it was through an accidental observance of its action upon a dead frog that galvanism was discovered. By connecting the muscles and nerves of recently-killed animals with the poles of a battery, many of the movements of life may be produced. Some remarkable experiments of this character were made some years since upon the body of a man recently executed for murder at Glasgow, in Scotland. The voltaic battery employed consisted of 270 pairs of plates, four inches square. On applying one pole of the battery to the forehead and the other to the heel, the muscles are described to have moved with fearful activity, so that rage, anguish, and despair, with horrid smiles, were exhibited upon the countenance.

781. Galvanic electricity is a powerful agent in effecting chemical decompositions, and in its application to such purposes, it is most practically useful.

Can galvanic electricity effect chemical decomposition?

When a current of galvanic electricity is made to pass through a compound conducting substance, its tendency is to decompose and separate it into its constituent parts.

How may water be decomposed?

Thus, water is composed of two gases, oxygen and hydrogen united together. When the wires connecting the poles of a galvanic battery are placed in water, and a sufficiently strong current made to pass through them, the water is decomposed, the hydrogen being given out at the negative pole of the battery, and the oxygen at the positive pole. Fig. 342 represents a form of apparatus by which this experiment can be performed in a very satisfactory manner. It consists of two tubes, O and H, supported vertically in a small reservoir of water, and two slips of platinum, *p p*, which can be connected with the poles of a voltaic battery, passing in at the open end of the tubes. When communication is effected between the platinum slips and a battery in action, gas rapidly rises in each tube and collects in the upper part. In that tube which is in connection with the positive pole of the battery oxygen accumulates, and in the other hydrogen. And it will be noticed that the quantity of the latter is equal to twice the quantity of the former gas, since water contains by volume twice as much hydrogen as it does oxygen.

FIG. 342.

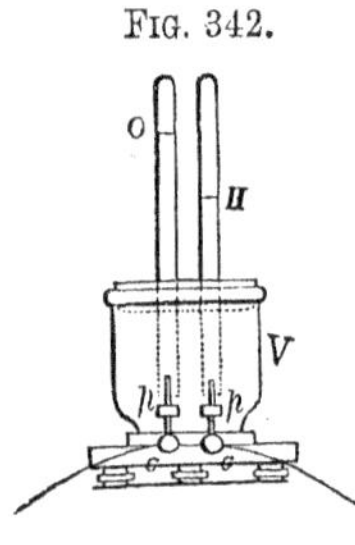

What is the theory of the decomposing action of galvanic electricity?

The explanation of this phenomenon may be briefly given as follows:—All atoms of matter are regarded as originally charged with either positive or negative electricity. In the case of water, hydrogen is the electro-positive element and oxygen the electro-negative element. It has been already shown that bodies in opposite electrical states are attracted by each other. Hence, when the poles of a galvanic battery are immersed in water, the negative pole will attract the positive hydrogen, and the positive pole the negative oxygen. If the attractive force of the two electricities generated by the battery is greater than the attractive force which unites the two elements, oxygen and hydrogen, together in the water, the compound will be decomposed. Upon the same principle other compound substances may be decomposed, by employing a greater or less amount of electricity. In this way Sir Humphrey Davy made the discovery that potash, soda, lime, and other bodies, were not simple in their nature, as had previously been supposed, but compounds of a metal with oxygen.

What quantity of electricity is necessary to decompose a substance?

782. Recent experiments have shown that the electricity which decomposes, and that which is evolved by the decomposition of a certain quantity of matter, are alike. Thus, water is composed of oxygen and hydrogen; now, if the electrical power which holds a grain of water in combination, or which causes a grain of oxygen and hydrogen to unite in the right proportions to form water, could be collected and thrown into a voltaic current, it would be exactly the quantity required to produce the decomposition of a grain of water or the liberation of its elements, oxygen and hydrogen.

What is an Electrode? 783. For convenience in certain experiments, the ends of the copper wires connecting the poles of the galvanic battery are frequently terminated with thin strips of platinum, which are called Electrodes. The platinum slip connected with the positive pole forms the positive electrode, and that with the negative pole, the negative electrode.

Platinum is used for the reason, that in employing the battery for effecting decompositions, it is frequently necessary to immerse the ends of the conducting wires in corrosive liquids, and this metal generally is not affected by them.

What is Electro-metallurgy? 784. Electro-metallurgy, or electrotyping, is the art or process of depositing, from a metallic solution, through the agency of galvanic electricity, a coating or film of metal upon some other substance.*

Upon what is the process based? The process is based on the fact, that when a galvanic current is passed through a solution of some metal, as of sulphate of copper (sulphuric acid and oxyd of copper), decomposition takes place; the metal is separated in a metallic state, and attaches itself to the negative pole, or to any substance that may be attached to the negative pole; while the oxygen or other substance before in combination with the metal, goes to, and is deposited on the positive pole.

In this way a medal, a wood-engraving, or a plaster cast, if attached to the negative pole of a battery, and placed in a solution of copper opposite to the positive pole, will be covered with a coating of copper; if the solution contains gold or silver instead of copper, the substance will be covered with a coating of gold or silver in the place of copper.

The thickness of the deposit, providing the supply of the metallic solution be kept constant, will depend on the length of time the object is exposed to the influence of the battery.

In this way, a coating of gold thinner than the thinnest gold-leaf can be laid on, or it may be made several inches or feet in thickness, if desired.

The usual arrangement for conducting the electrotype process is represented

* The general name of electro-metallurgy includes all the various processes and results which different inventors and manufacturers have designated as galvano-plastic, electro-plastic, galvano-type, electro-typing, and electro-plating and gilding.

by Fig. 343. It consists of a trough of wood, or an earthen vessel, containing the solution, the decomposition of which is desired—for example, sulphate of copper. Two wires, one connected with the positive, and the other with the negative pole of a battery, Q, are extended along the top of the trough, and supported on rods of dry wood, B and D. The medal, or other article to be coated, is attached to the negative wire, and a plate of metallic copper to the positive wire. When both of these are immersed in the liquid, the action commences—the sulphate of copper is decomposed—the copper being deposited on the medal, and the liberated oxygen on the copper plate. As the withdrawal of the metal from the solution goes on, the copper plate attached to the positive pole undergoes corrosion by the sulphuric acid which is liberated and attracted to it, and sulphate of copper is formed. This, dissolving in the liquid, maintains it at a constant strength. When the operator judges that the deposit on the medal is sufficiently thick, he removes it from the trough, and detaches the coating. The deposit is prevented from adhering to the medal by rubbing its surface in the first instance with oil, or black-lead, and if it is desired that any part of the surface should be left uncoated, that portion is covered with wax, or some other non-conductor.

Fig. 343.

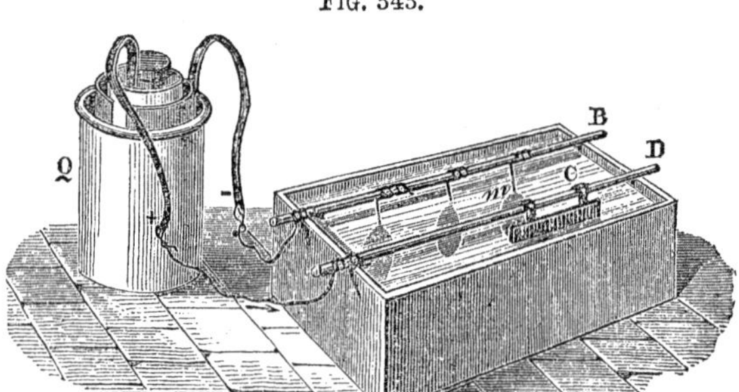

In this way a most perfect reversed copy of the medal is obtained,—that is, the elevations and depressions of the original are reversed in the copy. To obtain a fac-simile of the original, the electrotype cast is subjected to a repetition of the process.

In general, it is found more convenient to mold the object to be reproduced in wax, or Plaster of Paris. The surface of this cast is then brushed over with black-lead to render it a conductor, and the metal deposited directly upon it. The deposit obtained will then exactly resemble the original object.

The pages and engravings in the book before the reader are illustrations of the perfection and practical application of the electrotype process. The engravings were first cut upon wood-blocks, and then, with the ordinary type, formed into pages. Casts of the whole in wax were next made, and an electrotype coating of copper deposited upon them, and from the copper plates so formed the book was printed. The great advantage of this is, that the copper being harder than the ordinary type metal, is more durable, and resists the wear of printing from its surface for a longer period.

How has the electrotype process affected engraving?

The improvement effected by electro-metallurgy in engraving is very great. When a copper plate is engraved, and impressions printed off from it, only the first few, called "proof impressions," possess the fineness of the engraver's delineation. The plate rapidly wears and becomes deteriorated. But by the electrotype process, the original plate can at once be multiplied into a great many plates as good as itself, and an unlimited number of the finest impressions procured.

In this way the map plates of the Coast Survey of the United States, some of which require the labor of the engraver for years, and cost thousands of dollars, are reproduced—the original plate being never printed from.

One of the simplest illustrations of metallic deposit by electro-chemical action is afforded by the following experiment:—Put a piece of silver in a glass containing a solution of sulphate of copper, and into the same glass insert a piece of zinc. No change will take place in either metal so long as they are kept apart; but as soon as they touch, the copper will be deposited upon the silver, and if it be allowed to remain, the part immersed will be completely covered with copper, which will adhere so firmly that mere rubbing alone will not remove it.

How does the union of two metals affect their durability?

785. When two metals which are positive and negative in their electrical relations to each other, are brought in contact, a galvanic action takes place which promotes chemical change in the positive metal, but opposes it in the negative metal.

What are illustrations of this principle?

Thus, when sheets of zinc and copper immersed in dilute acid touch each other, the zinc oxydizes or rusts more, and the copper less rapidly, than without contact. Iron nails, if used in fastening copper sheathing to vessels, rust much quicker than when in other situations, not in contact with the copper. The reason is, that the contact of the two metals excites galvanic action, which causes the iron to rust speedily, but protects the copper.

What is galvanized iron?

What is called galvanized iron, is iron covered entirely, or in part, with a coating of zinc. The galvanic action between the two oxydizes the zinc, but protects the iron from rust.

How did Davy attempt to protect the sheathing of ships from corrosion?

Copper, when immersed in sea-water, rapidly wastes by the chemical action of the oxygen dissolved in sea-water; but if it be brought in contact with zinc, or some metal that is more electro-positive than itself, the zinc will undergo a rapid change, and the copper will be preserved. Sir Humphrey Davy attempted to apply this principle to the protection of the copper sheathing of ships, by placing at intervals over the copper small strips of zinc. The experiment was tried, and a piece of zinc as large as a pea was found adequate to preserve forty or fifty square inches of copper; and this wherever it was placed, whether at the top, bottom, or middle of the sheet, or under whatever form it was used. The value of the application was, however, neutralized by a consequence which had not been foreseen. The protected copper bottom rapidly acquired a coating of sea-weeds and shell-fish, whose friction on the water became a serious resistance to the motion of the vessel, and it was discovered that the bitter, poisonous taste of the copper surface, when corroded, acted in preventing the adhesion of living objects. The principle, however, has been applied with success to protect the iron pans used in evaporating sea-water.

CHAPTER XVII.

THERMO-ELECTRICITY.

What is Thermo-electricity?

786. If two dissimilar metallic bars be soldered together, and heated at the point of junction, an electric current will circulate through them, and may be carried off by connection with any good conductor. Electricity thus generated or developed is called Thermo-electricity.

Thus, if two bars, one of German silver and the other of brass, as represented in Fig. 344 (the dark one being the brass), be heated at their junction, an electric current will flow in the direction of the arrows from the German silver to the brass.

Fig. 344.

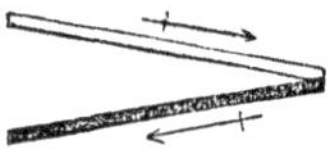

Different degrees of temperature, also, in the same metal, will occasion an electric current to flow from the colder to the warmer portions.

The properties of thermo-electricity are the same as those of ordinary electricity.

The metals best adapted for showing its effects are German silver, bismuth, brass, iron, and antimony.

How are thermo-electric batteries constructed?

Thermo-electric batteries of considerable power may be constructed by combining together alternate plates of German silver and brass, or bismuth and antimony, thick cards of pasteboard being so placed between the plates, that a contact of the metals is prevented, except at the ends. Such a battery, represented by Fig.

345, may be made to develop electricity by heating one end of the bundle, or pile of plates.

Fig. 345.

By binding together two bars of bismuth and antimony, an electric current can be proved to circulate with the slightest variation of temperature.

A series of slender bars of these two metals, arranged as a thermo-electric battery, is far more sensitive to heat than the most delicate thermometer; so that the heat radiated from the hand brought near to one end of the battery is sufficient to excite an appreciable amount of electricity.

Fig. 346 represents the construction of such a battery. It consists of thirty-six delicate bars of bismuth and antimony, alternately connected at their extremities and packed in a case, the ends of which are removed in the figure to show the bars. The area of such a battery is not quite one half an inch. A represents a conical reflector, used to concentrate rays of heat in experimenting.

Fig. 346.

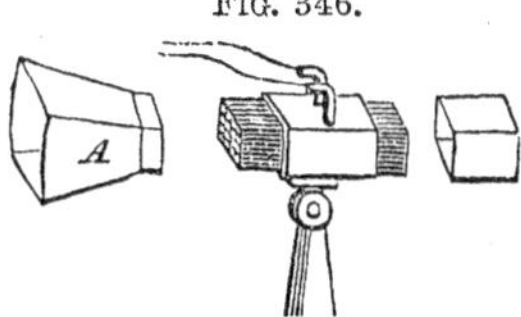

It has been also found that when hot water mixes with cold water, that electricity is produced; the hot liquor being positive and the cold negative.

CHAPTER XVIII.

MAGNETISM.

What is a natural magnet?

787. A NATURAL magnet, sometimes called a loadstone, is an ore of iron, known as the protoxyd of iron, or magnetic oxyd of iron, which is capable of attracting other pieces of iron to itself.

Fig. 347.

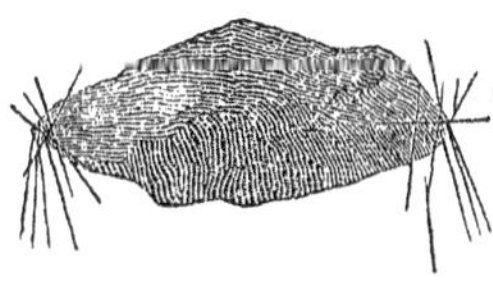

Natural magnets are by no means rare; they are found in many places in the United States, and in Arkansas, especially, an ore of iron possessing remarkably strong attractive powers is very abundant.

The magnetic ore is usually of a dark color, and possesses but little metallic luster. If a piece of this ore be dipped in iron filings, or brought in contact with a number of small needles, they will adhere to the extremities of the magnet, as is represented in Fig. 347.

Can a magnet communicate its properties?

When a natural magnet is brought near to, or in contact with a piece of soft iron or steel, it communicates its attractive properties, and renders the iron a magnet. In doing so, it loses none of its original attractive influence.

What are artificial magnets?

Bars of iron or steel which by contact with natural magnets, or by other methods, have acquired magnetic properties, are termed artificial magnets.

For all practical purposes, artificial magnets are used in preference to natural magnets, and can be made more powerful.

Define the meaning of the terms magnetic force and magnetism.

The attractive force of magnets has received the name of MAGNETIC FORCE, and that department of science which treats of magnets and their properties is denominated MAGNETISM.

This designation must not be confounded with Animal Magnetism, a term which has been adopted to designate a certain influence which one person may exercise over another by means of the will.

What are the poles of a magnet?

788. The attractive power of the magnet is not diffused uniformly over every part of its surface, but resides principally at opposite points or extremities of its surface. These points are termed *poles*.

Between the regions of greatest attraction, a point may be found where the attractive influence wholly disappears.

When a bar magnet is rolled in iron filings, the filings attach themselves to the magnet in the manner represented in Fig. 348, and in this way clearly indicate the location of the magnetic force.

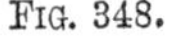

FIG. 348.

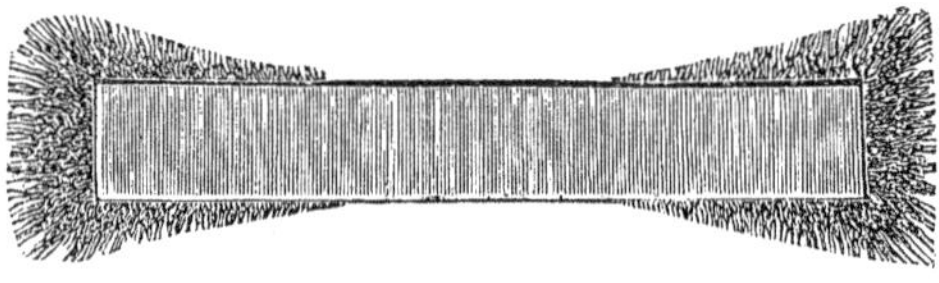

In a steel magnet, the actual poles, or points of greatest magnetic intensity, are not exactly at the ends, but at a distance of about one tenth of an inch from each extremity.

In what position will a magnet freely suspended rest?

789. When a magnet is supported in such a way as to move freely, it will rest only in one position, viz., with its poles, or extremities directed nearly north and south.

If drawn aside from this position, it will continue to vibrate backward and forward, until it again rests in the same position.

What are the north and south poles of a magnet?

The pole, or extremity of the magnet that constantly points toward the north, is called the North Pole, and the one that points toward the south, the South Pole of the magnet.

What is magnetic polarity or directive power?

790. That property of a magnet which will cause it, when suspended freely, to constantly turn the same part toward the north pole, and the opposite part toward the south pole of the earth, is termed magnetic polarity, or directive power.

When is a magnet said to traverse?

When a magnet, being free to move, places itself after deflection in a nearly north and south line, it is said to traverse.

The attractive force of the loadstone, or natural magnet, can not be considered as of any great amount. Native magnets, in their rude state, will seldom lift their own weight, and, with some rare exceptions, their power is limited to a few pounds.

What is the general law of magnetic attractions and repulsions?

791. When two bodies possessing magnetic properties are brought near, or in contact with each other, the like poles will repel, and the unlike attract each other.

Thus, the north or the south poles of two magnets repel each other; but the north pole of the one will attract the south pole of the other.

In what substances may magnetism be most easily excited?

792. Magnetism may be excited most readily in iron and steel. In steel the magnetic property, when induced, remains permanent; but soft iron loses its power as soon as it is removed from the influence of the exciting magnet. Brass, nickel, and cobalt may also be rendered magnetic.

Recent investigations have shown that the influence of magnetism, which was once supposed to be wholly restricted to iron and its compounds, is almost as pervading and wide-extended as that of electricity. The emerald, the ruby, and other precious stones, the oxygen of the air, glass, chalk, bone, wood, and many other substances, are more or less susceptible to magnetic influence. This influence, however, is perceptible only by the nicest tests, and under peculiar circumstances.

In what form are artificial magnets constructed?

Artificial magnets of iron or steel may be of any required form, or of almost any dimensions. For general purposes, they are limited to straight bars.

FIG. 349.

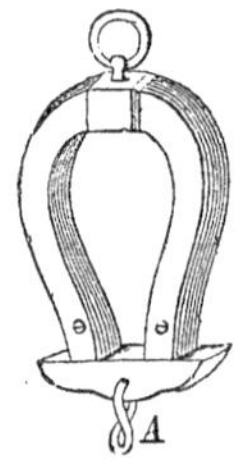

When a piece of iron not magnetic is brought in contact with a common magnet, it will be attracted by either pole; but the most powerful attraction takes place when both poles can be applied to the surface of the piece of iron at once. The magnetic bars are for this purpose bent somewhat into the shape of the letter U, and are termed horse-shoe magnets.

Several of these are frequently joined together with their similar poles in contact; they then constitute a compound magnet, and are very powerful, either for lifting weights or charging other magnets.

For the purpose of distinguishing between the two poles of an artificial magnet, the end of the bar which is designated as the north pole is generally marked with a + or with the letter N.

If we break an artificial magnet, what occurs?

If we break a magnet across the middle, each fragment becomes converted into a perfect magnet; the part which originally had a north pole acquires a south pole at the fractured end, and the part which originally had a south pole, gets a north pole.

Thus, if the bar N S, Fig. 350, be broken in the center, each of the fractured ends will exhibit a polar state, as perfect as the entire magnet; the fractional end *s* becoming a south and *n* a north pole, although at this middle point, where *n* and *s* join, no magnetism could, before the breaking, have been detected.

FIG. 350.

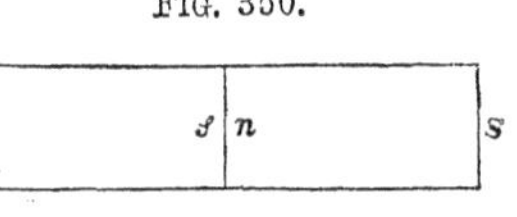

If we divide up a magnet to the extreme degree of mechanical fineness possible, each particle, however small, will be a perfect magnet.

The properties of a magnet are not at all affected by the presence or absence of air; but its influence is as great in a vacuum as in any other situation.

Heat weakens the power of a magnet, and a white heat destroys it entirely.

Where does the magnetic power of a body reside?

793. The magnetic power of an iron or steel magnet appears to reside wholly upon the surface, and to circulate about it.

How may steel be rendered magnetic?

To render a bar of steel magnetic, the north pole of a magnet is placed on the center of a bar of steel and repeatedly drawn over it toward one extremity; the other half is subjected to a similar treatment with the south pole of the magnet; the bar is thus rendered magnetic, and only loses this property when strongly heated.

How is soft iron magnetized?

A bar of soft iron becomes magnetic by simple contact with a magnet, but the effect, as before stated, is not permanent.

May iron be rendered magnetic by induction?

It is not necessary that absolute contact should take place between a bar of soft iron and a magnet, in order to render the iron magnetic; but whenever a magnet is brought near to a piece of iron in any shape, the latter is rendered magnetic by the influence of the former. To this phenomenon the name of induction has been given, and the distance through which this effect can take place is called the magnetic atmosphere.

FIG. 351.

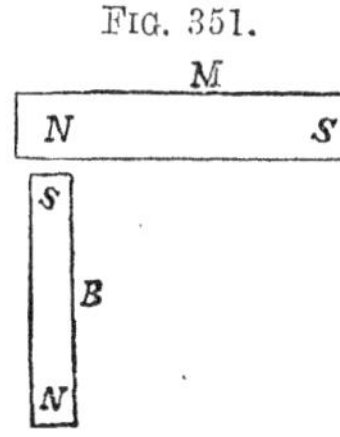

Thus, let a bar of soft iron, B, as in Fig. 351, be brought near to a magnet, M, whose poles, north and south, are indicated by N and S. By induction, the bar will be rendered magnetic, the end of the bar toward the north pole of the magnet constituting its south pole, and the other end the north pole.

In all cases, where either pole of a magnet is brought near to, or in contact with bodies capable of acquiring magnetism, the part which is nearest to the pole of the magnet acquires a polarity opposite, while the remote extremity becomes a pole of the same kind; hence the attraction of a magnet for iron, is simply the attraction of one pole of a magnet for the opposite pole of another.

How may the phenomena of magnetic induction be exhibited?

The general effect of magnetization by induction may be clearly exhibited by bringing a powerful magnet near to a piece of soft iron, as a large key, when it will be found that the large key will support several smaller ones; but as soon as the body inducing the magnetic action is removed, they all drop off.

Can the earth induce magnetism?

Magnetism may be also induced in a bar of iron by the action of the earth.

Most iron bars and rails, as the vertical bars of windows, that have stood for a considerable time in a perpendicular position, will be found to be magnetic.

What are illustrations of magnetism induced by the earth?

If we suspend a bar of soft iron sufficiently long in the air, it will gradually become magnetic; and although when it is first suspended it points indifferently in any direction, it will at last point north and south.

If a bar of iron, such as a kitchen poker, which has been found to be devoid of magnetism, is placed with one end on the ground, slightly inclined toward the north, and then struck one smart blow with a hammer upon the upper end it will acquire polarity, and exhibit the attractive and repellent properties of a magnet.

Does magnetic attraction extend through other bodies?

Magnetic attraction can be made to exert its influence through glass, paper, and solid and liquid substances generally which are not capable of acquiring magnetic influence in the ordinary manner.

FIG. 352.

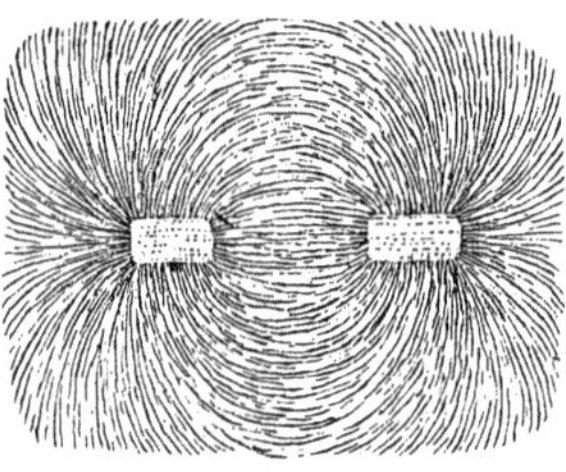

If a horse-shoe magnet be placed underneath a sheet of paper which has iron filings sprinkled over its surface, the filings, upon the approach of the magnet, will arrange themselves in great regularity in lines diverging from the poles of the magnet, in curves, and extending from the one pole to the other, as is represented in Fig. 352. The numerous fragments of iron, being rendered magnets by induction, have their unlike poles fronting each other, and they therefore attract one another, and adhere in the direction of their polarities, forming what are termed magnetic curves.

If a plate of iron is caused to intervene between the magnet and the under surface of the paper, the magnetic influence is almost entirely cut off.

Do artificial magnets lose their properties?

794. Magnets, if left to themselves, gradually, and in a space of time varying with the hardness of the metal composing them, lose their magnetic properties, from the recombination of their separate fluids.

This is prevented by keeping their poles united by

What is an Armature? means of a soft iron bar called an Armature, represented at A, Fig. 349.

This becoming magnetic by induction, reacts upon the magnetism in the poles of the magnetic bar, and tends to increase rather than diminish their intensity.

What is the power of artificial magnets? The lifting or sustaining power of magnets varies very materially. The most powerful that we are acquainted with are capable of sustaining twenty-six times their own weight.

How does the force of magnetic attraction and repulsion vary? The law of magnetic attraction and repulsion is the same as that of gravitation; that is, these forces increase in the same proportion as the square of the distance from the center of attraction or repulsion diminishes.

According to what theory are magnetic phenomena accounted for? 795. The various phenomena of magnetism have been accounted for by supposing that all bodies susceptible of magnetism are pervaded by a subtle imponderable fluid, which is compound in its nature, and consists of two elements, one called the austral, or southern magnetism, and the other the boreal, or northern magnetism. Each of these, like positive and negative electricities, repel their own kind, and attract the opposite kind.

When a body pervaded by the compound fluid is in its natural state and not magnetic, the two fluids are in combination and neutralize each other. When a body is magnetic, the fluid which pervades it is decomposed, the austral fluid being directed to one extremity of the body, and the boreal to the other.

Iron and steel are easily rendered magnetic, because the fluids which pervade them can be easily decomposed by the action of other magnets. In iron, the separation of the two kinds of magnetism may be easily, but only transitorily effected. The magnet, therefore, attracts it powerfully, converting it, however, into only a temporary magnet. In steel, the two kinds of magnetism are not so easily separated; hence the latter is but slightly attracted by the most powerful magnets. When once effected, however, the separation is permanent, and the steel becomes a perfect magnet.

As, according to this theory, the act of rendering a body magnetic consists simply in decomposing a fluid pervading it, we can easily understand how, by means of one artificial magnet, an infinite number of other magnets may be made, without the former losing any of its magnetic properties.

What is a Magnetic Needle? 796. The Magnetic Needle (Fig. 353) is simply a bar of steel, which is a magnet, balanced upon a pivot in such a way that it can turn freely in a horizontal direction.

FIG. 353.

Such a needle, when properly balanced, will be observed to vibrate more or less, until it settles in such a direction that one of its extremities, or poles, points toward the north, and the other toward the south. If the position of the needle be altered or reversed, it will always turn and vibrate again until its poles have attained the same direction as before.

It is this remarkable property of a magnetized steel bar, of always assuming a definite direction, that renders the compass of such value to the mariner, the engineer, and the traveler.

What is a Compass?

The ordinary compass consists of a magnetic needle, or bar balanced upon a pivot, and inclosed within a shallow box, or metallic case. Upon the bottom of the box is a circular card with the chief, or cardinal points of the horizon, north, south, east, west, marked upon it.

Fig. 354 represents the form and construction of the ordinary, or land compass. The term compass is derived from the card, which *compasses*, or *involves*, as it were, the whole plane of the horizon.

FIG. 354.

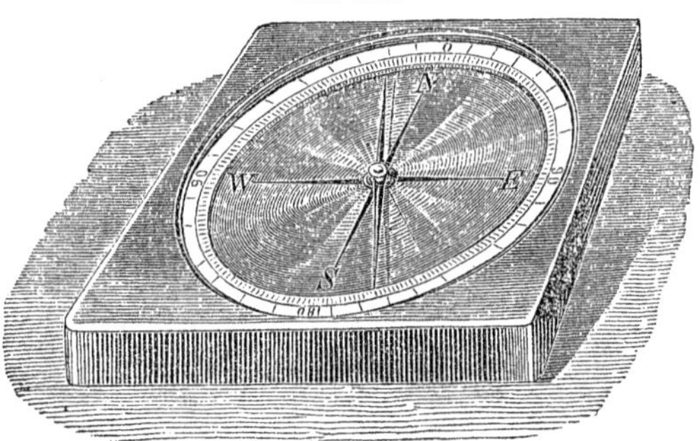

What is the construction of the Sea, or Mariner's Compass?

In the Sea, or Mariner's Compass, the needle is attached to the under side of the card, in such a way that both traverse together—the needle itself being out of sight. Upon the surface of the card is engraved a radiating diagram, dividing the whole circle of the horizon into thirty-two parts, called points. The compass-box is supported by means of two concentric hoops, called *gimbals*. These are so placed as to cross each other, and support the box immediately in the center of the two; so that whichever way the vessel may roll or lurch, the card is always in a horizontal position, and is certain to point the true direction of the head of the ship. Fig. 355 represents the construction and mounting of the Sea Compass.

FIG. 355.

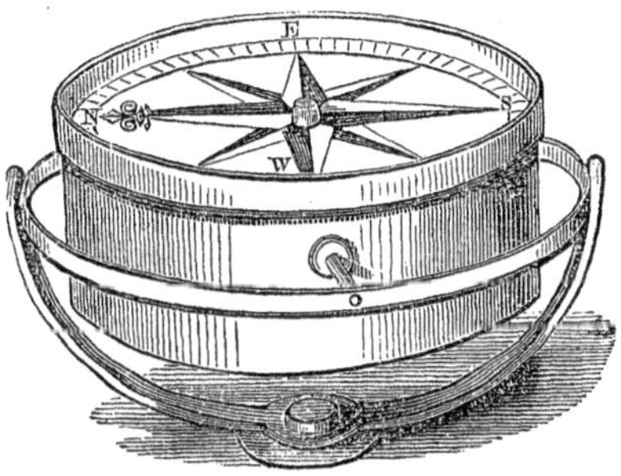

What is a Dipping Needle?

797. If a simple bar of unmagnetized steel, or an ordinary needle be suspended from a center, instead of being balanced upon a pivot beneath it, it will hang horizontally, and manifest no inclination to dip from a horizontal line, either on one side or the other of the center of suspension. But if the bar, or needle, be made a magnet, it will no longer lie in a horizontal direction, but one pole will incline downward and the other upward; the inclination in this latitude to the horizon being about 70°.

FIG. 356.

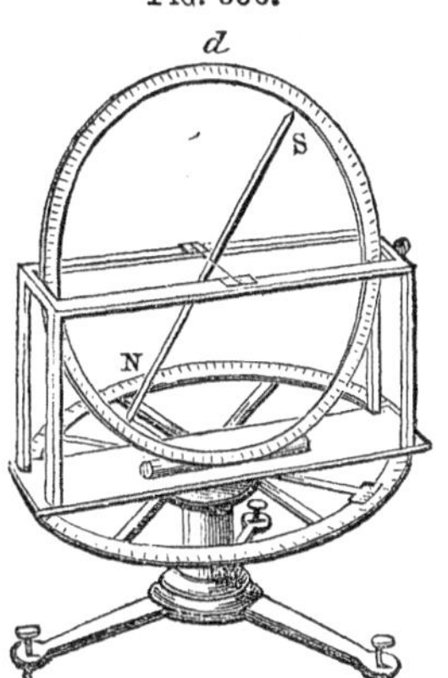

Such arrangement is called a Dipping Needle.

Fig. 356. represents the construction and appearance of the dipping needle.

Does the magnetic needle point due north and south.

798. Although the magnetic needle is said to point north and south, accurate observations have shown that it does not point exactly north and south except in a few restricted positions upon the earth's surface.

What is the magnetic meridian?

799. The direction assumed by a horizontal needle in any given place upon the earth's surface, is called the magnetic meridian.

What is a terrestrial meridian?

A terrestrial meridian, it will be remembered, is a great circle, supposed to be drawn around the earth, passing through both poles, and any given place upon its surface, and intersecting the equator at right angles. (See § 68, Fig. 6, page 36.) The direction of a needle which would point due north and south at any place, will be the true, or terrestrial meridian of that place.

What is the variation or the declination of the needle?

The deviation of the needle from the true north and south, or the angle formed by the magnetic meridian and the terrestrial meridian, is called the variation, or declination of the needle.

What are the lines of no variation?

There are two lines upon the earth's surface, along which the needle does not vary, but points to the true north and south. These lines are called the eastern and western lines of no variation, and are exceedingly irregular and changeable.

Their position is as follows:—The western line of no variation begins in latitude 60°, to the west of Hudson's Bay, passes in a south direction through the American lakes, to the West Indies and the extreme eastern point of South America. The eastern line of no variation begins on the north in the White Sea, makes a great semicircular sweep easterly, until it reaches the latitude of 71°; it then passes along the Sea of Japan, and goes westward across China and Hindoostan to Bombay; it then bends east, touches Australia, and goes south.

In proceeding in either direction, east or west from the lines of no variation, the declination of the needle gradually increases, and becomes a maximum at a certain intermediate point between them. On the west of the eastern line the declination is west; on the east it is east.

At Boston, in the United States, the declination of the needle is about 5½° west; in England it is about 24° west; in Greenland, 50° west; at St. Petersburg, 6° west.

How is the directive power of the needle accounted for?

800. As the directive property of the magnetic needle is observed everywhere in all parts of the world, on all seas, on the loftiest summits of mountains, and in the deepest mines, it is evident that there must be a magnetic force which acts at all points of the earth's surface, since magnetic needles can no more take up a direction of themselves than a body can acquire motion of itself. To explain these phenomena, the earth itself is considered to be a great magnet, and the points toward which the magnetic needle constantly turns are called the magnetic poles of the earth. These poles, by reason of their attractive influence, give to the needle its directive power.

Where are the magnetic poles situated?

The two poles of the great terrestrial magnet which are situated in the vicinity of the poles of the earth's axis, are termed respectively the magnetic north pole and the magnetic south pole. These contrary poles attract each other, and thus a magnetic needle will turn its south pole to the north, and its north pole to the south. Hence, what we generally call the north pole of a needle is in reality its south pole, and its south pole is its north pole.

The exact position of the northern magnetic pole is about 19° from the north pole of the earth, in the direction of Hudson's Bay. It was visited by Sir J. Ross in 1832, in his voyage of Arctic discovery. The south magnetic pole is situated in the antarctic continent, and has been approached within 170 miles.

If a compass needle be carried to the magnetic pole what will occur?

If the ordinary compass be carried to either of the magnetic poles, it will lose its power and point indifferently in any direction. If it is carried beyond the magnetic pole, to any point between it and the true pole, the poles of the needle become reversed, the end called the north pole pointing to the south, and the south to the north.

How does the position of the dipping needle vary?

The position assumed by the dipping needle varies in different latitudes. If it were carried directly to the north magnetic pole, its south pole would be attracted downward, and the needle would stand perfectly upright. At the south magnetic pole, its position would be exactly reversed.* If the dipping needle be taken to the equator of the earth, or to a point midway between the north and south magnetic poles, it will be attracted equally by both, and will remain perfectly horizontal, or cease to dip at all: as we go north or south, however, it dips more and more, until at the magnetic poles, as before stated, it becomes perpendicular—the end which was uppermost at the north being the lowest at the south.†

Fig. 357.

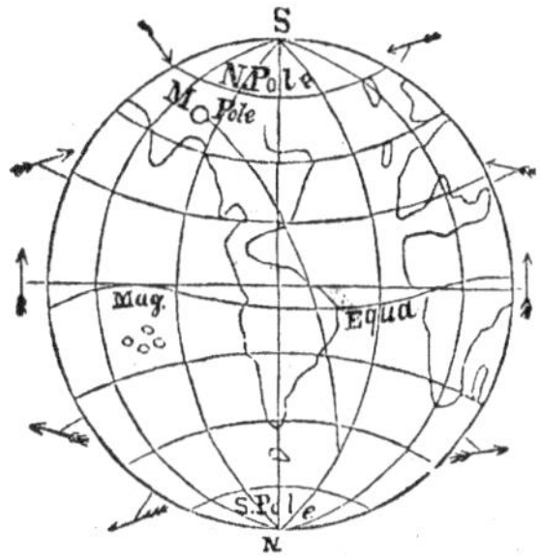

Fig. 357 represents the position assumed by the magnetic needle in various latitudes. The magnetic poles of the earth are not stationary, but change their position gradually during long intervals of time.

Observations on the temperature of the earth have afforded some reason for believ-

* Like the declination and dip, the intensity of the earth's magnetism varies very much in different parts of the earth; at the magnetic equator being the most feeble, and gradually increasing as we approach the poles. The intensity of terrestrial magnetism in different places may be measured by the number of vibrations made by a magnetic needle in a given time.

† As the directive tendency of the horizontal needle arises from its poles being attracted by those of the earth, it is evident from the rotundity of the earth, that its poles will not be attracted by those of the earth horizontally, but downward, so that the needle can not tend to be horizontal, except when it is acted upon by both poles equally—that is, when midway between them. When nearer the north magnetic pole than the south, its north end must be attracted downward, and the contrary when it is nearest the south pole. Accordingly, a needle which was accurately balanced on its support before being magnetized, will no longer balance itself when magnetized, but in this country its north pole will appear to dip, or appear to be the heavier end. This circumstance has to be corrected in ships' compasses by a small sliding weight attached to the southern half, which weight has to be removed on approaching the equator, and shifted to the other side of the needle when in the northern hemisphere.

ing, that the points upon the earth's surface where the greatest degree of cold is experienced, or where the yearly mean of the thermometer is lowest, coincides with the location of the magnetic poles.

What is the diurnal variation of the needle?

801. Beside the variation from the true north and south, the magnetic needle is subject to a diurnal variation. This movement, or variation, commences about seven in the morning, when the north end of the needle begins to deviate toward the west; it reaches its maximum deviation about two o'clock in the afternoon, when it begins to return slowly to its original position.

The magnetic needle is subject also to an annual movement, and a movement different in the winter months from that noticed in the summer months.

What is the supposed cause of the periodical variations of the needle?

The daily, monthly, and yearly variations of the needle are supposed to be occasioned by variations in the temperature of the earth's surface, depending upon the changes in the position and action of the sun.

Observations made for a great number of years seem to show that the entire magnetic condition of the earth is subject to a periodical change, but neither the cause or the laws of this change are as yet understood.

For most practical operations, as in navigation and surveying, the deviation of the magnetic needle from the true north and south, is carefully taken into account, and a rule of corrections applied. A knowledge of the amount of variation, east or west, for different localities upon the earth's surface, may be obtained from tables accurately arranged for this purpose.

The variation of the magnetic needle from the true north and south, is said to have been first noticed by Columbus in his first voyage of discovery. It was also observed by his sailors, who were alarmed at the fact, and urged it as a reason why he should turn back.

When was the compass discovered?

The compass is claimed to have been discovered by the Chinese: it was, however, known in Europe, and used in the Mediterranean, in the thirteenth century. The compasses of that time were merely pieces of loadstone fixed to a cork, which floated on the surface of water.

802. The resemblance between magnetism and electricity is very striking, and there are good reasons for believing that both are but modifications of

one force. Both are supposed to consist of two fluids, which repel their own kind, and attract the opposite. The fluid in both cases is supposed to reside upon the surface of bodies; the laws of induction in both are the same; and each can be made to excite or develop the other.

CHAPTER XIX.

ELECTRO-MAGNETISM.

What is Electro-magnetism?

803. MAGNETISM developed through the agency of electrical or chemical action, is termed Electro-magnetism.

Among the earliest phenomena observed which indicated a connection between magnetism and electricity, it was noticed that ships' compasses have their directive power impaired by lightning, and that sewing-needles are rendered magnetic by electric discharges passed through them.

What effect is produced when a magnetic needle is brought near a conducting wire?

In 1820, a discovery was made by Professor Oersted of Denmark, which established beyond a doubt the connection of electricity and magnetism. He ascertained that a magnetic needle brought near to a wire, through which an electric current was circulating, was compelled to change its natural direction, and that the new direction it assumed was determined by its position in relation to the wire and to the direction of the current transmitted along the wire.

Further experiments developed the following law:—

In what direction do electric currents exert their influence?

Electric currents exert a magnetic influence at right angles with the direction of their flow, and when they act upon a magnetic needle they tend to cause the needle to assume a position at right angles to the direction of the current.

FIG. 358.

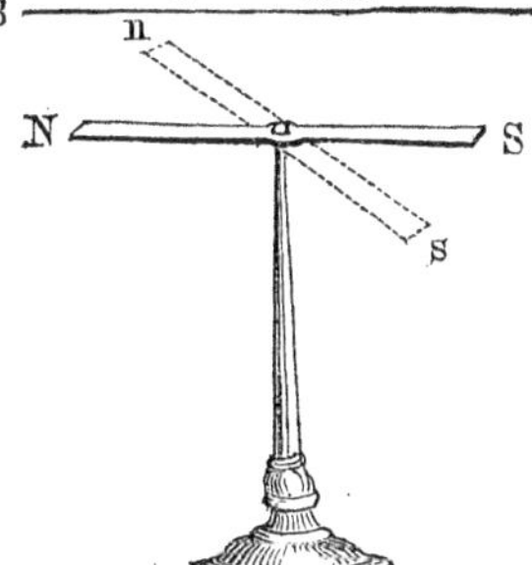

Thus, suppose an electric current to pass on the wire A B, Fig. 358, in the direction of the arrow; suppose a magnetic needle, N S, to be placed directly under the wire and parallel to it. By the action of the electric current flowing in the direction A B, the needle is caused to move from its north and south position and turn round, and if the current is sufficiently strong, it will place itself at right angles with the wire, as is represented in the figure.

If the current, however, had passed in the same direction below the needle, instead of above it as in the first instance, the deflection of the needle would have taken place as before, but in an opposite direction, the pole S standing where the pole N did previously, and N also in the place of S.

In like manner, if the needle be placed by the side of the wire, a like effect will be produced; on one side it dips down, and on the other it rises up; and in whatever other position the needle may be placed, it will always tend to set itself at right angles to the current. If the wire be bent in the form of a rectangle, as is represented in Fig. 359, so as to carry the current around the needle, above and below it in opposite directions, the opposite currents, instead of neutralizing, will assist each other, and the needle will move in accordance with the first direction of the current.

FIG. 359.

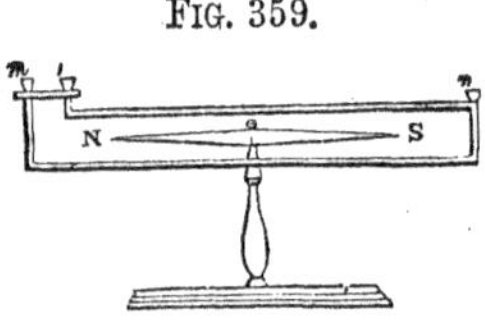

If the wire, instead of making a single turn, is bent many times around the needle, the magnetic force excited by the current of electricity traversing the wire, will be greatly increased, the increase being, within certain limits, proportional to the number of turns of the wire.

Describe the Galvanometer.

It is upon this principle that an instrument called the Galvanometer, for measuring the quantity of an electric current, is constructed. It consists of a rectangular coil of copper wire, N B S, Fig. 360, containing about 20 convolutions, the separate coils being insulated by winding the wire with silk thread. A magnetic needle, supported on a pivot, is placed in the center of the coil, and a graduated circle is fixed below it to measure the amount of the deflection; the two ends of the wire connect with two cups, C and Z, which contain mercury, and into which the poles of the battery transmitting the current dip.

FIG. 360.

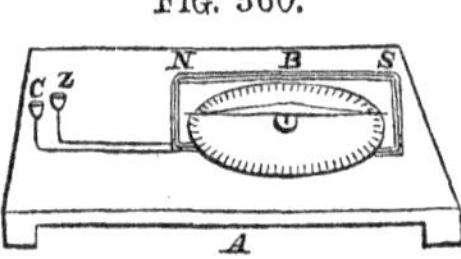

What is the Astatic Needle?

In this form of the instrument the transmitted current is obliged to contend with the influence of the earth's magnetism, which tends to hold the needle in its original position, and unless the former is more powerful than the latter, the needle is not moved. This difficulty has been overcome by means of an arrangement called the *Astatic Needle.* This consists essentially of two needles fastened together, one above the other, but with their poles in opposite directions, as is represented in Fig. 361. In this way the influence of the earth is almost entirely emoved, and the force of the transmitted current is rendered more effective.

FIG. 361.

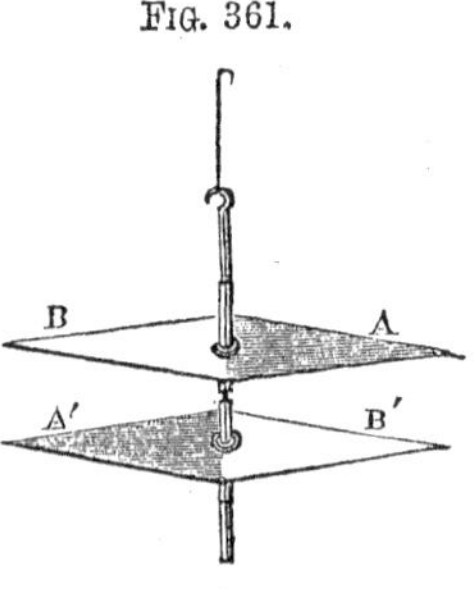

By means of the galvanometer, the most feeble traces of electricity can be detected; and electric currents which would fail to influence the most sensitive gold leaf electrometer can be made to affect perceptibly the magnetic needle. Galvanometers are sometimes called electro-multipliers.

In what manner does an electric current exert its magnetic force?

804. Electricity, unlike all other motive forces in nature, exerts its magnetic force laterally; all other forces exerted between two points act in the direction of a straight line connecting their points, but the electric current exerts its magnetic influence at right angles to the direction of its course.

When a magnetic pole is influenced by an electric current, it does not move either directly toward or directly from the conducting wire, but it tends to revolve about it.

By the application of these facts, it has been discovered that rotatory movements can be produced by magnets around conducting wires, and conversely, that conducting wires can be made to rotate around magnets.

FIG. 362.

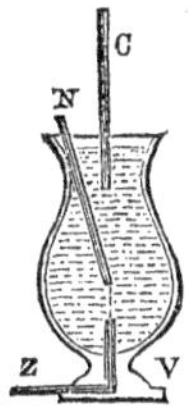

The rotation of the pole of a magnet around a fixed conducting wire may be shown by a piece of apparatus represented by Fig. 362. A small magnet, N, is fixed to the lower part of a vessel, V, by means of a silk thread; the vessel is filled with mercury nearly to the top of the magnet; G is a conducting wire dipping into the mercury, and Z is another conductor communicating with the mercury at the bottom of the vessel. Now, when the electric current is established, by connecting the extremities of the wires C and Z with the opposite poles of the battery, the pole N of the magnet revolves round the conducting wire C. If the current is descending, that is, if C be connected with the positive pole of the battery, and if N be a north pole, its motion round the wire will be direct, that is, in the direction of the hands of a watch; and so on, *vice versâ*.

FIG. 363.

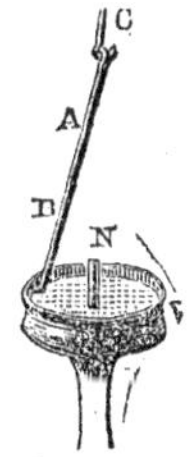

A different arrangement, by which a movable wire traversed by a current, may be made to revolve around the pole of a fixed magnet, is represented by Fig. 363. A wire, A B, is suspended from the wire C by a loop, and dips into the mercury in a vessel, V; when the circuit is established, by connecting C and N with the respective poles of the battery, the conducting wire revolves around the pole N of the magnet.

If the current be descending, and N be the north pole of the magnet, the rotation will be direct.

On similar principles, various kinds of reciprocating and rotatory movements may be produced.

In what manner can an electric current be made to excite magnetism?

805. If a piece of soft iron, entirely wanting in magnetism, be placed within a coil of wire through which an electric current is circulating, it will be rendered intensely magnetic, so long as the current continues; but the moment the current ceases, the iron loses its magnetism.

What is an Electro-magnet?

Magnets formed in this way, through the agency of electricity, are called Electro-magnets, and are more powerful than any others.

What is a Helix?

The coil, or spiral line of wire used for exciting magnetism in the iron by conducting a current of electricity about it, is called a Helix.

Fig. 364.

It is usually made of copper wire, coated with some non-conducting substance, such as silk wound round it. The coils of the wire are generally repeated one over the other, until the size of the helix is sufficient, since the magnetic action of an electric current upon a bar of iron increases to a certain extent with the number of revolutions it performs about it. Fig. 364 represents the appearance of a helix.

It is necessary for the induction of magnetism in iron bars by electricity, that the current should flow at right angles to the axis of the bars.

What determines the poles of an electro-magnet?

If the bar be steel, the magnetism thus induced will be permanent; and the direction in which the current moves round the helix determines which of its extremities shall constitute its north, and which its south pole.

When the current circulates in the direction of the hands of a watch, the north pole of the bar will be at the farthest end of the helix.

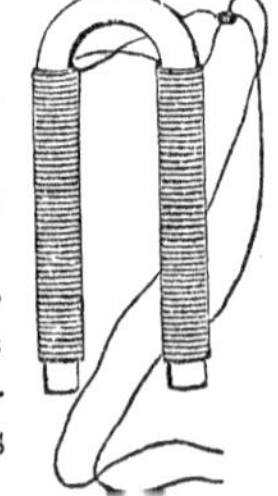

Fig. 365.

If a bar of soft iron, bent in the form of a horse shoe magnet, be wound with insulated wire, as is represented in Fig. 365, and a current of electricity transmitted through it, it becomes a most powerful magnet.

Electro-magnets of this character have been formed capable of supporting more than a ton weight. The magnetic power thus developed is wholly dependent upon the existence of the current, and the moment it ceases the weights fall away by the action of gravity.

Fig. 366.

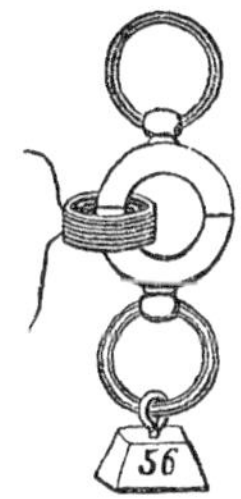

If two semicircular rings of soft iron be passed within a helical ring, as is represented in Fig. 366, they will become so strongly magnetic on passing the current of even a small battery, as to be separated with extreme difficulty. A rod of iron brought near to one of the extremities of a longitudinal helix, is not only attracted but lifted up into the center of the coil, where it remains suspended without contact or visible support, so long as the current continues in action. If the battery and helix be of sufficient size, a considerable weight may be suspended. In some experiments at the Smithsonian Institution at Washington, a few years since, a bar of iron weighing 80 pounds was raised and suspended in the air without being in contact with any body.

Has electro-magnetic force been applied to any practical purpose for propelling machinery?

806. Many attempts have been made to take advantage of the enormous force generated and destroyed, in an instant, by making or breaking an electric current, for propelling machinery, but thus far all efforts have failed to produce any practical results.

One of the reasons why this power can not be used to advantage is, that the rate at which the power diminishes as we recede from the contact point of the magnets, prevents our obtaining the full force of the magnets. Thus, a magnet whose force in contact would be sufficient to raise 250 pounds, would exert a force of only 90 pounds at the distance of 1-250th of an inch, and of only 40 pounds at the distance of 1-50th of an inch. It is also found that notwithstanding the loss of power with distance, a still greater loss takes place with motion. The moment any magnetic body is moved in front of either a permanent or an electro-magnet, it loses power, and this loss increases very rapidly with the increase of velocity. This obstacle stopped the progress of the very extensive researches of Professor Jacobi, after he had expended upward of $120,000 granted him for his experiments by the liberality of the Russian government.

Upon what does the construction of the Morse telegraph depend?

807. The construction of the Morse magnetic telegraph depends upon the principle, that a current of electricity circulating about a bar of soft iron temporarily renders it a magnet.

The construction and method of operating the Morse telegraph may be clearly understood by reference to Fig. 367. F and E are pieces of soft iron surrounded by coils of wire, which are connected at *a* and *b* with wires proceeding from a galvanic battery. When a current is transmitted from a battery located one, two, or three hundred miles distant, as the case may be, it passes along the wires, and through the coils* surrounding the pieces of soft iron, F and E, thereby converting them into magnets. Above these pieces of soft iron is a metallic bar, or lever, A, supported in its center, and having at one end the arm, D, and at the other a small steel point, *o*. A ribbon of paper, *p h*, rolled on the cylinder, B, is drawn slowly and steadily off by a train of clock-work, K, moved by the action of the weight, P, on the cord, C. This clock-work gives motion to two metal rollers, G and H, between which the ribbon of paper passes, and which, turning in opposite directions, draw the paper from the cylinder B. The roller H has a groove around its circumference (not represented in the engraving), above which the paper passes. The steel point *o* of the lever A is also directly opposite this groove. The spring, *r*, prevents the point from resting upon the paper when the telegraph is not in operation.

FIG. 367.

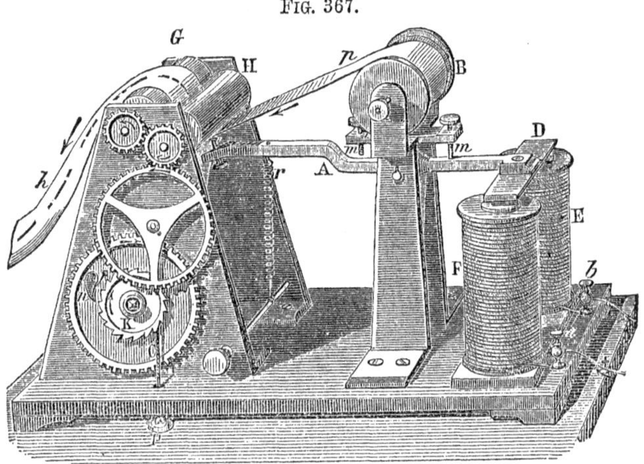

The manner in which intelligence is communicated by these arrangements is as follows: The pieces of soft iron, F and E, being rendered magnetic by the passage of a current of electricity transmitted from the battery through the coils of wire surrounding them, attract the metal arm D of the lever A. The end of the lever at D being depressed, the steel point *o* at the other extremity is elevated and caused to press against the paper ribbon and indent it. When the current from the battery is broken or interrupted, the pieces of soft iron F and E being no longer magnetic, cease to attract the arm D. The lever A is therefore drawn back to its former position by the action of the spring *r*, and the steel point *o* ceases to indent the paper. By letting the current flow

* These coils consist of very fine copper wire, some thousands of feet being generally contained in them. In this way a magnet of small size and great power may be obtained.

round the magnet for a longer or shorter time a dot, or a line is made, and the telegraphic alphabet consists of a series of such marks.*

Grove's battery (see Fig. 340) is generally used for working the telegraph, about thirty cups being required for a distance of 150 miles. These cups may be kept in one compact space, but operate the telegraph more successfully when distributed along the line. Such batteries will work for about two weeks without replenishing.

How many wires are necessary for working the telegraph?

Formerly two wires were required in telegraphing; one conveyed the current from the battery to the electro-magnet, at a distance, through which it passed, and then returned by another wire back to the battery. At present but one wire is generally used. It was found that the earth itself might be made to perform the function of the returning wire. To effect this all that is necessary is that one short wire from the battery at one end of a line, and from the electro-magnet at the other end, should be sunk into the moist earth, and there connected with a mass of conducting metal, from which the electricity passes to complete the closed circuit.

FIG. 368.

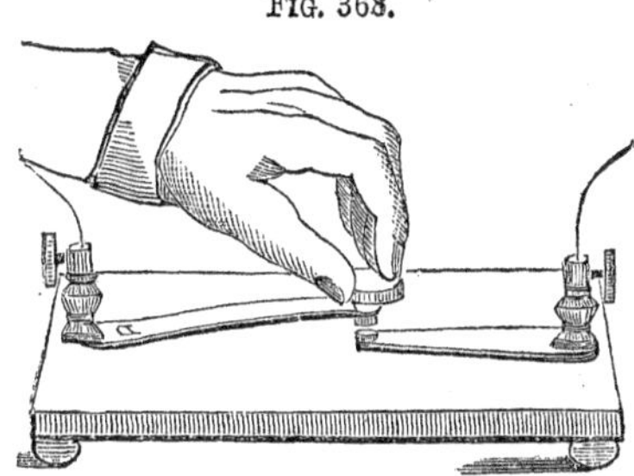

For interrupting the current and regulating the system of dots and lines, an instrument called the Signal-key, or, Break-piece, Fig. 368, is employed. This is placed near the battery, so as to be in the galvanic circuit. The operator, by pressing down the knob with the finger, closes the circuit and allows the current to pass, but when the pressure is removed communication is interrupted

* The following table exhibits the signs employed to represent letters in the Morse system of telegraphing:

ALPHABET.				NUMERALS.	
a	- —	n	— -		
b	— - - -	o	- -		
c	- - -	p	- - - - -	1	- — — -
d	— - -	q	- - — -	2	- - — - -
e	-	r	- - -	3	- - - — -
f	- — -	s	- - -	4	- - - - —
g	— — -	t	—	5	— — —
h	- - - -	u	- - —	6	- - - - - -
i	- -	v	- - - —	7	— — - -
j	— - — -	w	- — —	8	— - - - -
k	— - —	x	- — - -	9	— - - —
l	—	y	- - - -	0	—
m	— —	&	- - - -		

Experienced operators are often able to understand the message merely from the sounds, or clicks, of the lever.

What is the construction of the chemical telegraph?

808. In what is known as the "Bain," or chemical telegraph, there is no magnet created, but a small steel wire, connected with the wire from the line, presses upon a roll of paper, moved by clock-work. This paper, before being coiled on the roller, has been dipped in a nearly colorless chemical solution, which becomes colored when an electric current passes through it. By sending a current through the wire resting on the paper, we can stain it, as it were, in dots and lines in the same manner as the last instrument embossed it in dots and lines.

What is the printing telegraph?

809. The House's, or printing telegraph, differs from the others principally in an arrangement whereby the message as transmitted is printed in ordinary letters, at the rate of two or three hundred a minute.

What was the first telegraphic method proposed?

810. The method first proposed for communicating intelligence by electricity was by deflecting a compass needle by causing a current to pass along its length.

Thus, if at a given point we place a galvanic battery, and at a hundred miles from it there is fixed a compass needle, between a wire brought from, and another returning to the battery, the needle will remain true to its polar direction so long as the wires are free from the excited battery; but the moment connection is made, the needle is thrown at right angles to the direction of the current. The motion of the needle may thus be made to convey intelligence.

It is necessary, in conveying the wires from point to point, to support them on the poles by glass or earthen cylinders, in order to insure insulation, otherwise the electricity would pass down a damp pole to the earth, and be lost.

Does any principle or influence pass along the wires when a message is communicated?

811. The idea that many persons have, that some substance passes along the telegraphic wires when intelligence is transmitted, is wholly erroneous; the word current, as something flowing, expresses a false idea, but we have no other term to express electrical progression. We may, however, gain some idea of what really takes place, and of the nature of the influence transmitted, by remembering that the earth and all substances are reservoirs of electricity; and if we disturb this electricity at any given point, as at Washington, its pulsations may be felt at New York. Suppose the telegraphic wire a tube extending from Washington to New York perfectly filled with water; now, if one

drop more is forced into the tube at Washington, a drop must fall out at New York, but no drop is caused to pass from Washington to New York. Something like this occurs in the transmission of electricity.

Can electricity be made to measure time?

812. Electricity, through an electro-magnetic arrangement, can be made available for the measurement of time, and by its agency a great number of clocks can be kept in a state of uniform correctness.

The plan by which this is accomplished is substantially as follows:—A battery being connected with a principal clock, which is itself connected by means of wires with any number of clocks arranged at a distance from each other, has the current regularly and continually broken by the beating of the pendulum. This interruption is also experienced by all the clocks included in the circuit; and in accordance with this breaking and making of contact, the indicators or hands of the clock move over the dial at a constantly uniform rate.

What is the action of electrical currents upon each other?

813. The fundamental law of action in frictional electricity is, that bodies charged with like electricities at rest repel, and with unlike, attract each other. With electricity in motion the case is somewhat different, since currents of the same electricity moving in the same direction attract each other. The general law of this action may be stated as follows:

What is the general law of this action?

If electric currents flow in wires parallel to each other, and have freedom of motion, the wires are immediately disturbed. If the currents are moving in the same direction, the wires attract each other; if they are moving in opposite directions, they repel each other: or, like currents attract, and unlike repel.

How may a helix be converted into a magnetic needle?

814. When the wires connecting the positive and negative poles of a galvanic battery in action are coiled in the form of a helix, the helix becomes possessed of magnetic properties. If such a helix be suspended in a horizontal plane, it points, as a magnetic needle would, north and south; if it is suspended so as to move in a vertical plane, it acts as a dipping needle.

If two helices carrying currents are presented to each other, they attract and repel, precisely as if they were magnets, according as like or unlike poles are brought together. And, in short, all the properties of the magnetic needle may be imitated by a helix carrying a current.

What is Ampere's theory of magnetism?

815. From these, and other like phenomena, M. Ampere has propounded a theory which accounts for nearly all the phenomena of terrestrial magnetism.

He supposes that all magnetic phenomena are the result of the circulation of electrical currents. Every molecule of a magnet is considered to be surrounded with an atmosphere of electricity, which is constantly circulating around it, the difference between a magnet and a mere bar of iron being, that the electricity which exists equally in the iron, is at rest, whereas in the magnet it is in motion. The direction of these currents circulating in a magnet is dependent upon the position in which the magnet is held. If the opposite or unlike poles of two magnets be placed end to end, the electric currents of each will be found running the same way, and as currents moving in the same direction attract each other, the two poles will tend to come together. On the contrary, if the ends of like poles be presented, the course of the currents traversing each will be in opposite directions, and a repulsion will result.

Why does a magnetic needle tend to arrange itself at right angles to a current?

A magnetic needle tends to arrange itself at right angles with a wire transmitting an electric current, in order to bring the numerous currents circulating around its particles parallel with that of the wire.

How is the magnetism of the earth explained by this theory?

The magnetism of the earth is also explained by this theory on the same principles. If we take a metal ring and warm it at one point only by a spirit-lamp, no electrical effect ensues; but if the lamp is moved an electric current runs round the ring in the direction the lamp has taken. In a like manner, currents of electricity are known to be excited and kept in motion around the earth, by the sun, which heats in turn successive portions of its surface. They flow round it from east to west in a direction at right angles with a line joining the magnetic poles. A magnetic needle, therefore, points north and south, because that position is the one in which the electric currents in it are parallel to those of the earth, and this is the position, as has just been explained, that electric currents tend always to assume.

Fig. 369.

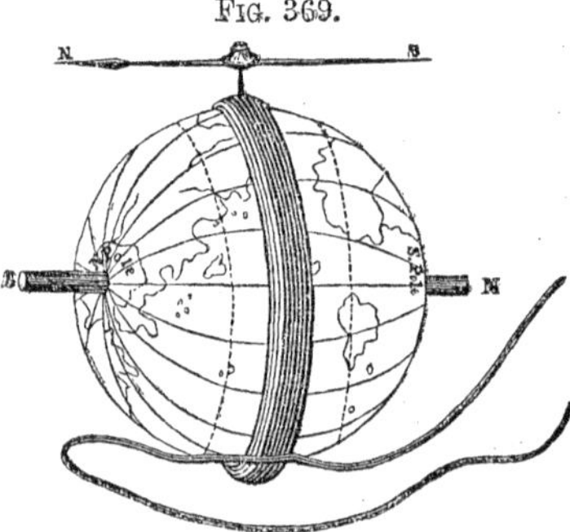

Fig. 369 represents an artificial globe, surrounded by a coil of insulated wire, surmounted by a magnetic needle. The needle will always point to the north pole of the globe, on transmitting the battery current.

The dip of the needle may be also readily accounted for in the same manner. At the polar re gions it dips vertically down in order that its currents may be parallel with those of the earth; for in those regions the sun performs his daily motion in circles parallel to the horizon. At the equator, the course of the sun is nearly at right angles to the horizon, and the needle maintains a horizontal position.

What is Magneto-electricity?

816. As an electric current passing round the exterior of a bar of soft iron induces magnetism in it, so on the contrary, a magnetized bar is able to generate an electric current in a conducting wire surrounding it.

Electricity thus produced by the agency of a magnet is called Magneto-electricity.

Fig. 370.

This may be shown by introducing one of the poles of a powerful bar magnet within a helix of fine insulated wire (see Fig. 370), the ends of which are connected with a delicate galvanometer. The deflection of the needle will indicate the flow of an electric current every time the magnet enters or leaves the coil—the direction of the current changing with the poles entered.

The same results will be obtained, if instead of introducing and removing a permanent steel magnet, we continually change the polarity of a soft iron bar. Thus, in Fig. 371, let *a b* be a bar of soft iron surrounding a helix, and N E S a horse-shoe magnet so arranged that it can revolve freely on a pivot at *c*, the poles in their revolution just passing by the terminations of the bar *a b*. On causing the magnet to revolve, the polarity of the bar *a b* will be reversed with every half revolution the magnet makes, and this reversal of polarity will generate electric currents in the wire.

Fig. 371.

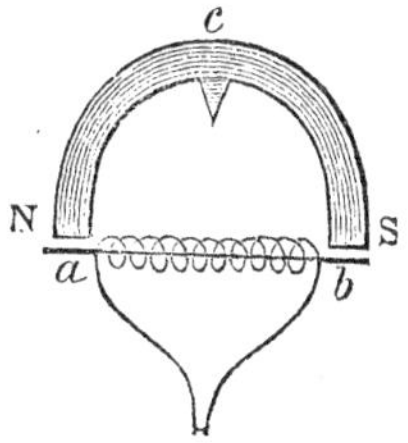

To instruments constructed on these principles the name of magneto-electric machines is given.

Can one electric current induce another?

817. Whenever an electric current flows through a wire it excites another current in an opposite direction, in a second wire held near to and parallel with it. Its duration, however, is only momentary. On stopping the primary current, induction again takes

place in the secondary wire ; but the current now arising has the same direction as the primary one.

The passage of an electrical current, therefore, develops other currents in its neighborhood, which flow in the opposite direction as the primary one first acts, but in the same direction as it ceases. Whenever a magnet, also, is moved in the neighborhood of a conducting wire, these currents are also induced.

What is the general construction of magneto-electric machines?

818. Magneto-electric machines, arranged for developing electricity by the reaction of a magnet, are constructed in a great variety of forms. In some, permanent steel magnets are used; in others, temporary soft iron ones, brought into activity by a galvanic current. A common form of magneto-electric machine is represented in Fig. 372.

FIG. 372.

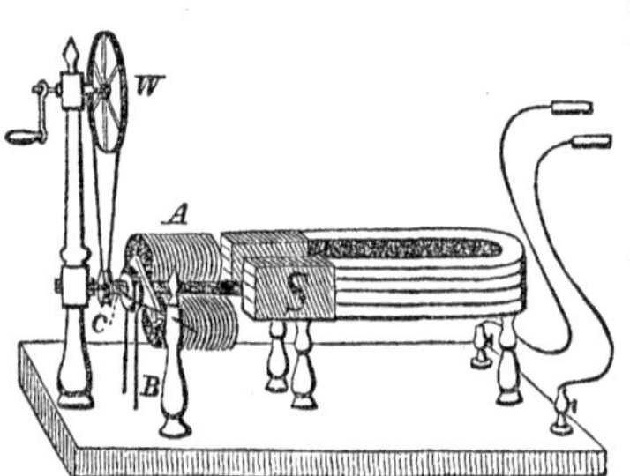

It consists of a compound horseshoe magnet, S, Fig. 372, bolted to a mahogany stand, arranged in such a manner that an electro-magnet, or armature, A B, mounted on an axis, revolves in front of its poles, by turning a multiplying wheel, W. This electro-magnet, or armature, consists of two cores of soft iron wound about with fine insulated copper wire. The ends of the wire in these coils are kept pressed, by means of springs, against a good conducting metal plate, which in turn is connected by wires with the screw-caps at the end of the base board. When the iron cores, or axes of the coils are in front of the poles of the magnet, they become magnetic by induction. This sets in motion the natural electricity of the coil, or helices, which flows in a certain direction, and is conveyed through the springs and wires to the screw-caps.

If the armature be turned half round, the magnetism of the iron is reversed, and a second current is excited in the opposite direction.

What effects may be produced by the action of electro-magnetic machines?

By turning the armature very rapidly, a constant current passes through the wires, and by connecting a small piece of platinum wire in the circuit, it is rapidly rendered red hot. By conveying connecting wires from the magneto-electric machine into acidulated water, its decomposition is effected; and many chemical compounds may in like manner be resolved into their ultimate constituents: machines also of this character may be used for electro-plating.

The effects of electricity thus generated on the human system are peculiar. If the two handles connected with the screw-caps of the machine are grasped by the hands, slightly moistened, and the armature is made to revolve rapidly, the muscles are closed so firmly, that the handles can not be dropped, and most powerful convulsive shocks are sent through the arms and body.

What is a diamagnetic body? 819. It has been demonstrated by Professor Faraday that bodies, not in themselves magnetic, may, when placed under certain physical conditions, be repelled by sufficiently powerful electro-magnets. Such substances have been termed diamagnetic, and the phenomena developed have received the general name of diamagnetism.

Bodies that are magnetic are attracted by the poles of a magnet; bodies that are diamagnetic are repelled by the poles of a magnet. Magnetism may be regarded as an attractive force, diamagnetism as a repelling one.

FIG. 373.

N S

Thus, if a bar of iron is suspended free to move in any direction, between the poles, N S, of a magnet, Fig. 373, the bar will arrange itself along a line which will unite the two poles; it places itself in the axial line, or along the line of force. Such is the condition of a magnetic body. If a substance of the diamagnetic class is placed in the same situation—as, for example, a bar of bismuth—between the poles, N S, Fig. 374, it places itself across, or at right angles to the axial line, or the line of force.

FIG. 374.

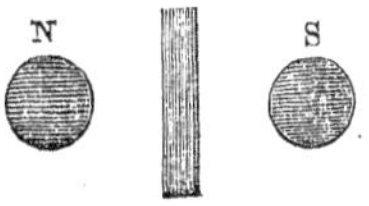

N S

Every substance in nature is in one or the other of these conditions. "It is a curious sight," says Dr. Faraday, "to see a piece of wood, or of beef, or an apple, or a bottle of water repelled by a magnet; or taking the leaf of a tree, and hanging it up between the poles, to observe it taking an equatorial position."

INDEX.

A

B

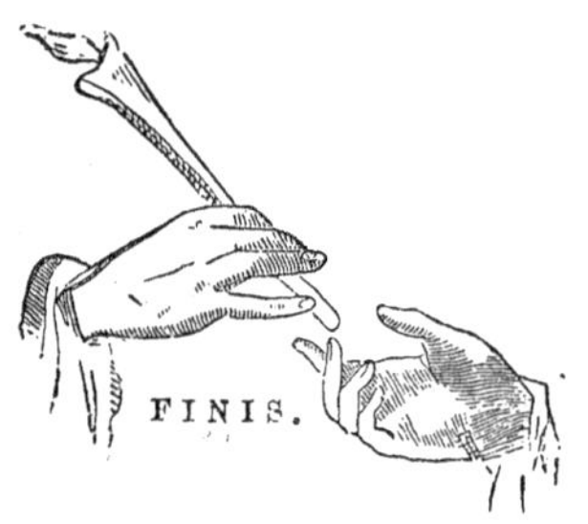
FINIS.

www.ingramcontent.com/pod-product-compliance
Lightning Source LLC
LaVergne TN
LVHW021127110826
845150LV00005B/959

* 9 7 8 1 4 2 5 5 5 1 0 7 0 *